Arsenic Remediation of Food and Water

Bhaskar Sen Gupta · Nadia Martínez-Villegas
Editors

Arsenic Remediation of Food and Water

Technological Interventions and Perspectives from Developing Countries

Springer

Editors
Bhaskar Sen Gupta OBE
School of Energy, Geoscience
Infrastructure and Society
Institute for Infrastructure and Environment
Heriot-Watt University
Edinburgh, Scotland, UK

Nadia Martínez-Villegas
Applied Geoscience Division
IPICYT
San Luis Potosí, Mexico

ISBN 978-981-97-4766-5 ISBN 978-981-97-4764-1 (eBook)
https://doi.org/10.1007/978-981-97-4764-1

This Springer imprint is published by the registered company Springer Nature Singapore Pte Ltd.
The registered company address is: 152 Beach Road, #21-01/04 Gateway East, Singapore 189721, Singapore

Foreword

Arsenic is naturally present in the groundwater. It is highly toxic in its inorganic form and exposure to high levels of arsenic poses a serious threat to the public health. Arsenic-contaminated water used for drinking purposes, food preparation and irrigation of crops are the major sources of exposure. Arsenic contamination of groundwater is widespread and there are a number of developing countries in which the arsenic contamination of drinking water is above the WHO provisional limit of 10 μg/L.

The signs and symptoms caused by long-term elevated exposure to arsenic differ between individuals, population groups and geographical areas. They are usually observed on the skin and may include hyperpigmentation, skin lesions and hard patches on the palms and soles of the feet (hyperkeratosis), and these may be a precursor to skin cancer. In addition, long-term exposure to arsenic may also cause cancers of the bladder and lungs. Other adverse health effects include developmental effects, diabetes, pulmonary and cardiovascular diseases. Exposure to arsenic in utero and early childhood has been linked to increases in mortality in young adults due to multiple cancers, lung disease, heart attacks and kidney failure. Also, numerous studies have demonstrated the negative impacts of arsenic exposure on cognitive development, intelligence and memory.

The most important action in the affected communities is the prevention of further exposure to arsenic by the provision of a safe supply of water for drinking, food preparation and irrigation of food crops. There are a number of options to reduce levels of arsenic in the drinking water: (a) substituting high-arsenic sources, such as groundwater, with low-arsenic safe sources such as rainwater and treated surface water, (b) discriminating between high-arsenic and low-arsenic sources, (c) mixing low-arsenic water with higher-arsenic water to achieve an acceptable arsenic concentration level, and (d) installing arsenic removal systems. Technologies for arsenic removal include oxidation, coagulation-precipitation, absorption, ion exchange and membrane techniques.

Awareness, education, and community engagement are the key factors for ensuring successful interventions for arsenic remediation. There is a need for the affected community to understand the risks of high arsenic exposure and its sources.

The United Nations' Sustainable Development Goals on good health and well-being (SDG 3), and safe and clean water and sanitation for all (SDG 6) cannot be achieved without monitoring and remediating arsenic-polluted groundwater. This book is a timely and useful contribution to arsenic remediation of food and water through technological interventions. It is hoped that such publications will contribute significantly in the dissemination of knowledge on the subject and will help in achieving these goals in a timely manner.

Professor Emeritus Mohd. Ali Hashim
Department of Chemical Engineering
Universiti Malaya
Kuala Lumpur, Malaysia

Preface

Arsenic is one of the ten chemicals of major public health concern. Arsenic contamination of groundwater from natural geogenic sources has long been regarded as a global problem by the World Health Organisation (WHO). A research team based at the Swiss Federal Institute (Podgorski and Berg, Science, 2020) has estimated by using a machine learning tool that 94 to 220 million people in the world that includes more than 94% in Asia, are exposed to unsafe levels of arsenic following the WHO provisional guideline value of 10 micro-grams per litre (μg/l). While the population affected in Africa is under 10 million, one may get little comfort from such data due to a relaxed drinking water standard (50 μg/l) for arsenic in countries such as Angola, Benin, Congo, Madagascar, Malawi, South Sudan and Gambia. Unfortunately, countries like Bangladesh, Myanmar, Nepal, Afghanistan, Cambodia and Pakistan have also set the same standard for drinking water risking the unabated occurrence of arsenicosis and cancer particularly amongst the rural population.

Environmentalists and health professionals working in the science and policy interface have been clamouring for a long time to tighten the current arsenic standard, where it exceeds 10 μg/l. At the same time, agronomists and soil scientists have been arguing against exploiting arsenic-contaminated groundwater for irrigation use to improve crop safety. There is a general reluctance among the affected nations to acknowledge the intensity of arsenic pollution and its effect on the population. While the arsenic problem is global and widespread in the Americas, it is the Global South that is bearing the brunt of the arsenic crisis due to poor infrastructure, lack of investment and enthusiasm from the local governments. The arsenic crisis emanating from the reckless use of contaminated groundwater for drinking and irrigation purposes in Asia, notably in the Ganga-Meghna-Brahmaputra delta in South Asia have far-reaching consequences for the population in the region and beyond. A maximum of 0.3 microgram of arsenic per kg bodyweight per day is often taken as the upper limit of exposure for an adult by international regulatory agencies, which may be imputed to 2 litres of water containing 10 micrograms of arsenic per litre. Such simple statistics would insinuate historicity of any claim on the safe limit of daily exposure from arsenic.

Accumulation of arsenic in soil in high concentrations due to contaminated groundwater use has resulted in translocation of arsenic from soil to the crop, predominantly turning rice into a toxic staple. This occurrence is global for arsenic-contaminated rice fields and is getting bad press for both importing and exporting nations. A recent example is the US rice export to Haiti, containing unsafe levels of arsenic (Reuters, 24 April 2024).

If we leave geopolitics aside, millions of people living in poverty and having little access to safe food and water, face a grim prospect of arsenic related cancer. It is unthinkable that the population living on the margins in scores of the affected countries could access safe food and water without help and support. Such responsibilities bestowed on 193 nations that have signed up for the UN Development Agenda.

In this context, we are delighted to present an edited book titled *Arsenic Remediation of Food and Water—Technological Interventions and Perspectives from Developing Countries* with general readers, policymakers, scientists and engineers in mind. It includes wide-ranging topics on arsenic-source identification, distribution, technologies for treating water for safe use, contamination of food and water and its impact on health. The book broadly covers the arsenic related issues in the Global South. It also includes two technical papers focusing on arsenic distribution in soil emanating from smelter waste in Mexico.

We would like to take this opportunity to thank the Centre for Science and Technology of the Non-Aligned and Other Developing Countries (NAM S&T Centre), New Delhi, and Springer Nature, Singapore for publishing this book. We hope this book serves the intended audience well and answers some critical questions on challenges for arsenic remediation of groundwater.

Edinburgh, UK — Bhaskar Sen Gupta, Ph.D.
San Luis Potosí, Mexico — Nadia Martínez-Villegas, Ph.D.

Introduction

Arsenic contamination of groundwater in many parts of the world is caused by the weathering of arsenic-bearing minerals, having high local concentration in the earth's crust.

In developing countries, groundwater provides a safe alternative source of drinking water to polluted surface water. However, the presence of arsenic in some groundwater sources has resulted in chronic arsenic poisoning around the world causing severe health hazards.

The problem became a serious health concern after mass poisoning of the population was detected due to arsenic contamination of groundwater in Bangladesh and West Bengal, India. Over the years, the same problem surfaced in many other parts of the world affecting the local populations. The World Health Organization (WHO) recommends limiting arsenic concentrations in water to 10 μg/L, although this is often an unattainable goal for many countries due to the difficult nature of removing arsenic from water sources, and the lack of appropriate, cost-effective and sustainable technologies for arsenic remediation of groundwater.

Considering the importance of the subject and to address the serious issue of arsenic contamination in developing countries, the Centre for Science and Technology of the Non-Aligned and Other Developing Countries (NAM S&T Centre), New Delhi has brought out this book titled *Arsenic Remediation of Food and Water—Technological Interventions and Perspectives from Developing Countries*. The book highlights the current status of arsenic contamination of water, related health hazards and the state of sustainable and cost-effective technological solutions for removing arsenic from groundwater in developing countries.

The book comprising 22 chapters contributed by leading experts and professionals from 9 countries has been divided into 5 parts. Part I and II of the book provides an understanding of the sources of arsenic contamination of groundwater and the process development such as nano-adsorbents for removal of arsenic and other heavy metals. Part III deals with technological interventions for the removal of arsenic, for example by filtration using locally manufactured ceramic membranes. Additionally, two chapters have been included in the Part III, which highlights the emerging and innovative technology like Subterranean Arsenic Removal (SAR) that can remove arsenic and iron from groundwater without using any adsorption bed. Part IV of the book includes a few chapters on arsenic contamination in the food materials and the food chain systems. Arsenic has long been associated with a variety of health complications in the human body. In order to address this issue, a chapter on arsenic contamination and its impacts on human health has been included in the Part V of the book.

I would like to express my sincere gratitude to Professor Emeritus Mohd. Ali Hashim, Department of Chemical Engineering, Universiti Malaya, Kuala Lumpur, Malaysia for writing the "Foreword" of the Book.

The leading international research and academic publisher—Springer Nature, Singapore has published this Book. We are thankful to Dr. Loyola D'Silva, the Executive Editor of Springer Nature for his encouragement and kind support; and Ms. Kavitha Palanisamy, Production Editor, Springer Nature, Chennai,India and Mr. Gowtham Chakravarthy V, Production Supervisor, Straive, Chennai, India for handling all the technical and production matters related to publication of this book. I am confident that our association with "Springer" would lead to such valuable collaborative endeavours in future.

I would like to express my gratitude to the Editors, Prof. Bhaskar Sen Gupta, OBE (UK) and Prof. Nadia Martínez-Villegas (Mexico) for sparing their valuable time for technical editing of the papers published in this book.

The publication of this book has been made possible through a Project Grant titled "*Reducing Arsenic Exposure from Food and Water in Developing Countries—A Road Map for Technological Solutions for the Future*" (Grant No. G-77/PGTF Project No. INT/20/K01 Project Code: 124567) from the G-77 Secretariat and project support from the UN Office for South-South Cooperation (UNOSSC), United Nations, New York, USA. We are extremely grateful to Dr. Donaldo Lopez from UNOSSC for his kind support and guidance in execution of the above project.

I will take this moment to express my sincere gratitude to Mr. Madhusudan Bandyopadhyay, Senior Adviser, NAM S&T Centre for his kind advice and guidance in various stages of planning and execution of this book project. Further, I am also thankful to Ms. Jasmeet Kaur Baweja, Programme Officer, NAM S&T Centre for her significant contributions in coordinating this book project.

I also acknowledge the help and support received from the entire team of the NAM S&T Centre. I am especially thankful to Mr. Pankaj Buttan, Data Processing Manager; Mr. Rahul Kumra, Assistant Administrative Officer; Mr. Sunil Kumar, Accounts Manager and Mr. Jayakumaran, Public Relations Manager, NAM S&T Centre for their support in bringing out this publication.

I am sure that this book would be a valuable reference material for the scientific community in developing countries working on community water supply and treatment, food safety, public health and policy.

Amitava Bandopadhyay, Ph.D.
Director General
NAM S&T Centre
New Delhi, India

Contents

Part I
Sources of Arsenic Contamination of Groundwater

Chapter 1
The Contamination of Water and Soil from the Dissolution of As-Bearing Mineral Waste in Matehuala, Mexico

Nadia Martínez-Villegas, Margaret Suárez, René Loredo-Portales, Francisco Castillo-Rivera, Patricia González-Hernández, Diana Meza-Figueroa, Mario Villalobos, and Bhaskar SenGupta

Abstract Extremely high concentrations of arsenic (As) in groundwater have been reported in central Mexico related to leachates from metallurgical wastes from an abandoned smelter. At the site, contaminated groundwater has been extensively used for maize cultivation for a long time resulting in very high soil pollution. However, the As-containing minerals' identity, concentration, and solubility remain unresolved. In the present work, highly contaminated soil samples from the area were studied to determine total As concentrations in shallow (0–5 cm) and deepsoils (5–30 cm) and to

N. Martínez-Villegas (✉)
Division de Geociencias Aplicadas, IPICYT, Instituto Potosino de Investigacion Cientifica y Tecnologica, Camino a La Presa San Jose No. 2055, Col. Lomas 4a Sec. 78216, San Luis Potosi, SLP, Mexico
e-mail: nadia.martinez@ipicyt.edu.mx

M. Suárez
INSTEC, Instituto Superior de Tecnologia y Ciencias Aplicadas, Av. Salvador Allende y Luaces S/N, Plaza de La Revolucion 10400, La Habana, Cuba

R. Loredo-Portales
Departamento de Ciencias Ambientales, Universidad de Guanajuato, Carretera Irapuato-Silao, Divisi0n de Ciencias de La VidaEx Hacienda El Copal Km 9, 36824 Irapuato, Guanajuato, Mexico
e-mail: r.loredo@ugto.mx

F. Castillo-Rivera
CONACYT-Instituto de Geologia, Universidad Autonoma de San Luis Potosi, Av. Manuel Nava #6, Zona Universitaria, 78290 San Luis Potosi, SLP, Mexico
e-mail: jjfcastillori@conacyt.mx

P. González-Hernández
Facultad de Quimica, de la Universidad de La Habana, Zapata y G. S/N, Vedado 10400, La Habana, Cuba

D. Meza-Figueroa
Departamento de Geología, Division de Ciencias Exactas y Naturales, Universidad de Sonora, Rosales y Encinas, 83000 Centro, Hermosillo, Sonora, Mexico
e-mail: diana.meza@unison.mx

B. Sen Gupta and N. Martínez-Villegas (eds.), *Arsenic Remediation of Food and Water*,
https://doi.org/10.1007/978-981-97-4764-1_1

identify and characterize As-bearing minerals and their solubility behavior. Results showed that soil samples contained up to 4.2% As and the mineralogy consisted mainly of calcite, gypsum, and quartz. Identified arsenic minerals included pharmacosiderite, bukovskýite, scorodite, beudantite, clinoclase, sodium arsenate, adamite, arsenolite, arsenopyrite, orpiment, and a mixture of calcium arsenates (guerinite, haidingerite and pharmacolite). Additionally, As was adsorbed on ferrihydrite. Soil fractionation analysis showed that up to 74% of total As was present in the most mobile fractions, e.g., soluble, exchangeable, phosphate absorbable, and slightly reducible. Furthermore, As solubility in water accounted for up to 60 mg/L at pH 7, explaining the high pollution observed in groundwater and highlighting the risk to humans and ecosystems. According to saturation indices calculations, As may derive from the dissolution of adamite, arsenolite, pharmacolite, and haidingerite, while clinoclase and ferrihydrite may precipitate, counteracting As solubility and mobilization. The results of this study increase knowledge on the identification and solubility of As-bearing minerals in calcicxerosols and semi-arid climatic conditions where considerable contamination is observed in groundwater from As-containing waste disposal.

Keywords Arsenic · Soil · Groundwater · Arsenic dissolution · Soluble metal arsenates · Environmental contamination

Introduction

In soils and groundwater contaminated by metallurgical and mining wastes, identifying arsenic-bearing minerals is challenging due to the variety of arsenic-containing solid wastes, which might include sulfide concentrates, roaster and calcine wastes, smelter wastes, waste rock, and their weathering products. Such identification is, however, of primary importance to devise ways to mitigate arsenic (As) dispersion across the wider environment, such as in the case of historical smelter sites in calcareous and semi-arid environments. In these cases, As geochemistry knowledge is limited compared with more humid environments typically reported in the literature.

M. Villalobos
Departamento de Ciencias Ambientales y del Suelo, LANGEM, Instituto de Geologia, Universidad Nacional Autonoma de Mexico, Ciudad Universitaria 04510, Mexico City, D.F, Mexico
e-mail: mariov@geologia.unam.mx

B. SenGupta
School of Energy, Geoscience, Infrastructure & Society, Heriot-Watt University. Water Academy, Scotland EH14 4AS, UK
e-mail: B.Sengupta@hw.ac.uk

Arsenic may be a significant contaminant in effluents from extractive hydrometallurgical processes of copper, lead, zinc, silver, or gold. In such cases, remediation methods such as oxidation–reduction, precipitation, adsorption, electrolysis, cementation, solvent extraction, ion exchange, membrane separation, and biological treatment are often adopted (Donahue & Hendry, 2003; Fernández-Martínez et al., 2008; Morin & Calas, 2006; Paktunc et al., 2003; Smedley & Kinniburgh, 2002; Vaughan, 2006, Robins et al., 2001). In the case of precipitation, the precipitated materials from the hydrometallurgical processes are supposed to produce As phases of very low solubility (Robins et al., 2001). However, these materials may be unstable and not suitable for direct disposal as they would produce leachates containing dissolved As (Martínez-Villegas et al., 2013; Nishimura et al., 1985; Robins, 1984; Robins and Tozawa, 1982). The most commonly adopted treatment methods for effluents from extractive hydrometallurgical processes involve precipitation of As(III) sulfide, calcium arsenates, or ferric arsenates.

In the case of calcium arsenates, several solid phases can be precipitated from As(V) solutions via lime additions. Historically, the removal of As in the presence of lime was attributed to the formation of $Ca_3(AsO_4)_2$ (Dutré & Vandecasteele, 1995; Robins, 1982; Sadiq et al., 1983; Zouboulis et al., 1993). However, further studies discarded this anhydrous arsenate compound (Bothe & Brown, 1999a, b; Zhu et al., 2006) in favor of hydrated, hydroxylated, and protonated forms as the most thermodynamically stable species under aqueous conditions (Guerin, 1941; Pierrot, 1964). So far, reported calcium arsenates responsible for As removal include an extensive suite of compositions and structures: calcium hydroxy-arsenate hydrates [$Ca_4(OH)_2(AsO_4)_2{\cdot}4H_2O$ and $Ca_5(AsO_4)_3OH$], calcium arsenate hydrates [$Ca_3(AsO_4)_2{\cdot}nH_2O$ where $n = 2.25, 3.0, 3.67, 4.0$ or 4.25], calcium dihydrogen arsenates [isomers of $Ca_5H_2(AsO_4)_4.9H_2O$ such as ferrarisite and guerinite], and calcium hydrogen arsenates [$CaHAsO_4{\cdot}nH_2O$ where $n = 1$ or 2] (Bothe & Brown, 1999a, b; Nishimura & Robins, 1998; Zhu et al., 2006). Differences in characteristics and solubility of these calcium arsenate minerals are reported (Bothe & Brown, 1999b). For example, the $Ca_4(OH)_2(AsO_4)_2{\cdot}4H_2O$ formed at pH 12 exhibits low solubility (0.0016 mmol As/L) compared to the $CaHAsO_4{\cdot}H_2O$ formed at pH 6.22, which exhibits a solubility that is five orders of magnitude higher (27.36 mmol As/L) than the former. Therefore, calcium arsenates of higher solubility may be more likely to prevail in circumneutral to slightly acidic aqueous environments. However, a gap exists in the identity and solubility of calcium arsenates in the environment as a function of environmental conditions after their disposal in soils. In the case of iron arsenates, these minerals can be precipitated with iron (III) from process solutions below pH 2 to form amorphous and/or crystalline phases, such as scorodite [$FeAsO_4{\cdot}2H_2O$]. However, scorodite is known to be thermodynamically unstable in the neutral to high pH region (Dove & Rimstidt, 1985), suggesting, therefore, a risk of dissolution after their disposal in soils. Yet, studies report the presence of scorodite at circumneutral pH values (Craw & Pacheco, 2002; Haffert et al., 2010) likely due to a slow dissolution rate (Bluteau & Demopoulos, 2007; Harvey et al., 2006) and a stability field that increases in the presence of high sulfate and As concentrations in the solution (Bluteau and Demopoulos, 2007; Craw et al., 2003; Langmuir et al., 2006).

For the specific case of mining wastes, neutralization of acid effluents leads to the precipitation, at room temperature, of Fe oxyhydroxides with adsorbed As (Demopoulos, 2005; Jia & Demopoulos, 2005; Langmuir et al., 1999), while maturation ($\geq$50 years), re-mobilization, and crystallization of ferrihydrite leads to the formation of pharmacosiderite [$KFe_4(AsO_4)_3(OH)_4{\cdot}6H_2O$] (Haffert et al., 2010). At a high concentration of sulfate and in the presence of moisture, bukovskýite [$(Fe)_2(AsO_4)(SO_4)(OH){\cdot}15H_2O$], a relatively rare secondary mineral found in medieval deposits in the Czech Republic (Novák et al., 1967) and mineral residues in New Zealand (Haffert et al., 2010), is formed (Majzlan et al., 2012).

The present investigation pertains to an old, abandoned metallurgical site, where limestone and iron ores were used in a smelting process (Manz & Castro, 1997; Martínez-Villegas et al., 2013), likely leading to the formation of calcium and iron arsenate wastes and other As-bearing mineral phases, which were deposited and/or precipitated on the smelter soils (Fig. 1.1a). Since the contamination of As in the underlying aquifer is very high, reaching up to 158 mg/L (Martínez-Villegas et al., 2013), As is expected to exist not only in stable As mineral phases but also in the form of highly soluble As-bearing phases. Therefore, the objectives of this study were: (i) to determine total As concentrations in both shallow (0–5 cm) and deep (5–30 cm) soils and (ii) to identify and characterize the As-bearing minerals as well as their solubility behavior in the contaminated soil samples.

Materials and Methods

Study Site

Soil samples were collected from the terrains of an old metallurgical site in Matehuala (Fig. 1.1b), which is an abandoned smelter that operated between 1905 and 1965 to process ores of Ag, Au, and Pb from Santa Maria de la Paz, a mining district located 8 km west of Matehuala (Castro Larragoitia et al., 1997; Gámez Rodríguez, 2014). Skarn deposits of Cu-Au with sulfides, such as pyrite (FeS_2), arsenopyrite (FeAsS), sphalerite (ZnS), galena (PbS), chalcopyrite ($CuFeS_2$) and bornite (Cu_5FeS_4) were investigated at Santa Maria de la Paz (Castro Larragoitia et al., 1997). In the terrains of the abandoned smelter, soils are classified as calcic xerosols, characterized by an accumulation of calcareous material (Cetenal, 1972).

A total of 16 soil samples were collected from eight different sampling points (namely 1, 3, 4, 9, 11, 13, 15, and 16 in Table 1.2) in the abandoned smelter site (Fig. 1.1b) for both shallow (0–5 cm) and deep (5–30 cm) soil using an exploratory sampling technique (SE, 2006), as conditioned by accessibility to reasonable bare soil, free of construction debris. Samples for surface soils (0–5 cm) were labelled 1, 3, 4, 9, 11, 13, 15 and 16 s and for sub-surface soils (5–30 cm) 1, 3, 4, 9, 11, 13, 15 and

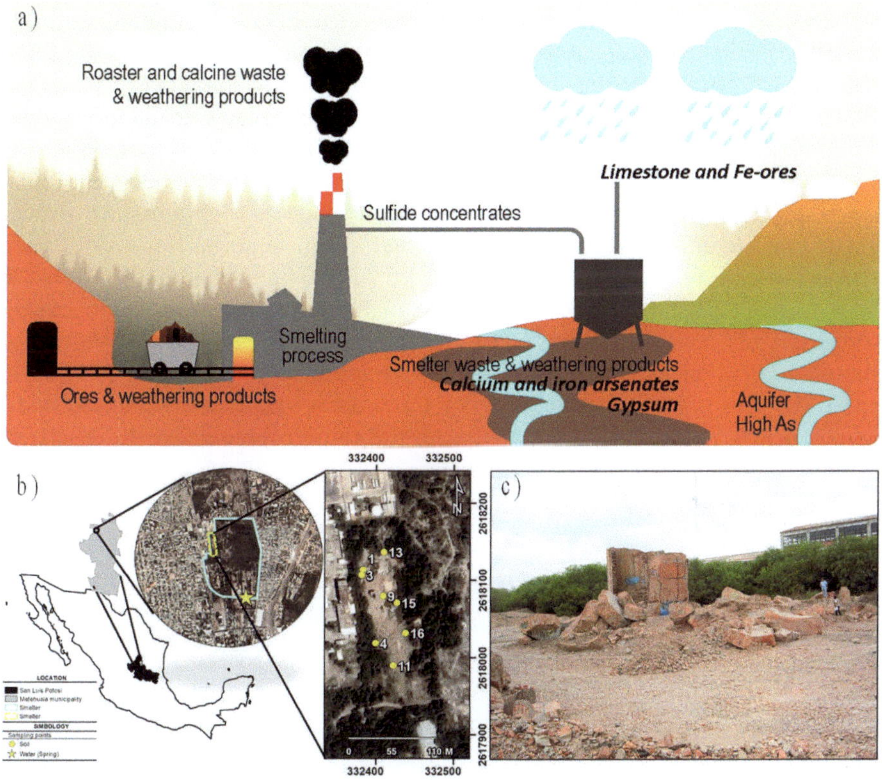

Fig. 1.1 Schematic of the Matehuala site, San Luis Potosi, Mexico showing **a** a conceptual model of the great variety of solid wastes that may either prevail or form in a (historical) metallurgical site, including ores, sulfide concentrates, roaster and calcine wastes, smelter wastes, waste rock, and weathering products, **b** the terrains of the old, abandoned smelter, the soil sampling points, and a nearby spring, contaminated with very high concentrations of As in water (up to 158 mg As/L) (Martínez-Villegas et al., 2013), and **c** the actual view of the abandoned smelter

16 d. All samples were stored in polyethylene bags and maintained at 4 °C until air-dried, ground, and homogenized for elemental and mineralogical analysis. The most As-contaminated soil samples (1, 15, and 16) were further chosen for fractionation and dissolution test analyses.

Determining Total As and Soil Parameters

Total As and Fe concentrations were determined with a portable XRF analyzer NitonFXL GOLDD from Thermo Scientific, according to the USEPA Method 6200 (USEPA, 2007). Detection limits for As and Fe were < 10 mg/kg. Duplicates of silicon dioxide (SiO_2) were used as a blank. Acceptable recovery for all elements was between 90 and 110% using seven replicates of three standard reference materials (SRM) from the National Institute of Standards and Technology (NIST) SRM 2709, SRM 2710, and SRM 2711. In order to verify the XRF outputs, 12.5% of the collected samples were sent to the ALS Laboratory in Toronto, Canada for analysis by plasma-atomic emission spectrometry (ICP-AES), after the EPA Method 3050B and 6010B (USEPA, 2007, 1996). Results agreed within a 95% confidence level.

Soil pH (potentiometric method), electrical conductivity (saturation extract method), and total iron-oxides were determined in soil samples. Iron-oxides were extracted by a dithionite-citrate digest, buffered with sodium bicarbonate (Mehra & Jackson, 1958). The Fe released during this extraction was determined by inductively coupled plasma optical emission spectrophotometry (ICP-OES) (EPA, 1994). The calibration curve ranged from 0.05 to 20 mg/L and showed good linearity ($r = 0.999$, $n = 6$). The detection limit was 0.10 mg/kg. Calibration with reference samples and a replicate analysis for quality control purposes was carried out to ensure the reliability of analytical data during ICP-OES analysis.

Determining Soil Bulk Mineralogy

The bulk mineralogy was determined by X-ray diffraction (XRD) using a Bruker D8 Advance X-ray diffractometer with Cu Kα radiation. All samples were previously sieved through a stainless steel 2 mm mesh to remove coarse debris and gravel. The analyses were carried out from 5 to 90° at a scan rate of 0.02° s^{-1}. Phase identification was carried out using X'pert High Score Plus software and the PDF-4 + database (International Center for Diffraction Data). Mineral semi-quantification was estimated by the Reference Intensity Ratio (RIR) method using the X'pert High Score Plus software.

Additionally, specimens of soil samples mounted on carbon tape in aluminium pins were analyzed by scanning electron microscopy (SEM) using a scanning electron microscope FEI-ESEM QUANTA FEG-250 coupled with a dispersive energy spectroscopy (EDS) system. The analyses were performed using a retro-scattered electron detector under low vacuum conditions (10–130 kPa).

Determining As Fractionation and Dissolution From Contaminated Soil Samples

Arsenic fractionation was carried out by sequential extractions using 1 g of soil for the fractionation of soluble, exchangeable, phosphate-adsorbable and slightly reducible, Fe and Mn oxyhydroxides (poorly crystalline and crystalline), organic, and sulfide phases (Table 1.4, Appendix). After each step, samples were centrifuged and decanted to determine the concentration of As in the supernatant using ICP-OES.

Arsenic concentration in the residual fraction was calculated by subtracting the sum of all previous phases from the total As concentration of each sample.

Moreover, As dissolution tests were carried out in triplicates by mixing 1 g of soil with 100 mL of deionized water in polypropylene containers for 4, 8, 20, 64, 128, 252, and 500 h, at 25 °C, in an orbital shaker. This experimental design permitted the study of As dissolution rates for each soil composition while testing the influence of other parameters on the final solubility of As. Changes occurring during the dissolution process were therefore, monitored over time to obtain information on potentially occurring additional reactions (e.g., gypsum and calcite dissolution). After equilibration, the suspensions were decanted, and the pH was determined for each supernatant using a multiparameter WE50050 from YSI. Each supernatant was then filtered through a 0.45 μm-pore membrane and divided into 4 subsamples to determine alkalinity, sulfate, chloride, major cations, and trace elements. Alkalinity was determined by titration using 0.02 M hydrochloric acid. Sulfate was determined by a turbidimetric method using barium as the precipitating agent, while chloride was determined by titration using silver nitrate. Although the precipitation of barium arsenate may have occurred during sulfate determination with barium as a precipitating agent, it was considered negligible as sulfate concentrations in the solution were orders of magnitude higher than those of arsenate (Table 1.5, Appendix). Major cations and trace elements were determined in acidified subsamples to $pH < 2$ using HNO_3 and stored at 4 °C until analysis. Total As, Cu, Fe, Pb, Zn, Ca, K, Mg, Na, Al, and Si were determined in the extracts by ICP-OES.

Determining As-Bearing Minerals in Soil Samples

Arsenic mineralogy was determined by X-ray Absorption near edge structure (XANES) using the XAFS Beamline of Elettra Synchrotron Trieste operated with-ring energy of 2 GeV and a current of 310 mA (Di Cicco et al., 2009). For doing so, soil samples less than 2 mm were homogenized using an agate mortar. Model compounds (see list below) and samples were prepared as pellets using polyvinyl pyrrolidone and according to suitable absorption length based on As content. Model compounds included: scorodite [$FeAsO_4{\cdot}2H_2O$], clinoclase [$Cu_3AsO_4(OH)_3$], pharmacosiderite [$KFe_4(AsO_4)_3(OH)_4{\cdot}6H_2O$], arsenolite [$As_2O_3$], beudantite [$PbFe_3(OH)_6SO_4AsO_4$], orpiment [$As_2S_3$], arsenopyrite [FeAsS], sodium arsenate [Na_3AsO_4], adamite [Zn_2AsO_4OH], bukovskýite [$Fe_2(AsO_4)(SO_4)(OH){\cdot}9H_2O$], and As adsorbed on ferrihydrite [$As{\equiv}Fe(OH)_3$]. Additionally, a mixture of calcium arsenates that comprised guerinite [$Ca_5H_2(AsO_4)_4{\cdot}9H_2O$], haidingerite [$CaHAsO_4{\cdot}H_2O$], and pharmacolite [$CaHAsO_0{\cdot}2H_2O$] was used as a standard for these minerals (Hernández-Bárcenas et al., 2017). The chemical structure of the standards was double-checked by Extended X-ray absorption fine structure (EXAFS) fitting to corroborate their identity.

For the XANES analysis, measurements were performed in fluorescence mode using a Silicon Drift Detector for diluted samples and transmission mode for model

Table 1.1 Dissolution reactions and solubility products available in the literature for the minerals found in this study

Mineral name	Dissolutionreaction	Log K_{sp}	References
Arsenic minerals			
Arsenolite	$As_2O_3 + 3H_2O \rightarrow 2H_3AsO_3$	–1.38	Ball and Nordstrom (1991)
Pharmacolite	$CaHAsO_4{\cdot}2H_2O + H^+ \rightarrow Ca^{2+} + H_2AsO_4^- + 2H_2O$	–4.68	Rodríguez-Blanco et al. (2007)
Haidingerite	$CaHAsO_4{\cdot}H_2O \rightarrow Ca^{2+} + HAsO_4^{2-} + H_2O$	–4.79	Bothe and Brown (1999b)
Beudantite	$(H_3O)_{0.68}Pb_{0.32}Fe_{2.86}(SO_4)_{1.69}(AsO_4)_{0.31}[(OH)_{5.59},(H_2O)_{0.41}] + 4.91H^+ \rightarrow 0.32Pb^{2+} + 2.86Fe^{3+} + 1.69SO_4^{2-} + 0.31AsO_4^{3-} + 6.68H_2O$	–13.94 ± 1.89	Forray et al. (2014)
Scorodite	$FeAsO_4{\cdot}2H_2O \rightarrow Fe^{3+} + AsO_4^{3-} + 2H_2O$	–20.24	Nordstrom et al. (2014)
Bukovskýite	$Fe_2(AsO_4)(SO_4)(OH){\cdot}9H_2O + H^+ \rightarrow 2Fe^{3+} + SO_4^{2-} + AsO_4^{3-} + 10H_2O$	–30.63	Majzlan et al. (2012)
Clinoclase	Cu3AsO4(OH)3 + 5H+ → 3Cu2+ + H2AsO4- +3H2O	10.10 ± 0.38	Magalhaes et al. (1988)
Adamite	$Zn_2AsO_4(OH) + 3H^+ \rightarrow 2Zn^{2+} + H_2AsO_4^- + H_2O$	5.71 ± 0.22	Magalhaes et al. (1988)
Guerinite	$Ca_5H_2(AsO_4)_4{\cdot}9H_2O \rightarrow 5Ca^{2+} + 2AsO_4^{3-} + 2HAsO_4^{2-} + 9H_2O$	–30.69	Bothe and Brown (1999b)
Pharmacosiderite	$KFe_4(AsO_4)_3(OH)_4{\cdot}6H_2O \rightarrow K^+ + 4Fe^{3+} + 3AsO_4^{3-} + 4OH^- + 6{\cdot}H_2O$	Not available on the literature	
Ferrihydrite	$Fe(OH)_3 \rightarrow Fe^{3+} + 3OH^-$	–39	Norvell and Lindsay (1982)
Orpiment	$As_2S_3 + 6H_2O \rightarrow 2H_3AsO_3 + 3HS^- + 3H^+$	–46.3	Ball and Nordstrom (1991)
Arsenopyrite	$FeAsS + 1.5\ H_2O + 0.5H^+ \rightarrow 0.5\ AsH_3 + 0.5H_2AsO_3^- + Fe^{2+} + HS^-$	–14.44	Johnson et al. (2000)

(continued)

Table 1.1 (continued)

Mineral name	Dissolutionreaction	Log K_{sp}	References
Bulk soil minerals			
Q uartz	$SiO_2 + 2H_2O \rightarrow H_4SiO_4$	–3.98	Ball and Nordstrom (1991)
Gypsum	$CaSO_4 \cdot 2H_2O \rightarrow Ca^{2+} + SO_4^{2-} + 2H_2O$	–4.58	Ball and Nordstrom (1991)
Calcite	$CaCO_3 \rightarrow CO_3^{2-} + Ca^{2+}$	–8.48	Ball and Nordstrom (1991)

Table 1.2 Sampling points and labels, coordinates, depth, pH, electrical conductivity (EC), total As, total Fe, Fe/As ratio, and Fe-oxides content in soil samples

Sampling		Coordinate (UTM)		Depth	pH	EC	As	Fe	Fe/As	Fe-oxides
Point	Label	X	Y	(cm)		(mS/m)	(mg/kg)	(mg/kg)		(%)
1	1 s	332,383	2,618,114	0–5	7.12	286	42,167	52,503	1.24	1.12
	1 d			5–30	6.96	287	29,657	24,369	0.82	0.51
3	3 s	332,381	2,618,107	0–5	8.32	149	1308	18,237	13.94	N.D.
	3 d			5–30	8.77	132	2283	21,058	9.22	N.D.
4	4 s	332,399	2,618,019	0–5	8.94	48	436	16,995	38.98	N.D.
	4 d			5–30*						
9	9 s	332,409	2,618,080	0–5	9.20	222	11,764	33,306	2.83	N.D.
	9 d			5–30	9.07	69	1291	20,098	15.16	N.D.
11	11 s	332,422	2,617,990	0–5*						
	11 d			5–30	8.68	68	1126	23,593	20.95	N.D.
13	13 s	332,410	2,618,137	0–5	9.17	140	3751	22,308	5.94	N.D.
	13 d			5–30	8.88	222	8535	30,148	3.53	N.D.
15	15 s	332,427	2,618,071	0–5	8.36	245	33,196	32,496	0.97	0.57
	15 d			5–30	7.59	223	40,567	34,780	0.85	0.70
16	16 s	332,438	2,618,032	0–5	8.85	111	15,552	38,720	2.48	0.33
	16 d			5–30	7.39	211	23,200	48,658	2.09	0.46

* = Lost sample. N.D. = Not determined

compounds and As concentrated samples. The As-K-edge (11,867 eV) was measured 100 eV below and 1000 eV above the edge to acquire XANES and EXAFS information for each sample. Two to four spectra were collected according to the quality of the spectra and merged into a final spectrum. Data analysis was performed using Athena's X-ray absorption spectroscopy (XAS) analysis software. A GaAs spectrum was collected for each measurement as a reference to align the energy scale. Linear combination analysis (LCA) was performed on all samples using Athena software (Ravel & Newville, 2005) with normalized data. LCA was done in the $-20 < E_0 < 30$ eV range, using all the references in groups of four standards first and then reducing this set to the main 4 components that explained each sample; all contributions between 0 and 1, without forcing them to add to 1 for better alignment. Final contributions were recalculated to obtain 100%.

Estimating Saturation Indices

Values of temperature, pH, alkalinity, total dissolved concentrations of major cations and anions, trace metals (Cu, Zn, Al, Pb), Si, Fe,and As were inputted (Table 1.5,

Appendix) to the chemical speciation code PHREEQC version 3.0 (Parkhurst & Appelo, 2013) in order to estimate aqueous solution speciation and saturation indices to help identify the metal arsenates controlling the solubility or immobilization of As. Model runs were performed for samples 1 s, 1 d, and 15 d at each sampling time, and in samples 15 s, 16 s, and 16 d at 500 h. All model runs were performed using the wateq4f database for PHREEQC as modified by Martínez-Villegas et al. (2013). Minerals in the database are listed in Table 1.1. Later, we plotted the different saturation indexes over time and identified the minerals likely to contribute to either the dissolution or the immobilization of As in our soil samples.

Results and Discussion

Total As in Soil Samples

Table 1.2 shows soil coordinates, depth, pH, electrical conductivity (EC), As concentration, Fe concentration, Fe/As ratio, and Fe-oxides content. Soil pH ranged from 6.96 to 9.20 indicating neutral to strongly alkaline conditions. Electrical conductivity ranged from 48 to 287 mS/m, likely due to the calcium sulfate present in the soil samples (CETENAL, 1972). Total As concentrations ranged from 436 to 42,167 mg/kg (Table 1.2), exceeding the maximum permissible levels of As in industrial soils (260 mg/kg) (DOF, 2007) and indicating a high risk of environmental contamination. Total Fe ranged from 16,995 to 52,503 mg/kg, revealing normal soil iron contents ($\leq$6%). However, because of the high As concentrations, some soil samples exhibited relatively low Fe/As ratios (<2.5), which, in turn, suggested a low potential for As retention in the iron phases (Vaca-Escobar et al., 2012). Arsenic and Fe concentrations by XRF agreed with previous analyses using ICP-OES in digest soil samples (Martínez-Villegas et al., 2013). Iron-oxides ranged from 0.33 to 1.12%, revealing a low percentage of Fe-oxide phases and suggesting the presence of other more recalcitrant phases of Fe, resistant to the dithionite-citrate digest buffered with sodium bicarbonate. This would lower further the above Fe/As ratio when Fe content is taken only from Fe-oxide phases, and confirms the low capacities of As retention by Fe-oxide minerals in the soils studied.

Bulk Soil Mineralogyin Soil Samples

Figure 1.2a shows the relative mineralogical composition of bulk soil samples. Calcite and quartz were present in all soils, while gypsum was identified only in 1 s, 1 d, 13 d, 15 s, 15 d, and 16 d samples. As mentioned above, soils at the abandoned smelter are classified as calcic xerosols (CETENAL, 1972) due to the significant presence of calcite and quartz. However, we found a direct relationship between gypsum and

As concentration, up to about 20–30% gypsym (Fig. 1.2a and b). In this study, we attributed the presence of gypsum to the metallurgical activities carried out on-site during smelting operations. Gypsum is a common industrial by product of many processes involving the neutralization of sulfuric acid and sulfur oxide fumes (Chen et al., 2014; Das et al., 2015; Fernández-Martínez et al., 2008; Jia & Demopoulos, 2005; Swash & Monhemius, 1994; Zhang et al., 2015). The neutralization of As-rich acidic solutions in the metallurgical industry generates As-rich gypsum sludge (Fernández-Martínez et al., 2008). The high percentages of gypsum in some soil samples may indicate larger disposal of As-rich gypsum sludge in these sampling points. This observation is consistent with lower concentrations of As in the soils with higher calcite and quartz (Fig. 1.2a and c). However, an As spatial pattern was difficult to establish due to protracted mechanical disturbance of the landscape by material movement (Fig. 1.7, Appendix). Such a disturbance might move As-bearing minerals, including gypsum, from their original deposition spots to other places within the smelter's terrains, making it difficult to draw an As spatial distribution at the study area.

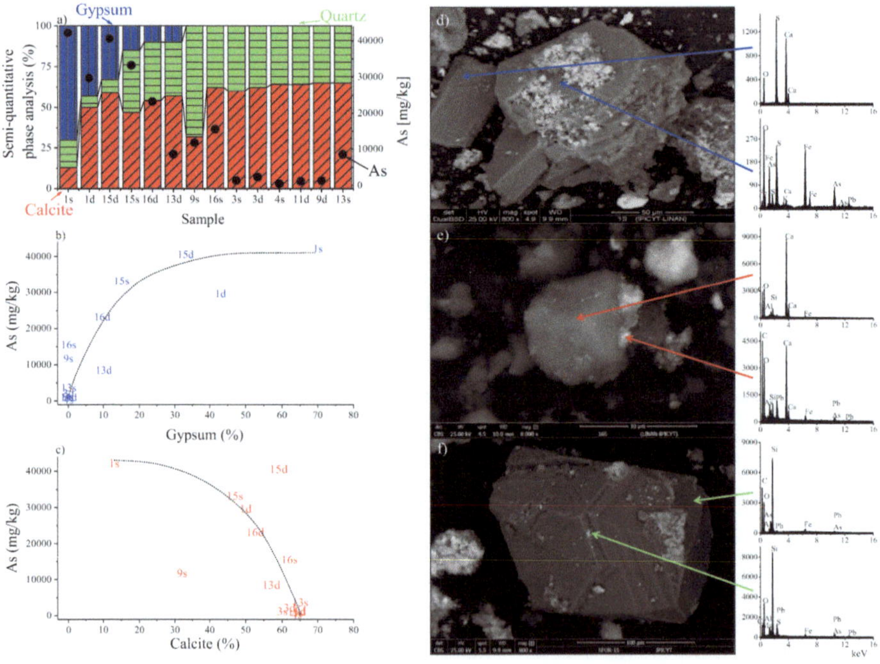

Fig. 1.2 **a** Major mineral soil composition as determined by XRD and RIR analysis showing the presence of calcite, gypsum, and quartz. **b** Arsenic as a function of gypsum concentration in soil samples showed a direct relationship. **c** Arsenic as a function of calcite content showed an inverse relationship between As and calcite. Dotted lines depict the relationships between As and gypsum and As and calcite. Sample 9 s (mostly comprised of construction debris) was not included in the fitting. Specimens of **d** gypsum, **e** calcite, and **f** quartz particles with probable metallic As minerals deposited on their surfaces

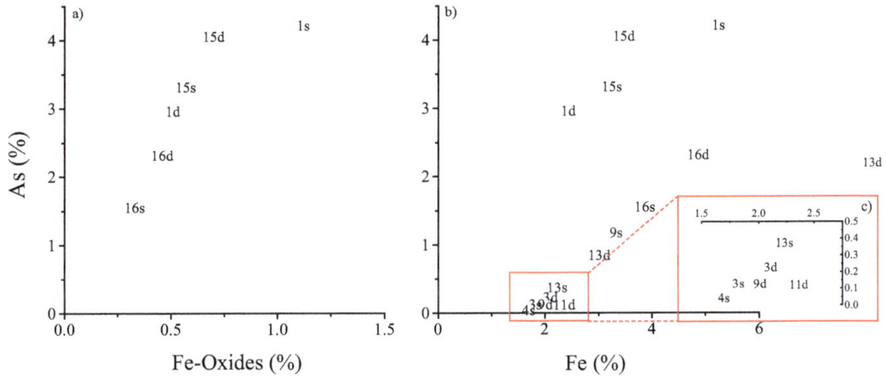

Fig. 1.3 **a** As versus Fe-oxides. **b** As versus total Fe, showing a direct relationship. **c** Close-up of samples 3 s, 3 d, 4 s, 9 d, 11 d, and 13 s

Specimens with the characteristic morphology and chemical composition of gypsum, calcite, and quartz were observed in bulk soil samples by SEM–EDS analyses (Fig. 1.2d–f). Furthermore, particles comprising Ca, Fe, Pb, Al, Mg, and As were deposited on their surfaces (Fig. 1.2d–f).

Figure 1.3a depicts a relationship between total soil As and Fe-oxides. However, more data points might be needed to show a stronger relationship between As and Fe-oxides. Given the extremely high concentration of As in the soil matrix compared to the relatively low Fe/As ratio, a saturation of Fe-oxide adsorption sites might occur at the abandoned smelter. In this type of scenario, precipitation of metal arsenate phases would be favored compared to the adsorption of As on Fe-oxides (Villalobos et al., 2010). A closer inspection of soil samples by SEM–EDS analysis revealed the presence of likely metal arsenates (Fig. 1.8, Appendix). If highly mobile, such phases could explain, in turn, the exceptionally high concentrations of dissolved As found in the adjoining aquifer (up to 158 mg/L) (Martínez-Villegas et al., 2013).

Arsenic Fractionation and Dissolution from Contaminated Soil Samples

Arsenic fractionation was determined (Fig. 1.9, Appendix). Arsenic concentration in the soluble fraction ranged from 9.9 to 6.31 × 10^3 mg/kg. This fraction was expected to represent As dissolved from highly soluble arsenate minerals and/or As (co) precipitated with gypsum. Precipitation of highly soluble calcium arsenates onto a gypsum lattice and isomorphic substitution of sulfate by arsenate in the gypsum structure (Fernández-Martínez et al., 2008; Rodríguez-Blanco et al., 2007) are likely to occur at the abandoned smelter due to geological and environmental conditions. The concentration of As in the exchangeable fraction was very low, ranging from 5.8 to 2.04 × 10^2 mg/kg. It might correspond to easily exchangeable As or loosely adsorbed As on Fe and Al oxides. For the phosphate-desorbable and slightly reducible fraction, As ranged from 1.26×10^2 and 1.53×10^4 mg/kg and might correspond to As adsorbed and/or (co) precipitated with calcite and Fe oxides as the phosphate might have displaced As. Arsenic adsorbs on calcite (and Fe oxides) by inner-sphere surface complexes through corner-sharing with Ca octahedral and coprecipitates with calcite through arsenate substitution in carbonate sites (Alexandratos et al., 2007; Sø, 2011). Arsenic concentrations in the Fe and Mn oxy-hydroxides fraction ranged from 12.0 to 5.65×10^3 mg/kg. Although Fe-oxides are a major immobilizing pool for As, the relatively low quantities of Fe-oxides found in this study and the relatively low concentrations of As found in the Fe and Mn oxy-hydroxides fractions, compared to the sum of the more mobile fractions (soluble, exchangeable, and phosphate-absorbable and slightly reducible), helped to explain the extremely high concentrations of As found in the adjoining aquifer and suggested that As contamination exceeded the retention capacity of Fe-oxides for immobilizing As effectively. The concentrations of As extracted from the organic matter and sulfides fractions ranged from 3.0 to 9.83×10^2 mg/kg, while As concentrations in the residual fraction ranged from 2.92×10^2 to 3.09×10^4 mg/kg. These results were in contrast with what is commonly reported in the literature. In this case, we found a significant amount of As in the most mobile soil-solid phases (i.e.73% in sample 1 d in Fig. 1.9, Appendix) that explains the mobilization of As in the study area.

Figure 1.4 shows the evolution of As, pH, and Ca and their relationships during the dissolution experiments of contaminated soil samples. We found that dissolved As increased with time until reaching steady state (Fig. 1.4a). At 500 h, total dissolved As was highest in test 1 d (60 mg/L), medium in tests 1 s, 15 s, and 15 d (30 mg/L average), and lowest in tests 16 s and 16 d (8 mg/L). In contrast, pH was highest in tests 16 s and 16 d (8.1), medium in tests 1 s, 15 s, and 15 d (7.7), and lowest in test 1 d (7.1) (Fig. 1.4b). This high release time to reach steady state explains why the soluble fraction in the extraction procedure at 30 min (Table S1) was inadequate and suggests that the Assoluble fraction in soil samples might be even higher than estimated. Similar to As, dissolved Ca increased with time (Fig. 1.4c), showing a concomitant dissolution of As and Ca that suggested either dissolution of As-rich gypsum wastes and/or the presence of highly soluble calcium arsenates. At 500 h, total dissolved Ca concentrations were highest in test 1 s (412 mg/L), medium in tests 1 d, 15 s, and 15 d (215 mg/L average), and lowest in tests 16 s and 16 d (54 mg/L average) (Fig. 1.4c). Effect of pH on As dissolution over time was unclear (Fig. 1.4d), probably due to other participating variables. However, As dissolution was observed to be lower at higher pH values (Fig. 1.4d). As mentioned above, As dissolution increased concurrently with calcium over time (Fig. 1.4e), within a constrained tendency, even when the concentration of dissolved calcium was displaced to the right, probably due to higher dissolution of gypsum in soil sample 1 s (Fig. 1.4e), which was supported by Fig. 1.4f where Ca decreased as pH increased, except for test 1 s due to gypsum dissolution. In fact, we suggest that sample 1 s was more a metallurgical waste than a contaminated soil sample due to high amounts of gypsum (70%), total As (42,167 mg/kg), total Fe (52,503 mg/kg), and Fe-oxides (1.12%), which could derive from the As immobilization processes and the neutralization of pH using limestone and iron ores (Fig. 1.1a). In contrast, sample 4 s was more a soil sample given the low As concentration (436 mg/kg), the high Fe/As ratio (38.98), and the absence of gypsum.

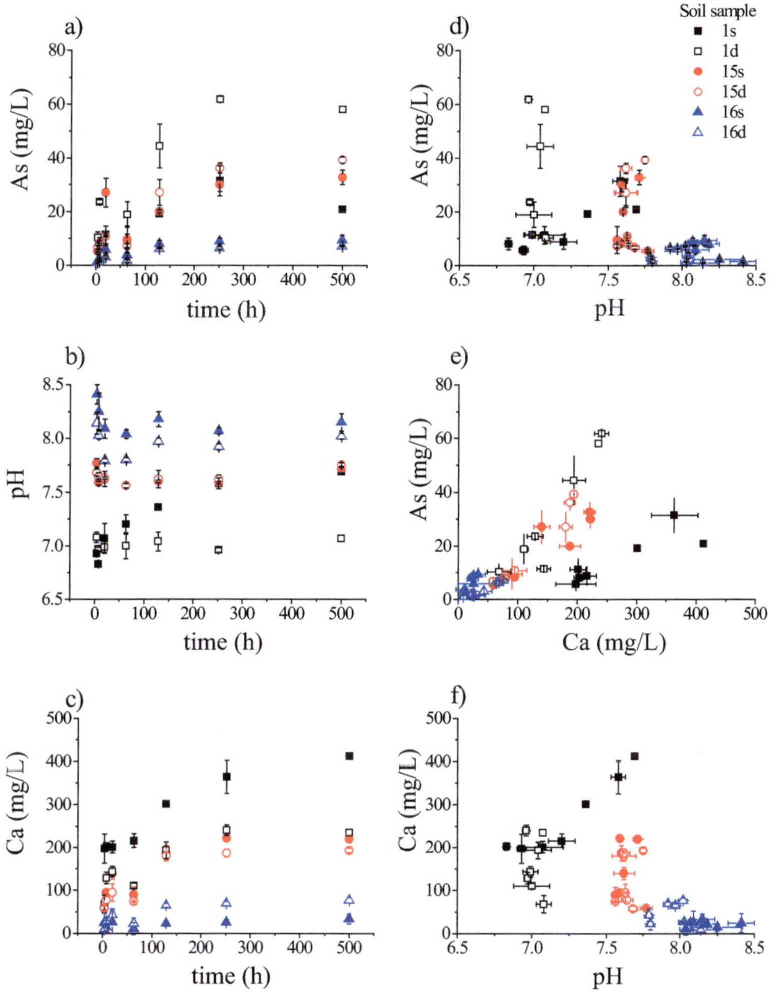

Fig. 1.4 **a** As concentration, **b** pH, and **c** Ca concentration as a function of time in water dissolution tests of As contaminated soil samples. **d** Dissolved As concentration as a function of pH, **e** dissolved As concentration as a function of Ca, and **f** dissolved Ca concentration as a function of pH. Error bars are standard deviations of datasets comprising triplicate experimental runs

As-Bearing Minerals in Soil Samples

Figure 1.5a shows the XANES spectra obtained for the soil samples collected at the abandoned smelter in Matehuala. Figure 1.5b shows the XANES spectra obtained for the different standards including Ca, Zn, Pb, Na, Cu, and Fe arsenates, As adsorbed on ferrihydrite, arsenolite, and As-sulfur minerals. Our results showed that soil samples shared similar XANES spectral features with arsenate standards and As(V) adsorbed on ferrihydrite, except for sample 11 that showed a shoulder at an As(III) position (Fig. 1.5). The major As(V) minerals found in the soil samples according to LCA were pharmacosiderite, bukovskýite, scorodite, calcium arsenates (guerinite, haidingerite and pharmacolite), beudantite, clinoclase, sodium arsenate, adamite, and As adsorbed on ferrihydrite (Table 1.2). Arsenic (III) minerals (arsenolite, realgar, and orpiment) were found in samples 3 s, 9 s, 13 s, 16 s, 9 d, and 11 d.

The presence of secondary calcium arsenate minerals, such as guerinite, pharmacolite, haidingerite, as well as scorodite might be the result of the neutralization of As-rich acidic solutions using lime and iron ores (Fig. 1.1a) (Bothe & Brown, 1999a, b; Frost et al., 2009; Nishimura & Robins, 1998; Rodríguez-Blanco et al., 2007; Swash & Monhemius, 1995; Zhu et al., 2014a) or a result of As precipitation over time. Pharmacolite is known to grow on gypsum at circumneutral pH values (Rodríguez-Blanco et al., 2007), while secondary calcium arsenate minerals (i.e.

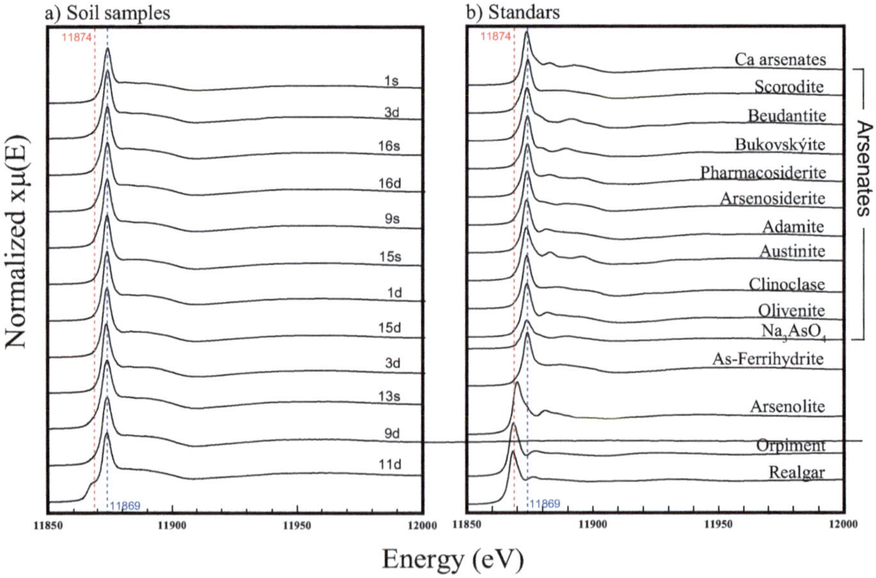

Fig. 1.5 **a** XANES spectra of soil samples and **b** model compounds show that most soil samples shared similar XANES spectral features with arsenate standards and As adsorbed on ferrihydrite, while sample 11 showed a shoulder at an As-sulfur position. Blue and red vertical dotted lines indicate similarities in peaks between samples and standards

guerinite, pharmacolite, and haidingerite) and scorodite are present in As contaminated sediments near the abandoned smelter (Castillo et al., 2015) and other contaminated sites (Castillo et al., 2015; Frost et al., 2009; Paktunc et al., 2003; Zhu et al., 2014b). Pharmacolite, haidingerite, and scorodite were likely to dissolve in the soluble fraction of the sequential extraction and the dissolution tests (Figs. 1.5, 1.6a and b) as determined by their high solubility products (as is the case of pharmacolite and haidingerite) (Table 1.1) and its thermodynamic instability at circumneutral pH values (as it is the case of scorodite). In the abandoned smelter, neutralization of acid effluents might have additionally led to the precipitation of different Fe compounds with adsorbed As, while maturation of As-ferrihydrite for more than 50 years after the closure of the smelter might explain the presence of pharmacosiderite in samples 1 s, 3 s, 9 s, 15 s 15 p, 16 s, and 16 p. The presence of bukovskýite may derive from the high concentrations of As in solution in the presence of gypsum and moisture (Julliot et al., 1999). The former from the neutralization of As-rich acidic solutions and the latter from the shallow aquifer underlying the abandoned smelter. The presence of As sulfide minerals such as orpiment, characteristic of low temperature hydrothermal veins (Zhu et al., 2005), such as those reported in Santa Maria de la Paz mining district (Castro Larragoitia et al., 1997; Manz & Castro, 1997), may derive from raw materials and/or mineral ores brought to the site for the smelting process, while scorodite might be the result of the oxidation of arsenopyrite (Filippi et al., 2015). The presence of scorodite in our soil samples was double-checked by comparison of the first and second shell of the EXAFS spectra against the scorodite standard and its theoretical spectra. In all cases, soil samples showed the characteristic structural arrangement of scorodite. Although scorodite is known to be thermodynamically unstable at circumneutral pH values (Dove & Rimstidt, 1985), high concentrations of As and S in the solution may lead to the presence of this mineral in mining sites as secondary phase or as a product of arsenopyrite weathering at circumneutral pH values. According to Craw et al., (2003), when the concentrations of As and S increased from 10^{-6} to 10^{-5} M, the stability field of scorodite shifts from pH 5 to 6 (Craw et al., 2003). Assuming that the same holds for an As concentration of 10^{-3} M, as is the case of the present study, the stability field of scorodite would extend to pH 8, supporting therefore the presence of scorodite in our study area. Furthermore, scorodite instability at the study site may be impeded by low humidity.

In semi-arid conditions, the presence of metastable phases, like ferrihydrite and schwertmannite, are also known to persist, in tailings, due to a low water through-flux that limits the translocation of sulfate and avoids its leaching out of the soil system (Hayes et al., 2014). The presence of arsenolite might be the result of incomplete oxidation of pyrite and arsenopyrite in previous metallurgical operations (Haffert et al., 2010). Adamite is an accessory mineral found in the oxidation zone of polymetallic deposits enriched with copper and zinc (Gołębiowska et al., 2006) like those occurring in Santa Maria de la Paz mining district, while beudantite might be the result of the oxidation of sulfide minerals (Botinelly, 1976). In this study, orpiment was identified in samples 3 s, 9 d and 11 d, while beudantite was found in sample 3 d. Similar to adamite, clinoclase is a mineral associated with areas of oxidation of Cu mineral deposits (Frost et al., 2009) like those found in Santa Maria de la Paz, from which the abandoned smelter used to process ores (Castro Larragoitia et al., 1997).

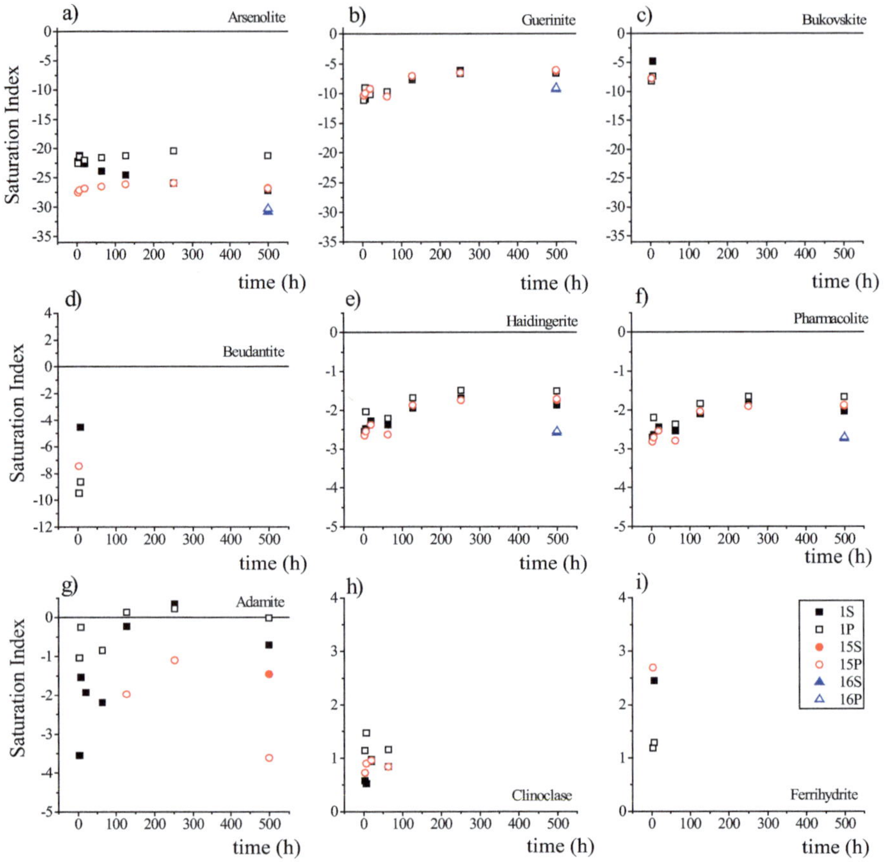

Fig. 1.6 Metal arsenate saturation indexes over time showed that **a** arsenolite, **b** guerinite, **c** bukovskýite, **d** beudantite, **e** haidingerite, **f** pharmacolite,and **g** adamite were undersaturated, except for samples 1 s and 1 d, where adamite reached equilibrium. In contrast, **h** clinoclase and **i** ferrihydrite were initially oversaturated. Since Fe and Cu were below detection limits, no saturation indexes could be calculated for bukovskýite, beudantite, clinoclase, and ferrihydrite at certain longer equilibration times

Minerals found in this study showed the complexity of As mineralogy at an abandoned metallurgical site and highlight the importance of mineral identification to gain insight into the phases that could explain the mobilization of As as well as for the development of suitable remediation strategies and technologies. The minerals responsible for As immobilization in this study comprised pharmacosiderite, bukovskýite, guerinite, and beudantite as determined by LCA of XANES spectra. Precipitation of these low solubility minerals might be an effective As immobilization technology on site. Precipitation of metallic arsenates of very low solubility has been proposed as a predominant control in an old Cu smelter and an As_2O_3 plant (Villalobos et al., 2010), which could be the case for sample 16 s, where the precipitation of metal arsenates such as bukovskýite and pharmacosiderite, might play a role in the immobilization of As as determined by the lower concentration of As extracted from the dissolution tests and the sequential extraction procedures compared with other samples. According to the literature, beudantite, bukovskýite, guerinite, ferrihydrite, and orpiment exhibit no potential dissolution (Ball & Nordstrom, 1991; Bothe & Brown, 1999a; Forray et al., 2014; Majzlan et al., 2012; Norvell & Lindsay, 1982). Yet, given enough time, orpiment can oxidize and therefore mobilize As in the soil environment. Ideally, the selective precipitation of the stable minerals over the soluble ones would decrease As mobilization on site.

Saturation Indices

Figure 1.6 shows the saturation indexes for the As minerals found in this study (except for orpiment, arsenopyrite, and scorodite as they were not found by LCA analysis) in samples 1, 15 and 16. Saturation indexes showed highly undersaturated conditions for arsenolite, guerinite, bukovskýite, beudantite, haidingerite, pharmacolite, and adamite; suggesting that, if they were initially present in the soils, these minerals would dissolve, releasing As from the soil to the solution. In contrast, clinoclase and ferrihydrite were oversaturated and therefore likely to precipitate, immobilizing As in the system.

Conclusions

The analysis performed in this study identifies that the As mineral phases present in the old, abandoned smelter site, as well as the presence of gypsum as a byproduct of metallurgical activity, are responsible for As groundwater and stream pollution. While some of the minerals are most certainly raw mining material brought to the former smelter, others might derive from the former smelting process or the weathering of the metallurgical wastes. We found that arsenic solubility is high at circumneutral pH values, perhaps due to the presence of highly soluble As-bearing minerals such as adamite, pharmacolite, and haidingerite, which explain the extremely high concentrations of As found in the underlying aquifer in the study area. On the other hand, less soluble minerals such as pharmacosiderite, bukovskýite, guerinite, beudantite, clinoclase, and arsenic adsorbed on ferrihydrite might contribute to limited As immobilization after cycles of dissolution and precipitation or weathering of the original As-bearing waste dumped on surface soils decades ago. The extent of these immobilizing phases is, however, very limited throughout the site, since, (1) the site is the source of the extremely high As concentration in the aquifer, and (2) soluble calcium arsenate phases were identified by the other techniques. Perhaps at the beginning of the operation of the smelter, there was much more abundance of the soluble calcium arsenate phases that released the dissolved As present now at the aquifer (Table 1.3).

Table 1.3 Arsenic mineral phases and bulk mineral composition in soil samples as determined by Linear Combination Analysis (LCA) performed on XANES spectra

		Soilsimple													
		1 s	3 s	16 s	16 d	9 s	15 s	1 d	15 d	3 d	13 s	9 d	11 d	4 s	13 d
Mineral name	Formula	LCA (%)													
Scorodite	$FeAsO_4 \cdot 2H_2O$	0	0	0	0	0	0	0	0	0	46.9	48	13.3	–	–
Ferrihydrite	$As{\equiv}Fe(OH)_3$	22.1	16.2	36.2	0	0	0	34.6	0	0	0	0	0	–	–
Calciumarsenate	$Ca_5H_2(AsO_4)_4 \cdot 9H_2O$ $CaHAsO_4 \cdot H_2O$ $CaHAsO_4 \cdot 2H_2O$	0	0	0	0	22.6	17.9	23.2	18.3	41.4	28.5	27.6	0	–	–
Adamite	Zn_2AsO_4OH	0	0	0	0	0	0	0	0	0	0	0	37.7	–	–
Clinoclase	$Cu_3AsO_4(OH)_3$	0	0	0	6.2	0	15.9	15.3	25.3	47.4	0	0	0	–	–
Sodiumarsenate	Na_3AsO_4	0	0	0	8.9	0	0	0	0	0	18.6	18	21.6	–	–
Beudantite	$PbFe_3(OH)_6SO_4AsO_4$	0	0	0	0	0	0	0	0	11.2	0	0	0	–	–
Bukovskýite	$Fe_2(AsO_4)(SO_4)(OH) \cdot 9H_2O$	34.8	36.9	23.6	29	47.9	45.2	26.9	38.7	0	0	0	0	–	–
Pharmacosiderite	$KFe_4(AsO_4)_3(OH)_4 \cdot 6H_2O$	43.1	43.4	37.5	55.9	20.7	21	0	17.7	0	0	0	0	–	–
Arsenolite	As_2O_3	0	0	2.7	0	8.8	0	0	0	0	0	0	0	–	–
Orpiment	As_2S_3	0	3.5	0	0	0	0	0	0	0	0	0	0	–	–
Arsenopyrite	FeAsS	0	0	0	0	0	0	0	0	0	6	6.4	27.4	–	–
	Total	100	100	100	100	100	100	100	100	100	100	100	100	–	–
	R Factor	6E–04	3E–04	3E–04	5E–04	1E–03	9E–04	3E–04	6E–04	9E–04	8E–04	4E–04	4E–04	–	–

Acknowledgements Thanks to Mary Carmen Rocha, Beatriz Rivera, Ana Iris Peña, Gerardo Hernández-Bárcenas, Esther Salazar Amador, Abraham Hernández, Christian Cordero, and Lucy Rose McKenna for laboratory work, sample collection or language corrections. We also thank LANBAMA, LINAN, and ELETRA for providing access to their facilities. This study was partially funded by grant NA140182 Newton Advanced Fellowship from The Royal Society awarded to Dr. Nadia Martinez-Villegas for working with Prof B. Sen Gupta at Heriot-Watt University. M.V. would like to thank the financial support of CONACyT CienciaBásica project 283416, as well the LANGEM (Laboratorio Nacional de Geoquímica y Mineralogía).

Appendix

See Table 1.4 and 1.5.

Table 1.4 Sequential extraction procedure used in this study (Salazar Amador, 2017)

Soil fraction	Extractant
(1) Soluble	Deionized H_2O, pH 7 (Bubbled N_2), 30 min, five repetitions
(2) Exchangeable	0.5 M $MgCl_2$, 20 ml, pH 8 (NaOH), 2 h
(3) Phosphate-adsorbable and slightly reducible	(1) 1 M NaH_2PO_4, 20 ml, pH 5 (HCl), 24 h (2) 1 M $NaC_2H_3O_2$, 20 ml, pH 5 (CH_3COOH), 5 h
(4) Fe and Mn oxyhydroxides (poorly crystalline and crystalline)	0.1 M NH_2OH–HCl, 20 ml, pH 2 (HCl), 2 h
(5) Organic matter and sulphides	(1) 30% (w/w) H_2O_2, 15 ml, pH 2 (HNO_3), 2 h (2) 3.2 M CH_3COONH_4, 20 ml, 20:1 ratio with HNO_3, 1 h

To determine the distribution of As in the different soil fractions, a sequential extraction procedure was carried out in 1 g of soil for each of the following fractions:

(1) Soluble: 100 ml of deionized water at pH 7 were added to soil samples. Deionized water was bubbled with N_2 to avoid water acidification by CO_2 dissolution. After water addition, each tube was stirred for 30 min, centrifuged at 4500 rpm, and decanted. Electrical conductivity was measured in each supernatant for decision making in order to pass to the next fractionation step. Electrical conductivity was taken as an indicative parameter of complete gypsum dissolution in soil samples when it reached low values and remained constant. A total of five washes were applied to each soil sample for the soluble fraction. Supernatants were analyzed for As using inductively coupled plasma optical emission spectrophotometry (ICP-OES)

(2) Exchangeable: 20 mL of 0.5 M $MgCl_2$ were mixed with the solid phase from the soluble fraction. Each tube was stirred for 2 h, centrifuged at 3000 rpm, and decanted. Supernatants were analyzed for As using ICP-OES

(3) Phosphate-adsorbable and slightly reducible: 20 mL of 1 M NaH_2PO_4 were mixed with the solid phase from the exchangeable fraction. Each tube was stirred for 24 h, centrifuged at 3000 rpm, and decanted. Then, 20 mL of 1 M $NaC_2H_3O_2$,acidified at pH 5 using CH_3COOH, were mixed with the solid phase from the previous extraction. Each tube was stirred for 5 h, centrifuged at 3000 rpm, and decanted. Supernatants were analyzed for As using ICP-OES

(4) Crystalline Fe and Mn oxyhydroxides: 20 mL of 0.1 M NH_2OH–HCl at pH 2 adjusted with 0.1 HCl were mixed with the solid phase from the phosphate-adsorbable and slightly reducible fraction. Each tube was stirred for 2 h, centrifuged at 3000 rpm, and decanted. Supernatants were analyzed for As using ICP-OES

(5) Organic matter and sulphides: 15 mL of 30% H_2O_2 at pH 2 adjusted with HNO_3 were mixed with the solid phase from the crystalline Fe and Mn oxyhydroxides fraction and stirred until the oxidation reaction was complete for approximately 2 h. After, 20 mL of 3.2 M CH_3COONH_4 in a 20:1 ratio with HNO_3 were added to each tube. Each tube was then stirred for 1 h, centrifuged at 3000 rpm, and decanted. Supernatants were analyzed for As using ICP-OES. Arsenic concentration in the residual fraction was calculated by subtracting the sum of all previous phases from the total As concentration of each sample

Table 1.5 Parameters used to simulate aqueous solution speciation and saturation indexes, over time, for dissolution tests 1 s, 1 d, 15 s, 15 d, 16 s, and 16 d in As contaminated soil samples. Numbers in parenthesis are standard deviations of triplicates. Hyphens were used for indicating anion missing data due to an accidental loss of the samples that prevented their acquisition

t (h)	Parameter[a]	Water dissolution test											
		1 s		1 d		15 s		15 d		16 s		16 d	
4	pH	6.93	(0.03)	7.08	(0.05)	7.77	(0.04)	7.68	(0.03)	8.41	(0.09)	8.14	(0.10)
	Ca	197.08	(34.11)	68.35	(19.93)	59.75	(4.07)	57.95	(5.85)	24.25	(21.88)	8.80	(4.59)
	Na	6.55	(1.29)	5.07	(0.24)	4.72	(0.35)	5.02	(0.20)	3.82	(0.12)	4.08	(0.07)
	K	2.35	(0.21)	2.88	(0.34)	2.73	(0.07)	2.97	(0.11)	2.26	(0.12)	2.06	(0.11)
	Mg	0.25	(0.12)	0.26	(0.08)	0.20	(0.02)	0.35	(0.06)	0.06	(0.03)	0.07	(0.04)
	SO_4^{2-}	453.10	(0.00)	276.11	(125.46)	–	–	196.46	(26.55)	–	–	–	–
	Alkalinity	31.11	(0.00)	33.70	(4.49)	–	–	28.52	(4.49)	–	–	–	–
	Cl^-	6.39	(0.00)	7.57	(3.91)	–	–	5.33	(0.00)	–	–	–	–
	As	5.71	(1.61)	10.37	(2.00)	5.47	(0.50)	6.70	(0.51)	1.57	(0.45)	0.95	0.22
	Al	0.15	(0.01)	0.18	(0.03)	0.28	(0.02)	0.32	(0.04)	0.31	(0.02)	0.21	0.04
	Si	0.19	(0.24)	0.00	(0.00)	0.13	(0.13)	0.30	(0.15)	0.92	(0.24)	0.16	0.14
	Cu	0.12	(0.01)	0.11	(0)	0.11	(0)	0.11	(0)	0.12	(0.01)	0.11	(0)
	Fe	bdl*	–	bdl	–	bdl	–	0.02	(0.01)	0.04	(0.01)	bdl	–
	Pb	2.26	(0.63)	1.23	(1.11)	1.68	(0.51)	1.94	(0.39)	2.08	(1.25)	1.78	(0.51)
	Zn	0.02	(0.03)	0.11	(0.05)	bdl	–	bdl	–	bdl	–	bdl	–
8	pH	6.93	(0.03)	7.00	(0.02)	7.59	(0.03)	7.64	(0.04)	8.25	(0.17)	8.03	(0.04)
	Ca	202.69	(8.59)	128.50	(13.45)	95.02	(3.49)	77.47	(7.69)	14.51	(8.22)	26.67	(3.19)
	Na	7.06	(0.72)	5.81	(0.43)	5.31	(0.16)	5.15	(0.10)	3.90	(0.22)	4.22	(0.24)
	K	2.40	(0.07)	3.73	(0.11)	3.14	(0.02)	3.12	(0.14)	2.39	(0.13)	2.45	(0.05)

(continued)

Table 1.5 (continued)

t (h)	Parameter[a]	Water dissolution test											
		1 s		1 d		15 s		15 d		16 s		16 d	
	Mg	0.36	(0.12)	0.60	(0.03)	0.44	(0.05)	0.50	(0.05)	0.09	(0.04)	0.22	(0.02)
	SO_4^{2-}	541.59	(53.83)	320.35	(38.57)	–	–	299.71	(58.92)	–	–	–	–
	Alkalinity	38.89	(7.78)	28.52	(2.38)	–	–	26.96	(6.29)	–	–	–	–
	Cl^-	4.97	(0.36)	4.97	(0.61)	–	–	6.39	(0.00)	–	–	–	–
	As	8.15	(2.25)	23.53	(0.80)	8.37	(0.84)	7.85	(0.41)	2.22	(0.39)	1.40	(0.09)
	Al	0.19	(0.11)	0.15	(0.03)	0.46	(0.28)	0.24	(0.03)	0.26	(0.02)	0.21	(0.01)
	Si	1.07	(0.60)	1.07	(0.08)	1.29	(0.71)	0.61	(0.14)	1.82	(0.04)	0.85	(0.05)
	Cu	0.13	(0.01)	0.11	(0)	0.13	(0.02)	0.12	(0.01)	0.13	(0)	0.12	(0)
	Fe	0.23	(0.44)	0.02	(0.07)	0.18	(0.26)	bld	–	0.03	(0.01)	bld	–
	Pb	7.38	(7.97)	3.07	(0.96)	2.15	(2.83)	1.31	(1.14)	2.52	(0.40)	2.18	(1.00)
	Zn	0.19	(0.30)	0.25	(0.05)	0.10	(0.17)	bld	–	bld	–	bld	–
20	pH	7.07	(0.14)	7.00	(0.05)	7.62	(0.07)	7.63	(0.01)	8.09	(0.09)	7.79	(0.02)
	Ca	200.98	(13.54)	143.49	(11.73)	139.78	(13.69)	95.04	(19.81)	25.77	(26.68)	42.73	(14.36)
	Na	6.66	(0.59)	6.05	(0.22)	6.13	(0.36)	5.52	(0.25)	4.20	(0.52)	4.36	(0.19)
	K	2.58	(0.03)	3.37	(0.34)	3.73	(0.34)	3.16	(0.45)	2.78	(0.10)	2.69	(0.09)
	Mg	0.56	(0.32)	0.54	(0.07)	0.62	(0.11)	0.54	(0.18)	0.36	(0.09)	0.39	(0.08)
	SO_4^{2-}	618.29	(87.31)	382.30	(38.57)	–	–	400.00	(15.33)	–	–	–	–
	Alkalinity	42.52	(3.92)	33.70	(4.49)	–	–	28.00	(4.12)	–	–	–	–

(continued)

Table 1.5 (continued)

t (h)	Parameter[a]	Water dissolution test											
		1 s		1 d		15 s		15 d		16 s		16 d	
	Cl^-	4.62	(0.00)	6.15	(0.89)	–	–	3.55	(0.00)	–	–	–	–
	As	11.24	(3.25)	11.40	(0.61)	27.05	(5.40)	10.90	(0.42)	5.83	(1.41)	3.06	(0.39)
	Al	0.13	(0.01)	0.18	(0.02)	0.14	(0.01)	0.16	(0.01)	0.23	(0.03)	0.17	(0.01)
	Si	3.19	(0.20)	2.10	(0.26)	2.13	(0.44)	1.18	(0.37)	4.29	(1.51)	3.16	(0.33)
	Cu	0.11	(0.01)	0.11	(0)	0.11	(0)	0.11	(0)	0.12	(0.01)	0.12	(0.01)
	Fe	bld	–	bld	–	bld	–	bld	–	0.03	(0.03)	bld	–
	Pb	2.48	(1.14)	2.01	(0.54)	0.98	(0.88)	2.27	(0.65)	1.91	(0.78)	2.14	(1.04)
	Zn	0.06		bld	–	0.22	(0.08)	bld	–	bld	–	bld	–
64	pH	7.20	(0.09)	7.00	(0.12)	7.56	(0.02)	7.56	(0.02)	8.04	(0.04)	7.80	(0.02)
	Ca	215.67	16.25	110.30	(6.74)	90.31	(16.86)	74.34	(1.58)	10.60	(14.27)	22.56	(13.47)
	Na	7.39	(0.57)	5.70	(0.13)	5.03	(0.40)	5.04	(0.42)	3.87	(0.29)	4.02	(0.22)
	K	2.60	(0.07)	3.37	(0.62)	3.07	(0.32)	3.03	(0.10)	2.57	(0.11)	2.43	(0.07)
	Mg	0.57	(0.36)	0.45	(0.13)	0.35	(0.10)	0.42	(0.05)	0.17	(0.06)	0.20	(0.05)
	SO_4^{2-}	848.38	(13.52)	473.75	(5.11)	–	–	379.35	(35.77)	–	–	–	–
	Alkalinity	48.74	(1.80)	41.48	(4.49)	–	–	35.26	(3.91)	–	–	–	–
	Cl^-	3.55	(0)	6.63	(1.25)	–	–	2.84	(0.00)	–	–	–	–
	As	8.84	(2.75)	18.91	(4.65)	9.56	(4.92)	7.22	(0.70)	3.71	(1.00)	1.61	(0.46)
	Al	0.12	(0.01)	0.13	(0.00)	0.18	(0.03)	0.20	(0.01)	0.19	(0.02)	0.17	(0.00)

(continued)

Table 1.5 (continued)

t (h)	Parameter[a]	Water dissolution test											
		1 s		1 d		15 s		15 d		16 s		16 d	
	Si	1.74	(0.32)	0.83	(0.22)	0.75	(0.18)	0.70	(0.05)	2.53	(0.89)	1.33	(0.50)
	Cu	0.11	(0)	0.11	(0.01)	0.11	(0)	0.11	(0)	0.12	(0.01)	0.12	(0)
	Fe	bld	–	bld	–	bld	–	bld	–	bld	–	bld	–
	Pb	2.07	(0.53)	2.25	(1.95)	1.71	(0.35)	1.48	(1.31)	1.52	(1.32)	1.51	(0.85)
	Zn	0.04	(0.03)	0.16	(0.14)	0.03	(0.05)	bld	–	bld	–	bld	–
128	pH	7.36	(0)	7.04	(0.09)	7.6	(0.01)	7.62	(0.08)	8.18	(0.07)	7.97	(0.03)
	Ca	300.69	(0)	193.58	(19.61)	187.12	(17.75)	179.87	(11.04)	22.55	(6.45)	64.78	(4.30)
	Na	12.87	(0)	7.90	(1.35)	7.35	(1.20)	7.18	(0.80)	1.57	(1.08)	2.38	(0.18)
	K	3.16	(0)	4.36	(0.50)	4.05	(0.30)	4.19	(0.02)	3.14	(0.13)	3.02	(0.06)
	Mg	0.98	(0)	1.02	(0.17)	1.01	(0.14)	1.49	(0.13)	0.56	(0.06)	0.69	(0.02)
	SO_4^{2-}	886.73	(30.66)	491.45	(20.44)	–	–	641.89	(13.52)	–	–	–	–
	Alkalinity	51.85	(4.49)	38.89	(0)	–	–	41.48	(4.49)	–	–	–	–
	Cl^-	3.55	(0)	6.86	(0.41)	–	–	6.04	(0)	–	–	–	–
	As	19.12	(0)	44.39	(8.19)	19.94	(0.59)	27.03	(4.75)	7.97	(1.18)	6.18	(0.09)
	Al	0.00	(0)	0.00	(0)	0.03	(0.01)	0.03	(0.03)	0.01	(0)	0.00	(0.00)
	Si	5.14	(0.00)	2.16	(0.50)	1.93	(0.32)	2.05	(0.36)	4.19	(0.46)	3.61	(0.12)
	Cu	bld	–	bld	–	bld	–	bld	–	bld	–	bld	–
	Fe	bld	–	bld	–	bld	–	bld	–	bld	–	bld	–

(continued)

Table 1.5 (continued)

t (h)	Parameter[a]	Water dissolution test											
		1 s		1 d		15 s		15 d		16 s		16 d	
	Pb	1.42	–	1.04	(0.34)	1.19	(0.63)	1.05	(0.31)	0.88	(0.34)	0.93	(0.90)
	Zn	0.19	–	0.33	(0.04)	0.02	(0.02)	0.01	(0.01)	bld	–	bld	–
252	pH	7.52	(0.00)	6.96	(0.02)	7.59	(0.01)	7.62	(0.04)	8.07	(0.03)	7.92	(0.03)
	Ca	363.80	(38.59)	240.29	(12.12)	221.54	(3.24)	186.78	(9.23)	25.71	(6.65)	68.74	(4.69)
	Na	16.03	(3.78)	8.53	(0.08)	8.95	(0.78)	7.22	(0.75)	1.09	(0.14)	2.51	(0.46)
	K	3.32	(0.11)	4.88	(0.21)	4.21	(0.05)	4.07	(0.15)	3.03	(0.09)	3.11	(0.16)
	Mg	0.92	(0.04)	1.25	(0.06)	1.30	(0.04)	1.66	(0.07)	0.58	(0.03)	0.72	(0.03)
	SO_4^{2-}	1028.32	(23.41)	479.65	(25.03)	–	–	650.74	(74.22)	–	–	–	–
	Alkalinity	52.37	(5.00)	46.67	(0.00)	–	–	41.48	(1.80)	–	–	–	–
	Cl^-	3.79	(0.41)	7.28	(0.25)	–	–	3.43	(1.02)	–	–	–	–
	As	31.38	(5.57)	61.85	(1.14)	29.99	(2.43)	36.16	(1.95)	8.93	(1.23)	6.48	(0.49)
	Al	0.00	(0.00)	0.00	(0.00)	0.02	(0.01)	0.00	(0.00)	0.02	(0.01)	0.01	(0.00)
	Si	8.97	(1.12)	3.68	(0.18)	3.46	(0.15)	2.72	(0.03)	4.62	(0.09)	3.84	(0.10)
	Cu	bld	–	bld	–	bld	–	bld	–	bld	–	bld	–
	Fe	bld	–	bld	–	bld	–	bld	–	bld	–	bld	–
	Pb	1.39	(0.03)	1.22	(0.13)	2.52	(0.94)	1.18	(0.26)	1.14	(0.62)	2.70	(0.88)
	Zn	0.19	(0.02)	0.40	(0.04)	0.03	(0.02)	0.02	(0.01)	bld	–	bld	–
500	pH	7.69	(0.00)	7.07	(0.00)	7.71	(0.03)	7.75	(0.02)	8.15	(0.08)	8.02	(0.03)

(continued)

Table 1.5 (continued)

t (h)	Parameter[a]	Water dissolution test											
		1 s		1 d		15 s		15 d		16 s		16 d	
	Ca	412.27	(0.00)	234.97	(0.00)	220.00	(9.12)	193.23	(6.02)	33.52	(11.92)	76.48	(5.52)
	Na	16.85	(0.00)	8.41	(0.00)	6.98	(0.45)	5.04	(1.64)	1.58	(0.65)	2.10	(0.27)
	K	3.33	(0.00)	4.76	(0.00)	4.19	(0.30)	4.39	(0.11)	3.22	(0.10)	3.20	(0.08)
	Mg	1.01	(0.00)	1.12	(0.00)	1.34	(0.08)	1.85	(0.06)	0.66	(0.08)	0.75	(0.04)
	SO_4^{2-}	1255.75	(88.35)	508.41	(13.27)	633.04	(76.30)	692.04	(107.30)	10.60	(0.00)	128.60	(129.00)
	Alkalinity	53.28	(3.45)	49.39	(3.68)	44.59	(1.80)	48.74	(1.80)	53.41	(9.11)	48.22	(5.61)
	Cl^-	3.55	(0.00)	7.19	(1.77)	3.55	(0.00)	6.39	(0.00)	3.91	(0.61)	3.91	(0.61)
	As	20.76	(0.00)	58.13	(0.00)	32.74	(2.70)	39.23	(1.32)	9.41	(1.81)	7.18	(0.15)
	Al	0.00	(0.00)	0.01	(0.00)	0.00	(0.00)	0.00	(0.00)	0.00	(0.00)	0.00	(0.00)
	Si	11.45	(0.00)	4.64	(0.00)	3.88	(0.30)	3.36	(0.11)	4.96	(0.06)	4.72	(0.11)
	Cu	bld	–	bld	–	bld	–	bld	–	bld	–	bld	–
	Fe	bld	–	bld	–	bld	–	bld	–	bld	–	bld	–
	Pb	1.77	–	1.39	–	1.43	(0.29)	1.65	(0.73)	1.63	(0.75)	1.18	(0.43)
	Zn	0.06	–	0.24	–	0.01	(0)	bld	–	bld	–	bld	–

*bdl: below detection limit

Fig. 1.7 Photo collection of the metallurgical site studied at Matehuala, San Luis Potosi over time showing the smelter during operation, after demolition and mechanical disturbance

See Figs. 1.7, 1.8 and 1.9.

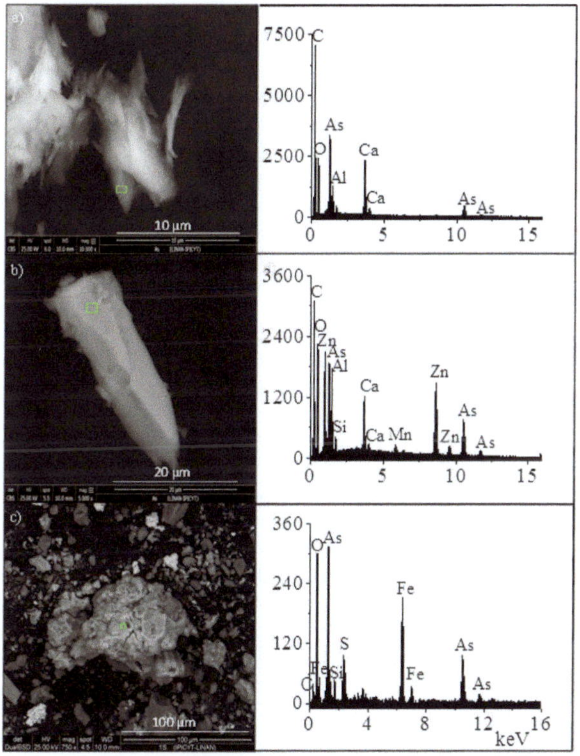

Fig. 1.8 SEM–EDS analysis of selected arsenic-bearing particles commonly found in contaminated soil samples. **a** Particles likely to be a calcium arsenate mineral as suggested by its fibrous morphology and chemical composition comprising Ca, As, and O. The C peak might come from calcite or the carbon tape used to mount the samples. **b** Particle likely to be a calcium and/or zinc arsenate mineral as suggested by its needle-like morphology and chemical composition comprised of Ca, Zn, As, and O. The C and Al peaks might come from other minerals such as calcite and aluminosilicates in soil samples. **c** Particle likely to be scorodite. All micrographs shown here were collected from the most contaminated soil sample (1 s)

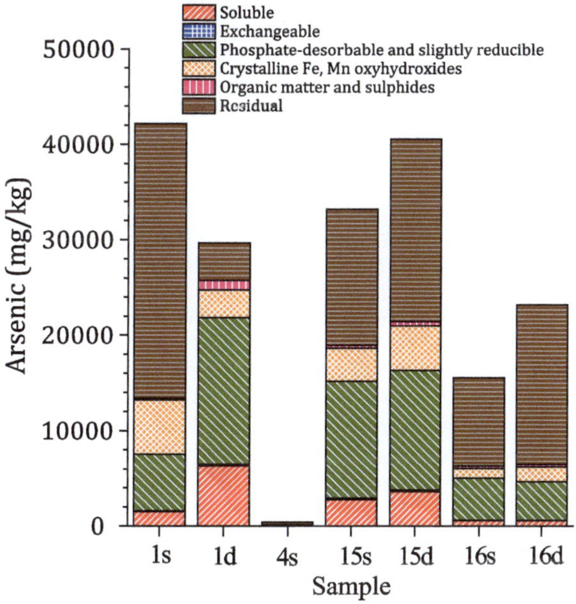

Fig. 1.9 Arsenic operational speciation in soil samples as determined by sequential chemical extractions

References

Alexandratos, V. G., Elzinga, E. J., & Reeder, R. J. (2007). Arsenate uptake by calcite: Macroscopic and spectroscopic characterization of adsorption and incorporation mechanisms. *Geochimica Et Cosmochimica Acta, 71*, 4172–4187. https://doi.org/10.1016/j.gca.2007.06.055

Ball, J. W., Nordstrom, D. K. (1991). *User's manual for WATEQ4F, with revised thermodynamic data base and text cases for calculating speciation of major, trace, and redox elements in natural waters (USGS Numbered Series No. 91–183), Open-File Report*. U.S. Geological Survey ; Books and Open-File Reports Section [distributor].

Bluteau, M.-C., & Demopoulos, G. P. (2007). The incongruent dissolution of scorodite—Solubility, kinetics and mechanism. *Hydrometallurgy, 87*, 163–177. https://doi.org/10.1016/j.hydromet.2007.03.003

Bothe, J. V., & Brown, P. W. (1999a). Arsenic immobilization by calcium arsenate formation. *Environmental Science and Technology, 33*, 3806–3811. https://doi.org/10.1021/es980998m

Bothe, J. V., & Brown, P. W. (1999b). The stabilities of calcium arsenates at 23+/–1 °C. *Journal of Hazardous Materials, 69*, 197–207. https://doi.org/10.1016/s0304-3894(99)00105-3

Botinelly, T. (1976). A review of the minerals of the alunite-jarosite, beudantite, and plumbogummite group. *J Res U Geol. Surv., U. S. Geological Survey 4.*

Castillo, F., Ávalos-Borja, M., Jamieson, H., Hernández-Bárcenas, G., & Martínez-Villegas, N. (2015). Identification of diagenetic calcium arsenates using synchrotron-based micro X-ray diffraction. *Bol Soc Geológica Mex, 67*, 479–491.

Castro Larragoitia, J., Kramar, U., & Puchelt, H. (1997). 200 years of mining activities at La Paz/San Luis Potosí/Mexico—consequences for environment and geochemical exploration. *Journal of Geochemical Exploration, 58*, 81–91. https://doi.org/10.1016/S0375-6742(96)00054-4

CETENAL. (1972). *Carta Edafológica Matehuala F-14-A-25*. Comisión de Estudios del Territorio Nacional.

Chen, X., Yang, L., Zhang, J., & Huang, Y. (2014). Exploration of As(III)/As(V) uptake from aqueous solution by synthesized calcium sulphate whisker. *Chinese Journal of Chemical Engineering, 22*, 1340–1346. https://doi.org/10.1016/j.cjche.2014.09.018

Craw, D., Falconer, D., & Youngson, J. H. (2003). Environmental arsenopyrite stability and dissolution: Theory, experiment, and field observations. *Chemical Geology, 199*, 71–82. https://doi.org/10.1016/S0009-2541(03)00117-7

Craw, D., & Pacheco, L. (2002). Mobilisation and bioavailability of arsenic around mesothermal gold deposits in a semiarid environment, Otago, New Zealand. *Scientific World Journal, 2*, 308–319. https://doi.org/10.1100/tsw.2002.101

Das, S., Essilfie-Dughan, J., & Hendry, M. J. (2015). Fate of adsorbed arsenate during phase transformation of ferrihydrite in the presence of gypsum and alkaline conditions. *Chemical Geology, 411*, 69–80. https://doi.org/10.1016/j.chemgeo.2015.06.031

Demopoulos, G. P. (2005). On the preparation and stability of scorodite. In *Arsenic metallurgy. TMS (The Minerals, Metals & Materials Society)* (pp. 25–50).

Di Cicco, A., Aquilanti, G., Minicucci, M., Principi, E., Novello, N., Cognigni, A., & Olivi, L. (2009). Novel XAFS capabilities at ELETTRA synchrotron light source. *Journal of Physics: Conference Series, 190*, 012043. https://doi.org/10.1088/1742-6596/190/1/012043

DOF. (2007). *Norma Oficial Mexicana NOM-147-SEMARNAT/SSA1–2004- Establece criterios para la caracterización y determinación de concentraciones de remediación de suelos contaminados por arsénico, bario, berilio, cadmio, cromo hexavalente, mercurio, níquel, plata, plomo, selenio, talio, vanadio.* (No. DOF 02/03/2007). SEMARNAT/SSA1, México.

Donahue, R., & Hendry, M. J. (2003). Geochemistry of arsenic in uranium mine mill tailings, Saskatchewan, Canada. *Applied Geochemistry, 18*, 1733–1750. https://doi.org/10.1016/S0883-2927(03)00106-9

Dove, P. M., & Rimstidt, J. D. (1985). The solubility and stability of scorodite, FeAsO4.2H2O. *American Mineralogist, 70*, 838–844.

Dutré, V., & Vandecasteele, C. (1995). Solidification/stabilisation of hazardous arsenic containing waste from a copper refining process. *Journal of Hazardous Materials, 40*, 55–68. https://doi.org/10.1016/0304-3894(94)00080-Z

EPA. (1994). *Method 200.7: Revision 4.4, Determination of metals and trace elements in water and wastes by inductively coupled plasma-atomic emission spectrometry*. United States Environmental Protection Agency, Cincinnati, Ohio.

Fernández-Martínez, A., Cuello, G. J., Johnson, M. R., Bardelli, F., Román-Ross, G., Charlet, L., & Turrillas, X. (2008). Arsenate incorporation in gypsum probed by neutron, X-ray scattering and density functional theory modeling. *Journal of Physical Chemistry A, 112*, 5159–5166. https://doi.org/10.1021/jp076067r

Filippi, M., Drahota, P., Machovič, V., Böhmová, V., & Mihaljevič, M. (2015). Arsenic mineralogy and mobility in the arsenic-rich historical mine waste dump. *Science of the Total Environment, 536*, 713–728. https://doi.org/10.1016/j.scitotenv.2015.07.113

Forray, F. L., Smith, A. M. L., Navrotsky, A., Wright, K., Hudson-Edwards, K. A., & Dubbin, W. E. (2014). Synthesis, characterization and thermochemistry of synthetic Pb–As, Pb–Cu and Pb–Zn jarosites. *Geochimica Et Cosmochimica Acta, 127*, 107–119. https://doi.org/10.1016/j.gca.2013.10.043

Frost, R. L., Sejkora, J., Čejka, J., & Keeffe, E. C. (2009). Vibrational spectroscopic study of the arsenate mineral strashimirite Cu8(AsO4)4(OH)4·5H_2O—Relationship to other basic copper arsenates. *Vibrational Spectroscopy, 50*, 289–297. https://doi.org/10.1016/j.vibspec.2009.02.002

Gámez Rodríguez, M. (2014). *Propiedad y empresa minera en la Mesa centro-norte de México.* Guanajuato, San Luis Potosí y Zacatecas, 1880–1910 (Doctorado). Universidad Autónoma de Barcelona.

Gołębiowska, B., Pieczka, A., & Franus, W. (2006). Olivenite-adamite solid solution from oxidation zone in Rędziny (West Sudetes, Poland). *Mineralogia Polonica, 37*, 101–110. https://doi.org/10.2478/v10002-007-0001-1

Guerin, H. (1941). On the alkali Earth arsenates: Study of the system As2O5-CaO-H_2O. *Ann Chim Paris, 16*, 101.

Haffert, L., Craw, D., & Pope, J. (2010). Climatic and compositional controls on secondary arsenic mineral formation in high-arsenic mine wastes, South Island, New Zealand. *New Zealand Journal of Geology and Geophysics, 53*, 91–101. https://doi.org/10.1080/00288306.2010.498403

Harvey, M. C., Schreiber, M. E., Rimstidt, J. D., & Griffith, M. M. (2006). Scorodite dissolution kinetics: Implications for arsenic release. *Environmental Science and Technology, 40*, 6709–6714. https://doi.org/10.1021/es061399f

Hayes, S. M., Root, R. A., Perdrial, N., Maier, R. M., & Chorover, J. (2014). Surficial weathering of iron sulfide mine tailings under semi-arid climate. *Geochimica Et Cosmochimica Acta, 141*, 240–257. https://doi.org/10.1016/j.gca.2014.05.030

Hernández-Bárcenas, G., Castillo, F., Ávalos-Borja, M., Martínez-Villegas, N.V. (2017). Síntesis de arseniatos de calcio (guerinita, haidingerita y farmacolita) morfológicamente similares a los encontrados en suelos contaminados. *Rev Int Contam Ambient 33*, 153–163. https://doi.org/10.20937/RICA.2017.33.01.14

Jia, Y., & Demopoulos, G. P. (2005). Adsorption of arsenite onto ferrihydrite from aqueous solutions: Influence of media (sulphate versus nitrate), added gypsum, and pH alteration. *Environmental Science and Technology, 39*, 9523–9527. https://doi.org/10.1021/es051432i

Johnson, J., Anderson, F., & Parkhurst, D. L. (2000). *Database thermo.com.V8.R6.230, Rev 1.11.* Lawrence Livermore National Laboratory, Livermore, California.

Julliot, F., Ildefonse, P., Morin, G., Calas, G., Kersabiec, A. M., & Benedetti, M. (1999). Remobilization of arsenic from buried wastes at an industrial site: Mineralogical and geochemical control. *Applied Geochemistry, 14*, 1031–1048.

Langmuir, D., Mahoney, J., MacDonald, A., & Rowson, J. (1999). Predicting arsenic concentrations in the porewaters of buried uranium mill tailings. *Geochimica Et Cosmochimica Acta, 63*, 3379–3394. https://doi.org/10.1016/S0016-7037(99)00259-8

Langmuir, D., Mahoney, J., & Rowson, J. (2006). Solubility products of amorphous ferric arsenate and crystalline scorodite (FeAsO4 $2H_2O$) and their application to arsenic behavior in buried mine tailings. *Geochimica Et Cosmochimica Acta, 70*, 2942–2956. https://doi.org/10.1016/j.gca.2006.03.006

Magalhaes, M. C. F., Pedrosa De Jesus, J. D., & Williams, P. A. (1988). The chemistry of formation of some secondary arsenate minerals of Cu(II), Zn(II) and Pb(II). *Mineralogical Magazine, 52*, 679–690. https://doi.org/10.1180/minmag.1988.052.368.12

Majzlan, J., Lazic, B., Armbruster, T., Johnson, M. B., White, M. A., Fisher, R. A., Plášil, J., Loun, J., Škoda, R., & Novak, M. (2012). Crystal structure, thermodynamic properties, and paragenesis of bukovskýite, Fe2(AsO4)(SO4)(OH)·9H2O. *Journal of Mineralogical and Petrological Sciences, 107*, 133–148. https://doi.org/10.2465/jmps.110930

Manz, M., & Castro, L. J. (1997). The environmental hazard caused by smelter slags from the Sta. Maria de la Paz mining district in Mexico. *Environmental Pollution, 98*, 7–13.

Martínez-Villegas, N., Briones-Gallardo, R., Ramos-Leal, J. A., Avalos-Borja, M., Castañón-Sandoval, A. D., Razo-Flores, E., & Villalobos, M. (2013). Arsenic mobility controlled by solid calcium arsenates: A case study in Mexico showcasing a potentially widespread environmental problem. *Environmental Pollution, 176*, 114–122. https://doi.org/10.1016/j.envpol.2012.12.025

Mehra, O. P., & Jackson, M. L. (1958). Iron oxide removal from soils and clays by a dithionite-citrate system buffered with sodium bicarbonate. *Clays and Clay Minerals, 7*, 317–327. https://doi.org/10.1346/CCMN.1958.0070122

Morin, G., & Calas, G. (2006). Arsenic in soils, mine tailings, and former industrial sites. *Elements, 2*, 97–101. https://doi.org/10.2113/gselements.2.2.97

Nishimura, T., Itoh, C. T., Tozawa, K., & Robins, R. G. (1985). The calcium-arsenic-water-air system. In *Proceedings of impurity control and disposal. presented at the 15th annual hydrometallurgical meeting* of CIM, Vancouver, Canada (pp. 2/1–2/19)

Nishimura, T., & Robins, R. G. (1998). A re-evaluation of the solubility and stability regions of calcium arsenites and calcium arsenates in aqueous solution at 25 °C. *Miner Process Extr Metall Rev Int J, 18*, 283–308. https://doi.org/10.1080/08827509808914159

Nordstrom, D. K., Majzlan, J., & Königsberger, E. (2014). Thermodynamic properties for arsenic minerals and aqueous species. *Reviews in Mineralogy and Geochemistry, 79*, 217–255. https://doi.org/10.2138/rmg.2014.79.4

Norvell, W. A., & Lindsay, W. L. (1982). Estimation of the concentration of Fe3+ and the (Fe3+)(OH–)3 ion product from equilibria of EDTA in soil. *Soil Science Society of America Journal, 46*, 710–715. https://doi.org/10.2136/sssaj1982.03615995004600040009x

Novák, F., Povondra, P., & Vtelensky, J. (1967). Bukovskýit, Fe3+2(AsO4)(SO4)(OH)7H2O, from Kank, near Kutná Hora, a new mineral. *Acta Univ Carol Geol, 4*, 297–325.

Paktunc, D., Foster, A., & Laflamme, G. (2003). Speciation and characterization of arsenic in Ketza River mine tailings using x-ray absorption spectroscopy. *Environmental Science and Technology, 37*, 2067–2074. https://doi.org/10.1021/es026185m

Parkhurst, D. L., Appelo, C. A. J. (2013). *PHREEQC (Version3)— A computer program for speciation, batch-reaction, one-dimensional transport and inverse geochemical calculations.* U.S. Geological Survey, Washinton, D.C.

Pierrot, R. (1964). Contribution to the mineralogy of natural calcium and calcium-magnesium arsenates. *Bull Soc Fr Miner Cristal, 87*, 169.

Ravel, B., & Newville, M. (2005). ATHENA, ARTEMIS, HEPHAESTUS: Data analysis for X-ray absorption spectroscopy using IFEFFIT. *Journal of Synchrotron Radiat., 12*, 537–541. https://doi.org/10.1107/S0909049505012719

Robins, R. G. (1984). The stability of arsenic in gold mine processing wastes. In V. Kudryk, D. A. Corrigan, & W. W. Lyan (Eds.), *Precious metals. The metallurgical society of AIME* (pp. 241–249).

Robins, R. G. (1982). *Hydrometallurgical, research, development and plant practice.* AIME.

Robins, R. G., Nishimura, T., Singh, P. (2001). Removal of arsenic from drinking water by precipitation, adsorption or cementation, In MF Ahmed, MA Ali, & Z. Adeel (Eds.), *Technologies for arsenic removal from drinking water*.

Robins, R. G., Tozawa, K. (1982). Arsenic removal from gold processing waste waters: the potential ineffectiveness of lime. *Can Inst Min Metall Pet, 75.*

Rodríguez-Blanco, J. D., Jiménez, A., & Prieto, M. (2007). Oriented Overgrowth of pharmacolite (CaHAsO4·2H2O) on gypsum (CaSO4·2H2O). *Crystal Growth & Design, 7*, 2756–2763. https://doi.org/10.1021/cg070222+

Sadiq, M., Zaidi, T. H., & Mian, A. A. (1983). Environmental behavior of arsenic in soils: Theoretical. *Water, Air, and Soil Pollution, 20*, 369–377. https://doi.org/10.1007/BF00208511

Salazar Amador, M. E. (2017.) *Fraccionamiento de arsénico en suelos contaminados con arseniatos en Matehuala, San Luis Potosí, México.* (Tesis de Licenciatura). Universidad Autónoma de San Luis Potosí, San Luis Potosi, S.L.P., México.

SE. (2006). *NMX-AA-132-SCFI-2001—Muestreo de suelos para la identificación y la cuantificación de metales y metaloides, y manejo de la muestra.* (No. CDU: 628.163). Secretaria de Economía, México.

Smedley, P. L., & Kinniburgh, D. G. (2002). A review of the source, behaviour and distribution of arsenic in natural waters. *Applied Geochemistry, 17*, 517–568. https://doi.org/10.1016/S0883-2927(02)00018-5

Sø, H. U. (2011). *Adsorption of arsenic and phosphate onto the surface of calcite as revealed by batch experiments and surface complexation modelling.* Technical University of Denmark.

Swash, P. M., Monhemius, A. J. (1995). Synthesis, characterization and solubility testing of solids in the Ca-Fe-AsO4 system. Sudbury '95—Mining Environ. *CANMET*, 17–28.

Swash, P. M., Monhemius, A. J. (1994). Hydrothermal precipitation from aqueous solutions containing iron(III), arsenate and sulphate. In *Hydrometallurgy '94* (pp. 177–190). Springer Dordrecht. https://doi.org/10.1007/978-94-011-1214-7_10

USEPA. (2007). *Test method 6200: Field portable x-ray fluorescence spectrometry for the determination of elemental concentrations in soil and sediment.* United States Environmental Protection Agency.

USEPA. (1996). *Method 3050B: Acid digestion of sediments, sludges, and soils. revision 2.* United States Environmental Protection Agency, Washington, DC.

Vaca-Escobar, K., Villalobos, M., & Ceniceros-Gómez, A. E. (2012). Natural arsenic attenuation via metal arsenate precipitation in soils contaminated with metallurgical wastes: III. Adsorption versus precipitation in clean As(V)/goethite/Pb(II)/carbonate systems. Appl. Geochem. *Geochemistry of Arsenic during Low-Temperature Water-Rock Interaction, 27*, 2251–2259. https://doi.org/10.1016/j.apgeochem.2012.01.011

Vaughan, D. J. (2006). Arsenic. *Elements, 2*, 71–75.

Villalobos, M., García-Payne, D. G., López-Zepeda, J. L., Ceniceros-Gómez, A. E., & Gutiérrez-Ruiz, M. E. (2010). Natural Arsenic attenuation via metal arsenate precipitation in soils contaminated with metallurgical wastes: I. Wet chemical and thermodynamic evidences. *Aquatic Geochemistry, 16*, 225–250. https://doi.org/10.1007/s10498-009-9065-4

Zhang, D., Yuan, Z., Wang, S., Jia, Y., & Demopoulos, G. P. (2015). Incorporation of arsenic into gypsum: Relevant to arsenic removal and immobilization process in hydrometallurgical industry. *Journal of Hazardous Materials, 300*, 272–280. https://doi.org/10.1016/j.jhazmat.2015.07.015

Zhu, X., Qi, X., Wang, H., Shi, Y., Liao, T., Li, Y., Liu, C., & Wang, X. (2014a). Characterization of high-arsenic sludge in copper metallurgy plant. *ResearchGate*, 173–186. https://doi.org/10.1002/9781118888056.ch21

Zhu, X., Qi, X., Wang, H., Shi, Y., Liao, T., Li, Y., Liu, C., & Wang, X. (2014b). Characterization of high-arsenic sludge in copper metallurgy plant. In *Characterization of minerals, metals, and materials 2014.*

Zhu, Y., Zhang, X., Xie, Q., Chen, Y., Wang, D., Liang, Y., & Lu, J. (2005). Solubility and stability of barium arsenate and barium hydrogen arsenate at 25oC. *Journal of Hazardous Materials, 120*, 37–44.

Zhu, Y. N., Zhang, X. H., Xie, Q. L., Wang, D. Q., & Cheng, G. W. (2006). Solubility and stability of calcium arsenates at 25 °C. *Water, Air, and Soil Pollution, 169*, 221–238. https://doi.org/10.1007/s11270-006-2099-y

Zouboulis, A. I., Kydros, K. A., & Matis, K. A. (1993). Arsenic(III) and arsenic(V) removal from solutions by pyrite fines. *Separation Science and Technology, 28*, 2449–2463. https://doi.org/10.1080/01496399308019748

Chapter 2
Arsenic Contamination in Indonesia

Ahmad Shoiful

Abstract Arsenic is one of the harmful substances that pose an adverse risk to human health. It is found naturally and is widely distributed in the Earth's crust in the form of inorganic and organic compounds; and therefore, it can be detected in water, soil, and air. Indonesia being one of the biggest mining nations in the world contributes significantly to the problem of arsenic contamination. Arsenic concentration in submarine mine tailings deposited in Buyat Bay, North Sulawesi, was 1.54 μg/L and 3.13 mg/kg wet weight, in bottom seawater and fish tissue, respectively. In another study, the concentration of arsenic in fish tissues obtained from a market in Ratatotok Sub-district and Manado City, North Sulawesi was 1.76 and 2.12 mg/kg wet weight, respectively. In 2014, arsenic was detected in foodstuff samples collected from the residents living near Gunung Pongkor, a gold mining area, at concentration of 2.27 mg/kg, 1.66 mg/kg, 2.71 mg/kg, 1.41 mg/kg, 2.71 mg/kg, in rice/grain, vegetable, cassava, banana, fish, respectively. Arsenic is also released from rock-water interactions in aquifer systems resulting in the deterioration of groundwater quality. Arsenic was also detected in coastal aquifers from big cities in North Coastal of Java, namely Jakarta, Semarang and Surabaya at concentrations of 15.47 ± 18.79, 1.25 ± 2.05 and 0.59 ± 0.26 μg/L, respectively. A study of arsenic concentration in foodstuff samples collected from Jakarta has been conducted. The highest concentration of arsenic was detected in fish and fish products (452.51 ng/g fresh weight of sample), followed by steamed white rice (85.53 ng/g fw), common food ready to be served (80.95 ng/g fw), home-made porridges (79.18 ng/g fw), while the concentration of arsenic in tap water (PDAM) was 0.72 ng/L. According to the above studies, arsenic is found in several cities in Indonesia. Hence, regular monitoring of arsenic in food stuffs or environmental samples in Indonesia is necessary to protect human health as well as to prevent global contamination of arsenic.

Keywords Arsenic · Indonesia · Mining activities · Food stuffs · Water

A. Shoiful (✉)
Research Center for Environmental and Clean Technology, National Research and Innovation Agency (BRIN), Building 820, KST. B.J. Habibie, Serpong, Tangerang Selatan, Banten 15314, Indonesia
e-mail: ahmad.shoiful@brin.go.id

B. Sen Gupta and N. Martínez-Villegas (eds.), *Arsenic Remediation of Food and Water*,
https://doi.org/10.1007/978-981-97-4764-1_2

Introduction

Arsenic is found naturally and is widely distributed in the Earth's crust in the form of inorganic and organic compounds (Yuan et al., 2022). Arsenic-contaminated groundwater has been reported in many countries and affects millions of people around the world (Podgorski & Berg, 2020). Around 108 countries and more than 230 million people are predicted to be affected by arsenic contamination in groundwater (Shaji et al., 2021).

Arsenic is a serious global concern due to its toxicity, persistence and bioaccumulation in biota. Exposure to arsenic can pose an adverse risk to human health, such as hyper pigmentation and keratosis, increased risk of skin, internal organ, and lung cancers, cardiovascular diseases and neuropathy (Kapaj et al., 2006). Arsenic contamination in groundwater is mainly associated with the geogenic origin (Alarcón-Herrera & Gutiérrez, 2022), while mining activity is a major source of arsenic contamination in surface water (Alonso et al., 2020).

The major pathway of arsenic exposure to humans is through food and drinking water. Many countries have established regulations concerning arsenic to protect public health. It has been reported that 176 countries including Indonesia have set regulations on arsenic in drinking water (Frisbie & Mitchell, 2022). A total diet study (TDS) to monitor exposure of the population to arsenic present in food stuffs and assess the health risk, has been carried out by many developed countries, such as France, Sweden, and Germany (Hackethal et al., 2021; Kollander et al., 2019; Sirot et al., 2018). Determination of arsenic in foods and health risk assessment has also been reported in many Asian countries, such as India, Bangladesh, Pakistan, and China (Joardar et al., 2022; Islam et al., 2017; Arain et al. 2009; Jiang et al., 2015).

This study aims to present the current status of arsenic contamination in Indonesia. Arsenic contamination in environmental samples, food stuffs and biota collected from papers are reported. In addition, Indonesian regulations regarding arsenic are also included.

Materials and Methods

Indonesian regulation on arsenic in water and food stuff are reviewed and collated in this study. Data concerning arsenic contamination in Indonesia presented in this paper were collected from the Scopus, the Elsevier Science Direct, and Google Scholar. In total, about 20 papers (published between 2010 and 2022) corresponding to the keywords "arsenic" and "Indonesia" were gathered. The results were expanded upon in the publication that followed.

Results and Discussion

Indonesian Regulations Regarding Arsenic

Arsenic contamination is one of global health concerns including Indonesia. The Indonesian government has issued several regulations for arsenic to protect human health and the environment (Table 2.1), which are described as follows.

a. Arsenic in River and Sea Water

The Government of Indonesia has set the maximum contaminant level in river water through Government Regulation (Peraturan Pemerintah) No. 22 Year (2021), Appendix VI regarding river water quality standard, and maximum concentration of arsenic is 0.05, 0.05, 0.05, and 0.1 mg/L in river water class 1, class 2, class 3, and class 4, respectively. In addition, in Appendix VIII, maximum concentration of arsenic is 0.025, and 0.012 mg/L, in sea water and marine biota, respectively. The river water quality class 1 can be utilized as raw drinking water, while river water quality class 2 is utilized for tourism infrastructure, aquaculture, and agriculture. The river water quality class 3 can be applied for aquaculture and agriculture. The class 4 is the lowest river water quality, which is only used for agriculture.

b. Arsenic in Food

The maximum level of arsenic in several groups of foodstuffis determined by the Indonesian government through Regulation of the National Agency of Drug and Food Control (Peraturan Badan Pengawas Obat dan Makanan) No. 5, Year (2018), regarding heavy metal contaminant in food. The maximum concentration of arsenic in food stuff varies from 0.01 mg/kg in mineral drinking water to 1 mg/kg in sugar.

c. Arsenic in Drinking Water and in Water for Sanitation Hygiene Purpose

The Ministry of Health defined the maximum level of arsenic in drinking water through the Ministry of Health (Permenkes) Regulation No. 492/MENKES/PER/IV/ 2010 regarding Quality Requirements for Drinking Water in order to safeguard the health of people from contaminants exposure through intake of water. The maximum level of arsenic in drinking water is 0.01 mg/L. This threshold limit value is in alignment with the provisional WHO guideline. While, the Regulation of Ministry of Health (Permenkes) No. 17 Year 2017 (regarding quality standard for environmental health, and health requirements for water for sanitation hygiene purpose, swimming pool water, solus per aqua, and public shower) has set the maximum concentration of arsenic in water for sanitation hygiene purpose (e.g., shower, washing food, washing cutlery, laundry, and raw drinking water) as 0.05 mg/L.

d. Arsenic in Cosmetic Products

Arsenic in cosmetic products is regulated by the Regulation of Head of the National Agency of Drug and Food Control (Peraturan Kepala Badan Pengawas Obat dan Makanan) No. 17 Year 2014 regarding requirements of microbes and heavy metals

Table 2.1 Regulatory limit of arsenic in various samples

Samples	Limit	References
River water	Class 1: 0.05 mg/L Class 2: 0.05 mg/L Class 3: 0.05 mg/L Class 4: 0.1 mg/L	Government Regulation No. 22/2021, Appendix VI
Sea water	0.025 mg/L	Government Regulation No. 22/2021, Appendix VIII
Marine biota	0.012 mg/L	Government Regulation No. 22/2021, Appendix VIII
Foodstuffs	Milk products: 0.1 mg/kg Fats, Oils and Oil Emulsions: 0.1 mg/kg Fat emulsions: 0.1 mg/kg Ice (edible ice): 0.2 mg/kg Fruits and vegetables: 0.15 mg/kg Jam, jelly: 1.0 mg/kg Candy and chocolate: 1.0 mg/kg Cerealia and cereal products: 0.1 mg/kg Bakery: 0.5 mg/kg Meats and meat products: 0.25 mg/kg Fish and fish products: 0.25 mg/kg Egg and egg products: 0.25 mg/kg Sweeteners: 1 mg/kg Sugar: 1.0 mg/kg Honey: 1.0 mg/kg Salt, spice, soup, sauce, salad: 0.15 mg/kg Baby foods: 0.1 mg/kg Dietary supplements: 0.02 mg/kg Mineral drinking water: 0.01 mg/kg Juice (fruit and vegetable): 0.10 mg/kg Coffee, tea, herbal tea: 1.0 mg/kg Alcoholic drink: 0.10 mg/kg Ready-to-eat snack: 0.25 mg/kg	Regulation of the National Agency of Drug and Food Control No. 5 Year (2018)
Cosmetic products	Less than 5 mg/kg or 5 mg/l	Regulation of Head of the National Agency of Drug and Food Control No. 17 Year 2014

contamination in cosmetics. The maximum level of arsenic is 5 ppm (5 mg/kg or 5 mg/L).

Arsenic in Environmental Samples

The occurrence of arsenic in environmental samples is associated with the mining activities. Buyat Bay, is a small bay, that is located in the north of Sulawesi which has been used as the tailing dumping area by production of gold mining since 1996, and was ceased in 2004. Arsenic concentration in submarine mine tailings deposited in Buyat Baywas found to be around 1.54 μg/L and 3.13 mg/kg wet weight, underneath seawater and in fish tissue, respectively (Shepherd et al., 2018). In a different study, in stream water, near Talawaan–Tatelu mining, North Sulawesi, concentration of arsenic was 10.00 μg/L (Palapa & Maramis, 2015).

The concentration of arsenic in water, sediment, and resident hair in the Ie Seu'um geothermal manifestation area, Aceh, have been assessed (Irnawati et al., 2021). The results show that the lowest arsenic concentration was 0.80–136.80 μg/L in river water, and the highest was 166.73 μg/L in hot spring water, while the concentration of arsenic in sediment were in the range of 2.56–6.86 mg/kg. In addition, arsenic in resident hair samples was in the range of n.d.–164.57 mg/kg. The detected arsenic in resident hair indicated that the community living near geothermal area may be exposed to arsenic.

Arsenic can also be released from rock-water interactions in aquifer systems resulting in deterioration of groundwater quality. Arsenic was detected in coastal aquifers from big cities in North Coastal of Java, namely Jakarta, Semarang and Surabaya at concentration of 15.47 ± 18.79, 1.25 ± 2.05 and 0.59 ± 0.26 μg/L, respectively (Suryono, 2015). A higher arsenic concentration was detected in other north coastal areas of Java Island, namely Rembang and Pati, at concentration of 0.173 and 0.265 mg/L in average, respectively (Rochaddi et al., 2020). The occurrence of arsenic in water from big cities might be influenced not only by rock weathering but also by human activities such as industrial. This conclusion is also supported by a study conducted in Sorong, West Papua, a pristine area, where a very low arsenic concentration was detected in river water and dug well at concentration of 0.01 and 0.01 mg/L, respectively (Mallongi et al., 2020). Concentration of arsenic in environmental samples, as indicated in various studies is summarized in Table 2.2 as follows.

Arsenic in Food Stuffs

Potential arsenic contamination in human is mainly through contaminated drinking water and food and dietary exposure. Monitoring of arsenic in rice is important because it is the main staple food in Indonesia. Since rice is the primary staple meal in Indonesia, it is crucial to monitor the presence of arsenic in it. Several studies have been conducted to identify arsenic contamination in rice. A study in Medan, North Sumatera, reported that arsenic concentration was 3.71 mg/kg; 3.40 mg/kg; 0.33 mg/kg; and 0.13 mg/kg, in inorganic red rice, organic brown rice, inorganic

Table 2.2 Concentration of arsenic in environmental samples

Sampling location	Concentration	References
Submarine mine tailings deposited, Buyat Bay, North Sulawesi	Bottom seawater: 1.54 µg/L, Fish tissue: 3.13 mg/kg wet weight	Shepherd et al. (2018)
Stream water near Talawaan–Tatelu mining North Sulawesi	Stream water: 10.00 µg/L	Palapa and Maramis (2015)
The Ie Seu'um geothermalarea, Aceh	Water: 0.80–166.73 µg/L Sediment:2.56–6.86 mg/kg Resident hair: n.d.–164.57 mg/kg	Irnawati et al. (2021)
Groundwater of big cities in North Coastal of Java	Jakarta: 15.47 ± 18.79 µg/L Semarang: 1.25 ± 2.05 µg/L Surabaya: 0.59 ± 0.26 µg/L Rembang: 0.229–0.310 mg/L Pati: 0.115–0.229 mg/L	Suryono (2015) Rochaddi et al. (2020)
Sorong, West Papua	River water: 0.01 mg/L Dug well: 0.01 mg/L	Mallongi et al. (2020)

white rice and organic black rice, respectively (Ginting et al., 2018). Rice grown and cultivated from artisanal gold mining site, Gunung Pongkor, Bogor contains arsenic at concentration of 2.27 mg/kg (Rahman et al., 2014). From the same area, arsenic was also detected in vegetable, cassava, banana, and fish at concentration of 1.66 mg/kg, 2.71 mg/kg, 1.41 mg/kg, and 2.71 mg/kg, respectively.

Various studies on arsenic contamination in fish have also been conducted. It was found that the concentration of arsenic in fish collected from West Java in 2017 was in a range of 2.89–36.05 mg/kg (Puspitasari et al., 2021). While, during the study in 2013, arsenic in fish collected from Jakarta fishing port was around 2.71 to 6.77 mg/kg (Koesmawati & Arifin, 2015). In another study, concentration of arsenic in fish tissues obtained from market in mine impacted area, Ratatotok Sub-district and Manado City, North Sulawesi was approximately 1.76 and 2.12 mg/kg wet weight, in the given order (Bentley & Soebandrio, 2017). The above results indicated that arsenic in fish exceeded the maximum level (0.25 mg/kg) set by Regulation of the National Agency of Drug and Food Control (Peraturan Badan Pengawas Obat dan Makanan) No. 5 Year (2018). Thus, indicating that arsenic may be bioaccumulated in fish tissues.

Arsenic was also detected at low level in edible bird's nest (EBNs). Indonesia is one the major producers of edible bird nests worldwide. As a result of its high protein content and nutritional value, edible bird's nest is also said to promote good health in Asian culture. The concentration of arsenic was 0.0145, 0.0061, 0.0009, and 0.0049 mg/kg in EBNs collected from Java, Sumatera, Sulawesi and Kalimantan, respectively (Wahyuni et al., 2022). Additionally, arsenic was detected in sea green turtle eggs and hawksbill turtle eggs collected at Venu Island, Kaimana, West Papua at concentration of 0.875 and 0.569 mg/kg, respectively (Tapilatu et al., 2020). Further

Table 2.3 Concentration of arsenic in different food stuffs (Suyanto et al., 2021)

Food group	Concentration (ng/g fresh weight)
Fish and fish product	452.51 ± 200.41
Steamed white rice	85.53 ± 6.45
Common food ready to be served	80.95 ± 15.50
Home-made of porridges	79.18 ± 2.83
Vegetable and vegetable products	54.24 ± 5.68
Legumes, nuts and seeds, and legume products	50.57 ± 0.72
Cereal products other than rice products	38.72 ± 4.70
Tuber and starchy root products	33.67 ± 1.56
Milk products	28.80 ± 3.67
Egg and egg products	26.32 ± 5.80
Fruit and fruit products	23.43 ± 1.31
Meat, poultry and their products	17.41 ± 1.01
Tap drinking water	0.72 ± 0.19
Refill drinking water	0.54 ± 0.06
Ground drinking water	0.32 ± 0.09

investigation on arsenic contamination is required to identify potential source of contaminations.

A complete diet study to assess the risk associated with arsenic in different food stuffs collected from Jakarta, has been conducted by Suyanto et al. (2021). Arsenic was detected at concentration of 0.32 in ground drinking water, 0.54 in refill drinking water, and 0.72 in ng/L tap drinking water (PDAM). Concentration of arsenic in different food stuffs is presented in Table 2.3 as follows.

Arsenic in Biota

Holothurians' has been widely used as a bio indicator of pollution in coastal sediment (Marrugo-Negrete et al., 2011). A study conducted at Totok Bay, nearby Buyat Bay, North Sulawesi presented that the concentration of arsenic was in the range of 10.38–118.72 μg/g in in body walls, and 24.65–219.68 μg/g in guts of Holothurian, while the concentration of arsenic in sediment was around 1.53 to 34.08 mg/g (Tamburini et al., 2022). A similar study was conducted to assess arsenic contamination in environment using *Pteris vittate. Pteris vittata*is is widely known as a bio indicator plant for heavy metal pollution. Concentration of arsenic in *Pteris vittate* was collected from gold mining areas along the Bone River in Gorontalo, Indonesia was in the range of 1.30–17,700 mg/kg, while the concentration of arsenic in soil samples of this area was in the range of n.d.–36,500 mg/kg (Gafur et al., 2022). These results indicated that

Table 2.4 Concentration of arsenic in biota

Sampling location	Sample and concentration	References
Totok Bay, nearby Buyat Bay, North Sulawesi	Holothurian Body walls: 10.38–118.72 μg/g Guts: 24.65–219.68 μg/g Sediment: 1.53 to 34.08 mg/g	Tamburini et al. (2022)
Bone River, Gorontalo	*Pteris vittate*: 1.30–17,700 mg/kg Soil: n.d.–36,500 mg/kg	Gafur et al. (2022)

bioaccumulation of arsenic in biota might have occurred. The summary of arsenic concentration in biota can be seen in Table 2.4 as follows.

Conclusion

This report provides a summary of Indonesian arsenic regulations and the most recent levels of arsenic pollution in Indonesia. In this research paper, the arsenic concentrations in foods, biota, and environmental specimens (river water, soil, and sediment) were examined. Arsenic has been found in food, river water, coastal aquifers, sea bottom sediment, and biota from many locations throughout Indonesia. Although mining and geothermal activity were the main sources of arsenic in Indonesia, arsenic pollution in urban areas was also discovered. Studies on arsenic contamination were mainly focused on mining impacted areas. Monitoring its presence in the environment, food stuffs, and drinking water from other areas such as urban area or industrial area should be the main focus for future studies. Further, a comprehensive diet research is required to assure Indonesians' health protection, especially considering their population growth, urbanization, and industrialization.

Concentration of arsenic in environmental samples (river water, soil, and sediment), food stuffs, and biota were presented in this paper. Arsenic was detected in river water, coastal aquifer, sea bottom sediment, food stuffs, and biota from several places in Indonesia. Even though the occurrences of arsenic in Indonesia were mainly associated with geothermal and mining activities, arsenic contamination in urban areas was also identified. Studies on arsenic contamination were mainly focused on mining impacted areas. Monitoring its presence in the environment, food stuffs, and drinking water from other areas such as urban and industrial areas should be the main focus for future studies. Due to increasing population, urbanization, and industrialization a comprehensive diet study is required to ensure health protection of Indonesian people.

References

Alarcón-Herrera, M. T., & Gutiérrez, M. (2022). Geogenic arsenic in groundwater: Challenges, gaps, and future directions. In *Current opinion in environmental science and health* (Vol. 27). Elsevier B.V. https://doi.org/10.1016/j.coesh.2022.100349

Alonso, D. L., Pérez, R., Okio, C. K. Y. A., & Castillo, E. (2020). Assessment of mining activity on arsenic contamination in surface water and sediments in southwestern area of Santurbán paramo, Colombia. *Journal of Environmental Management, 264*. https://doi.org/10.1016/j.jenvman.2020.110478

Arain, M. B., Kazi, T. G., Baig, J. A., Jamali, M. K., Afridi, H. I., Shah, A. Q., Jalbani, N., & Sarfraz, R. A. (2009). Determination of arsenic levels in lake water, sediment, and foodstuff from selected area of Sindh, Pakistan: Estimation of daily dietary intake. *Food and Chemical Toxicology, 47*(1), 242–248. https://doi.org/10.1016/j.fct.2008.11.009

Bentley, K., & Soebandrio, A. (2017). Dietary exposure assessment for arsenic and mercury following submarine tailings placement in Ratatotok Sub-district, North Sulawesi, Indonesia. *Environmental Pollution, 227*, 552–559. https://doi.org/10.1016/j.envpol.2017.04.081

Frisbie, S. H., & Mitchell, E. J. (2022). Arsenic in drinking water: An analysis of global drinking water regulations and recommendations for updates to protect public health. *PLoS ONE, 17*(4 April 2022). https://doi.org/10.1371/journal.pone.0263505

Gafur, N. A., Sakakibara, M., Komatsu, S., Sano, S., & Sera, K. (2022). Environmental survey of the distribution and metal contents of Pteris vittata in Arsenic–Lead–Mercury-contaminated gold mining areas along the Bone river in Gorontalo Province, Indonesia. *International Journal of Environmental Research and Public Health, 19*(1). https://doi.org/10.3390/ijerph19010530

Ginting, E. E., Silalahi, J., & Putra, E. D. (2018). Analysis of arsenic in rice in medan, north sumatera Indonesia by atomic absorption spectrophotometer. *Oriental Journal of Chemistry, 34*(5), 2651–2655. https://doi.org/10.13005/ojc/340557

Government Regulation (PP) No. 22 Year 2021. Implementation of Environmental Protection and Management.

Hackethal, C., Kopp, J. F., Sarvan, I., Schwerdtle, T., & Lindtner, O. (2021). Total arsenic and water-soluble arsenic species in foods of the first German total diet study (BfR MEAL Study). *Food Chemistry, 346*. https://doi.org/10.1016/j.foodchem.2020.128913

Irnawati, I., Idroes, R., Zulfiani, U., Akmal, M., Suhartono, E., Idroes, G. M., Muslem, M., Lala, A., Yusuf, M., Saiful, S., Suhud, K., Safitri, E., & Jalil, Z. (2021). Assessment of arsenic levels in water, sediment, and human hair around ie seu'um geothermal manifestation area, aceh, indonesia. *Water (Switzerland), 13*(17). https://doi.org/10.3390/w13172343

Islam, M. S., Ahmed, M. K., Habibullah-Al-Mamun, M., & Eaton, D. W. (2017). Arsenic in the food chain and assessment of population health risks in Bangladesh. *Environment Systems and Decisions, 37*(3), 344–352. https://doi.org/10.1007/s10669-017-9635-8

Jiang, Y., Zeng, X., Fan, X., Chao, S., Zhu, M., & Cao, H. (2015). Levels of arsenic pollution in daily foodstuffs and soils and its associated human health risk in a town in Jiangsu Province, China. *Ecotoxicology and Environmental Safety, 122*, 198–204. https://doi.org/10.1016/j.ecoenv.2015.07.018

Joardar, M., Das, A., Chowdhury, N. R., Mridha, D., Das, J., De, A., Majumder, S., Majumdar, K. K., & Roychowdhury, T. (2022). Impact of treated drinking water on arsenicosis patients with continuous consumption of contaminated dietary foodstuffs: A longitudinal health effect study from arsenic prone area, West Bengal, India. *Groundwater for Sustainable Development, 18*, 100786. https://doi.org/10.1016/j.gsd.2022.100786

Kapaj, S., Peterson, H., Liber, K., & Bhattacharya, P. (2006). Human health effects from chronic arsenic poisoning—A review. *Journal of Environmental Science and Health—Part A Toxic/Hazardous Substances and Environmental Engineering, 41*(10), 2399–2428. https://doi.org/10.1080/10934520600873571

Koesmawati, T. A., & Arifin, Z. (2015). Mercury and Arsenic content in seafood samples from the Jakarta fishing port, Indonesia. *Marine Research in Indonesia, 40*(1), 9–16. https://doi.org/10.14203/mri.v40i1.76

Kollander, B., Sand, S., Almerud, P., Ankarberg, E. H., Concha, G., Barregård, L., & Darnerud, P. O. (2019). Inorganic arsenic in food products on the Swedish market and a risk-based intake assessment. *Science of the Total Environment, 672*, 525–535. https://doi.org/10.1016/j.scitotenv.2019.03.289

Mallongi, A., Novitasari, D. A. T., Noor, N. B., Fatmawati, Saleh, M., Sehalyana, Apollo, & Nur, R. (2020). Health risk analysis of phenol and arsenic exposure among Kokoda communities village Sorong, West Papua. *Enfermeria Clinica, 30*, 123–126. https://doi.org/10.1016/j.enfcli.2019.10.055

Marrugo-Negrete, J., Pinedo-Hernández, J., Marrugo-Madrid, S., Navarro-Frómeta, E., & Sergi Díez, &. (2011). Sea cucumber as Bioindicator of trace metal pollution in coastal sediments. https://doi.org/10.1007/s12011-020-02308-3/Published

Palapa, T. M., & Maramis, A. A. (2015). Heavy metals in water of stream near an Amalgamation tailing ponds in Talawaan—Tatelu gold mining, North Sulawesi, Indonesia. *Procedia Chemistry, 14*, 428–436. https://doi.org/10.1016/j.proche.2015.03.058

Podgorski, J., & Berg, M. (2020). Global threat of arsenic in groundwater. *Science, 368*, 845–850. http://science.sciencemag.org/

Puspitasari, R., Purbonegoro, T., Oktaviyani, S., Rositasari, R., & Kaisupy, M. T. (2021). A preliminary study of metal accumulation in fish from Cimanuk and Cimandiri, West Java, Indonesia. *IOP Conference Series: Earth and Environmental Science, 919*(1). https://doi.org/10.1088/1755-1315/919/1/012040

Rahman, A., Hananto, M., Laelasari, E., Adi, H.K., Widodo, S., Lasut, D. (2014). Health risk estimates from exposure to minerals in artisanal gold mining site in Gunung Pongkor, Bogor, Indonesia. In *Full Paper Proceeding of International Trend in Multidisciplinary Academic Research* 2014; Global Illuminators (Vol. 1, pp. 467–490). Barcelona, Spain. ISBN 978-969-9948-24-4. www.globalilluminators.org

Regulation of Ministry of Health (Permenkes) No. 492/MENKES/PER/IV/2010 regarding quality requirements for drinking water.

Regulation of Ministry of Health (Permenkes) No. 17 Year 2017 regarding quality standard for environmental health, and health requirements for water for sanitation hygiene purpose, swimming pool water, solus per aqua, and public shower.

Regulation of the National Agency of Drug and Food Control (Peraturan Badan Pengawas Obat dan Makanan) No. 5 Year 2018 regarding heavy metal contaminant in food.

Rochaddi, B., Sabdono, A., & Zainuri, M. (2020). Heavy metal (As and Hg) contamination of shallow groundwater in the coastal areas of Pati and Rembang, Central Java, Indonesia. *IOP Conference Series: Earth and Environmental Science, 530*(1). https://doi.org/10.1088/1755-1315/530/1/012035

Shaji, E., Santosh, M., Sarath, K. v., Prakash, P., Deepchand, V., & Divya, B. v. (2021). Arsenic contamination of groundwater: A global synopsis with focus on the Indian Peninsula. *Geoscience Frontiers, 12*(3). https://doi.org/10.1016/j.gsf.2020.08.015

Shepherd, T., Rumengan, I., & Sahami, A. (2018). Post-depositional behaviour of mercury and arsenic in submarine mine tailings deposited in Buyat Bay, North Sulawesi, Indonesia. *Marine Environmental Research, 137*, 88–97. https://doi.org/10.1016/j.marenvres.2018.02.028

Sirot, V., Traore, T., Guérin, T., Noël, L., Bachelot, M., Cravedi, J. P., Mazur, A., Glorennec, P., Vasseur, P., Jean, J., Carne, G., Gorecki, S., Rivière, G., & Hulin, M. (2018). French infant total diet study: Exposure to selected trace elements and associated health risks. *Food and Chemical Toxicology, 120*, 625–633. https://doi.org/10.1016/j.fct.2018.07.062

Suryono, C. A. (2015). The toxic metal arsenic contamination of the coastal aquifers in the North Coast of Java, Indonesia. *Jurnal Kelautan Tropis, 18*(2), 76–81. ISSN 0853-7291.

Suyanto, Lioe, H. N., Giriwono, P. E., & Fardiaz, D. (2021). Total arsenic in complementary food and its exposure assessment for children aged 6–24 months. *Food Control, 122*. https://doi.org/10.1016/j.foodcont.2020.107786

Tamburini, M., Badocco, D., Ercadi, R., Turicchia, E., Zampa, G., Gasparini, F., Ballarin, L., Guerra, R., Lasut, M. T., Makapedua, D. M., Mamuaja, J., Pastore, P., & Ponti, M. (2022). Bioaccumulation of mercury and other trace elements in the edible Holothurian Holothuria (Halodeima) atra in relation to gold mining activities in North Sulawesi, Indonesia. *Frontiers in Marine Science, 9*. https://doi.org/10.3389/fmars.2022.863629

Tapilatu, R. F., Wona, H., Siburian, Rima H. S., (2020). Data on environmental contaminants in sea turtle eggs at Venu Island, Kaimana—West Papua, Indonesia. *Data in Brief, 31*(105778).

Wahyuni, D. S., Latif, H., Sudarwanto, M. B., Basri, C., & Thong, D. (2022). An investigation of heavy metals in edible bird's nest from Indonesia using inductively coupled plasma mass spectrometry. *Veterinary World, 15*(2), 809–816. https://doi.org/10.14202/vetworld.2022.509-516

Yuan, C., Li, Q., Sun, Z., Zhang, W., Chen, J., Chen, Z., Na, S., & Sun, H. (2022). Chemical oxidation of arsenic in the environment and applications: A mini review. *Pedosphere, 32*. https://doi.org/10.1016/j.pedsph.2022.06.033.

Chapter 3
Arsenic Contamination of Groundwater in Myanmar: A Review by the Irrigation and Water Utilization Management Department

Phyo Wai and Mya Thandar Khin

Abstract Myanmar, in Southeast Asia, is categorized as a Least Developed Country, facing substantial challenges towards meeting 2030 Sustainable Development Goal targets, including ensuring access to safe water and sanitation. IWUMD is currently conducting heavy metals pollution of both surface and ground water especially for irrigation and domestic usages. From 2004 to 2011 testing the Ayeyarwaddy Region, tested in 17 out of 26 township and ground water is saline remaining a township in general. Tested the 13 blanket and 4 selected testing township and a total of (1,239,692) samples tested for arsenic content. In Ayeyarwaddy Region, the most arsenic contamination is in shallow tube well about 50–150 ft depth water resources. To reduce high content of arsenic in groundwater, we can use very simple method-can used gravel sizes can be chosen as 15 mm, 12 mm, 10 mm and also coarse sand, medium sand and fine sand filter should be applied in this gravel sand filter.

Keywords Ayeyarwaddy region · Arsenic · Ground water · Shallow tube well · Gravel sand filter

Introduction

Water is essential to human life and the health of the environment water quality is commonly defined by its physical chemical biological and aesthetic (appearance and smell) characteristic. A healthy environment is one in which the water quality supports a rich and varied community of organism and protects public health In

P. Wai (✉)
IWUMD, MOALI, Yangon, Myanmar
e-mail: phyowai2004@gmail.com

M. T. Khin
Environmental Conservation Department, Ministry of Natural Resources and Environmental Conservation, Nay Pyi Taw, Myanmar

B. Sen Gupta and N. Martínez-Villegas (eds.), *Arsenic Remediation of Food and Water*,
https://doi.org/10.1007/978-981-97-4764-1_3

retrospect the discovery of Arsenic contamination in drinking water. In Myanmar was dated back to mid-1999 when traces of Arsenic less than 50 μg/L were defined by Myanmar. Science and Technology Research Departments Laboratory in water supply provided by Water Resources Utilization Department (WRUD) during its mission in collaboration with UNICEF conduct on sampling of water sample from selected shallow tube well (STW) Thabaung, Laymyatnar, Kyonpyaw and Hinthada township during March-May2000.

Arsenic is a natural element found widely in the earth's crust. It may be found in some drinking water supplies, including tube wells. Exposure to high level of arsenic can cause health effects. It is also a byproduct of copper smelling, mining and coal burning. Arsenic can enter the water supply from natural deposits in the earth or from industrial and agricultural pollution. Heating or boiling water will not remove arsenic because some of water evaporates during the boiling process arsenic concentration can actually increase slightly as the water is boiled. Additionally chlorine (bleach) disinfection will not remove arsenic. Arsenic may be enter lake, rivers or ground water naturally when mineral deposits or rock containing arsenic dissolve.

Background

Arsenic Mitigation Program in Myanmar was carried out by the collaboration among the Ministry of Health, Department of Rural Development and Irrigation and Water Utiliza-tion Management Department. From 2004 to 2011 testing the Ayeyarwaddy Region, tested in 17 out of 26 township and ground water is saline remaining a township in general. Tested the 13 blanket and 4 selected testing township and a total of (1,239,692) samples tested for arsenic content.

In that (123,962) sample arsenic content is included (19,183) samples from deep tube well (15.48%), (75,577) samples from shallow tube well (60.97%), (26,835) samples from Dug well (21.648%), (1486) samples from ponds (1.20%) and (881) samples from river and stream (0.71%).Therefore, in Ayeyarwaddy Region, the most arsenic contamination is in shallow tube well about 50–150 ft depth water resources.

Surface Geology

As a Geology map, Ayeyarwaddy Region, is mostly composed by Hollocence Younger Alluvial at Pathein, Maubin, Pyapon, Myaungmya and Eastern HinthadaDistrict. Tested the 13 blanket and 4 selected testing township and a total of (1,239,692) samples tested for arsenic content (Figs. 3.1, 3.2, 3.3 and 3.4).

But Thabaung Township and Ngaputaw Township in Pathein District is composed by Paleocene-Eocene, Mollase–type units Paunggyi formation. Eocene-unit and Paleocene Flysh type Ngapali formation Kwingu formation and their equivalent and

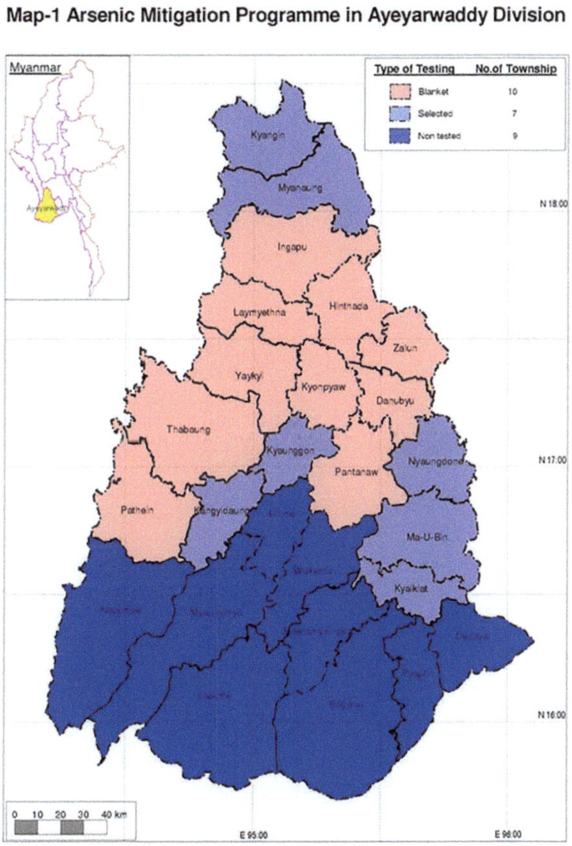

Fig. 3.1 Arsenic mitigation program in Ayeyarwaddy region

Kyangin Township western Hinthada Districts is composed by Oligocene, lower. Pegu group Minbu basin and its equivalents.

So that, in Ayeyarwaddy Region arsenic contamination of water source is mostly founded in Central Ayeyarwaddy Region and Thabaung, Ngaputaw and southern Pathein township are composed by hard rock and another place Myaungmya and Pyapon district are composed by saline water intrusion area.

Provided Safe Water Option in Ayeyarwaddy Region

For the rural people to have a good and clean drinking water, Irrigation and Water Utilization Management Department (Water Resources), Ayeyarwaddy is provided that according by Regional Government Budget from 2013–2014 to 2019–2020 fiscal year, (276) nos. of Ø4″ Drinking water deep tube well, (249) nos. of drinking water

Fig. 3.2 Arsenic contaminated water source by township

Sr	Township	No. of tested sample	Sample (As>10µ g/l)	Population (As>10 µgl)	Sample (As>50µ g/l)	Population (As>50 µgl)
1	Kyonpyaw	19,301	6,931	83,177	2,392	29,931
2	Hinthada	23,999	5,995	61,846	1,725	18,885
3	Zalun	10,415	4,316	50,722	888	10,259
4	Kyaungkone	6,996	3,484	50,034	1,508	20,874
5	Pantanaw	9,348	3,252	58,926	839	17,017
6	Nyaungdone	7,087	2,356	20,116	393	4,767
7	Yaykyi	9,587	2,300	41,148	529	10,578
8	Ingapu	8,918	1,794	17,765	342	3,470
9	Danubyu	4,785	1,539	30,716	367	6,124
10	Laymyathnar	6,053	1,255	22,660	195	2,922
11	Tharboung	4,088	1,041	19,987	562	10,385
12	Ma U Bin	4,576	959	16,487	240	3,147
13	Kangyidaunk	3,696	436	5,944	73	1,205
14	Kyeiklat	632	180	4,258	95	2,384
15	Pathein	3,342	170	5,059	4	77
16	Myanaung	1,002	137	425	9	35
17	Kyankin	137	33	763	0	
	Total	123,962	36,178	490,033	10,161	142,060

tank (257) nos. of pump house and (257) set of Engine and Compressor (Figs. 3.5 and 3.6).

Results

As a result of provided by Irrigation and Water Utilization Management Department (Water Resources) of Ayeyarwaddy Region in (276) tube wells, arsenic contamination are (93) tube wells are 0 ppb, (169) tube wells are less than 10 ppb, (12) tube wells are among the 10 and 50 ppb and (2) tube wells are above the 50 ppb in (28) townships (Figs. 3.7, 3.8, 3.9 and 3.10).

How to Do Safe Water Option?

To reduce high content of arsenic in groundwater, we can use very simple method can used gravel sizes can be chosen as 15 mm, 12 mm, 10 mm and also coarse sand, medium sand and fine sand filter should be applied in this gravel sand filter. In our country, this system can be installed low cost rate for rural area. This method can be

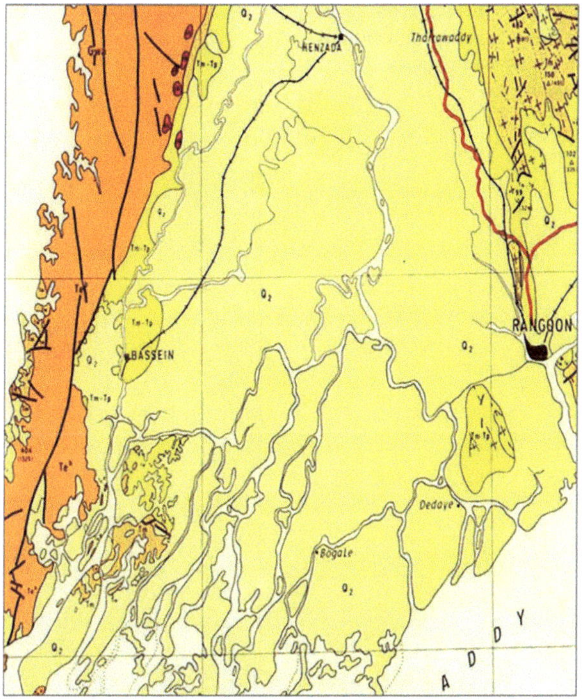

Fig. 3.3 Surface geology and arsenic contamination in drinking water sources

used not only to remove an arsenic contamination water sources but also the remove a high content iron water sources (Figs. 3.11, 3.12 and 3.13).

Conclusions

As a Irrigation and Water Utilization Management Department (IWUMD) Ayeyarwaddy Region according to Regional Government Budget, provided the safe water, option by drilling Ø4″ tube well is about 30–40 tube well Yearly. So this is not coverage for safe water option for Ayeyarwaddy all of us have responsibility to share the knowledge of arsenocosis in Ayeyarwaddy Region. Arsenic removal mechanism of gravel sand filter should be installed widely in these townships.

Regional Government budget is Limitation for department and need to increase public awareness especially for local people and regular monitoring on ground water system of this area must be carried out and need to strengthen cooperation and coordination among government and non-government agencies.

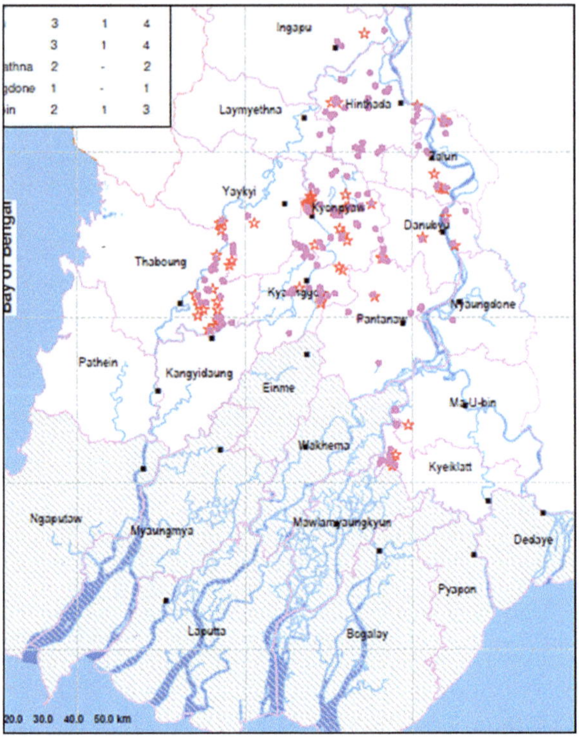

Fig. 3.4 Surface geology and arsenic contamination in drinking water sources by township

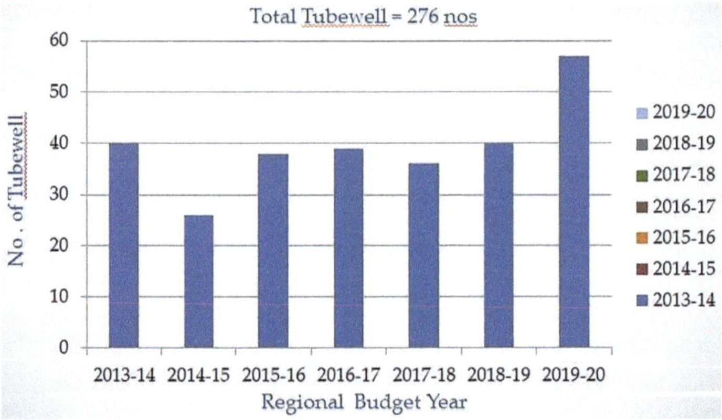

Fig. 3.5 No. of tested drinking water tube well by year

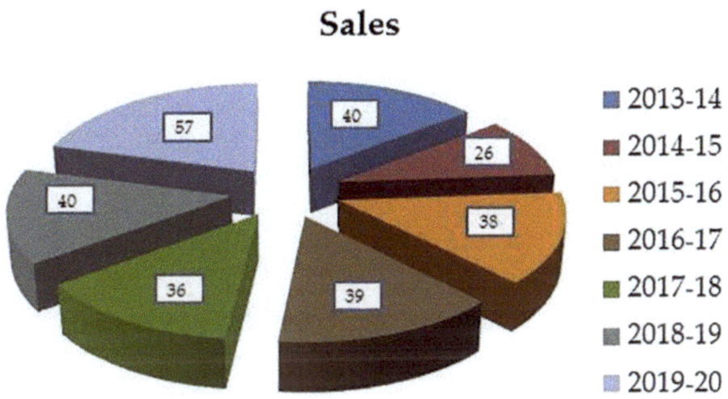

Fig. 3.6 No. of tested drinking water tube well by year

Fig. 3.7 Difference level of aresenic contamination by townships

Sr.	Tsp	No. of Tubewell					Remark
		As=O ppb	0<As<10 ppb	10< As<50 ppb	As_>50 ppb	Total Tubewell	
1	Pathein	6	8			14	
2	kangyidaunt	2	8			10	
3	Thaboung		10			10	
4	Kyonpyaw	17	16	3	1	37	
5	Kyaunggon	13	18	2	1	34	
6	Ngaputaw	2	2			4	
7	Yaykyi	5	6	2		13	
8	Hinthada	6	9	1		16	
9	Zalun	8	6			14	
10	Laymyethmar		5			5	
11	Ingapu	2	6	1		9	
12	Myanaung	1	4			5	
13	Kyankhein		2			2	
14	Myaungmya	3	11	3		17	
15	Warkhema		6			6	
16	Kyone mangee		1			1	
17	Shwe laung		1			1	
18	Eainme	2	7			9	
19	Lattputar		1			1	
20	Mawlamyinekyun		3			3	
21	Maubin		3			3	
22	Nyaungdone	2	3			5	
23	Danubyu	2	5			7	
24	Pantanaw	7	9			16	
25	Pyabon	8	5			13	
26	Bogalay	7	6			13	
27	Daydaye		3			3	
28	Kyiklat		5			5	
Total		93	169	12	2	276	

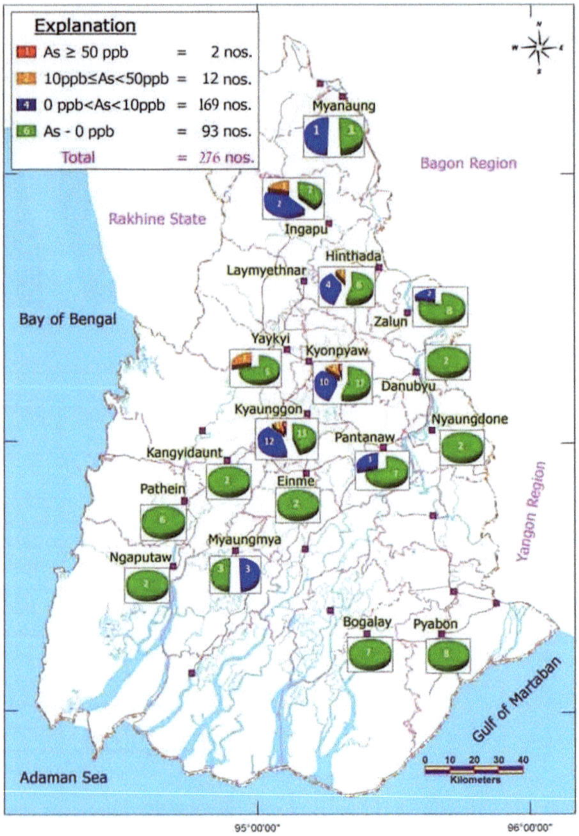

Fig. 3.8 Distribution of aresenic contamination

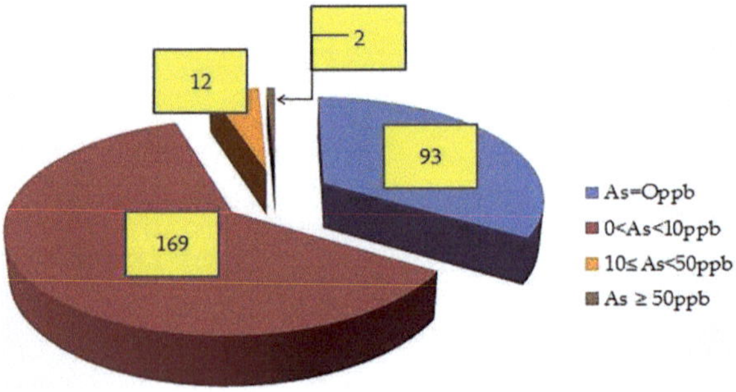

Fig. 3.9 Difference level of arsenic contamination

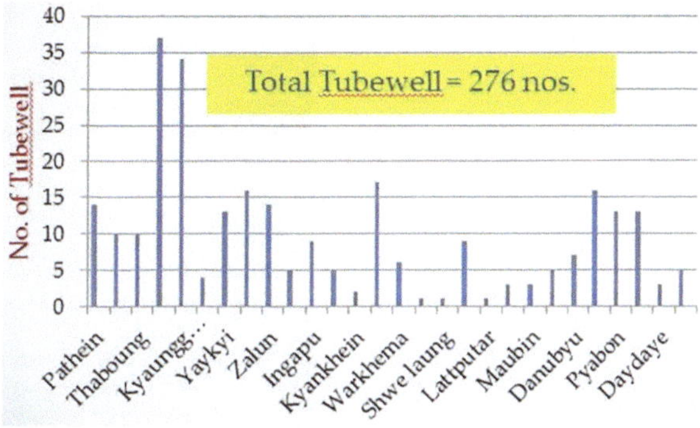

Fig. 3.10 No of tube well from difference township

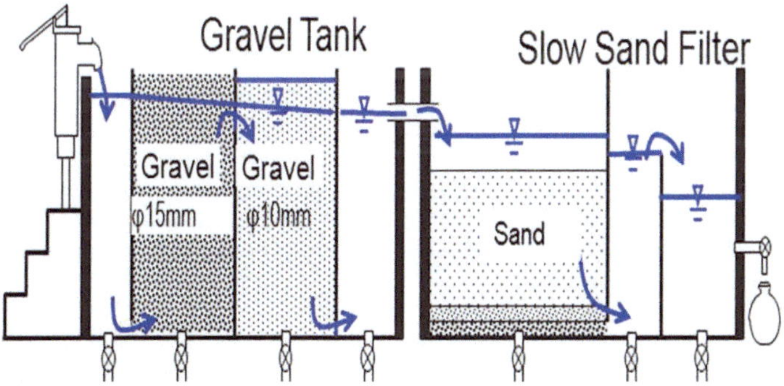

Fig. 3.11 Layout of filter

Fig. 3.12 Gravel sand filter demonstration

Fig. 3.13 Gravel sand filter demonstration

References

WRUD/UNICEF. (2006). *An analysis of arsenic contamination in drinking water sources at Ayeyarwaddy Division.* Water Resources Utilization Department and UNICEF report.

Htin, M. K. (2021). *Water resources management for arsenic contamination in Ayeyarwaddy Region, IWUMD, Ayeyarwaddy Region.*

Chapter 4
Assessing Hazards of Arsenic Leakage in Multi-layered Aquifer System in a Part of Middle Ganga Plains in Northern India

N. C. Mondal, L. Surinaidu, S. Ahmed, and V. M. Tiwari

Abstract In view of dwindling groundwater reserve and arsenic contamination in the shallow aquifers, indiscriminate pumping from the deeper aquifers is taking place at several parts of the Ganga river basin of Northern India. The severity of this uncontrolled pumping has disturbed the multi-layered aquifer hydrodynamics. The major concern is the inter-aquifer leakage through the intervening clay layers of varying thickness which has been deduced by dense geophysical measurements. The heavy pumping from the deeper aquifer (largely uncontaminated) and reduced thickness of the confining clay layer at places between the two aquifers offer a possible threat of leakage, and the arsenic-contaminated groundwater entering into the deeper aquifer, deteriorating the groundwater quality seriously. We present a groundwater flow model of a multi-layered aquifer system from a part of the Middle Ganga Plains (MGP). The simulation presented here is for understanding the hydrodynamics under stress conditions, and for assessing groundwater leakages. The model utilizes a previously 3-D aquifer geometry obtained from geophysical investigations and aquifer properties from in-situ measurements. The results reveal that the groundwater dynamics between two aquifers with increasing pumping from the deeper zone is leading to rapid depletion of the piezometric head, and possible mixing of arsenic-contaminated water to that from the shallow aquifer. This reversal in the inter-aquifer leakage has been simulated in both regional and micro-scales, and the areas vulnerable to the leakage reversal have been indicated as an indicator of arsenic hazard. These results provided safe pumping rates at a micro-scale for the marginal farmers and sites for the decision-makers to provide an alternate supply of water.

N. C. Mondal (✉) · L. Surinaidu · V. M. Tiwari
CSIR-National Geophysical Research Institute, Hyderabad, India
e-mail: ncmngri@gmail.com; mondal@ngri.res.in

N. C. Mondal · V. M. Tiwari
Academy of Scientific and Innovative Research (AcSIR), Ghaziabad, India

S. Ahmed
Maulana Azad National Urdu University, Hyderabad, India

B. Sen Gupta and N. Martínez-Villegas (eds.), *Arsenic Remediation of Food and Water*,
https://doi.org/10.1007/978-981-97-4764-1_4

Keywords Multi-layered aquifer system · Groundwater modeling · Hydrodynamics · Reversal leakages · Arsenic threats · A part of Ganga river basin · Northern India

Introduction

In the face of the rapid growth rate among the major economies in the world, India is undergoing a transformation with large-scale development in various sectors such as agriculture, industry, infrastructure facilities, energy resources, mining, etc. All these activities relate to Mother Earth in one way or the other and require precise geoscientific knowledge of the subsurface conditions at various scales of driving process-based parameters. For example, agriculture needs precise information on soil, and moisture conditions as well as assured water supply. Water itself is singularly the most vital resource for drinking, health, agriculture, industry, mining, urban development, etc. However, it is unevenly distributed not only in the Ganga river basin of Northern India but also globally.

The Ganga basin is one of the largest river basins in the world (area: ~8.6×10^5 km^2) and has been sustaining one of the largest and densest global populations (~10% of the global population) under increasing the stress for last few years in terms of freshwater availability and its deterioration water quality (GRBMP, 2015). The river is of high importance with an estimated 440 million people directly or indirectly depending on the water that the Ganga and its tributaries provide for agriculture, drinking, hydropower generation, navigation, and ecosystem services (Anand et al., 2018). It has been estimated that the average annual water potential and utilizable flow excluding groundwater are about 525.02 and 250.00 km^2, respectively (CWC, 2005). Its contribution comes only from monsoon precipitation but the storage of groundwater has declined at the rate of 2.0 cm/year between 2002 and 2013 (Asoka et al., 2017). In the evidence, the significant depletion of groundwater storage in many regions around the world including the Ganga basin in Northern India (Rodell et al., 2009; Tiwari et al., 2009) has been observed and attributed to groundwater pumping (abstraction) for the irrigation (Tiwari et al., 2009) caused the reduction of base flow triggering Gangs River being dried in summer (Mukherjee et al., 2018). People/marginal farmers are under pressure to trap deeper and deeper aquifers (depth: ~100 m or more, i.e., AQ-II) instead of shallower ones (Depth: < 100 m, i.e., AQ-I) for the groundwater extraction (Dabas et al., 2022; Mondal & Ajaykumar, 2022). The excessive pumping from deeper aquifers triggered an imbalance in the hydrodynamics and resulted in influencing the leakage of arsenic-contaminated water from the AQ-I to AQ-II (Muralidharan, 1998). Therefore, in the beginning, it has been encouraged for the construction of deeper wells only in areas with thick clay layers overlying the deeper aquifer to avoid any vertical leakage and also suggested the optimum groundwater pumping (Muralidharan, 1998).

In general, groundwater is extracted mainly from the AQ-I in the entire Middle Ganga Plains (MGP), India, mostly affected by arsenic contamination at several places along the river tracts (Saha, 2009; Saha & Sahu, 2016). Due to the indiscriminate over-exploitation from both the AQ-I and AQ-II, the detailed hydrodynamic is unknown due to the limited geoscientific knowledge (CGWB, 2012). Identifying and assessing the impact of irrigation and human activities on groundwater resources of the multi-aquifer system are the keys to finding the occurrence of groundwater resources, drinking water suitability, availability, and its dynamic. Thus, it is essential to explore the geoscientific knowledge of groundwater resources using hydrogeological, geophysical, and aquifer modeling for its sustainable management.

Our work gains the knowledge of the hydrodynamic systems made available with greater detail and precision, and the flow is used to simulate using the calibrated and validated regional and micro-scales groundwater flow model in a part of the MGP (Fig. 4.1). Further groundwater balance aquifer-wise is being estimated with the assigned zones of interest particularly taking the AQ-I and AQ-II separately to prepare an elaborative and sufficient number of scenarios. It provides the user enough freedom, and satisfaction in deciding with the help of variability of the natural facts. It could be attended using Visual Modflow which was initially developed by McDonald and Harbaugh (1984) of the United States Geological Survey (USGS) in 1984 and further updated at several stages (Harbaugh & McDonald, 1996; Harbaugh et al., 2000; McDonald & Harbaugh, 1988). There is a wide application not only India (Alam & Umar, 2013; Gupta et al., 1997; Maheswaram et al., 2016; Mondal & Singh, 2005, 2012; Mondal et al., 2009; Neupane et al., 2020; Patil et al., 2020; Pawar et al., 2022; Rejani et al., 2008; Sahoo & Jha, 2017), but also worldwide (Abd-Elaty et al., 2019; Alsharhan & Rizk, 2020; Lia et al., 2016; McKenzie et al., 2002; Shamsudduha et al., 2019; Weifang et al., 2010) for the simulation of groundwater flow and contaminant transport. Thus in this article, a regional groundwater flow model was developed in a part of MGP. The data generated from the field (CGWB, 2015; Mondal et al., 2014, 2015a, 2015b) was combined with other available secondary data related to hydrology, meteorology and irrigation (CGWB, 2012, 2014; Sahu et al., 2018) to estimate the groundwater flow dynamic among the two aquifers—AQ-I and AQ-II. Also the system response was projected to different extraction AQ-I and AQ-II regimes of a micro-scale for the decline of groundwater level and impact of As-free groundwater existing in the multi-aquifer system.

Description of the Study Area

The research area spreading over an extent of 1,102 km^2 covers parts of seven blocks of Patna district, Bihar in the MGP. It lies between longitudes: 84°44′34″ to 85°19′52″ E and latitudes: 25°22′55″ to 25°42′28″N (Fig. 4.1). It falls in Survey of India Toposheets 72G/2 and 72C/14, 15. Physiographically, the area represents a monotonously flat topography with a general slope from southwest to northeast and north with minor variations. Elevation varies from about 45.45 m to 69.00 m above mean sea level (amsl) with an average of 54.49 m (amsl). The mighty River Ganga forms the northern boundary and other rivers (i.e., Son and Punpun) are draining just outside

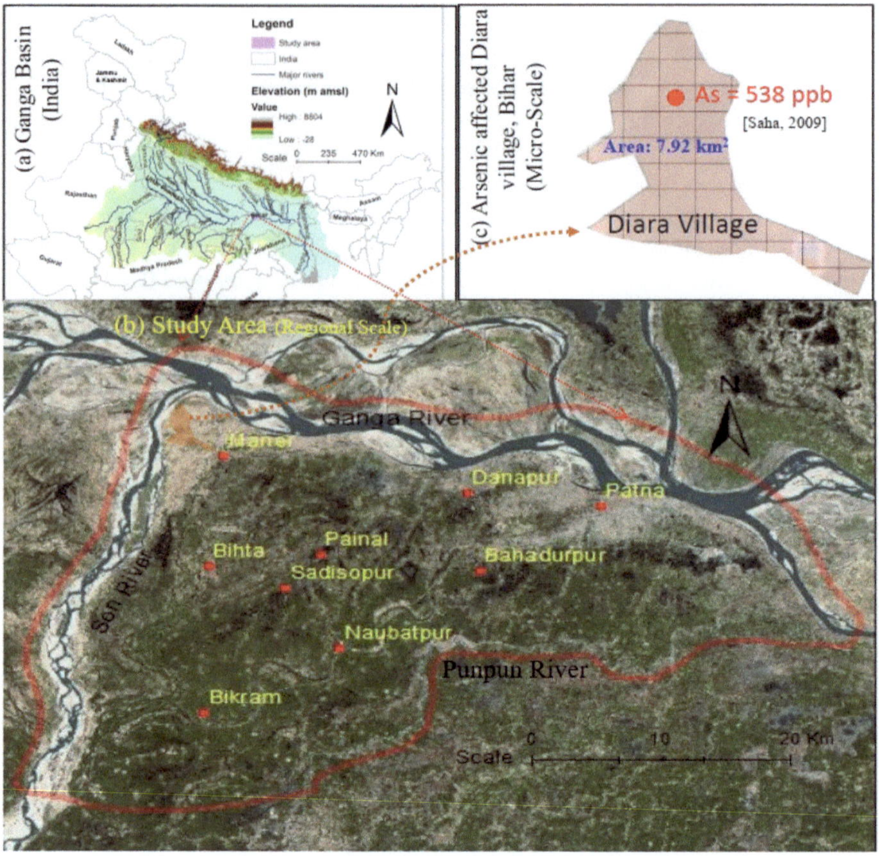

Fig. 4.1 Map showing the study area along with **a** Ganga basin, **b** regional boundary, and **c** arsenic-affected micro-scale area, Patna, Bihar as a part of Middle Ganga Plains (MGP), Northern India

the area. It enjoys a typical subtropical climate. Winter season starts in November and continues up to February. Summer season starts in March and continues up to April. The rainy season continues from May to October, which receives southwest monsoon and accounts for about 96% of the total rainfall. The area receives an average normal monsoon rainfall of about 960 mm/year (CGWB, 2012, Mondal & Ajaykumar, 2022).

Soils are predominantly sandy loams with clay loam at places with low to medium nutrient status. Typically, they are silty clay, clay, sandy clay, and sandy soil. The geomorphic map shows five different units (i.e., flood plain, active flood plain, natural levee, back swamp, and palaeo-channels, in Fig. 4.2a, CGWB, 2015, Mondal & Ajaykumar, 2022). The flood plain, basically part of the older flood plain of River Son, was observed throughout the south and western parts of the study area. Active flood plain formed by recent activities confined to the extreme north-western part. But the natural levees form the northern limits of Patna city boundary along the right bank of River Ganga extending 26 km length (east–west direction). The levee is well-exposed along the river Ganga (Fig. 4.2a). Back swamps are localized along the eastern and western margins of the Patna Urban area. The palaeo channel is the past channel of River Son which is flowing at present almost 35 km west of Patna city and its orientation is the same as the Son river flow.

The area constitutes a thick sequence of Ganga alluvium resting over the pre-Tertiary formations holding a multi-aquifer system. It represents sedimentary river basin environments with different stages and types of aquifers comprising various grades of clay, silt, and sand which constitute the groundwater reservoir due to the reworks of the Son and Ganga rivers. In the area immediately south of River Ganga the aquifers commence from an average depth of 40 m and continue downward up to 300 m as per the explored borehole information (Mondal et al., 2015b). However, the thickness of piled sediments is more than 700 m forming a multi-aquifer system and the aquifers are separated by clay beds of varying thicknesses deduced by the spare data (CGWB, 2012). It seemed to be unknown at a micro-scale. In addition, about 72 VES, 101 TEM, 8.32 LKM ERT, and 770 LKM SkyTEM dense data were collected under the AQUIM Project for the deciphering of the detailed aquifer disposition (CGWB, 2015; Dabas et al., 2022; Mondal et al., 2015b). The typical 2-D and 3-D aquifer dispositions are shown in Fig. 4.2b, c. Two main aquifers (namely AQ-I and AQ-II) are separated by intermediate clay with varying thickness with an average of 29 m. The top of AQ-I varied from 43.01 to 63.10 m (amsl), while the bottom from −119.02 to 52.00 m (amsl). Similarly, the top of AQ-II varied from −167.00 to −20.54 m (small), while the bottom from −316.00 to −83.86 m (amsl) in the modeled area. Broadly groundwater occurs under semi-confined to confined conditions in the AQ-I and occurs under the confined condition at deeper levels (AQ-II). The depth to the water level in the wells ranges between 1.32 and 9.69 m, bgl during pre-monsoon (May, 2012) whereas, during the post-monsoon (November 2012), it ranged between 0.61 and 4.96 m, bgl (CGWB, 2015). In general, groundwater flow is from southwest to north.

The quality of water in these aquifers is, in general, fresh with sporadic occurrences of higher concentrations of iron and nitrate. There is an occurrence of arsenic contamination at places in the AQ-I along the course of River Ganga in the extreme north-western and north-central part of the modelled area (Acharyya & Shah, 2007; Shah, 2008). An example, the highest AAS reading of arsenic level in government hand water is 724 ppb at Naikatola village in Maner block, 450 ppb at Kasimchak village at Danapur block, and 538 ppb at Kala Diara village (Saha, 2009). The source of the As-problem is geological in origin, which has been aggravated due to the excessive withdrawal of groundwater for paddy cultivation. The underlying deeper AQ-II is arsenic-free (Saha, 2009). Besides these, the palaeo-channels are localized reservoirs of fresh groundwater and form the near-surface aquifers (Mondal & Ajaykumar, 2022; Sahu et al., 2018).

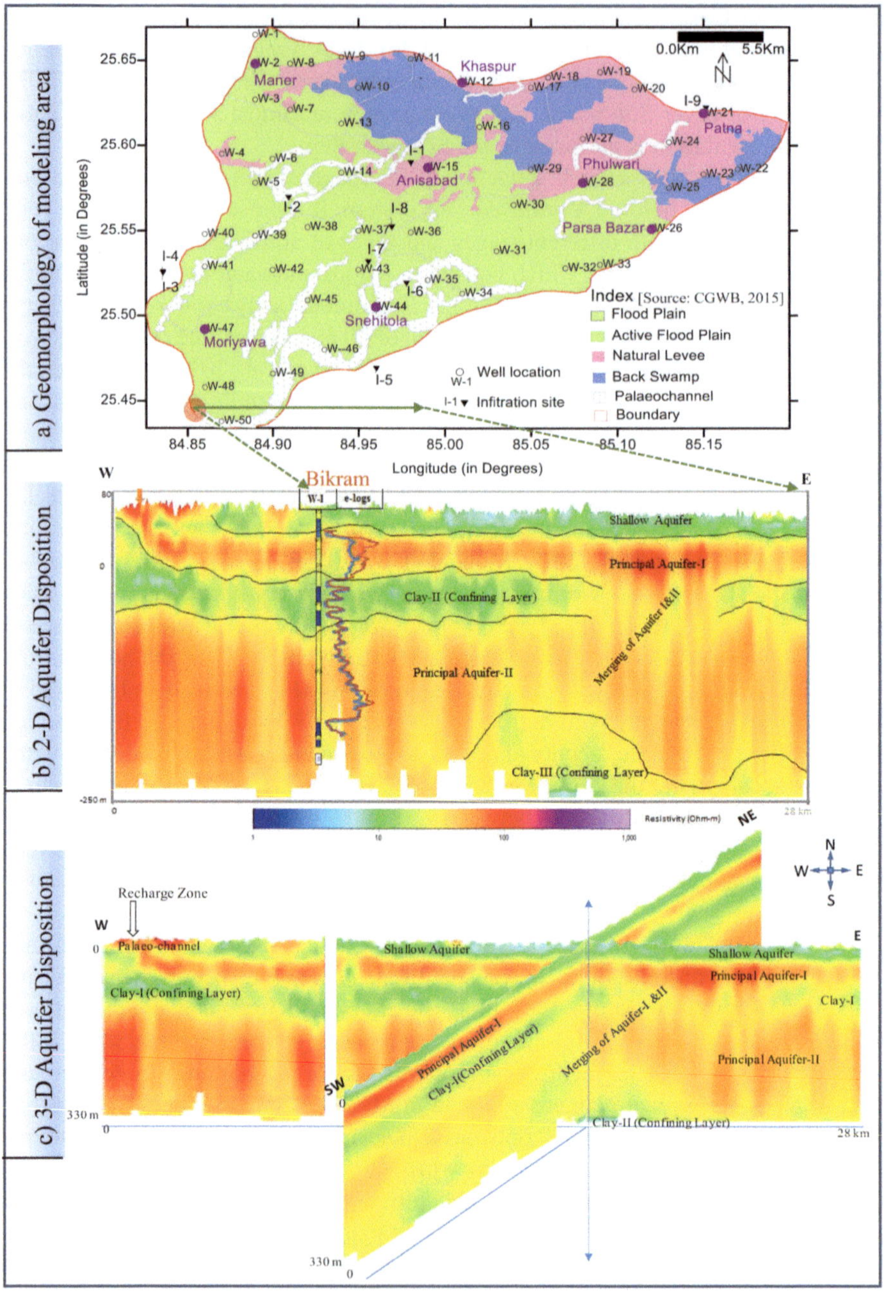

Fig. 4.2 Map showing **a** geomorphological setting, **b** 2-D, and **c** 3-D resistivity section along with the aquifer disposition of the multi-aquifer system, Patna, Bihar (deduced in the AQUIM Project, CGWB, 2015; Mondal & Ajaykumar, 2022; Mondal et al., 2015b)

Materials and Methods

Data Collection

Hydrogeological and geophysical surveys were carried out under the AQUIM Project (CGWB, 2015; Mondal et al., 2015b) for deciphering subsurface litho zones, understanding prevailing hydrogeological conditions, evaluation of aquifer parameters, such as natural recharge, storativity, and hydraulic conductivity, etc. Groundwater levels collected by CGWB, Patna from their monitoring wells were used. To determine aquifer properties, pumping tests were carried out (CGWB, 2012) and the estimated aquifer properties were utilized. The extensive geophysical survey carried out under the flagship program of aquifer mapping to delineate aquifer geometry (Mondal et al., 2015b) was incorporated into the groundwater model. The groundwater recharge has been reported by CGWB (2014). It varied from 32 to 40% of the annual rainfall, which was utilized for the modeling.

Groundwater Flow Model

The generated data was utilized for developing the groundwater flow model, which was also used for the prognostic model as discussed in this paper.

Conceptualization

Multi-aquifer system of the area (Fig. 4.1) consists mainly of two major aquifers (namely AQ-I and AQ-II) separated by an intermediate clay layer up to the explored depth of ~ 300 m, bgl (Dabas et al., 2022; Mondal et al., 2015b). An impermeable clay layer exists at the bottom of the second deeper aquifer (AQ-II). The average thicknesses of the AQ-I and AQ-II are 68 m and 64 m, respectively, with the intermediate clay of an average of 29 m thickness. But these two aquifers have different hydraulic characteristics and especially the top aquifer has a lower conductivity than the deeper AQ-II. To represent this, a four-layer model of the study area (Fig. 4.2) was constructed with the prevailing groundwater flow regime. The top of AQ-I varied from 43.01 to 63.10 m (amsl), while the bottom from −119.02 to 52.00 m (amsl). Similarly, the top of AQ-II varied from −167.00 to −20.54 m (amsl), while the bottom from −316.00 to −83.86 m (amsl) in the modeled area. Both the top and bottom elevations of the aquifers were defined in grid form with the aid of SURFER v.8.0 using the kriging method and converted into ASC format, which was imported into the model.

Analytical methods are rarely applicable for determining a closed-form solution of the partial differential equations of 3-D groundwater flow equation for the inhomogeneous anisotropic confined aquifer of the AQ-II (Rushton & Redshaw, 1979) as given below:

$$\frac{\partial}{\partial x}\left(K_{xx}\frac{\partial h}{\partial x}\right)+\frac{\partial}{\partial y}\left(K_{yy}\frac{\partial h}{\partial y}\right)+\frac{\partial}{\partial z}\left(K_{zz}\frac{\partial h}{\partial z}\right)=S_s\frac{\partial h}{\partial t}\pm W \tag{4.1}$$

The AQ-I was considered under an unconfined condition followed by the Boussinesg equation as below:

$$\frac{\partial}{\partial x}\left(K_{xx}h\frac{\partial h}{\partial x}\right)+\frac{\partial}{\partial y}\left(K_{yy}h\frac{\partial h}{\partial y}\right)=S_y\frac{\partial h}{\partial t}\pm -Q \tag{4.2}$$

where, K_{xx}, K_{yy} and K_{zz} are the hydraulic conductivities along the x, y, and z-directions, h is the hydraulic head, S_s is the specific storativity, S_y is the specific yield ($S_y >> S_s$), W is the leakage or groundwater volume fluxes per unit area (positive for outflow and negative for inflow), Q is the recharge rate, and x, y and z are the Cartesian coordinates.

Both the Eqs. (4.1 and 4.2) were solved using a finite difference approximation technique. The starting point for the application of this method was the discretization of small square sub-regions in a grid form (McDonald & Harbaugh, 1988). This leads to a set of simultaneous algebraic equations, which was solved using the Visual MODFLOW Classic Interface (Build: 4.6.0.168) modeling code. This code has been widely used and is accepted to produce numerically stable solutions. The method selected for the numerical solution of the algebraic equation set is WHS. The main stages of groundwater flow modeling were as follows (1) solving the groundwater flow equations using the finite difference method, (2) estimation of groundwater velocities at each node, and (3) zone budget was also run using the calibrated groundwater velocities in the flow domain.

The maximum recharge from rainfall takes place between June and September. The percentage of rainfall that becomes recharge was determined and varied from 32 to 40% of the average annual rainfall (CGWB, 2014; Mondal & Ajaykumar, 2022). The estimated recharge was adjusted during the model calibration. The reason why the recharge takes place between June and September and also less in summer monsoons during March–April, because it is low rainfall in conjunction with high temperature and evaporation. There is also surface water interaction with the adjoining rivers as the first aquifer is in hydraulic contact with the rivers. Groundwater interaction with adjacent rivers was considered in the boundary condition. It seems to be in interaction with the river and since the water table is higher than the river stage, the river is largely effluent in nature. In order to satisfy irrigation, domestic, and industrial needs, increasing groundwater abstractions take place in the area. Industrial and domestic abstractions are less than 20% of the total extracted volume (CGWB, 2014). Irrigation abstractions most likely occur during the dry period (January–May).

Model Grid Design

To set up the model in the MODFLOW set of codes, the area of interest was divided into a series of grid blocks or cells (Anderson & Woessner, 1992). The model simulated flow in multi-layered aquifers in a part of Ganga River Basin using a block-centered finite difference approach (Fig. 4.3). It was simulated as semi-confined and confined conditions. This simulation was considered with the help of a major program and a number of sub-routines called modules. The modules were grouped into various packages viz. basic, block-centered flow, river, recharge, evapotranspiration, wells, and general heads boundaries, etc. The modeled area was divided into 114 columns and 74 rows with a uniform grid size of 500 m × 500 m (total grid: 4409) and was grouped into 4 layers consisting of 2 main aquifers (AQ-I, Layer-1, and AQ-II, Layer-3) separated by intermediate clay, layer-2 (Figs. 4.2c and 4.4). The bottommost layer-4 was assumed as the base of the aquifer. The active grid design is illustrated in Fig. 4.3 along with the pumping and observation wells and natural boundary conditions.

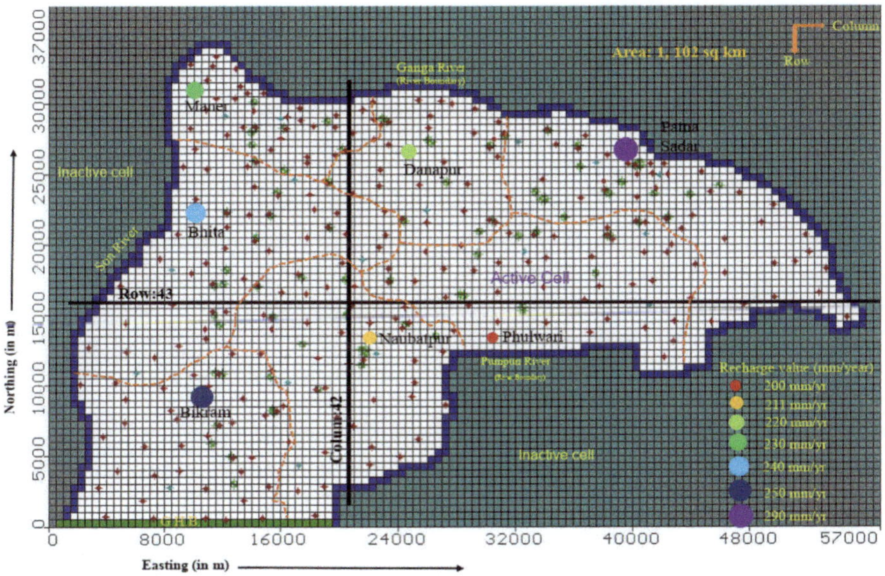

Fig. 4.3 Discretization of the modelling area showing the grids, active and inactive cells, and natural boundaries along with the assigned recharge value (in mm/year), pumping (red circle), and observation (green circle) wells in a part of Middle Ganga Plains (MGP), Northern India

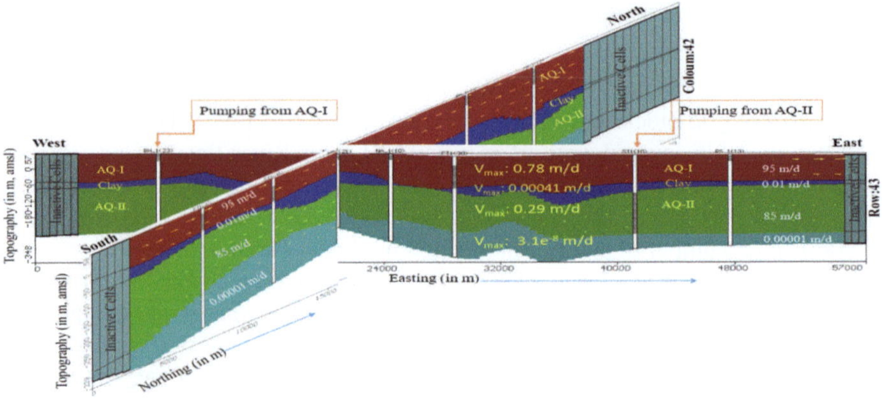

Fig. 4.4 Showing the assigned K-distribution of the multi-aquifer system along with the estimated maximum groundwater velocities and flow direction at Row: 43 and Colum: 42 (as shown in Fig. 4.3) in a part of the Ganga basin

Boundary Conditions

Head-Dependent Boundaries

Generalized head boundary: The generalized head boundary (GHB) was assumed in the form of continuous linear inflow /outflow. Flow through the boundary is the product of the conductance of the boundary and the difference between the head at or beyond the boundary and in the aquifer (Anderson & Woessner, 1992). The southern boundary of the modeled area was simulated through a GHB to represent the groundwater inflow to the modeled area (Fig. 4.3). The values that were used in the steady state simulation in this model were obtained by calibration and were kept constant throughout the transient simulation because there existed no data at all for this boundary. The values, that were finally used, were the boundary conductance range of 13.3–14.5 × 10^3 m^2/d and the head assigned at or beyond the boundary as 50.00–52.65 m (amsl) obtained from the groundwater hydraulic heads.

Rivers: Ganga, Son, and Punpun rivers are special cases of a head-dependent boundary (Fig. 4.3). They were simulated through the river boundary conditions. The properties were set for each river cell boundary as presented in Table 4.1. Because there were no data at all for these rivers, the properties were set through the conditional process except for the river stages which were obtained from the satellite altimetry data of the river stage.

Table 4.1 River properties assigned in the model

River (s)	Average river stage (m, amsl)	Average riverbed bottom (m, amsl)	Bed thickness (in m)	Vertical hydraulic conductivity (in m/d)	Average river width (in m)
Ganga	43.67	38.33	5.0	1.50	1333
Son	48.00	42.50	3.5	1.75	1050
Punpun	50.00	46.33	2.00	1.00	200

Aquifer Properties

Spatial and vertical variations in hydrologic characteristics in the aquifer framework were represented by discrete values in each of the model cells. Model cells extend vertically into the aquifer and divide the aquifer into discrete volumes of aquifer material that were assumed to have uniform hydrologic characteristics. For example, hydraulic conductivity data obtained from the pumping test were utilized in the preparation of the model. Vertical hydraulic conductivity had been taken as 10% of the horizontal hydraulic conductivity. The hydraulic conductivity of the AQ-I, which constitutes the first principal aquifer, varied from 80 to 115 m/day, which is available from the pumping test results. It had been assigned uniformly about 95 m/day as the location of the pumping test was unknown. For layer 2 (intermediate clay) which separates the AQ-I (layer 1: acting as first principal aquifer) and AQ-II (layer 3: acting as second principal aquifer), and layer 4, conductivity values had been assigned uniformly as 0.01 m/day and 0.00001 m/day, respectively (Fig. 4.4). The conductivity value of 85 m/day was also assigned to the AQ-II.

The final distribution of specific yield, as indicated by the model calibration, was assigned. The AQ-I was assigned a specific yield value of 0.25. Similarly, a storativity value of 0.000012 was assigned to the AQ-II uniformly. The effective and total porosities were assigned uniformly throughout the modeled area as 0.15 and 0.25 for the AQ-I and 0.20 and 0.22 for the AQ-II, respectively (Fetter, 2001; GWREC, 1997).

Applied Stresses

Recharge

Initially, as recharge, a percentage (32–40%) of the rainfall (CGWB, 2014) was used as a direct rainfall recharge in the model, although this approach proved to overestimate during transient calibration due to its spatial variability and the computed water levels were comparatively higher than the observed values. Therefore, a lower percentage of rainfall (22–30%) was used (Mondal & Ajaykumar, 2022). The recharge package was used to simulate distributed recharge to the groundwater system and assigned at the top of the first grid layer. Annual precipitation within the model area on an average of 959 mm, part of which infiltrated through the fine-grained material particularly the geomorphological setting overlying the aquifer to the water table condition. Hence, surficial geomorphic units were likely to represent a reasonable initial distribution of recharge. The recharge zones had been assigned considering the nature of the surficial material (Fig. 4.3). The recharge values of 240, 230, 200, 250, 211, 290, and 220 mm/year were assigned to the Blocks of Bhita, Maner, Phulwary, Bikram, Naubatpur, Patna Sadar, and Danapur, respectively, in the model area.

Abstraction

The main groundwater application in the area was irrigation using about 80% of the total abstracted volume. The remaining 20% of the abstraction was reserved for domestic and industrial uses (CGWB, 2014). Irrigation abstractions were spread out in the area, while the domestic/industrial abstractions were assumed to be concentrated around Patna City and the villages. However, all well locations were unknown. Therefore, the known location of 80 wells including unknown 185 wells (in Fig. 4.3), which were used for domestic, industrial, and irrigation abstractions, were used to simulate abstraction, because the total abstracted volume was about 5,09,723 m^3/day in 2011 from the AQ-I and about 1, 41, 244 m^3/day from the AQ-II. Abstractions from each irrigation and domestic/industrial well were used for steady, transient, and prognostic models.

Model Calibration

The purpose of the calibration was to estimate the actual parameters for the given boundary conditions and stresses within a certain established range of error (calibration targets). In this study, a trial and error calibration was used. Parameters were initially assigned to each node in the grid. These parameter values were then adjusted sequentially to match the calibration targets. The calibration parameters, set in the model, were the generalized head boundary, recharge, evapotranspiration, and hydraulic conductivity. During transient calibration, the main idea was to match the trends in the hydrographs reproduced at a few wells. But the numerical round-off error was evaluated by examining the water mass balance calculated by the model. This error should ideally be less than 0.1% (Konikow & Bredehoeft, 1978) and was also tested for the models.

Groundwater Flow Model Calibration

Steady State Calibration

The data was available for the period May 2012 to December 2017 in the modelled area. In the beginning of model was been calibrated under steady conditions. The steady-state model was calibrated against 40 head values observed in May 2012. The calibrated groundwater heads at the targets varied from 46.45 to 57.50 m (amsl) with a mean of 48.78 m (amsl). It showed that groundwater was flowing towards the river Ganga. The computed vs. observed heads with statistical analysis of the steady-state calibration results are shown in Fig. 4.5. The model was considered sufficiently calibrated as 95% of the targets. Three targets at the AQ-II showed comparatively low residual errors (Fig. 4.5) but these can be justified as located in Patna city. Finally, the numerical error in the model was minimal, since the mass balance error was 0.01%, as presented in Table 4.2. The table shows that groundwater abstraction was about 74.00% of annual groundwater recharge, but 9.31% of the annual recharge was being used up by groundwater evapotranspiration. There was a combination of inflow and outflow for river leakage. However, the outflow of river leakage was comparatively less than the inflow, which was about 1.22% of the annual groundwater recharge. In this steady state, the maximum groundwater velocities were estimated in the order of 0.78 and 0.29 m/d in the AQ-I and AQ-II, respectively. It is shown along with the direction of velocity in Fig. 4.4.

The zone budgets were also estimated aquifer-wise in the steady state. It could also be calibrated for any zone located aerially and vertically in the model area. The steady-state model indicates that there were vertical leakages between the AQ-I and AQ-II (in Table 4.2). In total, the AQ-I has lost a groundwater reserve of about 19.57% of the annual natural recharge to the AQ-II.

Transient State Calibration

The steady state calibrated parameter values were imported into the transient model for the period for June 2012 to December 2017 along with groundwater levels as an initial head. Each year in the transient model was divided into 2 stress periods: one dry period (March to April), which lasts 60 days and has 2 stress periods–one for each simulated month; four distinct 30 days for winter periods (i.e., November–February), and six distinct 30 days' wet periods, one for each recharge month (i.e., May–October). The predictive simulations were performed in 60 stresses and 600-time steps for the period between January 2018 and December 2022. This was done with the objective of better simulating the difference in rainfall and consequently in recharge between the three wet months. Each stress period was divided into 10 time steps in order to make it easier for the model to converge. In total, there are 68 stress periods with 680-time steps in the transient model. Computed versus observed heads

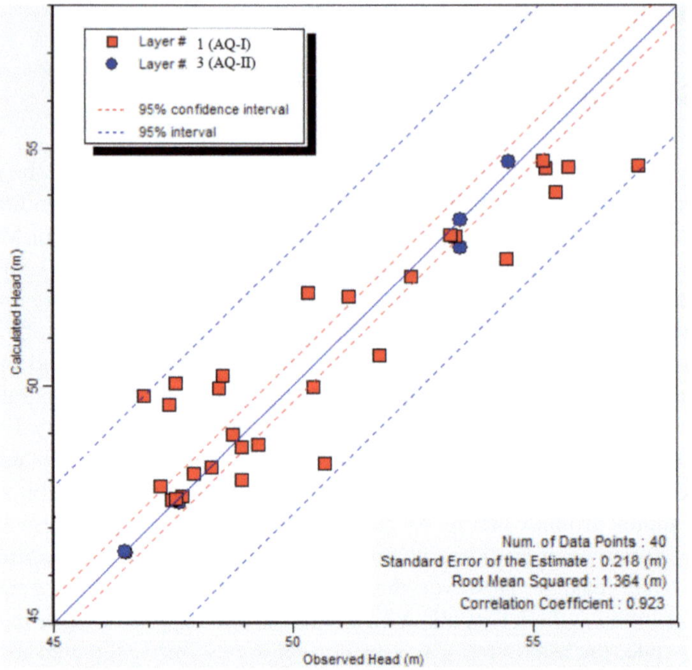

Fig. 4.5 Calculated versus observed heads for the AQ-I and II (May 2012)

Table 4.2 Steady-state calibration mass balance results (m^3/day) in the shallow aquifer (AQ-I)

In shallow aquifer (AQ-I)	Inflow (m^3/day)	Percentage (%) of recharge
Recharge	735,500	–
River leakage	514,180	–
General-Head	12,268	–
Zone AQ-II to AQ-I	114,660	–
Total IN	1,376,600	–
	Outflow (m^3/day)	
Wells	541,700	74.00
Groundwater evaporation (ET_{gr})	68,479	9.31
River leakage	505,220	–
General-head	2608	–
Zone AQ-I to AQ-II	258,650	–
Total OUT	1,376,700	–

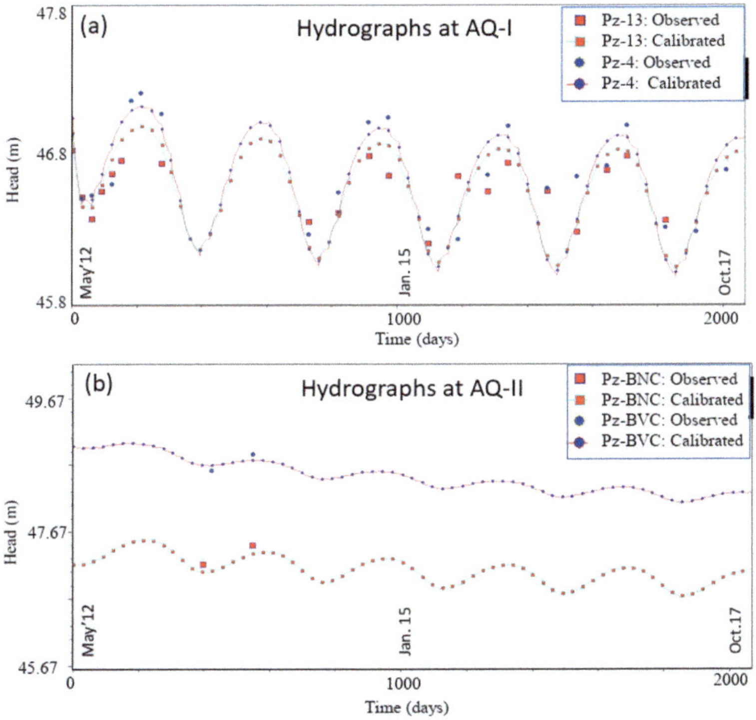

Fig. 4.6 Showing calibrated and observed well hydrographs at the **a** AQ-I and **b** AQ-II

for each calibration target, during all time steps considering a total of 579 observed data points in the entire simulation show that the maximum was −3.21 m at W-17 and minimum residual was −0.14 at W-20 in May 2012 with RMS = 1.635 m in the transient calibration statistics. The calibrated well hydrographs in both AQ-I and II were closely matched with the observed heads and a few typical examples are shown in Fig. 4.6a, b.

Predictive Model for Groundwater Management

The calibrated transient groundwater model was used to perform predictive simulations to assess the groundwater in both aquifers. Although a predictive simulation could last up to twice the calibrated period (Anderson & Woessner, 1992), it was simulated for the predictive model for 5 years in the future, because of the high uncertainty in predicting future hydraulic conditions, and stresses to the multi-aquifer system. Hence predictive simulations were performed in 60 stresses and 600-time steps for the period between January 2018 and December 2022.

Extrapolation of Past and Present Stresses to Future

Similar to the transient model, each year in the transient model was divided into one dry period (March–April), four distinct 30 days winter periods (i.e., November–February), and six distinct 30 days of wet periods (i.e., May–October). During wet periods the recharge occurred due to rainfall. Evapotranspiration and recharge from irrigation return flows were active during the dry stress period. Irrigation abstraction took place all year during the dry period. The extrapolation of rainfall and abstractions, to the future, is discussed.

Rainfall and Recharge

The generation of future rainfall values would require extensive statistical analysis. Since the objective of this study was not to predict exactly what will happen in the future but to examine the sustainability of the system, it was decided to generate future rainfall values by simply replicating the past and present rainfall patterns. Since rainfall for the year 2017 was known, the same value was repeated once more for the wet months until the year 2022. Then recharge from rainfall was calculated using the same percentages as in the transient model. In addition, the recharge had been changed by ±15% of the assigned value in the model for studying the leakages between two aquifers (Table 4.3b).

Abstraction

Projections of groundwater demand in the future would require a socio-economic study, which is beyond the scope of this study. Since the main objective was to examine the leakages among the multi-aquifers under various stress conditions, predictive simulations were performed at an abstraction rate of 141,244 m^3/day (in the year 2017) for domestic/industrial and irrigation purposes with the additional increasing pumping (at the rates of 10, 20, 30, 40, 50 and 100% of the discharge 141,244 m^3/day) at the AQ-II without addition load at the AQ-I. Because the AQ-I is already vulnerable in terms of arsenic contamination. The projection of the increased extraction is presented in Table 4.3a. Further, the pumping rate had been separately changed by ± 50% at the AQ-I & II under the diverse climatic conditions for studying the leakages in between two aquifers (Table 4.3b).

Predictive Well Hydrographs

It was presumed that the aquifer would receive the same recharge just like the previous year 2017, while the abstraction for irrigation, and domestic/industrial usage was kept in increasing order as the rates of 10, 20, 30, 40, 50, and 100% of the discharge 141,244 m^3/day.

Table 4.3 Projected extraction scenarios under the climatic condition from 2018 to 2022

(a) Projected extraction scenarios without additional pumping at the AQ-I with constant recharge from 2018 to 2022

Scenario (s)	Total extraction (m^3/day, AQ-II), 2017	Additional Pumping (from AQ-II)	Increased to Total extraction (m^3/d) (from AQ-II)	Increased extraction (m^3/d) (from AQ-II)
Scenario-I	141,244	10%	155,368	14,124
Scenario-II	–	20%	169,493	28,249
Scenario-III	–	30%	183,617	42,373
Scenario-IV	–	40%	197,742	56,498
Scenario-V	–	50%	211,866	70,622
Scenario-VI	–	100%	282,488	141,244

(b) Projected extraction scenarios in both the aquifers under the climatic conditions from 2018 to 2022

Scenario (s)	Extraction (in MCM/day)		Changing recharge (in %) at AQ-I
	In AQ-I	In AQ-II	
Scenario-I	Constant (0.509 MCM/d)	Constant (0.141MCM/d)	Constant
Scenario-II	Increased 50% (7.64 MCM/d)	Constant (0.141MCM/d)	Increased 15%
Scenario-III	Increased 50% (7.64 MCM/d)	Constant (0.141MCM/d)	Decreased 15%
Scenario-IV	Increased 50% (7.64 MCM/d)	Constant (0.141MCM/d)	Constant
Scenario-V	Constant (0.509 MCM/d)	Increased 50% (0.211MCM/d)	Constant

Computed groundwater heads in the both AQ-I, and AQ-II indicated a slow decline, which varied from 41.76 to 52.24 m (amsl) at the AQ-I and 43.08 to 49.34 m (amsl) at the AQ-II under the controlled condition in May 2017. Whereas these variations were from 41.76 to 52.24 m (amsl) at the AQ-I and 43.07 to 49.29 m (amsl) at the AQ-II under the controlled condition in May 2022. But in total, an average decline of groundwater head for the AQ-II is about 0.22 m from the AQ-I (Table 4.4), which initially indicates the leakage vulnerability of contaminated water from the AQ-I within the year 2022. The average head difference between the AQ-I and II varied from 0.23 m in the period May 2022 under the increased pumping of 10% to 0.40 m in the period of May 2022 under the increased pumping of 100%. It was observed that the condition of the pumping from the AQ-II at the rate of upto 30% was similar to the present condition and was abnormal beyond the total pumping increased to 183,617 m^3/day at the AQ-II.

Therefore, the computed groundwater heads were observed at the selective virtual wells. The computed well hydrographs at the selective targets indicated a slow decline of the groundwater heads under the stress conditions. It had been observed that the

Table 4.4 Statistics of the calibrated groundwater heads in both the AQ-I and AQ-II in the predictive modes

Sl. No	Period (s)	Condition (s)	AQ-I Heads (m, amsl)			AQ-II Heads (m, amsl)			Head difference (AQI-AQII) (in m)		
			Min	Max	Avg	Min	Max	Avg	Min	Max	Avg
1	May-17	Controlled	41.48	52.24	46.39	43.08	49.34	46.18	−1.62	2.92	0.21
2	May-22	Controlled	41.76	52.24	46.36	43.07	49.29	46.14	−1.62	2.96	0.22
Groundwater extractionincreased at the AQ-II											
3	May-22	10%	41.76	52.24	46.35	43.01	49.26	46.11	−1.61	3.00	0.23
4	May-22	20%	41.76	52.24	46.34	42.94	49.23	46.09	−1.59	3.02	0.25
5	May-22	30%	41.76	52.24	46.34	42.88	49.2	46.07	−1.58	3.05	0.27
6	May-22	40%	41.76	52.24	46.33	42.82	49.18	46.04	−1.57	3.08	0.29
7	May-22	50%	41.76	52.24	46.33	42.75	49.15	46.02	−1.56	3.11	0.31
8	May-22	100%	41.76	52.24	46.30	42.44	49.01	45.91	−1.50	3.32	0.40

heads of AQ-II were higher than the heads of AQ-I in both the well sites located at Maner and Danapur blocks under the 10% increased pumping at the AQ-II. When the pumping was increased to more than 10% and up to 50%, the groundwater heads of AQ-II were lowered down, which was being led to leakage of the contaminated water to the AQ-II (which is Arsenic free) at an alarming rate (Fig. 4.7a–k).

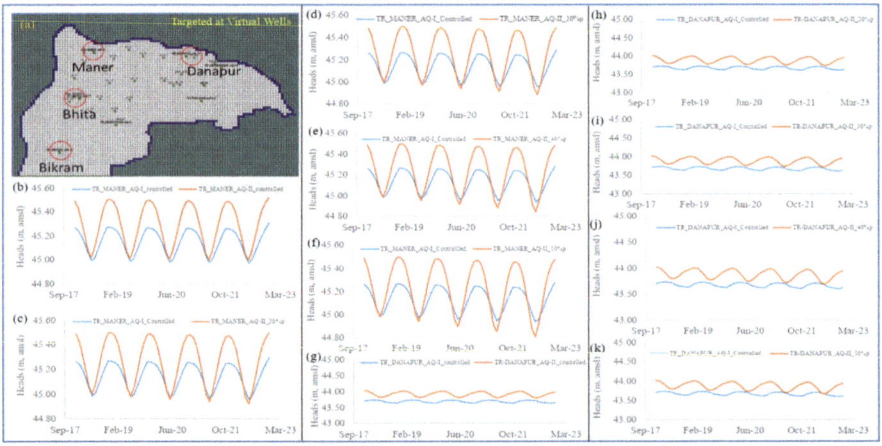

Fig. 4.7 Projected well hydrographs on (**a**) the targeted virtual wells under (**b**) both the controlled pumping condition in both the aquifers, (**c**) 20%, (**d**) 30%, (**e**) 40%, and (**f**) 50% pumping rate increase of the AQ-II at Maner Block during January 2018 to December 2022. And under (**g**) both the aquifers controlled pumping condition, (**h**) 20%, (**i**) 30%, (**j**) 40%, and (**k**) 50% pumping rate increase of the AQ-II at Danapur Block from January 2018 to December 2022

Predictive Groundwater Leakages and the Possibility of Arsenic Leakage

Based on the applied stresses on the AQ-II, the area was found for the upward and downward leakages due to the reversal of hydraulic heads almost throughout the study area. It was found that the model area covered for the downward leakage was about 771 km^2 under the controlled condition in the year 2022 with respect to the controlled condition of the year 2017 (Fig. 4.8a, b). It was observed that the downward leakage area was increased to about 810.5 km^2, but the upward leakage was reduced to about 12% of the controlled leakage area (~331 km^2), which is presented in Table 4.5. It indicated that groundwater heads at the AQ-I steeply declined particularly at Maner Block towards the confluence of Son and Ganga Rivers (Fig. 4.8d, e) under the stress conditions, if the extraction was increased up to 40% and 50% at the AQ-II. It was further observed that the upward leakage area was being reduced to 50% at the end of the year 2022, if the pumping increased to a total of 282,488 m^3/d from the AQ-II for the different uses (Fig. 4.8f). It had been observed that groundwater dynamics between two aquifers (AQ-I & II) in terms of vertical leakage is quite abnormal, if the extraction increased by more than 20% of the extraction, 141,244 m^3/day (Fig. 4.8). Therefore, there is a possibility of lowering the hydraulic head and possibly mixing arsenic safe water of the deeper aquifer (AQ-II) with that of the AQ-I with contaminated groundwater, if the AQ-II put under heavy pumping.

In addition, a few scenarios had been further estimated for observing leakage vulnerability by changing the parameters like recharge and the water extraction separately from both the aquifers. Scenario-I was considered as a base for the controlled recharge and pumping in both aquifers (Fig. 4.9). The upward leakage covered only 331 km^2 whereas the downward leakage was 771 km^2 of the modeled area. When the recharge was increased to about 15% and groundwater withdrawal around 50% without the changing abstraction at the AQ-II, the downward leakage covered more ~116 km^2 area than the upward leakage. In the third scenario, it was observed that the upward leakage had been considerably improved by the decreasing recharge of 15% and increasing pumping of 50% at the AQ-I keeping constant pumping at the AQ-II. Whereas the upward leakages were not improved by the ways of increased pumping of 50% separately at the AQ-I (scenario-IV) and AQ-II (Scenario-V) keeping the recharge constant (Fig. 4.9). It indicates that the upward leakage is only maintained by reducing the recharge or considerable extraction of groundwater from the AQ-I (Scenario-III) to reducing the aquifer vulnerability of As-contamination in the AQ-II.

Predictive Groundwater Leakage and the Possibility of Arsenic Leakage at the Village Level

The groundwater predictive flow model was used to understand the status of groundwater heads in the arsenic-affected area namely Diara village (area: ~7.92 km^2) where arsenic concentration was estimated at around 538 ppb (Saha, 2009), as shown in Fig. 4.10a. When the extraction had been increased year-wise at the AQ-II for the future demand, there is a reversal hydraulic head was observed. The extraction had

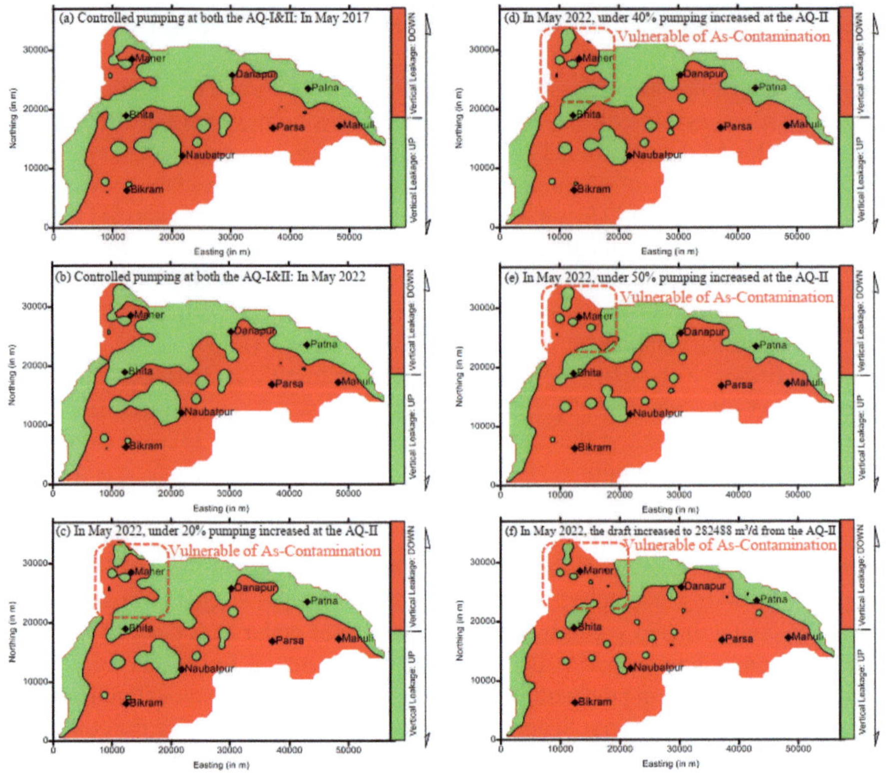

Fig. 4.8 Red color shows the zones for leakage from AQ-I to AQ-II (which is Arsenic free) whereas the green color for the leakages from the AQ-II to AQ-I under the controlled recharge and pumping for the months of **a** May 2017 and **b** May 2022, and under the pumping only increased to the AQ-II at the rates of **c** 20%, **d** 40%, **e** 50% and **f** 100% of the extraction 141,244 m^3/day

Table 4.5 Statistics of the vulnerable leakage area covered under the stress conditions (increasing pumping at the AQ-II)

Extraction at the AQ-II assigned as 141,244 m^3/d		Area (km^2) covered (at the alarming rate) under	
Extraction increased at the AQ-II for future demands (up to year 2022)	Increased	Downward leakage (from AQ-I to AQ-II)	Upward leakage (from AQ-II to AQ-I)
Controlled	N/A	771.00	331.25
10%	14,124	790.50	311.75
20%	28,249	810.50	291.75
30%	42,373	831.25	271.00
40%	56,498	850.75	251.50
50%	70,622	867.25	235.00
100%	141,244	938.25	164.00

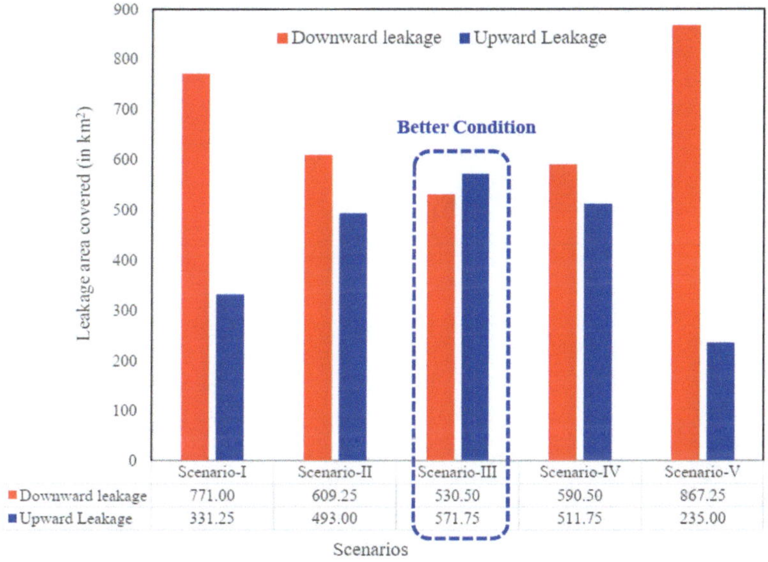

Fig. 4.9 Bar chart showing the leakage area of the Year 2022 under the stress conditions such as Scenario-I: All controlled conditions, scenario-II: recharge and pumping increased about 15% and 50%, respectively, at the AQ-I, scenario-III: decreased recharge of 15% and increased pumping of 50% at the AQ-I, scenario-IV: increased pumping of 50% at the AQ-I keeping the recharge constant, and Scenario-V: increased pumping of 50% at the AQ-II keeping the recharge and pumping constant at the AQ-I

increased to about 102, 203, 305, 406, 508, and 1015 m^3/day as the 10, 20, 30, 40, 50, and 100% of the extraction (the year 2017) at this village (Fig. 4.10b). The scenarios for the years 2018, 2019, 2020, 2012, and 2022 were been observed. It indicates that about 50% of the area will be under contaminated vulnerability zones in the year 2019 if the extraction is increased by about 30% of the present extraction. If we increase the extraction up to double, about 70% area of Diara village leads to the As-contamination at the AQ-II. The strategy of the As-contamination vulnerability under several stress conditions is shown in Fig. 4.10c.

Discussion and Concluding Remarks

The groundwater flow model of a multi-layered aquifer system from a part of Middle Ganga Plains (MGP), Patna, Northern India is simulated using the Visual MODFLOW Classic Interface (Build: 4.6.0.168) for understanding the hydrodynamics related to the threat of arsenic leakages in the stress conditions and also the projected ones. The specialty and objectives of the study have been to observe and handle the reversal of inter-aquifer leakages. The simulated results show the following.

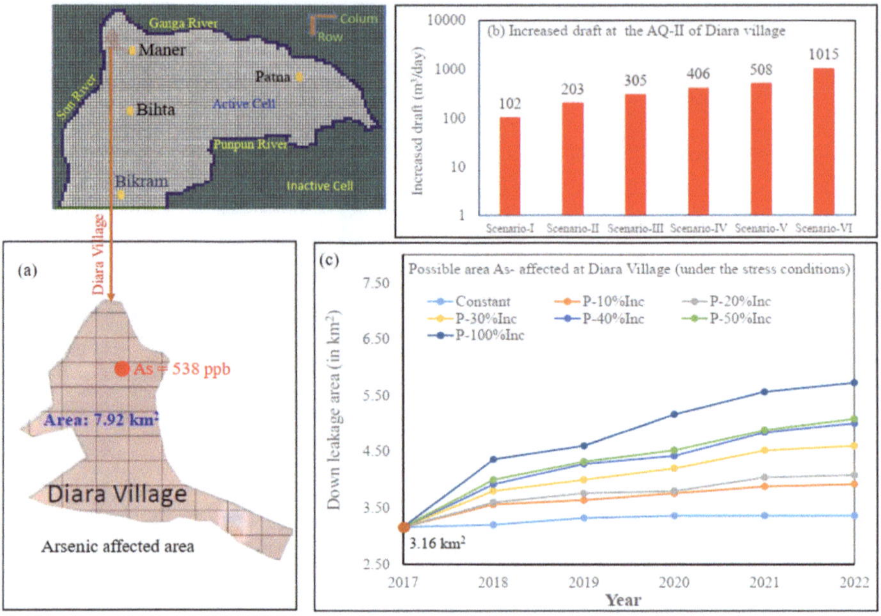

Fig. 4.10 Showing **a** grids of the Arsenic affected area, **b** increased extraction assigned at the AQ-II as the AQ-I is As-contaminated, and **c** predictive leakage areas at Diara Village, Patna from Middle Ganga Plains (MGP) under the different stress conditions

- Groundwater flow is in general towards the north and northeast in both the aquifers viz., shallow aquifer (AQ-I) and deeper aquifer (AQ-II).
- Estimated maximum groundwater velocities are ~0.78 and 0.29 m/d at the AQ-I & AQ-II, respectively. It is also estimated at ~0.0004 m/d at the intermediate clay layer.
- With increasing pumping of more than 20% from the AQ-II, groundwater dynamics between two aquifers (AQ-I & AQ-II) in terms of vertical leakage is quite disturbed, which leads to rapid depletion of water levels and quality concerns may arise in the presence of Arsenic contamination.
- Industrial & Irrigation allocation and deep tube wells especially at Maner Block towards the confluence of the Son and Ganga rivers should be discouraged, as this area has been found to affect As-contamination in the AQ-II. Any further increase in groundwater extraction (at the AQ-II, >20% of the recent) would lead to reversal hydraulic heads and possible mixing of arsenic-safe water of the deeper aquifer (AQ-II) with that of the AQ-I with contaminated groundwater.
- The upward leakage is only being maintained by reducing the recharge or considerably extraction of groundwater from the AQ-I to reducing the aquifer vulnerability of As-contamination in the AQ-II.

Figure 4.8 depicts the inter-aquifer leakage (an extremely important flux in this system in the present condition) showing clearly that there already exists a large area where considerable leakage is there from AQ-I to AQ-II. However, the majority of the area particularly away from the river Ganga where the reversal of leakage already has occurred, the aquifer AQ-I is also free from arsenic contamination. Thus the situation may affect the quantity of the groundwater but not the quality. Considering the northern and north-western parts of the study area, it has been shown that initially the majority of the villages have leakage from AQ-II to AQ-I (shown in green color). But the calibrated multi-layered model reveals the reversal in the leakage from AQ-II to AQ-I taking place once there is an increase in the pumping from the AQ-II. The red dotted rectangles in Fig. 4.8 indicate clearly the leakage reversal in 2022 and this is of much concern as the AQ-I in this area is contaminated by arsenic.

Of course, the groundwater withdrawals from the aquifer, AQ-II has to be reduced and optimized based on model response as it is safer to keep the piezometric head of the AQ-II higher than that of the AQ-I and particularly in the As-contaminated areas of the northern and north-western parts of the study area. As a demonstration, a scenario of a regional scale of the study area has been developed by the decreasing recharge of 15% and increased pumping of 50% at the AQ-I keeping constant pumping at the AQ-II (Fig. 4.9). It implies that the upward leakage is only maintained by reducing the recharge or considerable extraction of groundwater from the AQ-I to reduce the aquifer vulnerability of As-contamination in the AQ-II. But this issue certainly requires decision-making. Making a suitable and single decision may not be feasible for the entire area and hence a considerably small area e.g., a village named Diara has been demonstrated. Once the numerical model with appropriate real boundary conditions and satisfactorily calibrated hydrogeological parameters, a zone with a few grids (10 to 100; in the case of Diara village, it is 35, Fig. 4.10) can be taken and groundwater balance called as zonal balance may be analyzed. An appropriate decision for a sustainable scenario has to be taken in this regard adding a new dimension on the issue. This approach will be scientifically robust as the numerical model with appropriate boundary conditions could be calibrated keeping the boundary away from the decision-making area and hence reducing the errors. The targeted parameters may be different e.g., it could be water levels only in a water-scarce area but inter-aquifer leakage has been taken in the present study as the Arsenic free aquifer i.e., AQ-II has to be saved from the contaminations originating with the leakage from the contaminated shallow aquifer (AQ-I). Therefore, several scenarios of pumping from the AQ-II in this smaller area can be studied in the model and the best scenario that does not allow for the reversal of the leakage is adopted for sustainable groundwater and getting arsenic-free water. Of course, if that optimal withdrawal does not meet the demand of the village, other alternatives such as pumping from AQ-I with necessary treatment or supply from the river may be planned.

Acknowledgements Director of CSIR-National Geophysical Research Institute, Hyderabad has kindly accorded for the publication (Ref. No.: PME/CSIR-NGRI, e-mail, dated: 18/11/2019) and presentation (Ref. No: NGRI/Lib/2022/Pre-58). The Scientists and Technicians of CSIR-NGRI, Hyderabad, and CGWB-Patna have provided their scientific inputs and supported to collecting the necessary data and information. The DST-NGP (GoI), New Delhi (Ref. No.: NRDMS/GRACE/8342/NC Mondal/2020(G), January 21, 2021) has funded the first author for this research work.

The anonymous reviewer has suggested his constructive comments to improve this article. The authors are thankful to them.

References

Abd-Elaty, I., Eldeeb, H., Vranayova, Z. & Zelenakova, M. (2019). Stability of irrigation canal slopes considering the sea level rise and dynamic changes: Case study El-Salam canal, Egypt. *Water, 11*(5), 1046.

Acharyya, S. K., & Shah, B. A. (2007). Groundwater arsenic contamination affecting different geologic domains in India-a review: Influence of geological setting, fluvial geomorphology and quaternary stratigraphy. *Journal of Environment Science Health A, 42*(12), 1795–1805.

Alam, F., & Umar, R. (2013). Groundwater flow modelling of Hindon-Yamuna interfluve region, Western Uttar Pradesh. *Journal of the Geological Society of India, 82*(1), 80–90.

Alsharhan, A.S., & Rizk, Z. E. (2020). Application of modeling techniques for water-resource investigations in the UAE. In *Water resources and integrated management of the United Arab Emirates* (pp. 707–721). Springer.

Anand, J., Gosain, A. K., Khosa, R., & Srinivasan, R. (2018). Regional scale hydrologic modeling for prediction of water balance, analysis of trends in streamflow and variations in streamflow: The case study of the Ganga River basin. *Journal of Hydrology: Regional Studies, 16*, 32–53.

Anderson, M.P., & Woessner, W.W. (1992). *Applied ground water modeling: Simulation of flow and advective transport* (p. 381). Academic Press Inc.

Asoka, A., Gleeson, T., Wada, Y., & Mishra, V. (2017). Relative contribution of monsoon precipitation and pumping to changes in groundwater storage in India. *Nature Geoscience, 10*(2), 109.

Central Ground Water Board (CGWB). (2012). Background report on pilot micro level aquifer mapping projects, AQBHR area, Patna, Bihar, p. 95.

Central Ground Water Board (CGWB). (2014). Dynamic ground water resources of Bihar State as on 31st March 2011. Ground Water Directorate, Minor Water Resource Department, Patna, p. 276.

Central Ground Water Board (CGWB). (2015). Report on pilot project on aquifer mapping in Maner-Khagaul area, Patna District Bihar (Watershed -GNDK013), p. 260.

Central of Water Commission (CWC). (2005). Water and related statistics. Water Planning Project Wing, Central Water Commission, New Delhi.

Dabas, J., Sarah, S., Mondal, N.C., & Ahmed, S. (2022). Geostatistical spatial projection of geophysical parameters for practical aquifer mapping. *Nature-Scientific Reports, 12*, 4641, 1–14.

Fetter, C. W. (2001). *Applied hydrogeology (4th edn)*. Prentice Hall.

Ganga River Basin Management Plan—2015 (GRBMP) Report of Consortium of 7 "Indian Institute of Technology's (IITs), pp. 184 (2012).

Ground Water Resource Estimation Committee (GWREC). (1997). Ministry of Water Resources, Government of India, New Delhi.

Gupta, C. P., Thangarajan, M., & Rao, V. V. S. G. (1997). Assessment of groundwater pollution due to tannery effluents in the Upper Palar basin, South India. *Engineering Geology & The Environment, 1–3*, 1871–1875.

Harbaugh, A. W., & McDonald, M. G. (1996). User's documentation for MODFLOW96 an update to the US Geological Survey modular finite-difference groundwater flow model. *US Geological Survey Open-File Report, 96*, 485.

Harbaugh, A. W., Banta, E. R., Hill, M. C., & McDonald, M. G. (2000). MODFLOW 2000, the US Geological Survey modular groundwater mode-user guide to modularization concepts and the groundwater flow process. US Geological Survey Open-File Report 00-92, Reston, Virginia.

Konikow, L. F., & Bredehoeft, J. D. (1978). Computer model of two-dimensional solute transport and dispersion in groundwater: Techniques of water-resources investigations of the USGS; Chapter C2, Book 7, p. 90.

Lia, X., Hea, X., Yanga, G., et al. (2016). Study of groundwater using visual MODFLOW in the Manas River Basin, China. *Water Policy, 18*, 1139–1154.

Maheswaran, R., Khosa, K., Gosain, A. K., et al. (2016). Regional scale groundwater modelling study for Ganga River basin. *Journal of. Hydrology, 541* (Part B), 727–741.

McDonald, M. G., & Harbaugh, A. W. (1984). A modular three dimensional finite-difference groundwater flow model, U.S. Geological Survey. Techniques of Water Resources Investigations, Book 6.

McDonald, M. G., & Harbaugh, A. W. (1988). A modular three-dimensional finite-difference ground-water flow model: Techniques of Water-Resources Investigations of the United States Geological Survey, Book 6, Chapter A1, p. 586.

McKenzie, J. M., Siegel, D. I., Shotyk, W., Steinmann, P., & Pfunder, G. (2002). Heuristic numerical and analytical models of the hydrologic controls over vertical solute transport in a domed peat bog, Jura Mountains, Switzerland. *Hydrological Processes, 16*, 1047–1064.

Mondal, N. C., & Ajaykumar, V. (2022). Assessment of natural groundwater reserve of a morphodynamic system using an information-based model in a part of Ganga basin, Northern India. *Nature-Scientific Reports, 12*(1), 6191, 1–12.

Mondal, N. C., & Singh, V. P. (2012). Chloride migration in groundwater for a tannery belt in Southern India. *Environmental Monitoring and Assessment, 184*, 2857–2879.

Mondal, N. C., & Singh, V. S. (2005). Modelling for pollutant migration in the tannery belt, Dindigul, Tamil Nadu, India. *Current Science, 89*(9), 1600–1606.

Mondal, N. C., Singh, V. S., & Rangarajan, R. (2009). Aquifer characteristics and its modeling around an industrial complex, Tuticorin, Tamil Nadu, India: A case study. *Journal of Earth System Science, 188*(3), 231–244.

Mondal, N. C., Somvanshi, V. K., & Ramadevi, A. (2015a). Heliborne transient electromagnetic investigation in Middle Ganga Plains, Parts of Patna District, Bihar, Restricted Technical Report No.: NGRI-2015-GW-868, January 2015, p. 48.

Mondal, N. C., Somvanshi, V. K., & Ramadevi, A. (2015b). Hydrogeological Inferences using heliborne transient electromagnetic and ground geophysical data in AQBHR Area, Patna District, Bihar, India, Restricted Technical Report No.: NGRI-2015-GW-880, April 2015, p. 102.

Mondal, N. C., Somvanshi, V. K., & Tiwari, A. K. (2014). Mid Term Report, AQBHR Area, Patna, Bihar under the AQUIM Project. Aquifer Characterization using Advance Geophysical Techniques in Representative Geological Terrains of India, Technical Report No. NGRI-2014-GW-856, July 2014, p. 52.

Mukherjee, A., Bhanja, S. N., & Wada, Y. (2018). Groundwater depletion causing reduction of baseflow triggering Ganges river summer drying. *Scientific Reports, 8*(1), 12049.

Muralidharan, D. (1998). Protection of deep aquifers from arsenic contamination in Bengal Basin. *Current Science, 75*(4), 351–353.

Neupane, P. K., Mondal, N. C., & Manglik, A. (2020). Envisaging the sustainability of an aquifer by developing groundwater flow model for a part of Choutuppal Mandal, Nalgonda district, Telangana, India. *Nepal Journal of Science and Technology, 19*(1), 222–233.

Patil, N. S., Chetan, N. L., Nataraja, M., & Suthar, S. (2020). Climate change scenarios and its effect on groundwater level in the Hiranyakeshi watershed. *Groundwater for Sustainable Development, 10*, 100323.

Pawar, S. V., Mondal, N. C., & Ullagaddi, P. (2022). Aquifer response at the proximity of Patalganga and Shriganwadi micro-watersheds, Nanded district, Maharashtra using groundwater flow model. *Bulletin of Pure and Applied Sciences-Geology, 41F*(1), 12–23.

Rejani, R., Jha, M. K., Panda, S. N., & Mull, R. (2008). Simulation modeling for efficient groundwater management in Balasore Coastal Basin, India. *Water Resources Management, 22*, 23–50.

Rodell, M., Velicogna, I., & Famiglietti, J. S. (2009). Satellite-based estimates of groundwater depletion in India. *Nature, 460*, 999–1002.

Rushton, K. R., & Redshaw, S. C. (1979). *Seepage and groundwater flow*. Wiley.

Saha, D., & Sahu, S. (2016). A decade of investigations on groundwater arsenic contamination in Middle Ganga Plain, India. *Environmental Geochemistry and Health, 38*(2), 315-–337.

Saha, D. (2009). Arsenic groundwater contamination in parts of Middle Ganga Plain, Bihar. *Current Science, 97*(6), 753–755.

Sahoo, S., & Jha, M. K. (2017). Numerical groundwater-flow modeling to evaluate potential effects of pumping and recharge: Implications for sustainable groundwater management in the Mahanadi delta region, India. *Hydrogeology Journal, 25*, 2489–2511.

Sahu, S., Saha, D., & Shukla, R. R. (2018). Sone megafan: A non-Himalayan megafan of craton origin, forming a potential groundwater reservoir in marginal parts of Ganga Basin, India. *Hydrogeology Journal, 26*, 2891–2917.

Shah, B. A. (2008). Role of quaternary stratigraphy on arsenic-contaminated groundwater from parts of Middle Ganga Plain, UP–Bihar India. *Environmental Geology, 53*(7), 1553–1561.

Shamsudduha, M., Zahid, A., & Burgess, W. G. (2019). Security of deep groundwater against arsenic contamination in the Bengal Aquifer system: A numerical modeling study in southeast Bangladesh. *Sustainable Water Resources, 5*(3), 1073–1087.

Tiwari, V. M., Wahr, J., & Swenson, S. (2009). Dwindling groundwater resources in northern India, from satellite gravity observations. *Geophysical Research Letters, 36*, L18401.

Weifang, S., Weihua, Z., & Bin, C. (2010). Application of visual MODFLOW to assess the sewage plant accident pool leakage impact on groundwater in the Guanting reservoir area of Beijing. *Frontiers of Earth Science, 4*(3), 320–325.

Chapter 5
Systematic Review of Arsenic Contamination, Toxicity and Remediation Techniques in Malawi

Ibrahim Chikowe, Bonface Mwamatope, Ulemu Kankwatira, Henry Phiri, George Chirambo, and Collins Edward Jana

Abstract Arsenic (As) is one of the toxic contaminants in food and water. Excessive and long-term exposure to As causes harmful effects on humans, like arsenicosis, which manifests in disfiguring, disabling, and fatal diseases like cancer. People are mainly exposed to As through water and food intake, with a high exposure risk in developing countries like Malawi. The high prevalence of As exposure in many countries has led to studies to evaluate the As prevalence rate of exposure, toxic effects and consequences, and remediation measures and techniques. This review discusses an overview of the prevalence studies of As in foodstuffs and water, as well as their toxic effects and remediation measures employed to deal with the problem in Malawi. Although there are many studies conducted elsewhere about the presence of As in foodstuffs and water, there was only one (1) study on the presence of As in foodstuffs (0.040 mg/kg DW; inorganic arsenic) and one (1) study on herbal preparations (<0.045, undetectable). There were ten (10) studies on water (undetectable—0.5 mg/l; 0.4–14.5 lg/L) covering groundwater (7) (shallow wells, 3; boreholes, 4; hand-dug wells, 1 and households, 1), and surface water (3) (river, 1; lake, 2). Compared to the

I. Chikowe (✉)
Pharmacy Department, Kamuzu University of Health Sciences (KUHeS), Blantyre, Malawi
e-mail: ichikowe@medcol.mw

B. Mwamatope
Basic Science Department, Lilongwe University of Agriculture and Natural Resources, Bunda Campus, Lilongwe, Malawi

U. Kankwatira
Physics and Biochemical Sciences Department, Malawi University of Business and Applied Sciences (MUBAS), Blantyre, Malawi

H. Phiri
Pharmacy Department, Queen Elizabeth Central Hospital (QECH), Blantyre, Malawi

G. Chirambo · C. E. Jana
Biomedical Sciences Department, Kamuzu University of Health Sciences (KUHeS), Blantyre, Malawi

B. Sen Gupta and N. Martínez-Villegas (eds.), *Arsenic Remediation of Food and Water*,
https://doi.org/10.1007/978-981-97-4764-1_5

10 parts per billion Environmental Protection Agency (EPA) limit for As contamination in water and 0.5–2 parts per million in food, the estimated amounts ranged from safe to toxic limits in water and safe in foodstuffs. However, the number of studies was too small to determine the arsenic prevalence in food and water, and the coverage of the studies was skewed towards one region. There were no studies on toxic effects, remediation measures, and techniques. Therefore, more studies are needed in Malawi to determine the prevalence of As exposure in all media, its toxic effects and remediation measures.

Keywords Arsenic · Heavy metals · Foodstuffs · Herbal medicines · Water · Malawi

Abbreviations

As	Arsenic
DNA	Deoxyribonucleic acid
EPA	Environmental Protection Agency
EU	European Union
FAO	Food and Agriculture Organization
ICP–AES	Inductively coupled plasma–atomic emission spectroscopy
ICP–MS	Inductively coupled plasma–mass spectroscopy
ICP-OES	Inductively coupled plasma optical emission spectrometry
PPB	Parts per billion
PPM	Parts per million
US	United States
USA	United States of America
WHO	World Health Organization

Introduction

Arsenic (As) is a naturally occurring Group V element of the Periodic Table with atomic number 33, an atomic mass of 74.9, a specific gravity of 5.73, a melting point of 817 °C (at 28 atm), a boiling temperature of 613 °C, and vapour pressure of 1 mm Hg at 372 °C. Arsenic is a tasteless and odourless chemical with that can exist in two or more different forms (allotropy or allotropism) in the same physical state. It is a steel-grey metalloid that is brittle and crystalline. There are four different oxidation states of As that can be found in the environment: arsine (III), arsenite (+III), arsenic (0), and arsenate (+V). However, the most prevalent forms of arsenite found in natural water are the inorganic trivalent arsenite form (+III) and the pentavalent arsenate form (+V). Numerous minerals include arsenic, which can also be found as a pure elemental crystal. It typically occurs alongside sulphur and metals in nature (Asere et al., 2019; Fernanda et al., 2016; Pankaj et al., 2014; Vu et al., 2003). Arsenic is a naturally occurring element of the earth's crust and is found in large quantities in the air, water, and land (soils, rocks) (Vu et al., 2003; World Health Organisation [WHO], 2018). In its natural state, As may be found in both organic and inorganic forms. Examples of the organic form of arsenic in water include dimethylarsinic acid ($C_2H_7AsO_2$) and monomethylarsonic acid (CH_5AsO_3), while arsenic trioxide (As_2O_3) and sodium arsenate (NaH_2AsO_4) are examples of the inorganic forms of arsenic in water (Smedley & Kinniburgh, 2002).

Arsenic and its compounds are used in various industries, such as pharmaceuticals, wood, agricultural chemicals, mining, metallurgy, glassmaking, and semiconductor industries. For example, it is mostly used in lead alloys (for example, in car batteries and ammunition), and it is also a component in semiconductor electronic devices, pesticides, treated wood products, preservatives, herbicides, and insecticides. With the growing understanding of the toxicity of arsenic and related derivatives, these applications are waning (IARC, 2012).

In its inorganic form, As is extremely poisonous (WHO, 2018), and the trivalent form is approximately sixty times more poisonous than the oxidized pentavalent state. Consequently, the trivalent form of arsenic must be converted into the pentavalent form (Vu et al., 2003). Arsenic (As) is difficult to eradicate and can only be changed into new forms or made into insoluble compounds by combining them with other elements (Choong et al., 2007). These factors have led to identification, quantification, and remediation studies aimed at finding the species of As present as well as ways of removing or minimising their presence in their human exposure media.

People are exposed to high doses of inorganic arsenic through drinking contaminated water from natural causes such as erosion and human activities such as careless dumping of wastes containing As, using contaminated water for food preparations and crop irrigation, industrial activities, eating contaminated food, and smoking cigarettes (Darani et al., 2022). The source of arsenic's biggest threat to public health is contaminated groundwater. Many nations, including Argentina, Bangladesh, Chile, China, India, Mexico, and the United States of America, naturally contain high quantities of inorganic arsenic in their groundwater. The sources of exposure are drinking water, crops irrigated with polluted water, and food cooked with contaminated water. Other food sources of arsenic include fish, shellfish, pork, chicken, dairy products, and cereals, though exposure from these sources is often considerably lower than exposure from contaminated groundwater. Arsenic is primarily present in seafood in its less hazardous organic form (IARC, 2012).

Due to the toxic effects of As, several studies have been conducted worldwide to quantify the extent of exposure to As in people, routes of exposure, and epidemiological studies of its effects. Remediation studies have also been carried out to reduce the presence of As in water and foodstuffs, and the results have been used for the development of evidence-based interventions and policy formulation by local and international regulatory bodies (Darani et al., 2022). In Malawi, similar studies have been done, but there is sparse data about the extent to which the studies have been done and whether they are enough to formulate policies. This review is therefore aimed at describing human exposure to arsenic through food and water as well as remediation measures ever implemented in Malawi. In addition, a brief description of commonly assessed parameters worldwide is presented as a guide for the data collection focus in the current review. The study presents data from published literature only and therefore does not reflect the overall situation in Malawi.

Materials and Methods

Search Strategy

Databases from Google Scholar, PubMed, and Science Direct were searched for published laboratory and observational studies evaluating the presence, toxic effects, and remediation of As in water and food in Malawi. Search keywords comprising free texts Arsenic + Water + Food + Malawi, arsenic and food and Malawi, or arsenic and water and Malawi, arsenic Malawi, were used for studies up to November 2022. Figure 5.1 shows a summary of the procedure that was followed.

Each author was tasked with searching for information related to each subtopic, namely arsenic in water—Malawi, toxicity, and remediation of arsenic—Malawi, arsenic in food, arsenic toxicity and arsenic remediation—Malawi, arsenic removal—Malawi, and arsenic in medicines in Malawi. Then all authors also conducted confirmatory searches on all data generated by each author. All authors evaluated the papers and used the search criteria to select papers and extract information. The study targeted epidemiological studies on arsenic exposure and exposure outcomes,

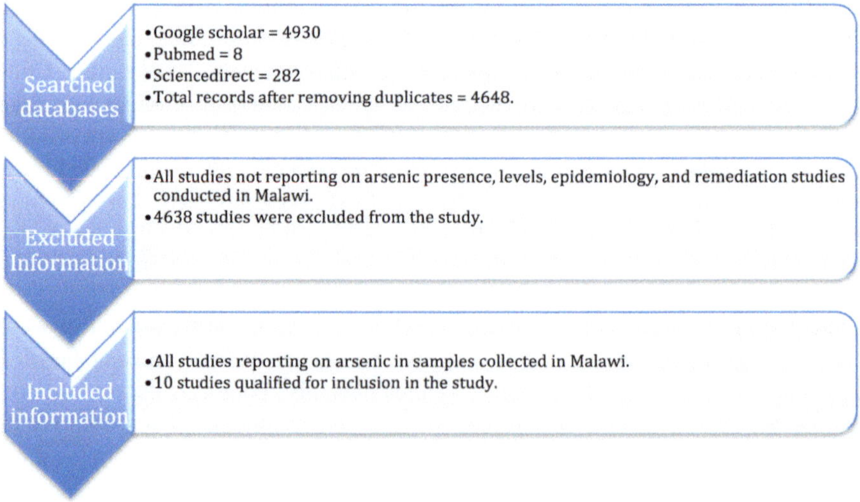

Fig. 5.1 Study data collection protocol

experimental studies, case reports, and case series on arsenic toxicity, quantity, and remediation. All studies that did not report on Malawi were excluded.

Data Analysis

Descriptive data analysis methods were used mostly using frequency on the number of studies conducted in food, water, toxicity, and remediation.

Results

There were twelve (12) studies comprising one (1) study on the presence of As in foodstuffs, particularly rice, whereby 0.040 mg/kg DW of inorganic arsenic was found; one (1) study on herbal preparations in which < 0.045, undetectable As was found; and ten (10) studies on water in which undetectable—0.5 mg/l; 0.4–14.5 lg/L levels of As were found. Seven of the water studies reported on groundwater, whereas three (3) of them covered shallow wells, four (4) covered boreholes, one (1) covered hand-dug wells, and one (1) covered households. Three (3) of the water studies covered surface water, whereas one (1) study evaluated river water, while the other two (2) evaluated lake water. Compared to the 10 parts per billion (ppb) US Environmental Protection Agency (EPA) exposure limit in water and 0.5–2 parts per million (ppm) limits in food, the amounts ranged from safe to toxic limits in water and safe in foodstuffs (Table 5.1). There were no studies on the toxic effects of As or epidemiology, as well as remediation measures and techniques.

Discussion

Arsenic in Water

Water contamination by human activity or due to natural causes has emerged as one of the most pressing environmental concerns worldwide due to its adverse socio-economic effects. The use of As-rich ground and surface water has subjected millions of people to high levels of exposure to arsenic (Smedley & Kinniburgh, 2002). Arsenic in its inorganic form is typically found in significant amounts in drinking water as arsenite, As (V), or arsenate, As (III). Water can have varying levels of arsenic pollution depending on where it is located, and studies have shed light on the connection between human activities and the presence of arsenic in water (Das & Murkhejee, 2022). Based on this background, studies have been conducted at various water sources worldwide. In our review, a good number of studies were done on

Table 5.1 Studies conducted in Malawi on Arsenic presence in water and food, toxicity impacts, and remediation

Study focus	Study purpose	Study parameters	Study place	Study sample	Source	Analytical method	Result	References
Food	Elemental composition of rice grain grown in Malawi	Concentrations of Ca, Cu, Fe, I, Mg, Mn, Se, Zn, As, Cd, and Pb inorganic arsenic (i-As) and phytic acid (PA)	18 extension planning areas (EPAs in Karonga, Nkhata Bay, Salima, Lake Chilwa, Shire River basin	Rice grain	Smallholder Production, government-sponsored irrigation schemes, irrigation schemes	Inductively coupled plasma-mass spectrometry (ICP-MS)	0.040 (mg kg^{-1} DW) /Inorganic arsenic	Edward et al. (2016)
	Bacterial and heavy metal contamination in selected commonly sold herbal medicine in Blantyre, Malawi	Heavy metals and bacteria	Blantyre	Herbal medicine	Herbal markets/ vendors	Atomic absorption spectrometry (AAS)	<0.045 mg/g (undetectable)	Kalumbi et al. (2020)
Water	Microbiological, chemical and physical quality of groundwater from shallow wells in selected villages in the rural areas in Blantyre	Arsenic	Lunzu—Blantyre	Groundwater	Shallow wells	Rapid arsenic test kit ('Quick Econo II)	3 ug/l	Mkandawire (2008)

(continued)

Table 5.1 (continued)

Study focus	Study purpose	Study parameters	Study place	Study sample	Source	Analytical method	Result	References
	Quality of water from shallow wells in three districts in southern Malawi namely, Balaka, Chikwawa, and Zomba districts	Arsenic	Chikwawa, Zomba and Balaka	Groundwater	Shallow wells	Rapid Arsenic Test Kit (Quick Econo II – ELE: 432–130)	$\leq$ 0.002 ug/l	Pritchard et al. (2008)
	Biological, chemical, and physical shallow well water quality	Arsenic	Blantyre, Chiradzulu and Mulanje	Groundwater	Shallow wells	Photometer	$\leq$ 3 ug/l	Pritchard et al. (2007)
	Safety of groundwater sources and to explore factors influencing water quality	Arsenic	Upper Limphasa River catchment in the Nkhata Bay District	Groundwater	Boreholes, hand-dug wells, and households	ICP-OES auto-analyser program	0 ug/l	Kanyerere et al. (2012)
	Suitability of the borehole water for human consumption	Arsenic	Phalombe plain	Groundwater	Boreholes	Flame atomic absorption spectrophotometer	Below detection limit	Vunain et al. (2019)

(continued)

Table 5.1 (continued)

Study focus	Study purpose	Study parameters	Study place	Study sample	Source	Analytical method	Result	References
	Levels and distribution of heavy metals in Lake Chilwa and its catchment, and the associated level of pollution	Arsenic	Lake Chirwa catchment area in Zomba	Surface water, lake	River, lake	Inductively coupled plasma optical emission spectroscopy	Below detection limit	Mussa et al. (2019)
	Trace element evolution in North Rukuru–Songwe alluvial aquifer of northern Malawi	Total arsenic	Karonga	Groundwater	Borehole	Inductively coupled plasma optical emission spectrometry (ICP-OES)	0.4–14.5 lg/L	Mapoma et al. (2017)

(continued)

Table 5.1 (continued)

Study focus	Study purpose	Study parameters	Study place	Study sample	Source	Analytical method	Result	References
	Quality of groundwater for drinking and irrigation in Blantyre, Malawi; geochemical processes governing mineralization of major and some minor elements	Arsenic	Machinjiri—Blantyre	Groundwater	Boreholes	Inductively coupled plasma optical emission spectrometry (ICP-OES)	0.0016–0.5 mg/l	Mapoma et al. (2016)

(continued)

Table 5.1 (continued)

Study focus	Study purpose	Study parameters	Study place	Study sample	Source	Analytical method	Result	References
	Address the scarcity of physicochemical datasets of water in Malawian artisanal gold mining hotspots	Arsenic	Lisungwi River in the Manondo area, within the Ntcheu-Neno-Balaka block	Surface water	River	Inductively coupled plasma–atomic emission spectroscopy (ICP–AES) and ICP–mass spectroscopy (ICP–MS)	<0.5 ug/l	Chikanda et al. (2022)
	Water, air, and precipitation samples for persistent organochlorines to determine inputs to the lake from the atmosphere, equilibrium state between air and lake water, and possible sources of the contaminants	Arsenic	Karonga, Nkhata-bay, Salima	Surface water	Lake	Flame atomic absorption	0.22–3.43 ug/g	Kidd et al. (1999)

various water media in different parts of the country, like those reported in the literature. There are also a variety of water sources that have been studied, such as shallow wells, boreholes, hand-dug wells and households, river, and lake water (Fig. 5.1; Table 5.1). The studied water parameters are similar to many studies conducted elsewhere that also studied the same parameters, such as boreholes in Nairobi City County in Kenya (Kiplangat et al., 2021) and groundwater in South West Ireland (McGrory et al., 2018). The results in this study were different from many reported elsewhere, as the levels of As in Malawian studies were mostly within acceptable limits (Table 5.1). This is different from many countries, such as Bangladesh, China, Pakistan, Cambodia, the USA, Indonesia, Canada, Hungary, Mexico, Nepal, India, and Greece, where higher than recommended amounts have been reported (Sharma et al., 2014). However, the studies in Malawi are not widespread enough for the country to guide policy direction because they cover a few areas not commensurate with the many sources of As the country has, such as irrigation areas, industries, and dumping sites for heavy metal-containing products such as cosmetics and batteries, wood, pharmaceuticals, metals, glass, alloys, pigments, leather, paints, poison baits, insecticides, weed killers, pesticides, soil sterilants, microelectronics, and other As-containing products (Chung et al., 2014), most of which are used in Malawi. For example, there are 27 districts in Malawi, but only 11 districts have been studied. Literature also shows that As can originate from natural sources in the ground; hence, sampling is supposed to be as wide as possible to identify the As hotspots, which would then be followed by routine monitoring studies. This would also be important in this time of continuous urbanisation and climate change, where many chemical dynamics are changing that can also change the status quo on the quality of water (Miller & Hutchins, 2017). This would also be important because As does not have a distinct colour or flavour, hence it is difficult to detect, and surveillance studies are needed periodically to know the present status of As levels (Asere et al., 2019; Bhowmick et al., 2018; Smith et al., 2000; Wang & Zhao, 2009).

The review has shown that As was present in some samples studied in lower amounts than the acceptable amounts set by many regulatory bodies (Table 5.1). The European Union (EU), the United States (US), and the World Health Organization (WHO) have all agreed on a maximum contamination level of ten grams per litre (10 g/L) for total arsenic in drinkable water. This number was defined as the highest allowable level of total arsenic (WHO, 2018). The presence of arsenic in freshwater poses a threat to public health since, even in trace amounts, this element can cause arsenicosis, which is responsible for several disorders (WHO, 2018). For example, prolonged consumption of As leads to the development of several dangerous ailments, including high blood pressure, skin pigmentation, diseases of the blood vessels, kidneys, and lungs, reproductive abnormalities, disruptions in the operations of the digestive tract, cardiovascular and brain systems, cancer, and ultimately death. Many countries have experienced substantial health issues because of the presence of arsenic in their drinking water (Asere et al., 2019; Bhowmick et al., 2018; Smith et al., 2000; Wang & Zhao, 2009). Therefore, the low presence of As is not a sign that there will be no toxic effects, as low levels also have the potential to cause negative health effects in the long run.

It has been estimated that at least one million people in fifty different nations are subjected to arsenic poisoning because they drink arsenic-tainted groundwater at concentrations greater than ten micrograms per litre (WHO, 2018). Therefore, it is important to scale up the As studies in Malawi to generate enough data for decision-making and evidence-based interventions on As presence in the water. This would also help with monitoring the As levels over time since the sources of As are continuously changing with the changes in development and technology, as well as climate change and disease incidences. For example, the increase in cancer incidences might also lead to the widespread use of As-based treatment options that might also add to the list of sources of As in water and food (Paul et al., 2022), while climate change and urbanisation have led to increases in flooding that have also affected water quality (Miller & Hutchins, 2017).

Remediation of Arsenic in Malawian Water

Numerous conventional and innovative treatment strategies have been proposed for removing As from groundwater, and these strategies have been tested in both the laboratory and the field (Mou et al., 2020). These include the use of chlorine, ozone, and other similar substances, as well as physical methods such as UV treatment, in-situ oxidation (Nicomel et al., 2015), arsenic mitigation by biological oxidation (Oremland et al., 2005), arsenic mitigation by coagulation-flocculation (Ucar et al., 2013), membrane filtration (Mou et al., 2020), reverse osmosis, and nanofiltration (Figoli et al., 2010), among others. However, our review showed that there has been no study ever conducted in Malawi to remediate the presence of As in water. Therefore, although the reported levels are low in Malawi based on the study (Table 5.1), the levels might change because there are no known remediation methods, either ongoing or deliberately implemented to deal with the As contamination since As levels can accumulate over time or can be reduced due to natural phenomena (Sandil et al., 2021). Therefore, in addition to identifying hotspots of As present in the country, remediation measures should be tested and implemented to minimise the presence of As in water, hence human exposure. It is important to test some technologies and methods applied elsewhere in Malawi because differences in climate, soil, microbiota, and water states can affect their efficacy. Therefore, optimisation and adaptation studies may be needed for them to work effectively.

The lack of data on remediation measures makes it difficult to implement exposure minimisation strategies since the levels help with decision-making. For example, due to the relatively higher irrigation needs of rice grains compared to other crops, the equivalent suggested limit by the Food and Agriculture Organization (FAO) in conjunction with WHO, currently only applies to rice grains and has been set at 0.2 mg/kg (FAO, 2014). It should be emphasized that neither the European Commission nor the US Environmental Protection Agency have yet approved this concentration level. This is one of the examples of the decision-making process that made use of prevailing data about commonly reported As levels in rice products, which helped to determine the safer levels in rice. Yet, in some areas, rice might not be their main agricultural produce; they may also need a different recommended limit of As exposure.

Arsenic in Malawian Food

Arsenic can enter the human body through drinking arsenic-contaminated water or eating arsenic-contaminated food. Arsenic has made its way into the food supply, primarily through rice, wheat, and potatoes. Numerous environmental transportation pathways have been identified, and rather high quantities (mostly in water sources) have been documented in numerous regions across the globe (Upadhaya et al., 2019). Differences in the way food is contaminated as well as commonly contaminated foods have been widely reported, such as rice, cereal grains (other than rice), sugar and sugar products, chocolate, vegetable fats and oils, vegetables, nuts, roots, potatoes, juices, soft drinks, bottled water, coffee, tea, alcoholic beverages, meat and meat products, milk, eggs, fish, and other products (Sharma et al., 2014). However, in Malawi, only two studies have reported on the presence of As in food, herbal medicines, and rice (Table 5.1). Therefore, studies are needed in Malawi to generate data on the presence of As in other foodstuffs for use in evidence-based interventions as well as use as a routine monitoring tool over time. The number of samples studied and the coverage of the food types are not enough to determine the prevalence of As in food. As such, it might be difficult to develop evidence-based interventions, let alone know whether there is a problem worthy of intervention. Furthermore, the changes in As contamination patterns in water may also affect the presence of As in foodstuffs because most foodstuffs are grown using irrigation, which is one of the ways As is transferred from water to food and people.

Toxic Effects of Arsenic on Human Health in Malawi

Several studies have been conducted worldwide on the epidemiology of As toxicity in people, and As consumption has been linked to many diseases. Inorganic arsenicals soluble in water are rapidly absorbed once they have passed through the gastrointestinal tract. The As(III) form of arsenic often enters the body by a straightforward diffusion route. Through an energy-dependent transport mechanism, a small amount of As (V) may pass through cell membranes and subsequently be converted to As (III), which would then attach to DNA or protein molecules (Jomova et al., 2011; Khairul et al., 2017). Arsenic poisoning is undetectable in the early stages, and depending on the individual's immune system and the amount of arsenic taken, it might take up to 8 years for symptoms to manifest. According to studies, arsenicosis results from human intakes of arsenic in the range of 0.05 mg/L. Arsenic has been linked to several ailments in the digestive, haematological, endocrine, renal, neurological, and reproductive systems, and these include pulmonary disease, cerebrovascular disease, cardiac disease, melanosis, hyperkeratosis, leucomelanosis, hepatic enlargement and cirrhosis, peripheral neuropathy, hypertension, chromosomal anomaly, heart failure, diabetes mellitus, goitre, and gangrene. It is also connected with disorders that affect the capillaries, arteries, and arterioles (Fontcuberta et al., 2011). These diseases, in the long run, raise the likelihood of developing cancers of the bladder, kidneys, liver, lungs, and lymphatic system, as well as diseases affecting the blood vessels in the legs and feet and excessive blood pressure (Khan et al., 2020; Santra & Samal, 2013). Other effects include children's cognitive and psychological development, a higher rate of foetal loss and neonatal mortality, genotoxicity, epidemiological toxicity, and cytotoxicity (Medda et al., 2021; Niaz & Nazmul, 2011).

These toxic effects have been found in *in vitro*, *in vivo*, and population or epidemiological studies worldwide, and their incidences and occurrences vary widely. In our review of similar studies in Malawi, there was no study found on these fronts. The results of the review have shown that there are a limited number of studies in Malawi on arsenic contamination in water and food, epidemiology, and remediation (Table 5.1). The lack of epidemiological studies of As toxicity in Malawi makes it difficult to estimate the toxicity burden, raise problem awareness, and develop public health interventions. According to research on the relationship between As and plants in soil, arsenic's toxicity to plants and humans varies depending on the kind of plant it occurs in (for example, arsenite, As (III), and arsenate, As (V)), as well as other soil conditions that affect how much As plants can accumulate. Different plant species have various absorption, toxicity, and detoxification strategies for As (III) or As (V) (Abbas et al., 2018). This may also affect the effective design of interventions in cases of problems because Malawi might have different disease burdens, widely consumed plants, and soil or water dynamics, which are all important in the development of interventions.

Conclusion

There are a very limited number of studies on the presence and quantities of As in food and water consumed in Malawi, toxicity epidemiology, and remediation studies. A good number of studies have been done on the presence of As in water, a few on food, and none on toxicity epidemiology. However, in those parameters, with whatever studies are available, the data is not widespread enough in the studied parameters and study areas to determine the prevalence levels of As in Malawi that would be needed for evidence-based interventions. Therefore, more studies are needed in all the research areas namely, contamination in water and food, as well as remediation and epidemiological studies (Mapoma et al., 2017).

References

Abbas, G., Murtaza, B., Bibi, I., Shahid, M., Niazi, N., Khan, M., et al. (2018). Arsenic uptake, toxicity, detoxification, and speciation in plants: Physiological, biochemical, and molecular aspects. *International Journal of Environment Research Public Health, 15*(1 Jan 2), 59. https://doi.org/10.3390/ijerph150

Asere, T. G., Stevens, C. V., Du Laing, G. (2019). Use of (modified) natural adsorbents for arsenic remediation: A review. *Science of the Total Environment, 676*, 706–720. https://doi.org/10.1016/j.scitotenv.2019, 04.237

Bhowmick, S., Pramanik, S., Singh, P., Mondal, P., Chatterjee, D., Nriagu, J. (2018). Arsenic in groundwater of West Bengal, India: A review of human health risks and assessment of possible intervention options. *Science of the Total Environment, 612*, 148–169. https://doi.org/10.1016/j.scitotenv.2017.08.216

Chikanda, F., Otake, T., Mwatseteza, J., & Sato, T. (2022). Water quality in the Lisungwi and Kaphamtengo Rivers in Manondo, Central Malawi: Assessment of the impact of artisanal small scale gold mining. *Malawi Journal of Science and Technology, 14*(1), 29–43.

Choong, T. S. Y., Chuah, T. G., Robiah, Y., Koay, F. L. G., & Azni, I. (2007). Arsenic toxicity, health hazards and removal techniques from water: An overview. *Desalination, 217*(1–3), 139–166. https://doi.org/10.1016/j.desal.2007.01.015

Chung, J. Y., Yu, S. D., Hong, Y. S. (2014). Environmental source of arsenic exposure. *Journal of Preventive Medicine and Public Health = Yebang Uihakhoe chi, 47*(5), 253–257. https://doi.org/10.3961/jpmph.14.036

Darani, K., Rehman, Y., Katsoyiannis, L., Kokkinos, E., & Zouboulis, A. (2022). Arsenic exposure via contaminated water and food sources. *Environmental Science Water, 14*(12). https://doi.org/10.3390/w14121884

Das, S., & Murkhejee, S. (2022). Implementation of biotechnological techniques in treatment of ground water contaminated with arsenic. *International Journal for Research in Applied Science & Enginneering Technology, 10*(II Feb), 45–98.

Edward, J. M., Ander, J. E. L., Broadley, M. R., Young, S. D., Chilimba, A. D. C., Hamilton, E. M., Watts, M. J. (2016). Elemental composition of Malawian rice. *Environmental Geochemistry Health*. https://doi.org/10.1007/s10653-016-9854-9.

FAO. (2014). Codex Alimentarius Commission. Retrieved November 15, 2022, from https://www.fao.org/news/story/en/item/%20238558/icode/

Fernanda, P. C., Amaral, C. D. B., Schiavo, D., & Nóbrega, J. A. (2016). Determination of arsenic in fruit juices using inductively coupled plasma tandem mass spectrometry (ICP-MS/MS). *Food Analytical Methods, 10*, 992–998.

Figoli, A., Cassano, A., Criscuoli, A. M., Mozumder, S. I. M., Uddin, T. M., Islam, A., Drioli, E. (2010). Influence of operating parameters on the arsenic removal by nanofiltration. *Water Research, 44*(1).

Fontcuberta, M., Calderon, J., Villalbí, J. R., Centrich, F., Portaña, S., Espelt, A., Duran, J., Nebot, M. (2011). Total and inorganic arsenic in marketed food and associated health risks for the Catalan (Spain) population. *Journal of Agricultural Food Chemistry, 59*, 10013.

IARC Working Group on the Evaluation of Carcinogenic Risks to Humans. Arsenic, Metals, Fibres and Dusts. Lyon (FR): International Agency for Research on Cancer and 2012. (IARC Monographs on the Evaluation of Carcinogenic Risks to Humans, No. 100C.).

Jomova, K., Jenisova, Z., Feszterova, M., Baros, S., Liska, J., Hudecova, D., Rhodes, C. J., & Valko, M. (2011). Arsenic: Toxicity, oxidative stress and human disease. *Journal of Applied Toxicology, 31*(2), 95–107. https://doi.org/10.1002/jat.1649

Kalumbi, M. H., Likongwe, M. C., Mponda, J., Zimba, B. L., Phiri, O., Lipenga, T., Mguntha, T., Kumphanda, J. (2020). Bacterial and heavy metal contamination in selected commonly sold herbal medicine in Blantyre, Malawi. *Malawi Medical Journal, 32*(3), 153–159.https://doi.org/10.4314/mmj.v32i3.8

Kanyerere, T., Levy, J., Xu, Y., & Saka, J. (2012). Assessment of microbial contamination of groundwater in upper Limphasa River catchment, located in a rural area of northern Malawi. *Water SA, 38*(4), 581–596.

Khairul, I., Wang, Q. Q., Jiang, Y. H., Wang, C., & Naranmandura, H. (2017). Metabolism, toxicity and anticancer activities of arsenic compounds. *Oncotarget, 8*(14), 23905–23926. https://doi.org/10.18632/oncotarget.14733

Khan, M. A., Hira-Smith, et al. (2020). Prospective cohort study of respiratory effects at ages 14 to 26 following early life exposure to arsenic in drinking water. *Environmental Epidemiology (Philadelphia, Pa.), 4*(2), e089. https://doi.org/10.1097/EE9.00

Kidd, K. A., Lockhart, W. L., Wilkinson, P., & Muir, D. C. (1999). Metals, pesticides and other persistent contaminants in water, sediments and biota from Lake Malawi. Water Quality Report, Lake Malawi/Nyasa Biodiversity Conservation Project. SADC/GEF, 243–276.

Kiplangat, A. S., Mwangi, H., Swaleh, S., Njue, W. M. (2021). Arsenic contamination in water from selected boreholes in Nairobi City County, Kenya. *European Journal of Advanced Chemistry Research, 2*(2), 1–6. https://doi.org/10.24018/ejchem.2021.2.2.50

Mapoma, H. W., Xie, X., Zhang, L., Nyirenda, M. T., Maliro, A., & Chimutu, D. (2016). Hydrochemical characteristics of rural community groundwater supply in Blantyre, southern Malawi. *Journal of African Earth Sciences, 114*, 192–202.

Mapoma, H. W. T., Xie, X., Liu, Y., et al. (2017). Hydrochemistry and quality of groundwater in alluvial aquifer of Karonga, Malawi. *Environmental Earth Science, 76*, 335. https://doi.org/10.1007/s12665-017-6653-2

McGrory, E., Holian, E., Alvarez-Iglesias, A., et al. (2018). Arsenic in groundwater in South West Ireland: Occurrence, controls, and hydrochemistry. *Frontiers in Environmental Science, 6*. https://www.frontiersin.org/articles/10.3389/

Medda, N., et al. (2021). Different mechanisms of arsenic related signaling in cellular proliferation, apoptosis and neo-plastic transformation. *Ecotoxicology and Environmental Safety, 208*. https://doi.org/10.1016/j.ecoenv.2020.111752

Miller, J. D., & Hutchins, M. (2017). The impacts of urbanisation and climate change on urban flooding and urban water quality: A review of the evidence concerning the United Kingdom. *Journal of Hydrology: Regional Studies, 12*, 345–362.

Mkandawire, T. (2008). Quality of groundwater from shallow wells of selected villages in Blantyre District, Malawi. *Physics and Chemistry of the Earth, Parts A/B/C, 33*(8–13), 807–811.

Mou, S. A., Kabir, M. H., Yasmin, S., & Ahmed, S. (2020). Arsenic mitigation technologies from ground water: A brief review. American *Journal of Pure Applied Science, 2*(5), 139–158. https://doi.org/10.34104/ajpab.020.01390158

Mussa, C., Biswick, T., Changadeya, W., Junginger, A., & Vunain, E. (2019). Levels and spatial distribution of heavy metals in Lake Chilwa Catchment, Southern Malawi. *ChemSearch Journal, 10*(2), 66–73.

Niaz, A. M., & Nazmul, C. (2011). Poisoning the mind: Arsenic contamination of drinking water wells and children's educational achievement in rural Bangladesh. *Economics of Education Review. Elsevier, 30*(5), 873–888.

Nicomel, N. R, Leus, K., Folens, K., Van Der Voort, P., Du Laing, G. (2015). Technologies for arsenic removal from water: Current status and future perspectives. *International Journal of Environmental Research and Public Health, 13*(1), ijerph13010062.

Oremland, R. S., Kulp, T. R., Blum, J. S., Hoeft, S. E., Baesman, S., Miller, L. G., & Stolz, J. F. (2005). A microbial arsenic cycle in a salt-saturated, extreme environment. *Science, 308*(5726), 1305–1308. https://doi.org/10.1126/science.1110832

Pankaj, V., Agarwal, A., & Singh, V. (2014). Arsenic removal from water through adsorption—A review. *Recent Research in Science and Technology, 6*, 219–226.

Paul, N. P., Galván, A. E., et al. (2022). Arsenic in medicine: past, present and future. *Biometals: An International Journal on the Role of Metal Ions in Biology, Biochemistry, and Medicine*. Advance Online Publication, 1–19.

Pritchard, M., Mkandawire, T., & O'Neill, J. G. (2007). Biological, chemical and physical drinking water quality from shallow wells in Malawi: Case study of Blantyre, Chiradzulu and Mulanje, *32*(15-18), 1177.https://doi.org/10.1016/j.pce.2007.07.013

Pritchard, M., Mkandawire, T., & O'Neill, J. G. (2008). Assessment of groundwater quality in shallow wells within the southern districts of Malawi, *33*(8-13), 0–823.https://doi.org/10.1016/j.pce.2008.06.036

Sandil, S., Óvári, M., Dobosy, P., Vetési, V., Endrédi, A., Takács, A., Füzy, A., & Záray, G. (2021). Effect of arsenic-contaminated irrigation water on growth and elemental composition of tomato and cabbage cultivated in three different soils, and related health risk assessment. *Environmental research, 197*, 111098. https://doi.org/10.1016/j.envres.2021.111098

Santra, S. S., Samal, A. C., et al. (2013). Arsenic in foodchain and community health risk: A study in gangetic West Bengal. *Procedia Environmental Sciences, 18*, 2–13. https://doi.org/10.1016/j.proenv.2013.04.002

Sharma, A. K., Tjell, J. C., Sloth, J. J., & Holm, P. E. (2014). Review of arsenic contamination, exposure through water and food and low cost mitigation options for rural areas. *Applied Geochemistry, 41*, 11–33. https://doi.org/10.1016/j.apgeochem.2013.11.012

Smedley, P. L., & Kinniburgh, D. G. (2002). A review of the source, behaviour and distribution of arsenic in natural waters. *Applied Geochemistry, 17*(5), 517–568. https://doi.org/10.1016/S0883-2927(02)00018-5

Smith, A. H., Lingas, E. O., & Rahman, M. (2000). Contamination of drinking-water by arsenic in Bangladesh: A public health emergency. *Bulletin of the World Health Organization, 78*(9), 1093–1103.

Ucar, C., Baskan, M. B., & Pala, A. (2013). Arsenic removal from drinking water by electrocoagulation using iron electrodes. *Korean Journal of Chemical Engineering, 30*, 1889–1895. https://doi.org/10.1007/s11814-013-0128-2

Upadhaya, M., Shukla, A., Yadav, P., & Srivastava, S. (2019). A review of arsenic in crops, vegetables animals and food products. *Food Chemistry, 276*, 607–618.

Vu, K. B., Kaminski, M. D., Nunez, L. (2003). Review of arsenic removal technologies for contaminated groundwaters. Report, May 2 and Illinois. Retrieved December 4, 2022, from https://digital.library.unt.edu/ark:/67531/metadc735691/

Vunain, E., Nkhuzenje, C., Mwatseteza, J., & Sajidu, S. (2019). Groundwater quality assessment from Phalombe plain, Malawi. *Chemsearch Journal, 10*(1), 1.

Wang, S., & Zhao, X. (2009). On the potential of biological treatment for arsenic contaminated soils and groundwater. *Journal of Environmental Management, 90*(8), 2367–2376. https://doi.org/10.1016/j.jenvman.2009.02.001

World Health Organisation (WHO). (2018). Arsenic. Retrieved November 7, 2022, from www.who.int/news-room/fact-sheets/detail/arsenic

Chapter 6
Groundwater Arsenic Contamination in Karimpur-I Block, District Nadia, West-Bengal and Investigation for Safe Water Option

Bhaskar Das, Somil Thakur, and Sanjana Chakraborty

Abstract The current chapter presents 18 years' of work (1989–2007) related to groundwater arsenic concentration in Karimpur I block in the Nadia District of West Bengal, India. The study was conducted by the School of Environmental Studies (SOES), Jadavpur University, Kolkata, and is taken from the Ph.D. thesis of the first author of this chapter. The work summarizes the results of groundwater samples collected from all 104 villages of 67 inhabited Mouzas (among a total of 74) from Karimpur I block. This block is one of the earliest identified arsenic-affected blocks from West Bengal and is considered to be severely arsenic-contaminated. Among the 3374 water samples collected from this block, 1896 (56.2%) and 1000 (29.6%) were having arsenic concentrations >10 μg/L and >50 μg/L, respectively. In addition to arsenic, depth information of 2825 samples was collected and 15 (0.5%) tubewells were found to have depths of >100 m. A total of 375 samples were analyzed for arsenic and iron. The highest iron concentration recorded was 34,730 μg/L with an average of 4870 μg/L. Analysis of urine samples indicated a substantial arsenic body burden as 78 samples (70.3%) were found to contain arsenic beyond the normal range (5–40 μg/1.5 L per day). In order to assess the current status of safe water options in this block, the existing arsenic mitigation strategies were investigated and analyzed for their usefulness and sustainability. This study concludes with recommendations for utilizing the available safe water options for mitigation efforts.

B. Das (✉)
School of Civil Engineering, Vellore Institute of Technology (VIT), Vellore, Tamil Nadu 632014, India
e-mail: bhaskar.ju@gmail.com

S. Thakur
Department of Civil Engineering, Institute of Technology, Nirma University, Ahmedabad, Gujarat 382481, India

S. Chakraborty
Department of Economics, Jadavpur University, Kolkata, West Bengal 700032, India

B. Sen Gupta and N. Martínez-Villegas (eds.), *Arsenic Remediation of Food and Water*,
https://doi.org/10.1007/978-981-97-4764-1_6

Keywords Arsenic · Groundwater · Health effects · Arsenic-safe water · Mitigation strategy

Introduction

Groundwater arsenic contamination has been a major public health concern in different parts of the world for the last few decades. Arsenic in the groundwater has been reported in 105 countries for affected populations of >200 million worldwide (Murcott, 2012; Naujokas et al., 2013; Shaji et al., 2021). Asian countries, especially the Ganga–Meghna–Brahmaputra (GMB) Plain of India and Bangladesh are the worst affected areas in the world. A significant part of the GMB plain of area and population (569,749 sq. km. and over 500 million respectively) are potentially at risk of groundwater arsenic contamination (Chakraborti et al., 2013). The source of arsenic in this region is geogenic and predominated in the shallow grey sand aquifers (GSA) in the floodplain constituted from the Holocene deposition from the rivers originating from the Himalayas and Tibet Plateau (Mukherjee et al., 2021; Pathak et al., 2022). The major mechanisms responsible for arsenic release in the groundwater are sulfide oxidation, reductive dissolution of iron, and alkali desorption, induced by microbial activities (Das et al., 2021; Jha & Tripathi, 2021). The arsenic contamination in the GMB plain of India was recorded back in 1976, from the Trans-Gangetic Plain Region of Chandigarh and its surrounding villages in North India (Datta, 1976). The next incident surfaced in the Lower Gangetic Plain of West Bengal, India (Garai et al., 1984). Both publications also addressed the presence of arsenic in the groundwater along with the sufferings of the people from the region. Over the next two decades, other Indian states in the Indo-Gangetic and GMB Plains were reported to be arsenic contaminated; i.e., Bihar and Uttar Pradesh (Ahamed et al., 2006; Chakraborti et al., 2003), Jharkhand (Nayak et al., 2008), Assam (Chakraborti et al., 2004; Singh, 2004) and Manipur (Chakraborti et al., 2008). School of Environmental Studies (SOES; http://www.jaduniv.edu.in/view_department.php?deptid=139) started surveying arsenic-affected districts in West Bengal in 1988, and since then they have contributed significantly towards the identification of arsenic-affected areas and creating awareness among the local population. Chakraborti et al. (2009), presented the most comprehensive depiction of arsenic contamination in West Bengal. A total of 140,150 groundwater samples were collected from 241 blocks (out of a total of 341) of all 23 districts of West Bengal. The analyses (Das et al.) showed that 48.1% of the samples had more than 10 μg/L (acceptable limit as per Indian Standard: IS10500:2012), 23.8% of samples contained more than 50 μg/L (the standard value of As in many developing countries) while 3.3% samples exceeded 300 μg/L (concentration predicting overt arsenical skin lesions). On the basis of the level of arsenic concentration, the districts were further classified as highly affected, mildly affected, and arsenic safe. The highly affected districts were mainly on the eastern side of the Bhagirathi River, such as Malda, Murshidabad, Nadia, North 24 Parganas, South 24 Parganas, Purba Bardhaman, Paschim Bardhaman, Howrah, Hooghly, and

Kolkata. A total of 135,555 water samples analysed from these districts had groundwater As concentrations more than 300 μg/L. The mildly affected districts were the districts from the northern part of West Bengal, such as Cooch Behar, Jalpaiguri, Alipurduar, Darjeeling, Kalimpong, North Dinajpur, and South Dinajpur. A total of 2923 water samples were analysed from these districts, with a few samples exceeding 50 μg/L. The remaining districts were arsenic-safe that include Bankura, Birbhum, Purulia, Jhargram, Medinipur East, and Medinipur West, where 1672 samples were analysed and none of them exceeded 3 μg/L.

The arsenic contamination scenario in West Bengal is complex in nature, not only because of the geochemistry but also because of the prevailing socio-economic status, awareness, education levels of the people, government intervention and the availability of freshwater (Bhowmik et al., 2022). These factors are also variable in micro to macro-level. In order to mitigate the problem, site specific solutions need to be adopted which must arrive from the knowledge acquired from comprehensive yet detailed investigations. Rahman et al. (2005a) investigated the past and present arsenic contamination scenario in a village called Rajapur in Murshidabad district of West Bengal and evaluated the efficiency of arsenic filtering devices. This paper also highlighted the consequence of food chain arsenic contamination, which was later found to be as catastrophic as the groundwater arsenic contamination. Rahman et al. (2005b) also studied the Jalangi block, one of the most severely arsenic-affected blocks in the Murshidabad district of West Bengal and evaluated the arsenic body-burden and associated health impacts. Rahman et al. (2003) provided the earliest and most convincing district level information on arsenic contamination. They investigated the groundwater contamination of North 24 Paraganas, another severely arsenic-affected district, and screened for suspected arsenic patients. This work provided a complete district wise information about the people's sufferings. This paper also reported one of the initial studies on the arsenic contamination in food chain, where samples (soil, rice and vegetables) from 10 fields were collected and analyzed for 21 months. Roy (2008) showed that the economic benefit that the households get by averting the arsenic problem supersedes the investment cost for arsenic-safe water provisioning. Similar study done in the Murshidabad district (Delaire et al., 2017) found that the choice for the alternative water sources are social, economic and aesthetic factors rather than health issues in the areas where awareness is minimum. Chatterjee et al. (2020) estimated the water level fluctuations in the shallow aquifers across the observation wells in Murshidabad districts and tried to establish the recharge-discharge disparity and induced pumping rate. Mishra et al. (2022) investigated groundwater arsenic contamination in all the 26 blocks of Murshidabad districts along with other ions like iron, manganese, cadmium and selenium in both pre and post monsoon season.

The groundwater arsenic contamination in Nadia district surfaced during the survey (1989–91) where 4 blocks were found to be contaminated. These blocks were Karimpur I, Chakdaha, Karimpur II and Nabadwip. In 1996, SOES published a paper (Mandal et al., 1996) on Nadia along with the other six affected districts of West Bengal. Based on 706 water samples from Nadia, 8 out of the 17 blocks of this district were identified as arsenic-affected. Chakraborti et al. (2009) analyzed,

28,794 groundwater samples from Nadia district and found 14,750 (51.2%), 4940 (17.2%), and 525 samples (1.8%) having concentrations more than 10 μg/L, 50 μg/L, and 300 μg/L, respectively. Rahman et al. (2014) surveyed all the 17 blocks in Nadia district and reported that an estimated 2.7 million people were drinking arsenic-contaminated water from this district alone. Upadhyay et al. (2019) studied two villages in Chakdah Block of Nadia district for arsenic in soil, water and rice grains and evaluated the risk associated with them. Pathak et al (2022) investigated the seasonal effects of dissolved arsenic mobilization in shallow aquifers in Nadia district. Sengupta et al (2022) analyzed 302 samples from Nakashipara and Tehatta Blocks of Nadia district for arsenic and other physico-chemical parameters and investigated the possible geochemical processes responsible for arsenic release in shallow aquifers. Verma and Sinha (2023) inspected both sides of River Ganga in Murshidabad district for groundwater arsenic contamination, source identification and associated health risk. However, no comprehensive study on arsenic concentration and possible mitigation options has been conducted on this block. Karimpur I block, which is one of the arsenic-affected blocks reporting arsenic in groundwater in early days of arsenic survey in West Bengal, from West Bengal have many suspected patients. Hence, the present study encompasses, (i) the groundwater arsenic situation in all the gram panchayats or cluster of villages (GPs) and Mouzas from this block, (ii) evaluates its relationship with depth and iron concentration, and (iii) investigates the sufferings of the people and (iv) evaluates the available safe water options in both the block and Mouza level. The data presented in this study is taken from the unpublished PhD thesis of the lead author of this chapter and the timeline of the study is 1989–2007 (Das, 2010).

Study Area

Karimpur I is one of the 17 blocks of Nadia district, which is in the Ganga-plain and situated between N22°53″ and N24°11″, and E88°09″ and E88°48″. According to the 2001 Government of India Census, the area and population of this district are 3927 sq. km. and 4,604,827 (according to Census 2001), respectively. Figure 6.1 shows the location of Karimpur I block in India. The district is linear in shape with an orientation of North–South and is approximately 14 m above the mean sea level. The Tropic of Cancer divides the district in two parts. Karimpur I block is one of the 17 blocks of Nadia district situated at the northern fringe of the district. It borders Bangladesh on the east and Murshidabad on the west. The area and population of Karimpur I block are 215.78 sq. km. and 166,725 respectively. There are 8 GPs (cluster of villages), 74 Mouzas and 104 villages in this block. A Mouza is a type of administrative unit, corresponding to a specific land area within which there may be one or more villages.

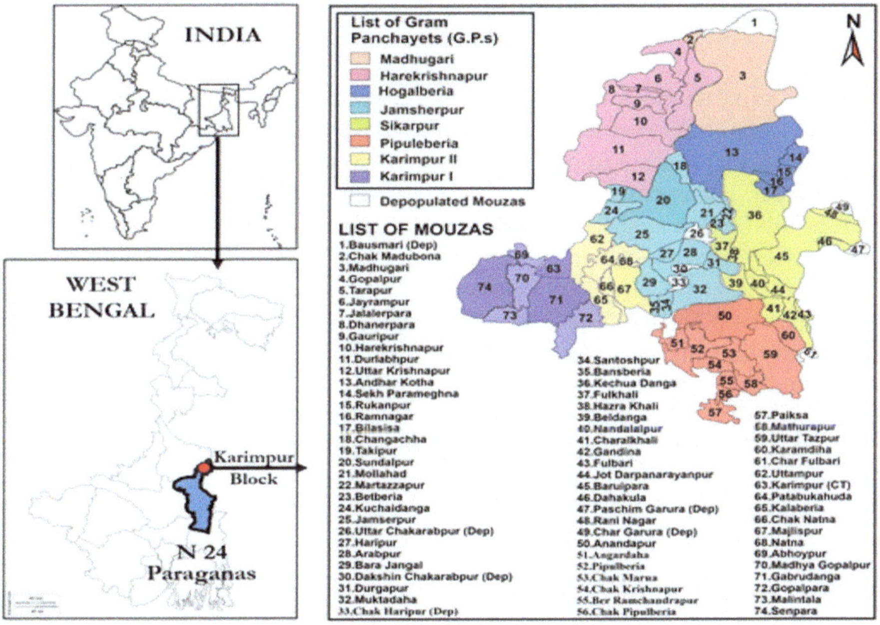

Fig. 6.1 Location of the Karimpur–I Block along with the list of GPs and Mouzas

Materials and Methods

Sampling Protocol

The sampling was conducted by skilled field workers, PhD students and medical teams following the protocol developed by SOES. During the initial study on arsenic in an area, the field workers collected water samples along with photographs of patients having arsenical skin lesions. They collected tubewell water samples from households following established protocols. The survey revealed the following: (a) the exact location of each tubewell through GPS, (b) the number of individuals drinking water from each tubewell, (c) the tubewell depth, (d) the date of installation, (e) whether it is a public or private tubewell, (f) and any preliminary information about arsenic patients. After the water samples were collected and acidified, they were transferred to SOES for analysis. There they were stored and refrigerated (at 4 °C) until analysis. From the analyses, the arsenic contamination in the area was assessed. Based on arsenic analysis and information received, villages suspected to be highly affected by arsenic contamination were selected for additional surveying efforts.

Sample Collection and Analysis

Water and urine samples were analyzed for arsenic by the flow injection-hydride generation atomic absorption spectrometry (FI-HG-AAS) method. Before the collection of the groundwater samples, the tubewells were purged in order to reach the aquifer water. High density polyethylene containers were used for collecting water samples. HNO_3 was added according to the standard protocol to stop precipitation and the adsorption of metal on the container surface (Rahman et al., 2009). Samples were preserved at 4 °C for future analysis. For urine samples, only inorganic As and its metabolites [(As(III), As(V), MMA(V) and DMA(V)] were measured without any chemical treatment. Under the experimental condition of FI-HG-AAS, arsenobetaine and arsenocholine do not produce a signal (Das et al., 2009). The details of the instrumentation techniques and quality assurance have been reported previously (Samanta et al., 1999). Among the samples collected for arsenic analysis, about 5% of the water samples were also analyzed for iron concentrations in spectrophotometry (Shimadzu double beam spectrophotometer Model UV-150-02).

Results and Discussion

Arsenic Concentration in the Hand Tubewell

All the Mouzas of Karimpur I Block were surveyed and groundwater samples along with other related information were collected during the sampling campaign. There are 74 Mouzas in the block. From the 67 inhabited Mouzas in eight GPs of the Karimpur I Block, a total 3374 water samples were analyzed. Table 6.1 shows the area, population (as of the 2001 census), estimated number of tubewells, and number of tubewell water samples examined from all eight GPs in this block. The approximate number of tubewells was estimated based on communicating with the people and visiting Panchayat offices during the collection of water samples, and an attempt was made to collect as many tubewell samples as possible from each Mouza. Where the direct information was not available the assumption of average number of people using a tubewell was taken as 34, from the previous publication on West Bengal (Chakraborti et al., 2009).

The average % of the sample collected from all the GPs are in the range of 27.0–78.9% with 39.8% in block level. The number and % of samples collected from various GPs are shown in Table 6.1. The overall sample size (3374) and its constituent sub-sample sizes (>30) are statistically relevant. Hence the result can be considered representative of the actual groundwater contamination levels. Table 6.2 shows the parametric presentation of the samples analysed. According to the result, all the Mouzas had water samples more than 10 μg/L (requirement for drinking water as per Indian Standard). At least 10% samples from 56 Mouzas (out of 67) had

Table 6.1 Area, population, approximate number of total tubewells (TW) and no of tubewell water samples collected from Karimpur I block

Name of GPs	Population	Area (km^2)	Approximate number of TW	No. of TW samples collected	% of samples collected
Hogalberia	17,808	18.46	937	301	32.1
Karimpur I	25,589	22.58	1587	439	27.7
Karimpur II	23,508	13.31	692	187	27.0
Shikarpur	27,136	30.46	1428	578	40.5
Harekrishnapur	22,492	30.70	1184	512	43.2
Jamserpur	26,702	33.87	1405	698	49.7
Pipulberia	18,064	26.97	951	434	45.6
Madhugari	5426	39.43	285	225	78.9
Total	166,725	215.7	8469	3374	39.8

exceeded 50 μg/L, i.e., the guideline value of arsenic in drinking water in most of the developing countries. The highest concentration found was 1363 μg/L.

When compared with the result from the previous study (Rahman et al., 2005b), on Jalangi Block of Murshidabad district, this block was not as contaminated as Jalangi Block, where out of the 1916 samples, 77.8%, 51.0%, and 17.2% samples had arsenic concentrations more than 10 μg/L, 50 μg/L, and 300 μg/L, respectively. But the number of samples having more than 300 μg/L arsenic was high, which indicated high arsenic exposure and sickness among the population (Chakraborti, 2002).

The status of arsenic pollution in each of the 8 GPs of the Karimpur I block is depicted in the form of a pie chart (Fig. 6.2). From the distribution, it can be seen that the Karimpur I GP is most severely affected in terms of contamination followed by Sikarpur.

Table 6.3 shows the arsenic distribution of the 3374 samples in all the Mouzas in the block. This study showed that all Mouzas had at least 10% of tubewells with > 10 μg/L arsenic. Only 11 Mouzas had arsenic in 10% of the samples between 10 μg/L to 50 μg/L. The remaining 56 have at least 10% of tubewells with arsenic contamination above 50 μg/L.

Arsenic Concentrations and Depth of Tubewells

The depth information for 2825 samples was also gathered along with the arsenic concentration and its findings have been shown in Table 6.4. Only 15 (0.5%) of the 2825 tubewell samples with depth information were deeper than 100 m, and none of them had arsenic levels greater than 100 μg/L. However, two of these tubewells contained arsenic at levels greater than 50 μg/L. The deeper aquifers are considered

Table 6.2 Parametric presentation of groundwater arsenic contamination status in Karimpur I block

Physical parameters	Karimpur I block
Area in km^2	215.78
Population (according to 2001 census)	166,725
Total number of Gram Panchayets (GPs)	8
Number of GPs surveyed	8
Number of GPs where groundwater contaminated >50 μg/L	8
Total number of Mouzas	74
Total number of Mouzas inhabited	67 (7 Mouzas are uninhabited)
Total number of Mouzas surveyed	67
Number of Mouzas where groundwater arsenic contamination >10 μg/L (at least 10% of samples)	67
Number of Mouzas where groundwater arsenic contamination >50 μg/L (at least 10% of samples)	56
Total number of hand tubewell water samples analyzed	3374
Number of samples >10 μg/L (%)	1896 (56.2%)
Number of samples >50 μg/L (%)	1000 (29.6%)
Number of samples >100 μg/L (%)	626 (18.6%)
Number of samples >300 μg/L (%)	150 (22.5%)
Number of samples >500 μg/L (%)	67 (2.0%)
Number of samples =>1000 μg/L (%)	8 (0.2%)
Highest concentration (μg/L)	1363

as one of the feasible arsenic safe water options. During the survey, very few tubewells were found to be more than 100 m deep. Those tubewells were publicly owned and were installed by the local panchayats (village councils) for the communities. None of the samples collected from the tubewells having depth more than 100 m contained arsenic more than 100 μg/L even in the most contaminated areas.

At least 10% of tubewells in all Mouzas had more than 10 μg/L of arsenic. Only 11 Mouzas, or 10% of tubewells, had arsenic at levels between 10 and 50 μg/L. At least 10% of the tubewells in the remaining 56 had arsenic contamination levels that were higher than the 50 μg/L.

Relation Between Arsenic and Iron Concentrations of the Tubewells Water

Arsenic and iron levels in 375 water samples were examined during this investigation. The distribution of iron and arsenic concentrations in the Karimpur I block has been shown in Table 6.5. Iron concentrations ranged from 34,730 μg/L to 4870 μg/L

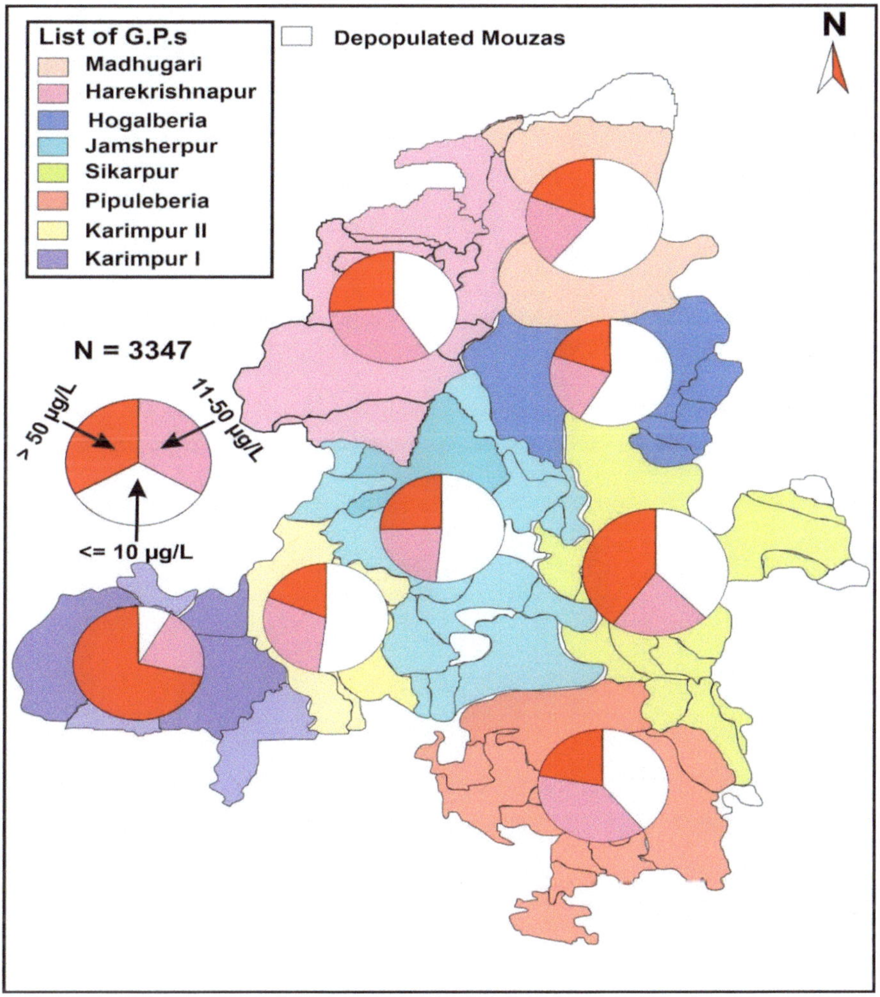

Fig. 6.2 Distribution of arsenic concentration in the tubewells of 8 GPs in Karimpur I block

on average across all samples, with 3694 μg/L serving as the median. However, 44 (11.7%) and 152 (40.5%) samples from this block included iron >10,000 μg/L and >5000 μg/L, respectively. Only 8 (2.1%) samples contained iron less than 300 μg/L (WHO Standard of iron in drinking water based on aesthetic consideration). This research indicated a high level of iron in the local groundwater. However, none of the samples with arsenic below 10 μg/L or below 50 μg/L had iron levels over 5000 μg/L.

Table 6.3 Details of the arsenic concentration situation in all the 69 inhabited Mouzas of 8 GPs in Karimpur I block

GPs	Mouzas	Arsenic concentration in μg/L								Total
		<=3	4–10	11–50	51–100	101–300	301–500	501–999	=>1000	
Harekrishnapur	Dhanerpara	7	7	22	13	16	3	4	0	72
	Durlabhpur	45	1	20	5	19	3	1	0	94
	Gauripur	22	5	0	1	0	0	0	0	28
	Gopalpur	14	6	34	9	10	4	1	0	78
	Harekrishnapur	28	16	20	1	3	1	0	0	69
	Jalaler Para	4	1	11	10	4	0	0	0	30
	Jayrampur	18	12	43	8	2	0	0	0	83
	Tarapur	1	0	3	3	7	5	0	0	19
	Uttar Krishnapur	8	13	17	1	0	0	0	0	39
	Total	*147*	*61*	*170*	*51*	*61*	*16*	*6*	*0*	*512*
Hogalberia	Andhar Kota	84	24	45	13	20	5	7	2	200
	Baliasisa	6	3	7	1	0	0	0	0	17
	Ramnagar	9	5	4	1	0	0	0	0	19
	Rukanpur	13	7	1	0	2	0	0	0	23
	Sekh Parameghna	18	7	8	7	2	0	0	0	42
	Total	*130*	*46*	*65*	*22*	*24*	*5*	*7*	*2*	*301*
Jamsherpur	Arabpur	25	11	7	1	4	0	0	0	48
	Bansberia	8	1	7	9	5	0	0	0	30
	Bara Jangal	4	1	5	5	3	0	0	0	18
	Betberia	1	0	1	0	10	6	8	0	26
	Changachha	34	3	12	12	8	0	0	0	69

(continued)

Table 6.3 (continued)

GPs	Mouzas	Arsenic concentration in µg/L								Total
		<=3	4–10	11–50	51–100	101–300	301–500	501–999	=>1000	
	Durgapur	7	1	3	1	0	0	0	0	12
	Haripur	75	27	42	15	9	0	0	0	168
	Jamserpur	31	0	4	5	7	0	0	0	47
	Kuchaidanga	24	2	10	1	6	0	0	0	43
	Mollahad	2	0	6	7	12	1	0	0	28
	Muktadaha	29	15	20	7	3	0	0	0	74
	Santoshpur	3	2	8	3	0	0	0	0	16
	Sundalpur	31	8	31	8	5	2	0	0	85
	Takipur	8	4	4	4	5	4	0	0	29
	Total	*284*	*75*	*160*	*78*	*78*	*14*	*8*	*0*	*698*
Karimpur I	Abhoypur	1	0	1	3	4	1	1	0	11
	Gabrudanga	2	6	8	12	23	2	1	0	54
	Gopalpara	3	2	6	8	6	2	4	0	31
	Karimpur CT	90	29	77	25	29	13	8	0	271
	Madhya Gopalpur	0	0	2	1	2	1	10	0	16
	Malintala	0	0	0	1	2	0	1	0	4
	Senpara	0	0	5	7	36	4	0	0	52
	Total	*96*	*37*	*99*	*57*	*102*	*23*	*25*	*0*	*439*
Karimpur II	Chak Natna	5	1	7	8	1	0	0	0	22
	Kalaberia	13	0	5	2	0	0	0	0	20

(continued)

Table 6.3 (continued)

GPs	Mouzas	Arsenic concentration in μg/L								Total
		<=3	4–10	11–50	51–100	101–300	301–500	501–999	=>1000	
	Majlispur	9	2	5	3	6	0	0	0	25
	Natna	4	0	5	3	10	0	0	0	22
	Patabukahuda	20	4	37	9	6	0	0	0	76
	Uttampur	13	6	2	0	1	0	0	0	22
	Total	*64*	*13*	*61*	*25*	*24*	*0*	*0*	*0*	*187*
Madhugari	Chak Madhubona	30	0	8	6	0	0	0	0	44
	Madhugari	103	4	37	14	16	7	0	0	181
	Total	*133*	*4*	*45*	*20*	*16*	*7*	*0*	*0*	*225*
Pipulberia	Anandapur	12	0	32	19	14	0	0	0	77
	Angardaha	2	0	16	9	17	1	1	0	46
	Chak Krishnapur	0	0	0	0	4	0	0	0	4
	Chak Marua	9	2	16	3	2	0	0	0	32
	Chawk Pipulberia	12	2	6	0	0	0	0	0	20
	Karamdia	3	1	4	0	0	0	0	0	8
	Mathurapur	9	6	27	0	0	0	0	0	42
	Paiksha	34	3	1	2	1	0	0	0	41
	Pipulberia	40	8	12	3	6	0	0	0	69
	Uttar Tazpur	15	1	28	7	0	0	0	0	51
	Total	*144*	*26*	*167*	*49*	*46*	*1*	*1*	*0*	*434*
Sikarpur	Baruipara	88	11	44	14	22	4	7	6	196

(continued)

Table 6.3 (continued)

GPs	Mouzas	Arsenic concentration in μg/L								Total
		<=3	4–10	11–50	51–100	101–300	301–500	501–999	=>1000	
	Beldanga	17	1	5	1	7	2	2	0	35
	Charal Khali	17	3	11	4	4	0	0	0	39
	Daha Khola	3	0	2	0	1	0	0	0	6
	Fulbari	0	1	5	5	6	0	0	0	17
	Fulkhali	0	0	7	4	13	2	1	0	27
	Gandhina	18	3	5	0	2	0	0	0	28
	Hazra Khali	0	0	3	4	7	2	0	0	16
	Joth Darpanarayanpur	19	6	9	1	0	0	0	0	35
	Kechua Danga	22	8	29	23	32	5	1	0	120
	Nandalalpur	0	0	8	14	28	0	0	0	50
	Raninagar	0	0	1	2	3	2	1	0	9
	Total	*184*	*33*	*122*	*69*	*112*	*15*	*11*	*6*	*578*
Total		1182	295	889	371	463	81	58	8	3374

Table 6.4 Comparison of arsenic distribution and tubewell depth

Depth in m	Arsenic concentration range in μg/L								Total
	<=3	4–10	11–50	51–100	101–300	301–500	501–999	=>1000	
<=10	112	15	27	4	5	2	0	0	165
10–30	837	218	666	263	341	61	44	3	2433
30–100	40	4	54	46	55	6	5	2	212
100–200	5	1	7	2	0	0	0	0	15
Total	994	238	754	315	401	69	49	5	2825

Table 6.5 Comparison of iron and concentration in Karimpur I block

Iron conc in μg/L	Arsenic concentration in μg/L								Total
	<=3	4–10	11–50	51–100	101–300	301–500	501–999	=>1000	
<=300	3	3	2	0	0	0	0	0	8
301–500	11	2	1	2	2	0	0	0	18
501–11,000	4	1	11	4	5	0	0	0	25
1001–3000	4	3	28	26	40	7	2	1	111
3001–5000	3	0	9	7	38	3	1	0	61
5001–10,000	0	0	7	12	61	10	17	1	108
>10,000	0	0	0	5	20	10	8	1	44
Total	25	9	58	56	166	30	28	3	375

Arsenic in Urine Samples

Total 112 urine samples were collected from the willing villagers from 19 Mouzas and 3 GPs in Karimpur I block. The urine-arsenic concentration was then compared with the arsenic concentration in their respective drinking water. The urine arsenic concentration represents the exposure to arsenic and the average range of arsenic in urine was 5 to 40 μg/1.5L per day (Farmer & Johnson, 1990). Table 6.6 displays the statistics of the arsenic concentration in urine.

The linear association between urinary arsenic levels and related drinking water samples has been depicted in Fig. 6.3. Arsenic levels in urine and water samples showed a moderately positive correlation ($R^2 = 0.412$). This indicated that local people were drinking water from multiple sources. Among the 122 samples, 78 (69.6%) of the urine samples had arsenic content above the average values. The results showed that the highest and lowest levels of arsenic in urine were 3 μg/L and 1886 μg/L, respectively. The standard deviation was 247.2 μg/1.5L, the median was 71 μg/1.5L, and the average concentration was 164.4 μg/1.5L.

Table 6.6 Results of urine sample analysis from Karimpur I block

Parameters	Arsenic concentration in urine (μg/1.5 L)
Normal level (range)*	5–40
Number of observations	112
Mean	162.34
Median	71
Range	<3–1886
Standard Deviation	247.2
% of samples having arsenic above normal level	69.6

* Normal excretion of arsenic in urine ranges from 5 to 40 μg/1.5 l (per day) (Farmer & Johnson, 1990)

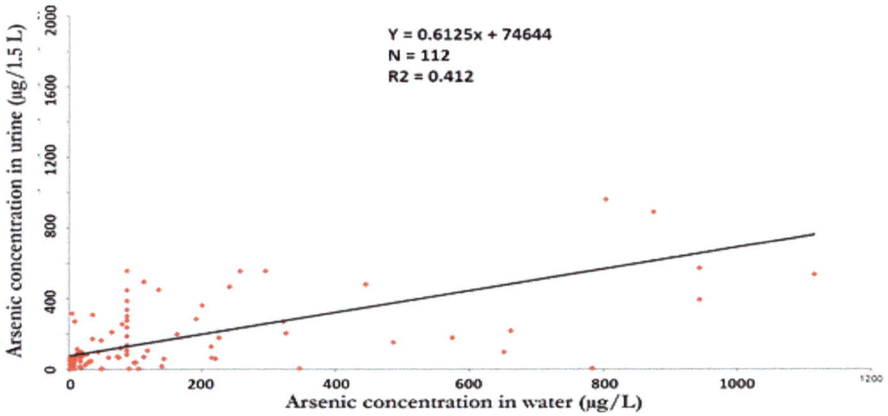

Fig. 6.3 Correlation between the arsenic concentration in urine and water samples used for drinking

Condition of Public Tubewells in Karimpur I Block

There were 324 Panchayat tubewells established by the government among the 3374 tubewell samples collected from Karimpur I block. Table 6.7 compares the distribution of this block's public and private tubewells by depth. Private and public tubewells had average depths of 18.0 and 44.7 m, respectively. The highest depths of tubewell found were 188.4 m for a public tubewell and 112.8 m for a private tubewell. The public tubewells in this block were often newer and deeper than private tubewells. Public and private tubewells that were sampled had ages averaging 4.95 and 6.16 years, respectively. Public and private tubewells each had an average of 86 and 12 users, respectively.

Table 6.7 Arsenic concentration distribution in the private and public tubewells in Karimpur I block

Type	Arsenic concentration in μg/L								Total
	<=3	4–10	11–50	51–100	101–300	301–500	501–999	=>1000	
Private tubewells	1096	271	805	315	414	75	53	7	3036
Public tubewells	87	24	91	59	62	8	6	1	338
Total	1183	295	896	374	476	83	59	8	3374

Possible Arsenic Mitigation Strategy in Karimpur I Block: Merits, Demerits, and Suggestions for Arsenic-Safe Water Provision

Some sources that were safe from arsenic contamination were located in practically every Mouza in Karimpur I block and each community could receive water free of arsenic from these sources. Ironically, many of these sources including dugwells and surface water were previously in use. However, with the advent of hand tubewells, as well as poor management and ignorance, many communities abandoned those sources. There were a lot of tubewells that were arsenic-free as well. These tubewells could also be used as a source of clean water depending on the depth, installation method, and regular monitoring for temporal fluctuations in arsenic levels. Arsenic from the water of polluted tubewells could be removed and the treated water could be used for drinking or other domestic applications. Locally available arsenic removal plants (ARP) could be used for this purpose. Backwashing and regular maintenance are essential for the success of ARPs. It is possible to develop community-specific programmes and offer clean water by taking all of these sources of arsenic-free water into account. Participation from every community and educational initiatives that emphasize the value of safe water, however, are essential to the success of these strategies.

Arsenic safe tubewells

Among the 3374 groundwater samples evaluated from Karimpur I block, 44.1% and 70.7% had arsenic levels less than 10 and 50 μg/L, respectively. Villagers could utilize the 44.1% (1476) of the examined tubewells that meet the WHO's safety standards as sources of safe water. Uncertainty over which tubewells were safe and which were contaminated existed as observed by the survey team. Safe tubewells might be painted green to indicate their safety, and the villages may be notified as a result. Similar to this, 70.7% (2236 tubewells) of them met the Indian Standard for arsenic in drinking water at that point in time (50 μg/L). The situation could be considerably improved by raising awareness and teaching the locals about the risks of consuming arsenic-contaminated water. This has been confirmed by our study on communities affected by a high arsenic contamination level (Chakraborti et al., 2009). Strong awareness campaign in Bangladesh's Sonargaon and Jhikargachi Upazilas (Jakariya & Bhattacharya, 2007), raised the use of safe tubewells by 50%.

Public water supply schemes

In Karimpur I block, there were 6 Ground Water Supply Schemes (GWSS). They were in Joyrampur, Durlabhpur, Senpara, Sundarpur, Rajapur and Patapokhra. All of them were visited and water samples were collected from them. Only one of the five tap water samples had arsenic at levels more than 50 μg/L. The arsenic levels in the other 4 samples were less than 50 μg/L.

In Karimpur I block, surface water-based public water delivery system could be another option besides groundwater. River Jalangi runs beside 11 Mouzas and 4 GPs on the block's western side. Some of these Mouzas, namely, Gopalpur, Tarapur, Dhanerpara, Takipur, Karimpur CT, Abhoypur Mouzas recorded greater than 50 μg/L As in 35.8% of samples. These areas would benefit from surface water treatment and supply provision.

Dugwell

Hundreds of dugwells were identified during field trips in the Karimpur I block; these wells were formerly in service but have since been abandoned as tubewell technology has advanced. Most of them had a diameter of 2–3 m and are water-filled year-round. All of the samples from the seven dugwells in this block that were collected (but not included in this report) and examined for arsenic had values below 10 μg/L, but they were also polluted bacteriologically. They may prove to be a reliable source of safe water if they are restored to serviceable condition and maintained properly.

Arsenic Removal Plants (ARP)

Hundreds of ARPs were found to be inoperative on trips to various blocks that had been seriously impacted by arsenic. They were often commissioned next to abandoned or contaminated tubewells in an effort to remove arsenic from the water; but over time they fell into poor state of repair and rejected by users. These technologies were sound in theory but were unproductive and functioned poorly for a variety of reasons, including inadequate maintenance, sand leaking issues, poor usability, and a lack of community management. According to SOES research, 80% and 95% of the installed ARPs in two distinct blocks, Domkol in Murshidabad and Swarupnagar in North 24 Parganas, respectively, were not usable (Hossain et al., 2005). Block I of Karimpur was not an exception. Almost every ARP was discovered to be inoperable when inspecting the communities.

Conclusion

The presence of arsenic in Karimpur I block was identified in 1989, and the present study started at the same time and continued till 2007. All of the Mouzas were surveyed and 3374 samples were analyzed from different locations of this block.

The overall arsenic contamination status of this block is moderate to severe. No Mouzas are arsenic safe (<10 μg/L) and 84% of the inhabited Mouzas are having at least 10% water samples >50 μg/L. Even the people were identified to have arsenic body burden. During several field visits, some potential arsenic patients were also identified. Nevertheless, this block had several safe water options. At least 20% of the Mouzas now have access to piped water supplies. A total of 59 Mouzas out of 67 were found to have public tubewells that were inspected throughout the course of my research (2004–2007). It also had plenty of surface water with the presence of perennial river and seasonal lakes (bills). They could be considered as drinking options after proper treatment. Even the dugwells could be rejuvenated for drinking purposes. It was also evident that the situation did not considerably improve despite the construction of hundreds of arsenic removal plants in this block to provide the communities with safe water. During the recurrent visit in the area, the lack of awareness and education was observed among the villagers. The general education level was quite low and there was little knowledge about the harmful effects of arsenic. In an interview with 241 households during 2005, it was found that 228 of them (94.6%) had heard of arsenic but that only 159 (66.0%) were aware of the negative consequences of arsenic in drinking water. This problem was evident in other arsenic-affected areas in West Bengal and Bangladesh. Proper awareness, coupled with switching to safe water options that can be adopted by the villagers may be the long term solution to the arsenic contamination in the area.

Acknowledgements The authors would like to express sincere gratitude to Late Dr. Dipankar Chakraborti, Former Director, School of Environmental Studies (SOES), Jadavpur University, Kolkata under whose guidance the survey and analyses was conducted. Thanks to all the researchers and field samplers who were involved in the field sampling and analysis. The authors also thank VIT for its support during the preparation of the manuscript.

Data Availability Data will be made available on request.

Declaration of Competing Interest The authors declare that they have no known competing financial interests or personal relationships that could have appeared to influence the work reported in this paper.

References

Ahamed, S., Kumar Sengupta, M., Mukherjee, A., Amir Hossain, M., Das, B., Nayak, B., Pal, A., Chandra Mukherjee, S., Pati, S., Nath Dutta, R., Chatterjee, G., Mukherjee, A., Srivastava, R., & Chakraborti, D. (2006). Arsenic groundwater contamination and its health effects in the state of Uttar Pradesh (UP) in upper and middle Ganga plain, India: A severe danger. *Science of the Total Environment, 370*, 310–322. https://doi.org/10.1016/j.scitotenv.2006.06.015

Bhowmik, T., Sarkar, S., Bhattacharya, A., & Mukherjee, A. (2022). A review of arsenic mitigation strategies in community water supplies with insights from South Asia: Options, opportunities and constraints. *Environmental Science: Water Research & Technology, 8*, 2491–2520. https://doi.org/10.1039/D1EW00958C

Chakraborti, D. (2002). Arsenic calamity in the Indian subcontinent What lessons have been learned? *Talanta, 58*, 3–22. https://doi.org/10.1016/S0039-9140(02)00270-9

Chakraborti, D., Mukherjee, S. C., Pati, S., Sengupta, M. K., Rahman, M. M., Chowdhury, U. K., Lodh, D., Chanda, C. R., Chakraborti, A. K., & Basu, G. K. (2003). Arsenic groundwater contamination in Middle Ganga Plain, Bihar, India: A future danger? *Environmental Health Perspectives, 111*, 1194–1201. https://doi.org/10.1289/ehp.5966

Chakraborti, D., Sengupta, M. K., Rahman, M. M., Ahamed, S., Chowdhury, U. K., Hossain, M. A., Mukherjee, S. C., Pati, S., Saha, K. C., Dutta, R. N., & Quamruzzaman, Q. (2004). Groundwater arsenic contamination and its health effects in the Ganga-Meghna-Brahmaputra plain. *Journal of Environmental Monitoring, 6*, 74N-83N.

Chakraborti, D., Singh, E. J., Das, B., Shah, B. A., Hossain, M. A., Nayak, B., Ahamed, S., & Singh, N. R. (2008). Groundwater arsenic contamination in Manipur, one of the seven North-Eastern Hill states of India: A future danger. *Environmental Geology, 56*, 381–390. https://doi.org/10.1007/s00254-007-1176-x

Chakraborti, D., Das, B., Rahman, M. M., Chowdhury, U. K., Biswas, B., Goswami, A. B., Nayak, B., Pal, A., Sengupta, M. K., Ahamed, S., Hossain, A., Basu, G., Roychowdhury, T., & Das, D. (2009). Status of groundwater arsenic contamination in the state of West Bengal, India: A 20-year study report. *Molecular Nutrition & Food Research, 53*, 542–551. https://doi.org/10.1002/mnfr.200700517

Chakraborti, D., Rahman, M. M., Das, B., Nayak, B., Pal, A., Sengupta, M. K., Hossain, M. A., Ahamed, S., Sahu, M., Saha, K. C., Mukherjee, S. C., Pati, S., Dutta, R. N., & Quamruzzaman, Q. (2013). Groundwater arsenic contamination in Ganga–Meghna–Brahmaputra plain, its health effects and an approach for mitigation. *Environment and Earth Science, 70*, 1993–2008. https://doi.org/10.1007/s12665-013-2699-y

Chatterjee, R., Samadder, S., Mondal, D., & Adhikari, K. (2020). Analysis of spatio-temporal trend in groundwater elevation data from arsenic affected alluvial aquifers—Case study from Murshidabad district, West Bengal, Eastern India. *Journal of Earth System Science, 129*, 228. https://doi.org/10.1007/s12040-020-01489-8

Das, B. (2010). *Investigation for safe water options in groundwater arsenic contaminated rural and urban settlements in Lower Ganga Plain, West Bengal –India*. Jadavpur University.

Das, B., Rahman, M. M., Nayak, B., Pal, A., Chowdhury, U. K., Mukherjee, S. C., Saha, K. C., Pati, S., Quamruzzaman, Q., & Chakraborti, D. (2009). Groundwater arsenic contamination, its health effects and approach for mitigation in West Bengal, India and Bangladesh. *Water Quality Exposure and Health, 1*, 5–21. https://doi.org/10.1007/s12403-008-0002-3

Das, A., Joardar, M., De, A., Mridha, D., Chowdhury, N. R., Bin Kashim Khan, M. T., Chakrabartty, P., & Roychowdhury, T. (2021). Pollution index and health risk assessment of arsenic through different groundwater sources and its load on soil-paddy-rice system in a part of Murshidabad district of West Bengal, India. *Groundwater Sustainable Development, 15*, 100652. https://doi.org/10.1016/j.gsd.2021.100652

Datta, D. (1976). Arsenic and non-cirrhotic portal hypertension. *Lancet, 307*, 433. https://doi.org/10.1016/S0140-6736(76)90282-8

Delaire, C., Das, A., Amrose, S., Gadgil, A., Roy, J., & Ray, I. (2017). Determinants of the use of alternatives to arsenic-contaminated shallow groundwater: An exploratory study in rural West Bengal, India. *Journal of Water Health, 15*, 799–812. https://doi.org/10.2166/wh.2017.321

Farmer, J. G., & Johnson, L. R. (1990). Assessment of occupational exposure to inorganic arsenic based on urinary concentrations and speciation of arsenic. *Occupational and Environmental Medicine, 47*, 342–348. https://doi.org/10.1136/oem.47.5.342

Garai, R., Chakraborty, A. K., Dey, S. B., & Saha, K. C. (1984). Chronic arsenic poisoning from tubewell water. *Journal of the Indian Medical Association, 82*, 34–35.

Hossain, M. A., Sengupta, M. K., Ahamed, S., Rahman, M. M., Mondal, D., Lodh, D., Das, B., Nayak, B., Roy, B. K., Mukherjee, A., & Chakraborti, D. (2005). Ineffectiveness and poor reliability of arsenic removal plants in West Bengal, India. *Environmental Science Technology, 39*, 4300–4306. https://doi.org/10.1021/es048703u

Jakariya, M. D., & Bhattacharya, P. (2007). Use of GIS in local level participatory planning for arsenic mitigation: A case study from Matlab Upazila, Bangladesh. *Journal of Environmental Science Health Part A, 42*, 1933–1944. https://doi.org/10.1080/10934520701567221

Jha, P. K., & Tripathi, P. (2021). Arsenic and fluoride contamination in groundwater: A review of global scenarios with special reference to India. *Groundwater Sustainable Development, 13*, 100576. https://doi.org/10.1016/j.gsd.2021.100576

Mandal, B. K., Roy Chowdhury, T., Samanta, G., Basu, G. K., Chowdhury, P. P., Chanda, C. R., Lodh, D., Karan, N. K., Dhar, R. K., Tamili, D. K., Dipankar Das, K. C. S., & Chakraborti, D. (1996). Arsenic in groundwater in seven districts of West Bengal, India—The biggest arsenic calamity in the world. *Current Science, 70*, 976–986.

Mishra, D., Sen, K., Mondal, A., Kundu, S., & Mondal, N. K. (2022). Geochemical appraisal of groundwater arsenic contamination and human health risk assessment in the Gangetic Basin in Murshidabad District of West Bengal, India. . *Environmental Earth Science, 81*, 157. https://doi.org/10.1007/s12665-022-10273-4

Mukherjee, A., Sarkar, S., Chakraborty, M., Duttagupta, S., Bhattacharya, A., Saha, D., Bhattacharya, P., Mitra, A., & Gupta, S. (2021). Occurrence, predictors and hazards of elevated groundwater arsenic across India through field observations and regional-scale AI-based modeling. *Science of the Total Environment, 759*, 143511. https://doi.org/10.1016/j.scitotenv.2020.143511

Murcott, S. (2012). Arsenic contamination in the world: An international sourcebook 2012. *Water Intelligent Online, 11*. https://doi.org/10.2166/9781780400396

Naujokas, M. F., Anderson, B., Ahsan, H., Aposhian, H. V., Graziano, J. H., Thompson, C., & Suk, W. A. (2013). The broad scope of health effects from chronic arsenic exposure: Update on a worldwide public health problem. *Environmental Health Perspectives, 121*, 295–302. https://doi.org/10.1289/ehp.1205875

Nayak, B., Das, B., Chandra Mukherjee, S., Pal, A., Ahamed, S., Amir Hossain, M., Maity, P., Dutta, R. N., Dutta, S., & Chakraborti, D. (2008). Groundwater arsenic contamination in the Sahibganj district of Jharkhand state, India in the middle Ganga plain and adverse health effects. *Toxicological and Environmental Chemistry, 90*, 673–694. https://doi.org/10.1080/02772240701655486

Pathak, P., Ghosh, P., Mukherjee, A., Ghosal, U., Liang, M.-C., Sikdar, P. K., & Kaushal, R. (2022). Impact of differential surface water mixing on seasonal arsenic mobilization in shallow aquifers of Nadia district; western Bengal Basin, India. *Journal of Hydrology, 612*, 128270. https://doi.org/10.1016/j.jhydrol.2022.128270

Rahman, M. M., Mandal, B. K., Chowdhury, T. R., Sengupta, M. K., Chowdhury, U. K., Lodh, D., Chanda, C. R., Basu, G. K., Mukherjee, S. C., Saha, K. C., & Chakraborti, D. (2003). Arsenic groundwater contamination and sufferings of people in North 24-Parganas, one of the nine arsenic affected districts of West Bengal, India. *Journal of Environmental Science Health Part A, 38*, 25–59. https://doi.org/10.1081/ESE-120016658

Rahman, M. M., Sengupta, M. K., Ahamed, S., Chowdhury, U. K., Hossain, M. A., Das, B., Lodh, D., Saha, K. C., Pati, S., Kaies, I., Barua, A. K., & Chakraborti, D. (2005a). The magnitude of arsenic contamination in groundwater and its health effects to the inhabitants of the Jalangi—one of the 85 arsenic affected blocks in West Bengal, India. *Science Total Environment, 338*, 189–200. https://doi.org/10.1016/j.scitotenv.2004.06.022

Rahman, M. M., Sengupta, M. K., Ahamed, S., Chowdhury, U. K., Lodh, D., Hossain, A., Das, B., Roy, N., Saha, K. C., Palit, S. K., & Chakraborti, D. (2005b). Arsenic contamination of groundwater and its health impact on residents in a village in West Bengal, India. *Bulletin World Health Organization, 83*, 49–57.

Rahman, M. M., Mondal, D., Das, B., Sengupta, M. K., Ahamed, S., Hossain, M. A., Samal, A. C., Saha, K. C., Mukherjee, S. C., Dutta, R. N., & Chakraborti, D. (2014). Status of groundwater arsenic contamination in all 17 blocks of Nadia district in the state of West Bengal, India: A 23-year study report. *Journal of Hydrology, 518*, 363–372. https://doi.org/10.1016/j.jhydrol.2013.10.037

Rahman, M. M., Das, B., & Chakraborti, D. (2009). Sampling and analysis of arsenic in Groundwater in West Bengal, India, and Bangladesh. In *Handbook of water purity and quality* (pp. 95–130). Elsevier. https://doi.org/10.1016/B978-0-12-374192-9.00005-4

Roy, J. (2008). Economic benefits of arsenic removal from ground water—A case study from West Bengal, India. *Science of the Total Environment, 397*, 1–12. https://doi.org/10.1016/j.scitotenv.2008.02.007

Samanta, G., Chowdhury, T. R., Mandal, B. K., Biswas, B. K., Chowdhury, U. K., Basu, G. K., Chanda, C. R., Lodh, D., & Chakraborti, D. (1999). Flow injection hydride generation atomic absorption spectrometry for determination of arsenic in water and biological samples from arsenic-affected districts of West Bengal, India, and Bangladesh. *Microchemical Journal, 62*, 174–191. https://doi.org/10.1006/mchj.1999.1713

Sengupta, D., Mukherjee, P. K., Mukhopadhyay, B., Toppo, S. M., & Patel, J. M. (2022). Multivariate statistics on groundwater geochemical data to identify arsenic release mechanism and related microenvironments: A case study from West Bengal, India. *Environment Earth Science, 81*, 439. https://doi.org/10.1007/s12665-022-10571-x

Shaji, E., Santosh, M., Sarath, K. V, Prakash, P., Deepchand, V., & Divya, B. V. (2021). Arsenic contamination of groundwater: A global synopsis with focus on the Indian Peninsula. *Geoscience Frontiers, 12*, 101079. https://doi.org/10.1016/j.gsf.2020.08.015

Singh, A. (2004). Arsenic contamination in the groundwater of North Eastern India. In: *Proceedings on national seminar on hydrology with focal theme on water qaulity* (pp. 255–262). National Institute of Hydrology.

Upadhyay, M. K., Majumdar, A., Barla, A., Bose, S., & Srivastava, S. (2019). An assessment of arsenic hazard in groundwater–soil–rice system in two villages of Nadia district, West Bengal, India. *Environmental Geochemistry and Health, 41*, 2381–2395. https://doi.org/10.1007/s10653-019-00289-4

Verma, S., Sinha, A. (2023). Appraisal of groundwater arsenic on opposite banks of River Ganges West Bengal India and quantification of cancer risk using Monte Carlo simulations. *Environmental Science and Pollution Research, 30*(10), 25205–25225. https://doi.org/10.1007/s11356-021-17902-8

Chapter 7
Arsenic Contamination of Water Sources in Southern Africa: Role of Artisanal and Small Scale Mining Sector

Xavier Poshiwa

Abstract Arsenic (As) is an element that occurs either as a component of many minerals, sulphur, and metals or as a pure element as a crystal. Elemental arsenic and arsenic compounds are toxic to the environment and threaten human health. Arsenic is mutagenic, carcinogenic, and teratogenic. It naturally exists as inorganic arsenic compounds, organic arsenic compounds, or arsenic gas. Inorganic arsenic is more toxic than organic arsenic compounds. This paper reviews the state of arsenic contamination in water sources in Southern Africa. Research in Africa has reported elevated levels of As in both surface and groundwater systems. The World Health Organization has set a standard of 10 μg/L. However, concentrations of arsenic of up to 1760 μg/L were reported in groundwater in South Africa, 188 μg/L reported in surface water in Botswana, 96 μg/L from Eastern Zimbabwe, and 119 μg/L in surface water from South Africa. Studies reported low levels of As in rivers from Mozambique compared to rivers in South Africa. However, studies have reported that only 13% of the groundwater in South Mozambique was suitable for drinking purposes. High As levels in surface water were generally linked to mining activities, agricultural activities through herbicides and pesticides, disposal, and incineration of municipal and industrial waste. Whereas high concentrations of As in groundwater are generally related to local geology, mineralization, and geothermal water. Furthermore, in Southern Africa, soil and groundwater contamination could be a result of the legacy effects of the application of inorganic arsenic-based compounds used for cattle dipping. It can be summarized that anthropogenic and geogenic sources are the major causes of elevated As levels, particularly in soils and water within the Southern Africa region. More efforts should be directed towards the removal of the metal from both surface and groundwater systems to curb the effects of arsenic toxicity.

X. Poshiwa (✉)
Great Zimbabwe University, P.O. Box 1235, Masvingo, Zimbabwe
e-mail: xposhiwa@gzu.ac.zw

B. Sen Gupta and N. Martínez-Villegas (eds.), *Arsenic Remediation of Food and Water*,
https://doi.org/10.1007/978-981-97-4764-1_7

Introduction

Arsenic (As) is a semi-metal or metalloid (group VA) that is stable in several oxidation states (−III, 0, +III, and +V). Naturally, arsenic exists as an inorganic As(III) or As(V) state as well as organic As(e.g. monomethylarsonic acid, dimethylarsinic acid, trimethylarsine oxide, arsenobetaine, and arsenicholine) (Hussain et al., 2021). It is classified as a toxic element that threatens human health with inorganic states being more toxic than organic compounds (Bhattacharya et al., 2021). The +III and + V states are most common in the natural systems. Arsenic occurs ubiquitously in the environment, particularly in the soil, water, and air. The semi-metal is a natural constituent of the Earth's crust and ranks twentieth in abundance in relation to other elements (Mandal & Suzuki, 2002; Rahman & Hasegawa, 2011).

Arsenic is a well-known poison, with a lethal dose in humans of about 125 mg (Mandal & Suzuki, 2002). Most of the ingested Arsenic is excreted from the body through urine, stools, skin, hair, nails, and breath. In cases of excessive intake, some arsenic is deposited in tissues, causing the inhibition of cellular enzyme activities. In addition to consumption through drinking water, arsenic can also be taken up via the food chain. Direct consumption of cereals irrigated with arsenic-rich waters is a significant source of arsenic exposure (Rahman & Hasegawa, 2011).

Water is recognized as the most fundamental and indispensable of all natural resources. Neither social and economic development nor environmental diversity can be sustained without water. Today, virtually every country faces severe and growing challenges in their effort to meet the rapidly escalating demand for water, driven mainly by growing populations (Ashton et al., 2001). Water supplies continue to dwindle because of resource depletion and pollution, whilst demand is rising fast because of population growth coupled with industrialization, mechanization and urbanization (Gleick, 1998). This situation is particularly acute in the more arid regions of the World where water scarcity and associated increases in water pollution, limit social and economic development and are linked closely to the prevalence of poverty, hunger and disease.

In regions that are characterized by climatic aridity with low rainfall occurrence, groundwater from weathered and fractured crystalline aquifers is a primary source of water supply for various activities (Abiye & Bhattacharya, 2019). Surface water resources have been impacted most worldwide, consequently causing communities to depend more on groundwater resources (Dohare et al., 2014; Misi et al., 2018). Approximately 75% of Africa and 62% of Southern African Development Community (SADC)'s population depend on groundwater for domestic use (Clough et al., 2010; Malzbender & Earle, 2007; Misi et al., 2018; Pietersen et al., 2009). Generally, the world's water is progressively becoming poor in quality thus threatening the ecosystem and human health (Makaya & Maphosa, 2023; Panneerselvam et al., 2021, 2023). While groundwater use has helped increase food security, it should be noted that increased use of this fresh subsurface water has also created several issues e.g. groundwater pollution and over-extraction (Atangana & Oberholster, 2021; Brindha et al., 2017). Another severe environmental problem is caused by wastewater from anthropogenic activities. A major issue facing communities in Africa and the world over is that of harmful heavy metals from sewage contact with groundwater and surface water or human industrial activities e.g. mining, industrial waste from chemical companies, or even airborne particle matter (Atangana & Oberholster, 2021; Oberholster et al., 2017). These heavy metals affect groundwater, leading to groundwater pollution, a potential risk to several households where this groundwater is used as drinking water. Excessive exposure to heavy concentrations could lead to serious health complications. In Africa in general and Southern Africa in particular, contaminants from municipal, agricultural, and industrial sources are the main concern for water supply managers (Abiye & Bhattacharya, 2019). In this paper, we review the effects, sources, and levels of arsenic contamination in both surface and groundwater sources in Southern Africa. Arsenic is highly toxic to consumers and its prevalence in both surface and groundwater in Southern Africa has not been given sufficient attention.

Causes of Arsenic Contamination

Studies done so far in Africa indicated that both anthropogenic and geogenic activities as major sources of high arsenic levels in water sources (Ahoule et al., 2015; Irunde et al., 2022). Anthropogenic processes such as mining, agricultural pesticide application, industrial discharge, and land-filling of sewage sludge are potential sources of arsenic (Mandal & Suzuki, 2002; Rukh et al., 2017). Additionally, coal-burning activities can significantly contribute to arsenic mobilization in the environment (Finkelman, 2004). Anthropogenic sources also include the treatment of wood, tanneries, the pharmaceutical and glass industries, and pesticides (Ramudzuli & Horn, 2014). In gold, uranium, and coal-producing countries in the world, the potential release of arsenic into the water supply is through oxidation–reduction processes (Abiye & Bhattacharya, 2019). See the graphical abstract of arsenic contamination below.

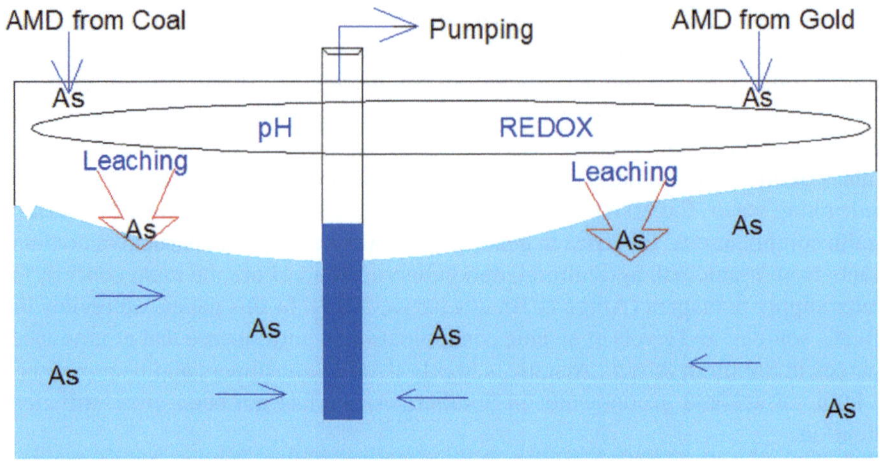

Graphical abstract of arsenic contamination (Abiye & Bhattacharya, 2019)

Coal is primarily mined through an open cast method where the overlying less mineralized rocks are dumped by exposing the sulphide rich coal layers to be oxidized, which could potentially mobilize arsenic in groundwater (Abiye & Bhattacharya, 2019) (see the graphical abstract of arsenic contamination below). Acid mine drainage (AMD) comes from the oxidation of pyrite (FeS_2) and other sulfide-based minerals in rock material that has been exposed from mining activity (Thomas et al., 2022). The biologically mediated chemical reactions lead to the generation of significant quantities of sulphate (SO_4), protons (H^+), and ferric iron (Fe^{3+}) that are released into the solution (Nordstrom et al., 2015) the latter of which function as an oxidizing agent and can perpetuate the reaction in oxygen-poor environments. The resulting, highly acidic solution dissolves other metal(loid)s contained within the surrounding mineral assemblage, often reaching toxic levels (Johnson, 2003; Johnson & Hallberg, 2005).

The release of arsenic into the environment is caused by the oxidation of arsenic-bearing sulphide (pyrite, arsenopyrite, and chalcopyrite) minerals (Irunde et al., 2022), weathering of arsenic-bearing rocks, arsenic adsorption from the surface of mineral iron or manganese oxides. The release of arsenic naturally can also be caused by reductive dissolution of iron oxides and evaporative concentration in arid and semi-arid regions (Ahoule et al., 2015; Irunde et al., 2022).

Effects of Arsenic Toxicity

Arsenic and its related compounds are classified as "toxic" and "dangerous for the environment" (Gupta et al., 2017). Arsenic toxicity is carcinogenic, mutagenic, and teratogenic (Gupta et al., 2017).

Arsenic is a "known human carcinogen" and can accumulate in the body (Genthe et al., 2018). Millions of people in several countries are exposed to high levels of arsenic via intake of arsenic-rich groundwater. Excessive and long-term (such as 5–10 years) human intake of toxic inorganic arsenic from drinking water and food may result in arsenicosis, a common name generally used for arsenic-related health problems that include skin disorders, skin cancers, internal cancers (bladder, kidney, and lung), diseases of the blood vessels of the legs, feet, diabetes, increased blood pressure, and reproductive disorders (Shankar et al., 2014). Acute effects lead to death from upper respiratory, pulmonary, gastrointestinal, and cardiovascular failure (Genthe et al., 2018). Nerve damage and sensory loss in the peripheral nervous system area are the primary symptoms of arsenic poisoning (Genthe et al., 2018). Arsenic hinders a series of gene proliferation processes (e.g. DNA damage and repair, cell cycle and differentiation) and distorts signal transduction pathways (Sinha et al., 2013; Wang et al., 2012). Trivalent arsenicals can induce diabetes via distortions of glucose metabolism (Shankar et al., 2014).

Arsenic toxicity is mutagenic, meaning that it damages genetic information inside the cell, resulting in genotoxicity which leads to mutations. Arsenic induces genotoxicity by generating Reactive oxygen species (ROS), which when present in excess amounts within the cell reacts with cellular components and result in genotoxicity (Lu et al., 1995; Shankar et al., 2014; Valdiglesias et al., 2010). Genotoxicity occurs since ROS reacts with both deoxyribose and bases in DNA causing base lesions and strand breaks. Additionally, ROS are also involved in the oxidation of DNA, alteration of DNA repair, gene regulation mechanism, and threatening of gene stability (Ramana et al., 1998). Arsenic interacts with DNA repair proteins which are involved in transcription, leading to breaks of single-strand and double-strand DNA and oxidative DNA damage (Ho, 2004).

Arsenic toxicity causes cell abnormalities referred to as cytotoxicity. Arsenic toxicity causes cytotoxicity or cell anomalies through several pathways. Arsenic can induce cytotoxicity by generating ROS (Sies & de Groot, 1992). ROS levels inside the cell increase dramatically when the cell is exposed to elevated levels of arsenic. Arsenic results in ROS production by inducing NADPH oxidase (Chou et al., 2004). Excess ROS causes damage to lipids and proteins as well as facilitating mitochondrial damage and its functions (Eun et al., 2007; Kim et al., 2002; Shanker et al., 2014). ROS can also act as modulators of signal transduction pathways, affecting various biological processes including cell growth, apoptosis, and cell adhesion (Apel & Hirt, 2004; Suzuki et al., 1997). Arsenic toxicity results in cytotoxicity by affecting the status of tumor-suppressor protein 53 (Huang et al., 1999; Yih & Lee, 2000). Protein 53 plays a very crucial role in a wide range of cellular functions by modulation of transformation and regulation of cell growth and control of cell cycle, DNA synthesis, DNA repair and differentiation, and apoptosis (Amundson et al., 1998; Ryan et al., 2001). Arsenic may induce protein 53 accumulation in human fibroblasts, which ultimately results in cell apoptosis.

Arsenic Contamination of Water Sources in Southern Africa

The anthropogenic and geogenic sources are the major causes of the elevated arsenic levels in soils and water within the southern African region (Irunde et al., 2022). The levels of arsenic vary considerably among countries and within the same country, exhibiting high spatial variability. Studies on arsenic in African waters show that arsenic can be found in high concentrations in both surface and groundwater (Ahoule et al., 2015). Arsenic concentrations in African groundwater ranged between 0.02 and 1760 μg/l, whilst the levels of arsenic in surface water ranged up to 10000 μg/l (Ahoule et al., 2015). High levels of arsenic in surface water is related to mining operations, agricultural drains, sediments, industrial and municipal disposals and incineration of municipal and industrial waste. However, mining activities remain the main source of surface water pollution. Mining, therefore, has a strong impact on arsenic concentrations in the environment, whereas, high arsenic concentrations in groundwater are due to natural conditions.

Table 7.1 shows that seven (7) of the 12 SADC states with high contributions to GDP by mining. Interestingly, these seven states show high arsenic levels in both surface water and groundwater. Those countries with low mining activity like Malawi, Lesotho and Mozambique and Swaziland have low mining contribution to GDP and low levels of arsenic contamination.

The majority of Southern African countries are endowed with a variety of economic minerals that also host minerals responsible for mobilizing arsenic and other toxic metals in water supply aquifers. Botswana, Namibia, DRC, Angola, South Africa, Zambia, and Zimbabwe are well known for producing minerals ranging from

Table 7.1 Relative importance of social and economic contributions made by the mining and mineral processing sector in 12 SADC states. Adapted from Hounsome and Ashton (2006)

SADC country	Population (millions)	Mining contribution to GDP (%)	Mining employment (%)	Mining share of foreign earnings (%)
Angola	10.145	52.3	9.0	90.0
Botswana	1.576	38.0	5.5	70.0
D.R. Congo	51.965	28.0	4.0	70.0
Lesotho	2.143	0.5	1.0	0.3
Malawi	10.386	0.9	0.4	0.5
Mozambique	19.105	2.0	1.3	2.0
Namibia	1.771	20.0	4.5	48.0
South Africa	43.421	8.0	9.0	28.6
Swaziland	1.083	2.0	3.5	2.0
Tanzania	35.306	2.8	2.7	22.0
Zambia	9.582	12.1	9.0	80.0
Zimbabwe	11.343	8.0	7.0	40.0

gold, coal, uranium, and platinum group elements (PGE). This means the vast areas of Southern Africa have great potential to release arsenic into the water supply aquifer through oxidation–reduction processes. Some of the minerals that are enriched in rocks are arsenopyrite (FeAsS), which is a high-grade arsenic ore, arsenical oxide, sulpharsenide, arsenopyritical reefs, leucopyrite, lÖllingite ($FeAs_2$) and scorodite ($FeAsO_4.2H_2O$) within the gold-bearing ores (Abiye & Bhattacharya, 2019).

Studies from the Okavango delta showed arsenic concentrations ranging from 1.1 to 188 μg/l in surface water (Huntsman-Mapila et al., 2006; Mladenov et al., 2014) and arsenic concentrations ranging from 3.2 to 185 μg/l in groundwater (Mladenov et al., 2014) recorded arsenic concentrations ranging between 87.5 and 185 μg/l in wells and associated its occurrence with geogenic processes such as sulfidic mineral weathering, and high alkalinity evapotranspiration. The high pH, dissolution of orpiment, and presence of dissolved organic matter led to arsenic mobility in groundwater through complexation reactions where the hydrological, biological, and chemical processes control such mobilization together with the water types.

In Zimbabwe, surface water arsenic concentrations of 1–96 μg/l were reported in the Eastern plains of Zimbabwe (Jonnalagadda & Nenzou, 1996). The high levels in a few studies reported in Zimbabwe could be attributed to mining. Zimbabwe's mining sector is predominantly dominated by artisanal and small-scale gold mining. This sector provides an income to people in many developing countries rich in gold resources. It is estimated that Zimbabwe has over 1.5 million artisanal and small-scale gold miners (Mkodzongi & Spiege, 2020; Mudzwiti et al., 2015; Rakete et al., 2022). Zimbabwe is also into heavy coal mining. Mining associated with gold and coal is the main process that mobilizes arsenic due to the oxidation process associated with mine waste (Abiye & Bhattacharya, 2019). Rakete et al. (2022) reported high levels of arsenic in urine samples of people from Shurugwi and Kadoma due to high arsenic exposure through drinking water and consumption of contaminated fish due to mining activities.

More studies on arsenic contamination in Southern Africa are from South Africa, which reported levels of 0.62–119 μg/l in surface water in Koekemoespruit and Cape Town (Akinsojiet al., 2013; Dzoma et al., 2010). Groundwater arsenic levels ranged between 12.3 and 830 μg/l in Mafikeng and Limpopo (Dzoma et al., 2010; Nephaloma & Muzerengi, 2016). Groundwater arsenic concentrations from Namaqualand ranged between <20 and 253 μg/l (Abiye & Bhattacharya, 2019), 17–318 μg/l from Johannesburg (Abiye & Bhattacharya, 2019), 1033–1369 μg/g of soil from cattle dip soil in Limpopo (Okonkwo, 2007), 0.001–46.76μg/g from cattle dip soil from Limpopo (Ramudzuli & Horn, 2014). While high levels of arsenic in surface water and groundwater in South Africa can be attributed to mining, the high levels of arsenic in soil could be due to the legacy effects of using arsenic-based dipping compounds to combat East Coast fever (ECF) among cattle (Ramudzuli & Horn, 2014). It can be summarized that the release of arsenic in the South African environment mainly depends on the sulfidic minerals oxidation, agricultural and industrial practices (glass manufacturing, semiconductors, steel melting, dye manufacturing, wood preservations, and insecticide (Abiye & Bhattacharya, 2019).

In Tanzania, arsenic levels ranged between <1 and 82 μg/l in surface water in Geita, Rwamagasa (Nyanza et al., 2014), 30–324 μg/l in mine drainage water in Serengeti National Park (Bowell et al., 1995), 5–70 μg/l in drinking water sources (wells) in Lake Victoria basin (Kassenga & Mato, 2008). Studies from Central Tanzania groundwater reported levels of 0.5–123μg/l (Nwankwo et al., 2020). Human activities (mining, small industries, and agricultural activities) cause contamination of arsenic in soil, surface water, and groundwater (Irunde et al., 2022; Kassenga & Mato, 2008; Nyanza et al., 2014; Taylor et al., 2005). The concentration of arsenic in the Tanzanian environment is expected to increase in the future due to mining activities such as coal, gold, iron, gas, and diamond mining that reduce the quality of drinking water sources (Irunde et al., 2022) The oxidation of exposed arsenopyrite is the main source of arsenic in the environment, particularly in drinking water sources during the mining process (Irunde et al., 2022; Kassenga & Mato, 2008).

Low levels of arsenic contamination of groundwater of <2–3 μg/l were reported from Malawi from Balaha, Chikwawa, and Zomba (Pritchard et al., 2007), Blantyre, Chiradzulu and Mulanje (Pritchard et al., 2007) and Blantyre (Mkandawire, 2008). Low levels of mining activities in Malawi could explain the low levels of arsenic contamination reported in the different studies. Low levels of arsenic contamination are also expected for countries like Lesotho, Swaziland, and Mozambique due to low mining activities as indicated by the low contribution of mining to GDP. However, for Mozambique, little information is available on the quality of groundwater in the aquifers of Mozambique. Those aquifers that may be vulnerable to arsenic contamination are likely to be areas of mineralization associated with gold, including alluvial gold deposits. Areas of young alluvium, particularly along the river valley, the Zambezi delta, and the coastal marshes may also be at increased risk (Smedley, 2002).

Conclusion

Arsenic concentrations of up to 10000 μg/l were reported in surface water systems in Africa. These high levels of arsenic concentrations are due to mining, industrial effluent, agriculture and municipal solid waste disposal. Arsenic concentrations of up to 1760 μg/l were reported in many groundwater systems which occur for approximately 60% of drinking water in rural areas (Irunde et al., 2022). These levels of contamination suggest that a large part of Southern Africa population suffers from arsenicosis. The accumulation of arsenic above the permissible limit in the food consumed has led to environmental contamination. Worse still, in Southern Africa region, elevated levels are found only irregularly, more as a result of the lack of research than absence of the problem. The contribution of artisanal and small scale mining in most of Southern African countries' economic activities suggests elevated levels of arsenic contamination to both surface and groundwater sources, posing a serious health risk to majority of the populations in the region.

Acknowledgements The author is grateful to the support given by NAM S&T Centre, India, and the Academy of Scientific Research and Technology (ASRT), Egypt.

References

Abiye, T. A., & Bhattacharya, P. (2019). Arsenic concentration in groundwater: Archetypal study from South Africa. *Groundwater Sustainable Development, 9.*

Ahoul´e, D. G., Lalanne, F., Mendret, J., Brosillon, S., Maïga, A. H. (2015). Arsenic in African waters: A review. *Water Air Soil Pollution, 226.*

Akinsoji, O., Fatoki, O. S., Ximba, B. J., Opeolu, B. O., & Olatunji, O. S. (2013). Assessment of arsenic levels in Guguleu and Langa rivers in Cape Town, South Africa. *International Journal of Physical Science, 8*, 1334–1340.

Amundson, S. A., Myers, T. G., & Fornace, A. J., Jr. (1998). Roles for p53 in growth arrest and apoptosis: Putting on the brakes after genotoxic stress. *Oncogene, 17*(25), 3287–3299.

Apel, K., & Hirt, H. (2004). Reactive oxygen species: Metabolism, oxidative stress, and signal transduction. *Annual Review of Plant Biology, 55*, 373–399.

Ashton, P., Love, D., Mahachi, H., & Dirks. (2001). An overview of the impact of mining and mineral processing operations on water resources and water quality in the Zambezi, Limpopo and Oliphants catchment in Southern Africa. Contract report on the mining, minerals and sustainable development (Southern Africa) project, by CSIR-Environmentek, Pretoria, South Africa and Geology Department, University of Zimbabwe, Harare, Zimbabwe. Report No. ENV-P-C 2001-042. Xvi +336pp

Atangana, E., & Oberholster, P. J. (2021). Using heavy metal pollution indices to assess water quality of surface and groundwater on catchment levels in South Africa. *Journal of African Earth Sciences, 182*, 104254.

Bhattacharya, S., Das, S., Das, S., Kalashetty, M., & Warghat, S. R. (2021). An integrated approach for mapping groundwater potential applying geospatial and MIF techniques in the semiarid region. *Environment, Development and Sustainability, 23*, 495–510.

Brindha, K., Pavelic, P., Sotoukee, T., Douangsavahn, J., & Elanga, L. (2017). Geochemical characteristics and groundwater quality in the vientiane plain. *Laos Expo Health.*

Bowell, R. J., Warren, A., Minjera, H. A., & Kimaro, N. (1995). Environmental impact of former gold mining on the Orangi river, Serengeti N.P., Tanzania. *Biogeochemistry, 28*, 131–160.

Chou, W.-C., Jie, C., Kenedy, A. A., Jones, R. J., Trush, M. A., & Dang, C. V. (2004). Role of NADPH oxidase in arsenic-induced reactive oxygen species formation and cytotoxicity in myeloid leukemia cells. *Proceedings of the National Academy of Sciences of the United States of America, 101*(13), 4578–4583.

Clough, J. et al. (2010). *Africa's water quality: A chemical science perspective.* A report by the Pan Africa Chemistry network.

Dohare, D., Deshpande, S., & Kotiya, A. (2014). Analysis of ground water quality parameters: *A review. Research Journal of Engineering Sciences, 3*(5), 26–31.

Dzoma, B. M., Moralo, R. A., Motsei, L. E., Ndou, R. V., & Bakunzi, F. R. (2010). Preliminary findings on the levels of five heavy metals in water, sediments, grass and various specimens from cattle grazing and watering in potentially heavy metal polluted areas of the North West province of South Africa. *Journal of Animal and Veterinary Advances, 9*, 3026–3033.

Eun, H. K., Sohn, S., Hyuk, J. K., et al. (2007). Sodium selenite induces superoxide-mediated mitochondrial damage and subsequent autophagic cell death in malignant glioma cells. *Cancer Research, 67*(13), 6314–6324.

Finkelman, R. B. (2004). Potential health impacts of burning coal beds and waste banks. *International Journal of Coal Geology, 59*, 19–24. https://doi.org/10.1016/j.coal.2003.11.002

Genthe, B., Kapwata, T., Le Roux, W., Chamier, J., & Wright, C. Y. (2018). The reach of human health risks associated with metals/metalloids in water and vegetables along a contaminated river catchment: South Africa and Mozambique. *Chemosphere, 199.*

Gleick, P. H. (1998). The World's water 1998–1999. Biennial report on freshwater resources. Island press, Washington DC. 217 pp.

Gupta, D. K., Tiwari, S., Razafindrabe.B. H. N., & Chatterjee, S. (2017). Arsenic contamination from historical aspects to the present. *Arsenic Contamination in the Environment: The Issues and Solutions* (pp. 1–12).

Ho, E. (2004). Zinc deficiency, DNA damage and cancer risk. *Journal of Nutritional Biochemistry, 15*(10), 572–578.

Hounsome R, & Ashton, P. (2006). *Sustainable development for the mining & minerals sector in Southern Africa.* http://www.mining.wits.ac.za/Peterob.doc

Huntsman-Mapila, P., Mapila, T., Letshwenyo, M., Wolski, P., & Hemond, C. (2006). Characterization of arsenic occurrence in the water and sediments of the Okavango Delta, NW Botswana. *Applied Geochemistry, 21*(8), 1376–1391.

Huang, C., Ma, W.-Y., Li, J., & Dong, Z. (1999). Arsenic induces apoptosis through a c-Jun NH2-terminal kinase- dependent, p53-independent pathway. *Cancer Research, 59*(13), 3053–3058.

Hussain, M. M., Bibi, I., Niazi, N. K., Shahid, M., Iqbal, J., Shakoor, M. B., Ahmad, A., Shah, N. S., Bhattacharya, P., Mao, K., Bundschuh, J., Ok, Y. S., & Zhang, H. (2021). Arsenic biogeochemical cycling in paddy soil-rice system: Interaction with various factors, amendments and mineral nutrients. *Science of the Total Environment, 773*, 145040.

Irunde, R., Ijamalane, J., Ligate, F., Maity, J. P., Ahmed, A., Mtamba, J., Mtalo, F., & Bhattacharya, P. (2022). Arsenic in Africa: Potential sources, spatial variability and the state of the art for arsenic removal using locally available materials. *Groundwater for Sustainable Development, 18.*

Johnson, D. B. (2003). Chemical and microbiological characteristics of mineral spoils and drainage water at abandoned coal and metal mines. *Water, Air, and Soil Pollution, 3*, 47–66.

Johnson, D. B., & Hallberg, K. B. (2005). Acid mine drainage remediation options: A review. *Science of the Total Environment, 338*(1–2 SPEC. ISS.), 3–14. https://doi.org/10.1016/j.scitotenv.2004.09.002

Jonnalagadda, S. B., Nenzou, G. (1996). Studies on arsenic rich mine dumps: III. Effect on the river water. *Journal of Environmental Science & Health Part A 31*(10), 2547–2555.

Kassenga, G., & Mato, R. (2008). Arsenic contamination levels in drinking water sources in mining areas in Lake Victoria Basin, Tanzania, and its removal using stabilized ferralsols. *International Journal of Biological and Chemical Science, 2*, 389–400.

Kim, T. S., Jeong, D.-W., Byung, Y. Y., & Ick, Y. K. (2002). Dysfunction of rat liver mitochondria by selenite: Induction of mitochondrial permeability transition through thiol-oxidation. *Biochemical and Biophysical Research Communications, 294*(5), 1130–1137.

Lu, J., Jiang, C., Kaeck, M., et al. (1995). Dissociation of the genotoxic and growth inhibitory effects of selenium. *Biochemical Pharmacology, 50*(2), 213–219.

Mandal, K. B., & Suzuki, T. K. (2002). Arsenic round the world: A review. *Talanta, 58*, 201–235.

Malzbender, D., & Earle, A. (2007). *The impact and implications of the adoption of the 1997 UN watercourse convention for countries in Southern Africa.* WWF.

Makaya, E., & Maphosa, M. (2023). Spatial analysis of groundwater quality in Bulawayo, Zimbabwe: Geogenic and anthropogenic contributions. *Physical Chemical Earth, 131.* Article 103436.

Misi, A., Gumindoga,W., & Hoko, Z. (2018). An assessment of groundwater potential and vulnerability in the upper manyame sub-catchment of Zimbabwe. *Physics and Chemistry of the Earth, 105*, 72–83.

Mkandawire, T. (2008). Quality of groundwater from shallow wells of selected villages in Blantyre District, Malawi. *Physics and Chemistry of the Earth, Parts A/B/C, 33*(8–13), 807–811.

Mkodzongi, G., & Spiegel, S. J. (2020). Mobility, temporary migration and changing livelihoods in Zimbabwe's artisanal mining sector. *ExtrIndSoc, 7*, 994–1001.

Mladenov, N., Wolski, P., Hettiarachchi, G. M., Murray-Hudson, M., Enriquez, H., Damaraju, S., Galkaduwa, M. B., McKnight, D. M., & Masamba, W. (2014). Abiotic and biotic factors influencing the mobility of arsenic in groundwater of a through-flow island in the Okavango Delta, Botswana. *Journal of Hydrology, 518*, 326–341.

Mudzwiti, P., Mukwakwami, N., Mungoni, M., & I.M. (2015). A golden opportunity: Scoping study of artisanal and small scale mining in Zimbabwe. Harare, Zimbabwe.

Nephalama, A., & Muzerengi, C. (2016). Assessment of the influence of coal mining on groundwater quality: Case of Masisi village in the Limpopo Province of South Africa. Mining Meets Water-Conflicts and solutions (pp. 430–438).

Nordstrom, D. K., Blowes, D. W., & Ptacek, C. J. (2015). Hydrogeochemistry and microbiology of mine drainage: An update. *Applied Geochemistry, 57*, 3–16.

Nwankwo, C. B., Hoque, M. A., Islam, M. A., Dewan, A. (2020). Groundwater constituents and trace elements in the basement aquifers of Africa and sedimentary aquifers of Asia: Medical hydrogeology of drinking water minerals and toxicants. *Earth System and Environment*.

Nyanza, E. C., Dewey, D., Thomas, D. S. K., Davey, M., & Ngallaba, S. E. (2014). Spatial distribution of mercury and arsenic levels in water, soil and cassava plants in a community with long history of gold mining in Tanzania. *Bulletin of Environment Contamination and Toxicology, 93*, 716–721.

Oberholster, P. J., Botha, A. M., Hill, L., & Strydom, W. F. (2017). River catchment responses to anthropogenic acidification in relationship with sewage effluent: An ecotoxicology screening application. *Chemosphere, 189*, 407–417.

Okonkwo, J. O. (2007). Arsenic status and distribution in soils at disused cattle dip in South Africa. *Bulletin of Environment Contamination and Toxicology, 79*, 380–383.

Panneerselvam, B., Muniraj, K., Thomas, M., Ravichandran, N., & Bidorn, B. (2021). Identifying influencing groundwater parameter on human health associated with irrigation indices using the Automatic Linear Model (ALM) in a semi-arid region in India. *Environmental Research, 202*, 111778.

Panneerselvam, B., Muniraji, K., Duraisamy, K., Ponde, C., Karuppaman, S., & Thomas, M. (2023). An integrated approach to explore the suitability of nitrate-contaminated groundwater for drinking purposes in a semi-arid region of India. *Environmental Geochemistry and Health, 45*, 647–663.

Pietersen, K., Beekman, H., Abdelkader, A., Ghany, H., et al. (2009). Environmental state and trends: 20-year retrospective. In: *Africa environment outlook 2.*

Pritchard, M., Mkandawire, T., & O'neill, J. G. (2007). Biological, chemical and physical drinking water quality from shallow wells in Malawi: Case study of Blantyre, Chiradzulu and Mulanje. *Physics and Chemistry of the Earth, Parts A/B/C, 32* (15–18), 1167–1177.

Rakete, S., Moonga, G., Wahl, A. M., et al. (2022). Biomonitoring of arsenic, cadmium and lead in two artisanal and small-scale gold mining areas in Zimbabwe. *Environmental Science and Pollution Research, 29*, 4762–4768.

Ramana, C. V. I., Boldogh, T. I., & Mitra, S. (1998). Activation of apurinic/apyrimidinic endonuclease in human cells by reactive oxygen species and its correlation with their adaptive response to genotoxicity of free radicals. *Proceedings of the National Academy of Sciences of the United States of America, 95*(9), 5061–5066.

Rahman, M. A., & Hasegawa, H. (2011). High levels of inorganic arsenic in rice in areas where arsenic-contaminated water is used for irrigation and cooking. *Science of the Total Environment, 409*, 4645–4655. https://doi.org/10.1016/j.scitotenv.2011.07.068

Ramudzuli, M. R., & Horn, A. C. (2014). Arsenic residues in soil at cattle dip tanks in the Vhembe district, Limpopo Province, South Africa. *South Africa Journal of Science, 110*, 1–7.

Rukh, S., Akhtar, M. S., Mehmood, A., Hassan, S., Khan, K. S., Naqvi, S. M. S., & Imran, M. (2017). Arsenate and arsenite adsorption in relation with chemical properties of alluvial and loess soils. *Journal of the Serbian Chemical Society, 82*, 943–954.

Ryan, K. M., Phillips, A. C., & Vousden, K. H. (2001). Regulation and function of the p53 tumor suppressor protein. *Current Opinion in Cell Biology, 13*(3), 332–337.

Smedley, P. (2002). *Groundwater quality: Mozambique.* British Geological Survey, 4 pp.

Shankar, S., Shanker, U., & Shikha. (2014). Arsenic contamination of groundwater: A review of sources, prevalence, health risks, and strategies for mitigation. *Science World Journal.*
Sies, H., & de Groot, H. (1992). Role of reactive oxygen species in cell toxicity. *Toxicology Letters, 64–65*, 547–551.
Sinha, D., Biswas, J., & Bishayee, A. (2013). Nrf2-mediated redox signaling in arsenic carcinogenesis: A review. *Archives of Toxicology, 87*(2), 383–396.
Suzuki, Y. J., Forman, H. J., & Sevanian, A. (1997). Oxidants as stimulators of signal transduction. *Free Radical Biology and Medicine, 22*(1–2), 269–285.
Taylor, H., Appleton, J. D., Lister, R., Smith, B., Chitamweba, D., Mkumbo, O., Machiwa, J. F., Tesha, A. L., & Beinhoff, C. (2005). Environmental assessment of mercury contamination from the Rwamagasa artisanal gold mining centre, Geita District, Tanzania. *Science of the Total Environment, 343*, 111–133.
Thomas, G., Sheridan, C., & Holm, P. E. (2022). A critical review of phytoremediation for acid mine drainage-impacted environments. *Science Total of the Environment,, 811*. Article 152230.
Valdiglesias, V., Pásaro, E., Méndez, J., & Laffon, B. (2010). In vitro evaluation of selenium genotoxic, cytotoxic, and protective effects: A review. *Archives of Toxicology, 84*(5), 337–351.
Wang, L., Kou, M.-C., Weng, C.-Y., Hu, L.-W., Wang, Y.-J., & Wu, M.-J. (2012). Arsenic modulates heme oxygenase-1, interleukin-6, and vascular endothelial growth factor expression in endothelial cells: Roles of ROS, NF -κB, and MAPK pathways. *Archives of Toxicology, 86*(6), 879–896.
Yih, L. H., & Lee, T.-C. (2000). Arsenite induces p53 accumulation through an ATM-dependent pathway in human fibroblasts. *Cancer Research, 60*(22), 6346–6352.

Chapter 8
Source Apportionment of Heavy Metal(loid)s in the Surface Soils of Cerrito Blanco, Mexico: A Comparative Study of Three Receptor Models (APCS-MLR, PMF and UNMIX Model)

Arnab Saha, Bhaskar Sen Gupta, Sandhya Patidar, and Nadia Martínez-Villegas

Abstract It is essential for the monitoring and conservation of soil environments to have a comprehensive awareness of the source characteristics including that of heavy metal (loid)s. This study focuses on a semi-arid mining area in Mexico where heavy metal (loid)s originated from past mining and rapid industrialisation activities, which have degraded the soil environment as well as the security of agricultural products. In this study, Cerrito Blanco, which is part of an abandoned mining region in Matehuala, Mexico was identified as the principal source of contaminants. It was also observed that the contributions of the contaminants to the overall pollution could be calculated by using a combination of multivariate statistics and receptor models. The three receptor models such as APCS-MLR, PMF, and UNMIX were used and mutually compared to improve the accuracy and quantitative assessment of source contributions. A total of eleven heavy metal(loid)s were selected for this study, out of which the mean concentrations level of As, Fe, Mn, and Zn exceeded their reference limit

A. Saha · B. S. Gupta (✉) · S. Patidar
School of Energy, Geoscience, Infrastructure and Society, Institute of Infrastructure and Environment, Heriot-Watt University, Edinburgh EH14 4AS, UK
e-mail: b.sengupta@hw.ac.uk

A. Saha
e-mail: as2059@hw.ac.uk

S. Patidar
e-mail: s.patidar@hw.ac.uk

N. Martínez-Villegas
Applied Geosciences Department, IPICyT, 78216 San Luis Potosi, Mexico
e-mail: nadia.martinez@ipicyt.edu.mx

B. Sen Gupta and N. Martínez-Villegas (eds.), *Arsenic Remediation of Food and Water*,
https://doi.org/10.1007/978-981-97-4764-1_8

values. The spatial distribution mapping revealed the distribution patterns and significant effects on concentrations of heavy metal(loid)s in surface soil. APCS-MLR identified three potential sources with contribution rates of 18.16% (groundwater source), 57.33% (past mining and industrialisation), and 24.51% (natural source) respectively. Two models, namely, PMF and UNMIX were employed to establish the contributions from common pollution sources. The contributions from four common sources (groundwater, past mining and industrialisation, natural source, and human activity) contributed 15.57, 42.86, 36.06, and 5.51% according to the PMF model, but 14.73%, 45%, 31.91%, and 8.36% respectively by the UNMIX model. The results of three receptor models showed that heavy metal(loid)s concentrations were mostly influenced by past mining and industrialisation activities. As a result, the identification of the potential sources of heavy metal(loid)s performed better using the APCS-MLR model than PMF and UNMIX model due to its higher R^2 value (0.90–1) and lower P/M error (1–1.15). To achieve more reliable and objective conclusions of source apportionment, it was recommended that multiple receptor models be applied.

Keywords Heavy metal(loid)s · Source apportionment · APCS-MLR · PMF · UNMIX

Introduction

Various sectors especially industrial and agricultural activities create substantial amounts of heavy metals as a result of the population increase and the corresponding demand for consumer items as a part of the economic development during the previous few decades (Massoud et al., 2019; Shi et al., 2022; Wu et al., 2015). Heavy metals have caused significant soil pollution resulting in considerable public concern (Wang et al., 2022; Xiang et al., 2021). Due to high toxicity, chemical instability, and bioavailability, heavy metals have the potential to harm not just human health but also the quality of soil, water, and air (Lu et al., 2018; Wang et al., 2020a; Yang et al., 2022). Heavy metals may accumulate in soil from both natural and anthropogenic causes, including soil parent material, industrial effluent, pesticides and fertiliser use, transportation activities, mineral extraction, and smelting (Gimeno-García et al., 1996; Saha et al., 2022a; Zhou et al., 2019). However, inefficient heavy metal extraction increases the accumulation of respective metal(s) in soil, water, and biota contaminating the ecosystem (Shi et al., 2022). For contamination prevention and management, the accurate identification of contaminant sources and quantitative assessment of their contributions are essential (He et al., 2019; Wang et al., 2022).

Arsenic and other heavy metals pollution in Mexican soil accentuated by mining activities have been covered extensively in several publications over the recent decades. A huge proportion of this pollution is caused by the atmospheric accumulation of metal oxides or tailings, and it is frequently found close to smelters or industrial activities (De Koning, 1974; Manz & Castro, 1997; Martínez-Villegas et al., 2013; Puchelt, 1992). The concentration of heavy metals and their sources and mobility have been studied for surface soil in the states of Baja California, Chihuahua, Guerrero, Hidalgo, and Sonora (Bundschuh et al., 2021). The area of Matehuala, located in the semi-arid zone of San Luis Potosi State in Mexico, is threatened by heavy metals contamination for the use of surface soil for maize cultivation purposes and artesian hydraulic complex for irrigation, recreation, or protection of aquatic ecosystems (Martínez-Villegas et al., 2018; Saha et al., 2022b). One of the most heavily mined regions in central Mexico is Villa de la Paz, located approximately 500 km NNW of Mexico City in San Luis Potosi. This region has had extensive Au, Ag, Cu, Pb, and Zn extraction for the past 250 years (Castro-Larragoitia et al., 1997; Chapa-Vargas et al., 2010; Razo et al., 2004). Therefore, there are a variety of mining waste dumps and old slag loads, as well as an abandoned smelter alongside operational tailings. Some of the urban areas of Mexico are beset with existing heavy metal contamination and is worth investigating the impact of the contaminants and geological environment on the pollution (Aguilera et al., 2022). However, the development of open pit mines, mining waste dumps, tailings ponds, and other mineral extraction or smelting infrastructure can significantly affect land use changes in addition to increasing environmental contamination through heavy metals (Gómez-Hernández et al., 2020; Martínez-Villegas et al., 2013; Ruíz-Huerta et al., 2017; Saha et al., 2022c). However, the scientific community has turned its attention to the identification of sources, prevention, and management of soil heavy metal contamination.

The sources of heavy metal contaminations in soil are identified using a variety of techniques. The multivariate statistical approach and receptor models are two of the most popular techniques among these (Huang et al., 2015; Wang et al., 2022; Zhao et al., 2019). Principal component analysis, correlation analysis, and cluster analysis are the three basic components of multivariate statistical techniques, which are effective tools for analysing the structure of datasets and detecting probable sources of contamination (Facchinelli et al., 2001; Hou et al., 2017; Xu et al., 2022; Yang et al., 2014). The principal component analysis is a method for reducing the number of dimensions in datasets that are often employed, while cluster analysis is frequently used to validate principal component analysis results (Goswami & Kalamdhad, 2022; Wang et al., 2022). Receptor models are frequently used to locate the sources of a certain contaminant and quantitatively estimate the effects of each source at different receptor sites (Guan et al., 2019; Li et al., 2021a). Over the past few years, a variety of receptor models, including absolute principal component analysis-multiple linear regression (APCA-MLR), UNMIX, factor analysis with non negative constraints (FA-NNC), positive matrix factorization (PMF), and chemical mass balance (CMB), etc., have been used extensively (Lang et al., 2015; Shi et al., 2014; Vega et al., 2000; Wang et al., 2010; Zhang et al., 2021). The initial purpose of using the UNMIX and PMF models was to locate the sources of air pollutants including PM2.5, PM10, and volatile organic contaminants. The sources of heavy metals in soil and water have recently been assigned using UNMIX and PMF successfully (Lang & Yang, 2014; Lang et al., 2015; Shi et al., 2012). The standard procedure to evaluate the contributions of contaminant emission sources to receptors using the Chemical Mass Balance (CMB) approach. The equation that defines the mass balance between the ambient concentrations of each contaminant and the linear sum of the components of the source characteristics and source contributions is solved using weighted least-squares solutions (Đorđević et al., 2013; Lee et al., 2009). Principal component analysis (PCA) is used in APCS-MLR to get the absolute true score (APCS) of a factor. The contribution of the common factor to the index is determined in association with a multiple linear regression (MLR) model (Paatero, 1997; Wu et al., 2020). Another popular kind of receptor-oriented modelling is factor analysis. Moreover, the loadings and scores of the factor derived from the conventional factor analysis might be negative from the perspective of matrix factorization, which complicates the evaluation of the sources extremely challenging (Lee & Seung, 1999; Zhao et al., 2004). An updated approach called factor analysis with non-negative constraints (FA-NNC) was developed to address this flaw (Wang et al., 2010). Nonetheless, the outcomes of a specific model are subject to source contribution uncertainty. As a result, in order to increase the reliability of the analyses, it is required to evaluate the results of several models (Gholizadeh et al., 2016; Wu et al., 2020; Zhang et al., 2020). Furthermore, spatial distribution maps of heavy metals and contamination sources can present strong substantiation of source identification. Many recent studies have demonstrated that combining several techniques is important to increase the objectivity of source apportionment (Dong et al., 2019; Wang et al., 2022).

In the present work, a past mining city of Matehuala in San Luis Potosi state, Mexico was identified as a case study area, and the APCS-MLR, PMF, and UNMIX models were implemented for identification and apportionment of the contamination sources of potentially harmful substances in the surface soils. Over the last 250 years, mining activities, inadequate human activities, and natural geological formation have resulted in the accumulation of heavy metal(loid)s in surface soils, which subsequently affected soil quality, human health, and the sustainability of agricultural products. The main objectives of this study were to: (a) determine the concentrations of heavy metal(loid)s like As, Cd, Co, Cr, Cu, Fe, Mn, Ni, Pb, Sb, and Zn in the surface soil and calculate the geo-accumulation index (I_{geo}) for investigating the contamination risk assessment; (b) evaluate the spatial distribution patterns of heavy metal(loid)s in soil based on the GIS interpolation techniques; (c) identify and apportion the contamination sources of heavy metal(loid)s in soil based on the APCS-MLR, PMF, and UNMIX models; and (d) compare and assess the effectiveness of APCS-MLR, PMF, and UNMIX to offer relevant information on the advantages and limitations of these techniques. The results would be important in confirming the quantitative sources of heavy metal(loid)s and also in managing and restoring polluted soil in past mining regions in the future.

Study Area

Soil samples were obtained from Cerrito Blanco, Matehuala municipality, San Luis Potosi, Mexico, located near an abandoned mining area surrounded by uncultivated cropland. The Joya Verde soccer sports club was identified as the study spot to enable the execution of this research activity. The club is around eighteen years old and contains three half-hectare grounds for soccer. Between 1974 and 2003, it was built on land with access rights for agricultural purposes (Martínez-Villegas et al., 2018). The surrounding areas are described as non-irrigated and sparsely vegetated underdeveloped agricultural land (Martínez-Villegas et al., 2018; Saha et al., 2022a, 2022b). The study site is located between 23°40′30″ N latitude and 100°35′27″ W longitude and has a total area of approx. 4.84 ha, which is 6.8 km to the east of the most populated locality of Matehuala municipality. The area is located next to a mining region where gold, silver, copper, lead, and zinc concentrations have been exploited from El Fraile hill for more than 240 years, and are still being exploited at the current time (Castillo-Nieto & Carranza-Alvarado, 1996; Castro-Larragoitia et al., 1997; Martínez-Villegas et al., 2013; Razo et al., 2004; Saha et al., 2022b). It is believed that slags, wastes, and construction debris have accumulated on the site from a dormant metal ore smelter that once operated within Matehuala city until the 1960s (Manz & Castro, 1997; Martínez-Villegas et al., 2018; Saha et al., 2022c). Previous studies have revealed that pollution of water, soils, agricultural croplands, and adverse impacts on the population in an area over 100 km^2 surrounding a mining region caused by mining and metallurgical waste (Chapa-Vargas et al., 2010; Martínez-Villegas et al., 2018; Mendoza-Chávez et al., 2021; Saha et al., 2022b; Yáñez et al., 2003). These areas often have dry and semi-arid climates with a mean yearly temperature of 20.3 °C (INEGI, 2009; Martínez-Villegas et al., 2018). The soil types in this area are Calcisol and Gypsisol, and the area receives a limited amount of precipitation every year, usually in the range of 300 to 500 mm (INEGI, 2009; Ruíz-Huerta et al., 2017). This area has a high concentration of heavy metals, several sources of pollution, and distinctive geographical patterns, providing an ideal location to investigate heavy metal contamination (Fig. 8.1).

Materials and Methods

Soil Sample Collection and Testing

To analyse the concentrations and source identification of heavy metal(loid)s in the surface soil, a total of 39 soil samples were collected using an auger at the depth of 0–5 cm. In this study, a systematic sampling approach based on a statistical procedure (NMX-132-SCFI-2001) was applied, and the guidelines from Mexican standard for topsoil at the depth of 0–5 cm was adopted for the identification and measurement of heavy metal(loid)s contamination in soil (Martínez-Villegas et al., 2018; Secretaria de Economía, 2006). Moreover, soil samples were equally distributed over a distance of approximately 10 50 m. The grid layout was adopted, and a global positioning system (GPS) was used to log the positions of each sample site. The sample sites were focused on the comparatively central plot of each sampling grid, and the distribution points were consistent and informative. After that, the samples were collected in

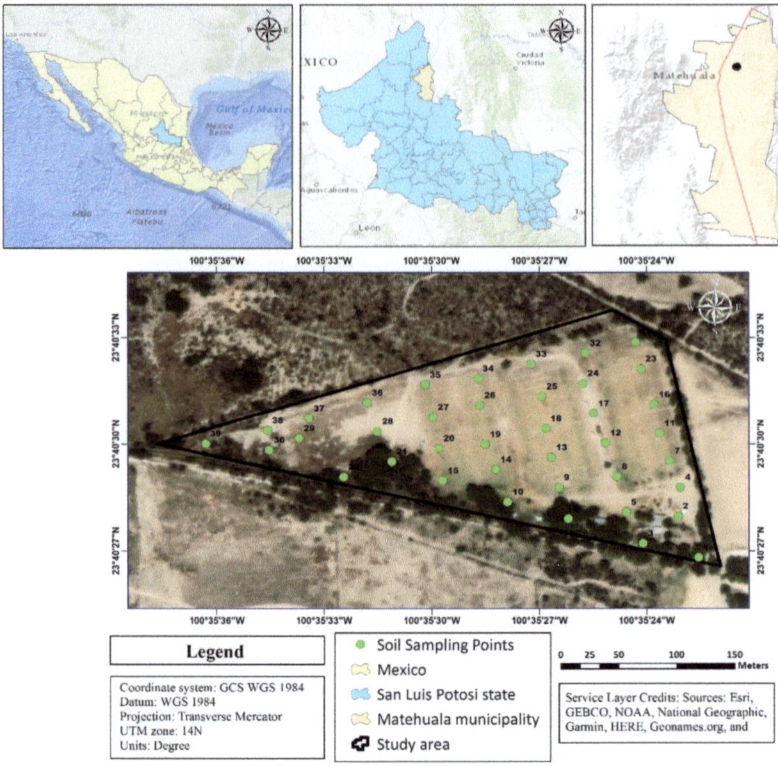

Fig. 8.1 Study area map of soil sampling locations for heavy metals extraction

polythene bags and taken back to the lab, where samples were air-dried at ambient temperature and sieved through a 2 mm stainless steel mesh filter to get rid of any stones or wastes (Martínez-Villegas et al., 2018). For chemical analysis, soil samples were digested in accordance with a slightly modified version of ISO 11466:1995. To perform this, 1.0 g of soil sample and 10 mL of aqua regia (HNO_3:HCl, 3:1) were stirred in a beaker (Martínez-Villegas et al., 2018; Vercoutere et al., 1995). After being properly dissolved with deionized water, aliquots of the digested samples were analysed for heavy metal(loid)s using inductively coupled plasma optical emission spectroscopy (ICP-OES) (Martínez-Villegas et al., 2018; USEPA, 1994).

Geo-accumulation Index (I_{geo})

A quantitative criterion for analysing contamination in the soil environment is the geo-accumulation index (I_{geo}), which is widely used to evaluate the level of soil

Table 8.1 The classifications of geo-accumulation index (I_{geo})

Category	Description
$I_{geo} < 0$	No contamination
$0 \leq I_{geo} < 1$	Slight to moderate contamination
$1 \leq I_{geo} < 2$	Moderate contamination
$2 \leq I_{geo} < 3$	Moderate to high contamination
$3 \leq I_{geo} < 4$	High contamination
$4 \leq I_{geo} < 5$	High to extreme contamination
$I_{geo} \geq 5$	Extreme serious contamination

contamination (Li et al., 2018; Wei et al., 2023; Yang et al., 2022). Using Muller's (1969) equation, the I_{geo} index was used to compare the overall measured metal concentration with background values or reference levels of concentrations in order to assess the harmful metal concentrations in the soil. This index was used to assess the levels of contamination for all heavy metals present in the topsoil obtained from inside the Joya Verde soccer sports club in Cerrito Blanco. This index classifies the potential level of contamination into seven levels (Table 8.1) as a basis for analysing soil contamination (Saha et al., 2022b). The following equation is used to calculate the I_{geo}.

$$I_{geo} = \log_2\left(\frac{C_n}{1.5B_n}\right)$$

where the concentration value of nth heavy metal in the observed soil sample is expressed as C_n; the background or reference value of nth heavy metals in soil is represented as B_n, and factor 1.5 is used to reduce the impact of any background value fluctuations that might be caused by lithologic variances in the soils (Abrahim & Parker, 2008; Stoffers et al., 1986).

Spatial Distribution of Heavy Metal(loid)s Based on Interpolation Technique

Spatial distribution mapping was used to create a geographical visualisation to identify hot spots, or areas with high levels of hazardous metal concentrations (Radomirović et al., 2020). Utilising measured data from existing sampling points, the spatial interpolation approach estimates the values of a specific variable at unsampled points (Arslan & Ayyildiz Turan, 2015; Jin & Lv, 2020). An analysis may be made for the unsampled region in this study utilising a spatial interpolation method based on the measured data from selected sampling sites called the Inverse distance weighted (IDW) interpolation approach. One of the commonly used stochastic interpolation methods in soil research is the IDW. It is worth noting that some previous studies in this area have proven that the IDW interpolation approach is the best interpolation approach to identify the spatial distribution patterns of heavy metal(loid)s (Martínez-Villegas et al., 2018; Saha et al., 2022a, 2022b). IDW determines values at unsampled points by combining values at sampled points that have been linearly weighted by an inverse function of the distance between the sampled points and the point of interest (Arslan & Ayyildiz Turan, 2015). IDW was used with a weighting power of two to simplify spatial variability of heavy metal(loid) concentration levels and source distribution factors (Saha et al., 2022a, 2022b; Wang et al., 2020b). The following equation can be used to determine the value of the estimated point.

$$Z = \sum_{i=1}^{n} \left(Z_i / d_i^p \right) / \sum_{i=1}^{n} \left(1 / d_i^p \right)$$

where, Z denotes the predicted value at an interpolated point, Z_i represents the observed value at point i, n indicates the total number of observed values used for interpolation, d_i denotes the distance between interpolated value Z and observed value Z_i, and p represents the weighting power that defines how weight decreases with distance.

Receptor Models for Source Analysis

Absolute Principal Component Score-Multiple Linear Regression (APCS-MLR)

The source identification of heavy metal(loid)s in soil was implemented using absolute principal component scores-multiple linear regression (APCS-MLR). The statistical analytic approach called principal component analysis (PCA) combines a large number of indicators into a smaller number of substantial indicators with the purpose of minimising data redundancy and uses the concept of dimensionality reduction (Jin et al., 2019). Thurston and Spengler (1985) developed the APCS/MLR receptor model, which evolved from conventional principal component analysis (PCA) and used it to calculate the distributions of various sources to certain elements. In APCS-MLR, a variable with zero concentration is introduced, and its factor fraction of the principal component is subtracted from that variable. Hence, the factor fraction of the principal component is transformed into an absolute principal component factor fraction (identifying the pollution sources and determining each source's contribution further using a multiple linear regression model) (Chen et al., 2016; Jia et al., 2020; Li et al., 2021a). Additionally, the model calculates the average contribution rate of each pollution source to its heavy metals and the contribution rate at each sampling point, based on both qualitative and quantitative measurements (Jin et al., 2019; Lv, 2019; Meng et al., 2018). Initially, the following equation was used to normalise the raw data.

$$Z_{ij} = \frac{C_{ij} - \overline{C_j}}{\delta_j}$$

where C_{ij} represents the concentration amount of the jth heavy metal(loid) in the ith soil sample, $\overline{C_j}$ is the average amount of concentration of the jth heavy metal(loid) for all soil samples, and δ_j is the value of standard deviation of jth heavy metal(loid).

Then, determine the factor fraction of the artificial sample introduced with a concentration of 0 using the following formula:

$$Z_0 = \frac{0 - \overline{C_j}}{\delta_j} = \frac{-\overline{C_j}}{\delta_j}$$

The PCA can be used to determine standardized factor scores of heavy metal concentration samples, and APCS is obtained by subtracting the factor score of each heavy metal element from the factor score of a sample with zero concentration. After that, multiple linear regression (MLR) was performed with APCS as the independent variable, the heavy-metal concentration as the dependent variable, and the regression coefficient to determine the source apportionment of each heavy metal(loid). The following equation was used to calculate the contributions to C_j using multiple linear regression.

$$C_j = b_0 + \sum_{k=1}^{p} b_k APCS_k$$

where, C_j denotes the estimated concentration value of heavy metal(loid), b_0 represents the intercept of regression for jth heavy metal(loid), b_k denotes the regression coefficient, p denotes the number of factors, b_k represents the MLR-derived regression coefficient of heavy metal source p, $APCS_k$ is the transformation value of the rotation factor for the processed sample, the concatenated phrase $b_k APCS_k$ indicates the contribution of source k to C_j. The average value of $b_k APCS_k$ for all sample points is used to reflect the contribution of heavy metal(loid) sources.

Positive Matrix Factorization (PMF) Model

The PMF model is a factor-based receptor model for estimating the quantity and contribution of the individual source based on the circumstances of the source of contamination (Chen et al., 2016; Li et al., 2021a; Zhang et al., 2020). The United States Environmental Protection Agency (USEPA) recommended the positive matrix factorization (PMF) model for heavy metal(loid) source apportionment based on a compositional dataset (Paatero & Tapper, 1994; United States Environmental Protection Agency, 2014; Wu et al., 2020; Xue et al., 2014). The PMF is differentiated by the concept that no source profiles are necessary, and uncertainty is applied to evaluate each data point respectively (Guan et al., 2018; Wu et al., 2020). The PMF model combines the source matrix into two parts: the source component spectrum matrix and the source contribution rate matrix (Deng et al., 2020; Li et al., 2021a; Saha et al., 2022d), which are represented as:

$$X_{ij} = \sum_{k=1}^{p} (G_{ik} \times F_{kj}) + E_{ij}$$

where, X_{ij} denotes the quantitative matrix for the concentration of jth heavy metal(loid) measured at the ith sampling location, G_{ik} represents the contribution matrix of the kth source factor to ith sample, F_{kj} signifies the amount of concentration of jth heavy metal(loid) from the kth source factor, E_{ij} refers to the residual error matrix for the jth heavy metal(loid) over i number of samples, and p denotes the specific sources. The residual error matrix E_{ij} derived by minimising the object function Q, which can be represented as:

$$Q = \sum_{i=1}^{n} \sum_{j=1}^{m} \left(\frac{X_{ij} - \sum_{k=1}^{p} G_{ik} \times F_{kj}}{U_{ij}} \right) = \sum_{i=1}^{n} \sum_{j=1}^{m} \left(\frac{E_{ij}}{U_{ij}} \right)^2$$

where, E_{ij} denotes the residual error matrix, U_{ij} indicates the uncertainty in the jth heavy metal(loid) for the ith sampling point, m and n indicate the number of metals and the number of samples.

The metal(loid)-specific method detection limit (MDL) is used to compute the uncertainty, and the percentage of error is assessed using standard reference materials. The uncertainty is determined by applying the following formula if the metal(loid)'s concentration is less than or equal to the respective MDL:

$$U_{ij} = \frac{5}{6} \times MDL$$

The following formula can be used to calculate the uncertainty if the metal concentration exceeds the MDL value.

$$U_{ij} = \sqrt{(\sigma \times c)^2 + (MDL)^2}$$

where, σ indicates the relative standard deviation of each metal or error fraction and c denotes the concentration of metals (Saha et al., 2022a, 2022d).

UNMIX Model

The UNMIX model is a multivariate receptor model focussed on factor analysis that evaluates the number of source identifications by implementing a single-value reduction approach to minimise the data dimensionality (Lang et al., 2015; Zhang et al., 2012). Henry (2003) developed the mathematical computational technique for

UNMIX model and launched the software on the official website of the United States Environmental Protection Agency. The model offers a highly apparent graphical interface, and the outputs might include pictorial diagnostic features. The concentrations of the samples are used as variables in this model, avoiding the normalisation procedure and difficult parameter modifications (Li et al., 2022). The rigorous data analysis procedure used by the UNMIX model is based on the analysis of eigenvalue, which eliminates lost data or data that is below the minimal detection level (Zang et al., 2022). In this study, the input file of the model was created in accordance with the UNMIX version 6 user manual guidelines (United States Environmental Protection Agency, 2007). The model does not require a source profile, and self-modelling curve resolution is applied to ensure the outcomes correspond to the non-negative limitations on source contributions (Hong et al., 2020; Wang et al., 2022; Zang et al., 2022). The computational equation is as follows:

$$C_{ij} = \sum_{k=1}^{m} F_{jk} S_{ik} + E$$

where, C_{ij} is the concentration of the jth heavy metal(loid) in the ith soil sample, F_{jk} represents jth heavy metal(loid) concentration in source k, S_{ik} indicates the contribution of kth source to the ith soil sample, and E denotes the standard deviation of analysis. Moreover, the influence of significant variations in heavy metal(loid) in soil was reduced by using the standard deviation normalisation approach (Liao et al., 2021; Wang et al., 2022; Zang et al., 2022). The calculation equation is as follows:

$$X_k = \frac{X_i - X_{imin}}{X_{imax} - X_{imin}}$$

where, X_k indicates the value after standard deviation, X_i indicates the sample's first analytical value, X_{imin} represents the minimum analytical value, and X_{imax} represents the maximum analytical value.

Processing of Data and Statistical Analysis

Microsoft Excel 2022 was used in this study to analyse the data and generate descriptive statistics for the presence of heavy metal(loid)s in soils. The APCS-MLR model and Pearson correlation analysis were performed using IBM SPSS 28.0 software. PMF and UNMIX models were carried out using the USEPA-recommended software PMF 5.0 and UNMIX 6.0, respectively. The soil sampling sites of the study area and spatial distribution maps of heavy metal(loid)s were generated by ArcGIS Pro software. Some data relationship diagram was created by Origin 2022 software.

Results and Discussion

Description of Heavy Metal(loid)s Concentration in Soil

The statistical descriptions of eleven heavy metal(loid)s and the related concentrations observed in 39 samples of soils obtained around the Joya Verde soccer sports club in Matehuala are summarised in Table 8.2. The average concentration of heavy metal(loid)s in soil with a decreasing order as: Fe (3700.52 mg/kg) > As (119.44 mg/kg) > Mn (137.82 mg/kg) > Zn (58.93 mg/kg) > Pb (36.95 mg/kg) > Cu (20.65 mg/kg) > Ni (3.20 mg/kg) > Cr (2.96 mg/kg) > Sb (1.62 mg/kg) > Cd (0.95 mg/kg) > Co (0.76 mg/kg). The reference values of heavy metal(loid)s were As (22 mg/kg) (Martínez-Villegas et al., 2018), Cd (37 mg/kg) (Secretaría de Medio Ambiente y Recursos Naturales, 2007), Co (40 mg/kg) (Kosiorek & Wyszkowski, 2020), Cr (100 mg/kg) (Ogundele et al., 2015), Cu (36 mg/kg) (Saha et al., 2022c), Fe (300 mg/kg) (Iyama et al., 2021), Mn (85 mg/kg) (Ashraf et al., 2021), Ni (35 mg/kg) (Ogundele et al., 2015), Pb (85 mg/kg) (Saha et al., 2022c), Sb (36 mg/kg) (Guo et al., 2009), and Zn (50 mg/kg) (Ogundele et al., 2015). The observed concentrations of As, Fe, Mn, and Zn exceeded their respective reference values by 5.43, 12.34, 1.62, and 1.18 times, respectively. The highest concentration of As reached 591.31 mg/kg reportedly, posing a significant threat to the environment. The coefficient of variation (CV) values of As, Co, Cr, and Sb in the soil samples was close to each other, whose value was more than 75%. Moreover, the percentage of CV values of the heavy metal(loid)s followed as a decreasing order: As (91.71%) > Co (85.77%) > Sb (82.43%) > Cr (77.22%) > Cd (68.06%) > Pb (67.10%) > Zn (58.90%) > Ni (58.32%) > Cu (47.22%) > Fe (35.32%) > Mn (29.60%). The concentrations distribution of all heavy metal(loid)s was in a highly dynamic environment under the effect of extrinsic events, except for Mn; and this heavy metal(loid)s displayed high variability with CVs over 35%, showing a highly variable condition (Zhao et al., 2022).

However, the detailed evaluation revealed that the soil around the Cerrito Blanco area contained high concentrations of As, Fe, and Mn which were greater than the reference values and it is a contamination risk and cause for concern for these metal(loid)s.

Evaluation of Contamination Risk Based on I_{geo}

In this study, I_{geo} was used to describe the contamination risk levels of heavy metal(loid)s in considerable detail. In terms of the I_{geo} classification criteria for heavy metal(loid)s (Table 8.1); Cd, Co, Cr, Cu, Ni, Pb, Sb, and Zn can be classified as contamination-free or no-contamination (Fig. 8.2). The average values of I_{geo} in the surface soils were in the decreasing order of: Fe (2.94) > As (1.33) > Mn (0.05) > Zn (−0.51) > Cu (−1.52) > Pb (−2.03) > Ni (−4.32) > Sb (−4.53) > Co (−5.28) > Cd (−5.38) > Cr (−6.01). The average value of I_{geo} in the soil sample for Fe ranged

Table 8.2 Descriptive statistics of heavy metal(loid)s

Metal(loid)s	As	Cd	Co	Cr	Cu	Fe	Mn	Ni	Pb	Sb	Zn
Mean (measured)	119.44	0.95	0.76	2.96	20.65	3700.52	137.82	3.20	36.95	1.62	58.93
Standard error	17.54	0.10	0.10	0.37	1.56	209.28	6.53	0.30	3.97	0.21	5.56
Median	90.51	0.94	0.69	2.49	18.10	3868.07	137.81	3.07	30.86	1.54	54.57
Standard deviation	109.54	0.65	0.65	2.28	9.75	1306.97	40.79	1.87	24.79	1.34	34.71
Sample variance	11,998.65	0.42	0.43	5.21	95.04	1,708,161.39	1663.99	3.49	614.63	1.78	1204.94
Kurtosis	8.37	−1.09	−0.74	4.85	3.63	0.44	0.77	0.93	5.73	6.57	15.38
Skewness	2.43	0.21	0.53	2.11	1.68	0.40	0.81	0.93	2.12	1.81	3.27
Range	578.17	2.18	2.19	10.82	47.85	5981.05	175.25	8.13	126.30	7.15	209.81
Minimum	13.14	0.00	0.00	0.28	7.88	1260.65	69.47	0.24	8.99	0.00	20.53
Maximum	591.31	2.18	2.19	11.10	55.73	7241.70	244.72	8.37	135.29	7.15	230.34
Sum	4658.01	37.12	29.73	115.30	805.17	144,320.22	5374.87	124.90	1440.99	63.17	2298.24
Coefficient of variation (CV) (%)	91.71	68.06	85.77	77.22	47.22	35.32	29.60	58.32	67.10	82.43	58.90
Count	39	39	39	39	39	39	39	39	39	39	39
Confidence level (95.0%)	35.51	0.21	0.21	0.74	3.16	423.67	13.22	0.61	8.04	0.43	11.25
Permissible limits (mg/kg)	22	37	40	100	36	300	85	35	85	36	50

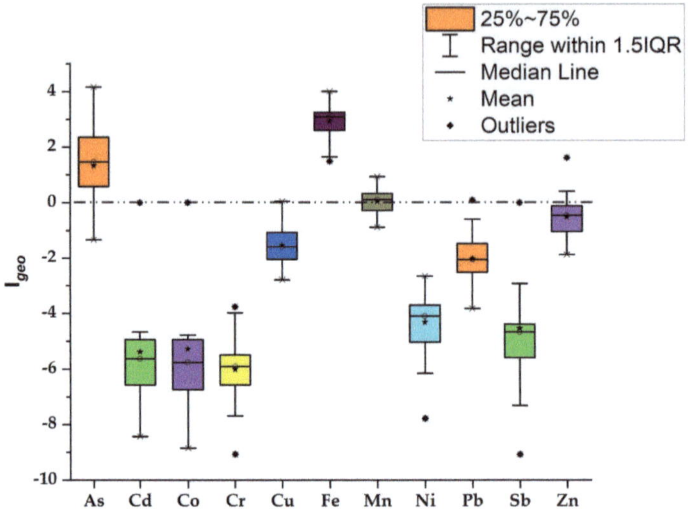

Fig. 8.2 Boxplots of I_{geo} of heavy metal(loid)s in soil

from 1.49 to 4.01, this indicates a moderate to a high level of contamination in the studied region. According to the classification category of Moderate contamination, where $1 \leq I_{geo} < 2$, As was found to have levels of 1.33. And Mn was found in the slight to moderate contamination category of I_{geo}. However, the results of this I_{geo} calculation revealed that the geological formation of rocks and past mining activities in Matehuala and the surrounding area of Cerrito Blanco were significantly contaminated by Fe and in addition to the transitional contamination carried on by the presence of As and Mn.

As a result, it is important to note that although the study area's economic growth is at the upper-middle level and contamination levels are relatively high with respect to As and Fe, indicating that the management of the area's soil environment can serve as a model for other surrounding locations.

Spatial Distribution of Heavy Metal(loid)s in Soil

The spatial distribution of heavy metal(loid)s concentration in surface soils of the Joya Verde soccer sports club is shown in Figs. 8.3 and 8.4. The graphical depiction of the concentrations of each heavy metal(loid) was presented through 39 different soil sample sites around the study area. The distribution maps were classified into six different categories based on the geometrical interval classification method. The IDW interpolation approach was used to map the spatial distribution of heavy metal(loid)s in the study area. Overall, the distribution patterns of heavy metal(loid)s were skewed. The high concentration values of As were mostly distributed in the soccer field area, indicating the significant influence of past mining activities and regular use of As-contaminated groundwater for irrigation purposes (Martínez-Villegas et al., 2018; Saha et al., 2022b). Soccer fields are irrigated twice a week with As-contaminated water, which means that 9.5 kg of As per year are added to the fields when no As is leaching from the soil in the worst case scenario. Based on an average As concentration in the irrigation water of 6 mg/L over the last 13 years, nearly 150 kg of As have been added to these soils (Martínez-Villegas et al., 2018). According to the spatial distribution map of As, it can be seen that the concentration of As is very high in a section towards the southeast which is denoted as soil sample point 3. This result was consistent with sample point 3 having 591 mg/kg of As. A detailed analysis of sample 3 revealed that there were also many white precipitates present inside the soil pores in addition to the sample's high level of organic material (Martínez-Villegas et al., 2018). Due to geological formation and increasing mining activities in this region, the component of Fe was dispersed over the area. The highest concentration of Fe was found in the centre and the lowest was in the soccer field area. The concentration of Mn in the soil on the left side of the soccer pitches area, was close to the high-value area in the centre and southern part; this suggests that Mn does not have much of an effect where the players are playing. Hence, it is concluded that anthropogenic activity is not affected by Mn. Cu, Pb, and Zn concentrations, which were substantial and an indication of a scattered distribution. For remaining heavy metals, there were no apparent spatial distribution patterns, indicating that their spatial distribution was influenced by natural origin such as soil parent material, and are more or less affected by past mining activities. Due to the slags, wastes, and debris that have accumulated in the surrounding areas, it is possible that Cd, Co, Cr, Ni, and Sb did exhibit a noticeable minimum value in the studied area and had a consistent spatial distribution pattern. However, heavy metal(loid)s have been continuously transported into the surrounding area considered in this study as a result of atmospheric sedimentation, sewage irrigation, transportation, and other causes.

Source Apportionment of Heavy Metal(loid)s

Correlation Analysis of Heavy Metal(loid)s

The correlation with heavy metal concentrations can reveal information regarding sources of heavy metal(loid)s. The Pearson correlation analysis results of eleven heavy metal(loid)s in soil were shown in Fig. 8.5. Heavy metal(loid)s in the study area showed several strong correlations, which indicated that their potential contamination sources may be comparable (Wang et al., 2022; Xia et al., 2018). Since there is a significant correlation between the Cd, Co, Cr, Cu, Cu, Fe, Mn, Ni, and Pb concentrations in the soils, it may be assumed that their sources are similar or identical. Additionally, at the 0.05 level, there were significant positive correlations between Cu and Zn ($r = 0.88$), Pb and Zn ($r = 0.54$), and Sb and Zn ($r = 0.66$) as well. This revealed that these heavy metal(loid)s may have originated from similar

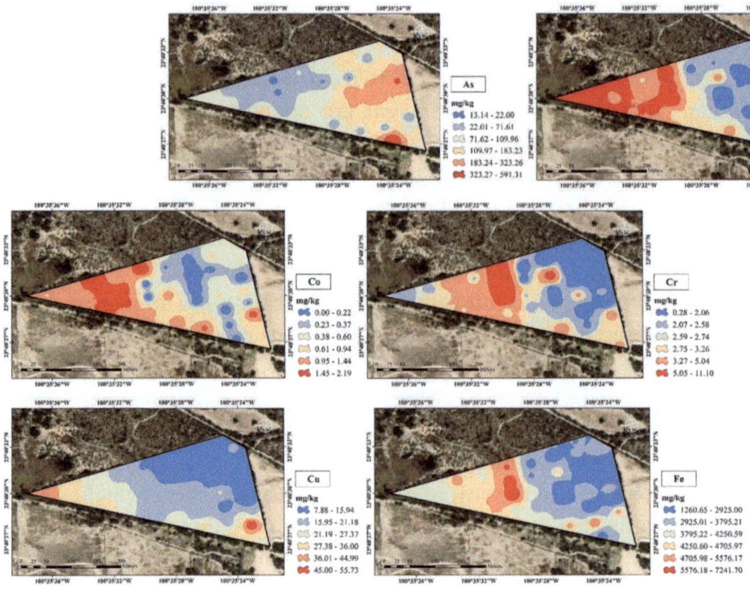

Fig. 8.3 Spatial distribution of heavy metal(loid) concentrations in soil (for As, Cd, Co, Cr, Cu, and Fe)

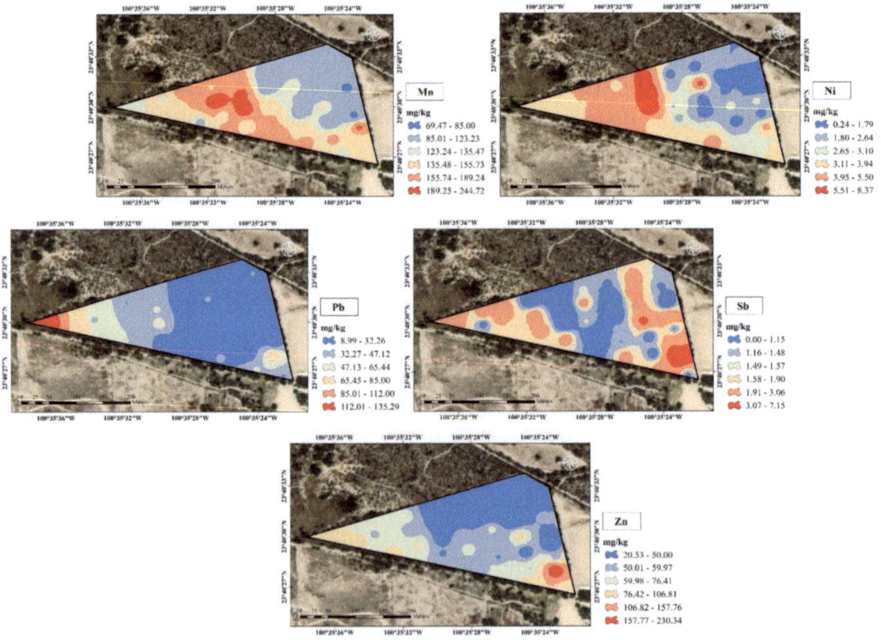

Fig. 8.4 Spatial distribution of heavy metal(loid) concentrations in soil (for Mn, Ni, Pb, Sb, and Zn)

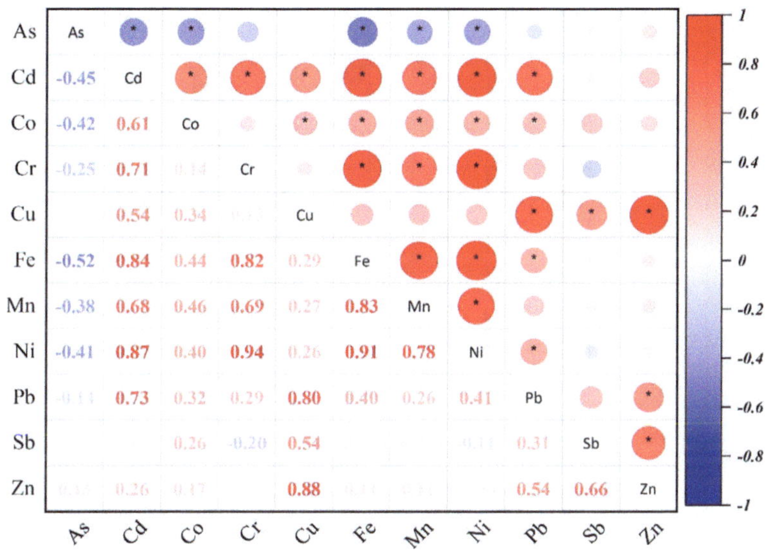

Fig. 8.5 Pearson correlation coefficients plot between heavy metal(loid)s

sources. However, a negative substantial correlation was found between Sb–Cr, Sb–Fe, Sb–Mn, and Sb–Ni. Also, As and other heavy metals had no correlation with one another. The concentrations of As in the soil showed a weak and negative indigenous correlation indicating that a particularly different source was mostly influenced by irrigational As-contaminated groundwater. The Sb, Cu, and Pb concentrations showed a strong correlation and may have been influenced by natural activities. For assessments of heavy metal sources, Pearson's correlation coefficient may be considered only as a reference, but it may not be used to draw any definite conclusions (Zang et al., 2022). The results of receptor models can further verify the source and correlation of each metal(loid)s.

Source Apportionment Results Based on the Three Receptor Models

To estimate the quantitative contributions of the various sources of contamination, three conventional receptor models were used. Three contamination sources were identified by the APCS-MLR model (Fig. 8.6), i.e., groundwater, past mining activities and industrialization, and natural sources. The PMF (Fig. 8.7) and UNMIX models (Fig. 8.8) identified four sources of contamination, i.e., groundwater, past mining activities and industrialisation, natural sources, and human activities. Detailed information on source apportionment using the various receptor models is provided below.

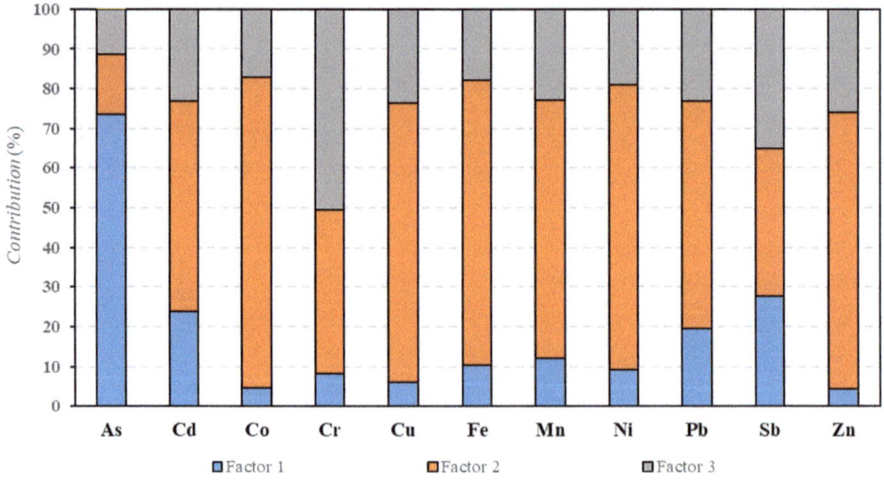

Fig. 8.6 Source apportionment of heavy metal(loid)s based on the APCS-MLR model

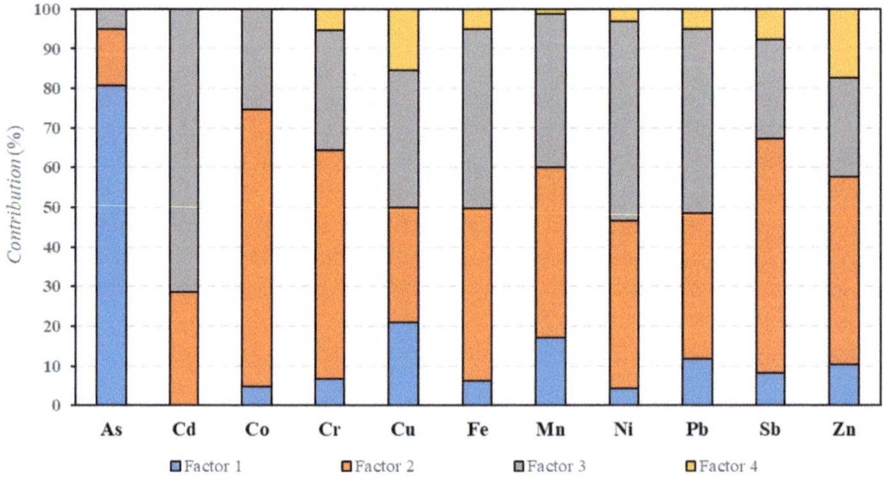

Fig. 8.7 Source apportionment of heavy metal(loid)s based on the PMF model

To ascertain if the data were appropriate for analysis, the Kaiser–Meyer–Olkin test (KMO) and evaluation of the model's Bartlett sphericity were carried out before completing PCA. The results indicated that the data were acceptable for PCA because the KMO value was 0.688 and the significance level of the Bartlett test was 0.001. This means the data may be used for factor analysis. Three principal components were identified using the APCS-MLR approach in accordance with the assumption of an eigenvalue greater than 1. The measured value of the heavy metal(loid)s concentration and the predicted value of the heavy metal(loid)s by the receptor models are listed in

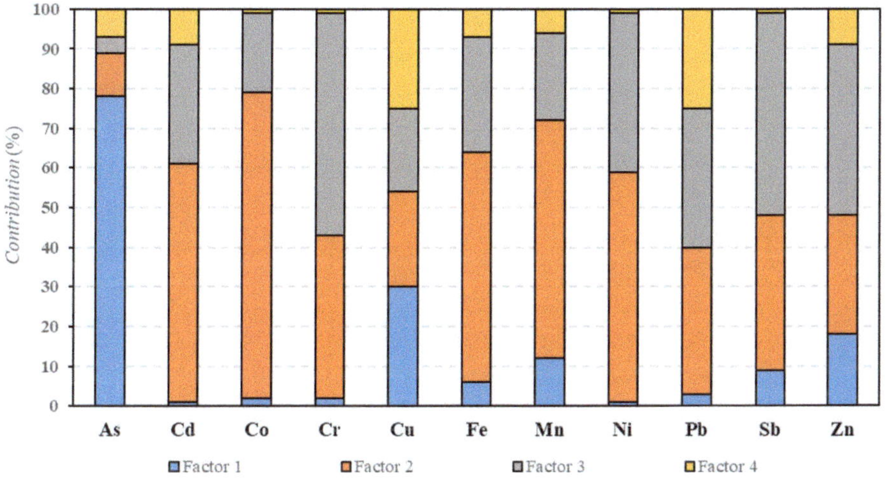

Fig. 8.8 Source apportionment of heavy metal(loid)s based on the UNMIX model

Table 8.3. The P/M value (predicted value/measured value) is close to 1, reflecting a high fitting level and significant reliability for the APCS-MLR analysis (Hingorani et al., 2020; Zhang et al., 2022). Factor 1 was accountable for 73.5% of As, showing that it almost fully dominated the amount of As. Factor 2 was heavily loaded with all heavy metal(loid)s in the APCS-MLR model, except for As, contributing for Cd (53%), Co (78%), Cr (41%), Cu (70%), Fe (72%), Mn (65%), Ni (72%), Pb (57%), Sb (37%), and Zn (69.6%), respectively. Factor 3, which was significantly influenced by Cr and Sb, with contributions of 50.5% and 35%, respectively, and the rest of heavy metal(loid)s contributing between 11 and 26%. According to these results, factor 1, factor 2, and factor 3 were groundwater sources, past mining activities & industrialisation sources, and natural sources, respectively. APCS-MLR was unable to identify factor 4, hence factor 1, factor 2, and factor 3 had comparatively high contribution rates.

The PMF model was used to confirm that the sources of heavy metal(loid)s contamination in soil were accurate. Before determining the PMF result, several factors were considered (ranging from 3 to 5) and runs were executed 20 times using a random selection in the PMF base model until lower conformed Q_{robust} values and higher R^2 values were found (Shi et al., 2022). As a result, the appropriate factor number was determined and could be easily interpreted. The optimal approach was decided upon using four factors. The outcomes mostly accord with the APCS-MLR model. Moreover, bootstraps were also performed 100 times to confirm maximum quantitative result precision. Compared with the APCS-MLR model, the PMF model evaluation identified an additional factor that had an influence on the heavy metal(loid)s contamination in the study area. The results of the four factors are shown in Fig. 8.7. Factor 1 is the groundwater source, which is shown to have As with a contribution of 80.6%. Factor 2 is the source of past mining activities and

Table 8.3 Comparison of the fitting degree of receptor models

Heavy metal(loid)s	Measured (mg/kg)	APCS-MLR			PMF			UNMIX		
		Predicted (mg/kg)	P/M	R^2	Predicted (mg/kg)	P/M	R^2	Predicted (mg/kg)	P/M	R^2
As	119.44	119.52	1.00	0.99	121.42	1.02	0.99	121.00	1.01	0.99
Cd	0.95	0.96	1.01	1.00	1.23	1.30	0.74	1.06	1.12	0.94
Co	0.76	0.85	1.12	0.99	0.96	1.27	0.64	1.01	1.32	0.96
Cr	2.96	3.15	1.06	0.96	3.66	1.24	0.40	3.05	1.03	0.91
Cu	20.65	20.81	1.01	0.97	22.55	1.09	0.96	20.76	1.01	0.85
Fe	3700.52	3700.55	1.00	0.99	3701.89	1.00	1.00	3548.48	0.96	0.79
Mn	137.82	137.95	1.00	0.98	159.21	1.16	0.70	131.80	0.96	0.58
Ni	3.2	3.51	1.10	0.90	3.76	1.18	0.78	3.21	1.00	0.96
Pb	36.95	37.06	1.00	0.99	37.41	1.01	0.99	37.63	1.02	0.68
Sb	1.62	1.87	1.15	0.94	2.23	1.37	0.61	1.80	1.11	0.79
Zn	58.93	59.15	1.00	0.95	64.91	1.10	0.95	59.42	1.01	0.82

industrialisation, which was highly contributed by Co (70%), Cr (57%), Fe (43.5%), Mn (43%), Ni (42%), Pb (37%), Sb (59%), and Zn (47%), respectively. Factor 3 is the source of natural activities and had a large contribution of Cd, Cr, Cu, Fe, Mn, Ni, and Pb with values of 71, 30, 34, 45, 39, 50, and 46%. The contribution of factor 4 is very less, which was dominated by Cu with a value of 15.5%; moreover, Factor 4 is denoted by human activities. On the whole, the average contributions from the identified sources were as follows: factor 2 > factor 3 > factor 1 > factor 4. As a result of PMF, factor 2 and factor 3 were the main contributors to heavy metal(loid)s contamination.

The UNMIX model revealed four primary sources of heavy metal(loid)s in the investigated soils; their source characteristics and contributions of a percentage of various sources were shown in Fig. 8.8 and Table 8.3. The proposed exclusion feature may be used to examine the data after it has been imported into the UNMIX model. The findings reveal that all eleven metal(loid)s meet the software operating criteria. As soon as the formal process is complete, the model normally compares the three sources and adjusts the balance of each source. As a next step, the number of factors is set to 4 manually and run the model. The least R^2 value for any metal(loid)s in the UNMIX model is called Min R_{sq}. The results indicate that the least R^2 value was 0.90, which is greater than the minimum value needed by the model (min $R^2 > 0.8$), and the minimum S/N (signal/noise) value in multiple operations was 2.05, which was greater than the required value needed by the model (min S/N > 2). Factor 1 was the dominant contributor of As (78%), which indicated groundwater source. Factor 2 was contributed by Cd (60%), Co (77%), Cr (41%), Fe (58%), Mn (60%), Ni (58%). It also had a specific contribution of Cu (24%), Pb (37%), Sb (39%), and Zn (30%), which was denoted by past mining activities and industrial sources. Factor 3 is the source of natural activities, which is highly contributed by Cr (56%), Ni

(40%), Pb (35%), Sb (51%), and Zn (43%), respectively. Similar to the PMF model, the results of the UNMIX model also show that the contribution of factor 4 is very less and mainly dominated by Cu and Pb with a similar value of 25%. The average contributions from the identified sources were as follows: factor 2 > factor 3 > factor 1 > factor 4.

Identification of Heavy Metal(loid)s Contamination Sources

Factor 1 showed a significant load value of As in the component of source distribution through three receptor models and the contribution rate exceeded 70% in three models. In the past, the mining and smelting operations severely contaminated the soil in the surroundings of this region. In the identification of factor 1 distribution, a very small role was dominated by Cd, Cu, Mn, Pb, and Sb contamination. Because the study area is on the border of an urban area of Matehuala city, there are several industrial sites in northeast of the studied area, that released wastes, contaminated water, and chemical residues into the natural environment. The contribution rates of the APCS-MLR, PMF, and UNMIX models to factor 1 were 18%, 15.5%, and 15%, respectively (Fig. 8.9). The paedogenic activities, ore mining and mineral processing operations, waste disposal, and As-contaminated groundwater are some of the potential sources of As bonded to other heavy metals. And earlier research has demonstrated that the higher arsenic levels in these soils only relate to the As-contaminated irrigational groundwater (Martínez-Villegas et al., 2018; Saha et al., 2022a, 2022d). Therefore, compared with the spatial distribution map of As and the results of three receptor models, it was revealed that factor 1 might be interpreted as groundwater sources.

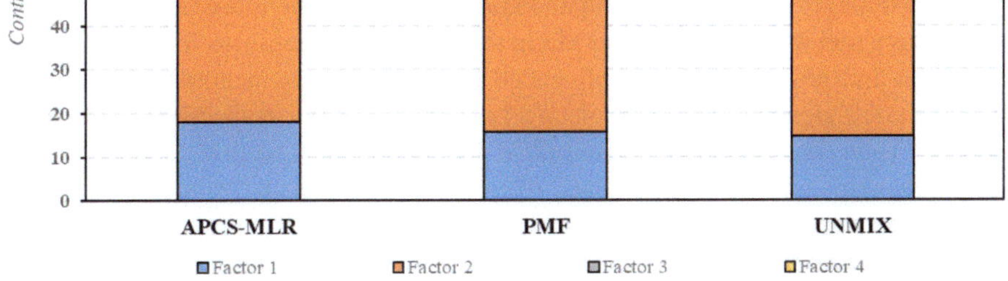

Fig. 8.9 Comparison of source appointment results by different receptor models

Factor 2 contributed the most to all heavy metal(loid)s in the studied soil, with three receptor models contributing more than 42%. The contribution rates of the APCS-MLR, PMF, and UNMIX model to factor 2 were 57%, 43%, and 45%, respectively (Fig. 8.9). Since this region is located in a mining area with more than 240 years of mining history, all are aware of its existence (Martínez-Villegas et al., 2013; Razo et al., 2004). Other industrial operations, such as metal casting and soil chemical formation, have the possibility of releasing heavy metals into the environment. Furthermore, Cu, Pb, and Zn smelting operations were one of the major sources of pollution in this region. Thus, metals such as Cd, Co, Cr, Fe, Mn, Ni and Sb, which were released into the environment during smelting and processing, and related metals to accumulate in the surface soils by dry and wet deposition. Numerous scientific research demonstrated that heavy metals such as Cd, Co, Cr, Fe, Mn, Ni, and Sb were produced over a long period of time in the mining and other industrialisation sectors, such as ore smelting, steel manufacture, metal processing, metallurgical industry, etc. (Guan et al., 2019; Li et al., 2021b; Ming-Kai et al., 2013; Wang et al., 2022). The interpretation of the spatial distribution of heavy metal(loid)s concentration (Figs. 8.3 and 8.4) in this region revealed that all studied heavy metal(loid)s concentrations in recreational surface soil were comparatively uniform throughout the entire study area, and provided additional evidence that heavy metals accumulation was primarily influenced by past mining activities and increasing industrialisation. As a result, factor 2 is considered to be past mining activities and industrialisation sources.

Factor 3 was significantly loaded with Cr, Fe, Mn, and Ni, and moderately loaded with Cd, Cu, Pb, Sb, and Zn. The contribution rates of the APCS-MLR, PMF, and UNMIX models to factor 3 were 24.5%, 36%, and 32%, respectively (Fig. 8.9). The weathering of rocks and their pedogenesis lead to the formation of soil (Wang et al., 2022). The parent material for the soil defines the primary element content and generates the background value of heavy metal(loid)s in the soil (He et al., 2017; Wang et al., 2022). The previous studies revealed that Cd, Cu, Cr, Fe, Mn, Ni, Pb, Sb, and Zn were frequently affected by soil minerals and were associated with natural and lithogenic sources such as soil parent material (Saha et al., 2022a, 2022d). All the heavy metals obtained from the surface soil are near to or below the reference limit requirements, except for the Fe and Mn. But the substantial contributions to Mn and Fe may indicate that this source is lithogenic or crustal based on the existence of those elements (Jose & Srimuruganandam, 2021). Furthermore, the distribution of the heavy metals, namely Fe and Mn in the entire study area was found to be very consistent and not significantly influenced by other activity, according to the low variation coefficients of these two metals, i.e., 35% and 25%, respectively. However, Cd, Cu, Cr, Ni, Pb, Sb, and Zn were partially obtained from natural sources; Fe and Mn components were mostly obtained from natural sources and were influenced by soil parental materials that formed rocks.

Factor 4 was only identified by PMF and UNMIX models. It contributed substantially to Cu, with rates of 15.5 and 25% in the PMF and UNMIX models, respectively, and all other heavy metal(loid)s were negligible. Additionally, factor 4 had a small impact on Pb and its contribution rate was around 25%, according to the result of UNMIX model. The contribution rates of PMF and UNMIX model to factor 4 were 5.5% and 8.36%, respectively (Fig. 8.9). In addition to the spatial distribution patterns, the moderate value of Cu and Pb regions were primarily found near the road and surrounding agricultural area. The accumulation of heavy metal(loid)s might have been caused by the processing of some chemical products in smelting plant areas, including fertiliser, cement, and sulphuric acid (Ran et al., 2021; Wang et al., 2022). Moreover, heavy metal(loid)s also accumulated from anthropogenic sources, such as vehicle emissions and machine wear (Wang et al., 2022; Yan et al., 2018). Therefore, compared with the spatial distribution map and the lower contribution rate of heavy metal(loid)s, it was revealed that factor 1 might be interpreted as human activity source.

Comparison of Three Receptor Models

The source apportionment results of the three receptor models were comparatively reliable, indicating that they may be independently validated. Using the APCS-MLR model, only three principal components were retrieved using PCA and limiting the number of contamination sources compared to PMF and UNMIX models. The contribution of various contamination sources was calculated by using multiple linear regression (MLR). All receptor model's associated factors may be compared using Pearson correlation analysis, which can also help to identify common sources (Jin et al., 2019). Due to its consistency and ease of implementation, the APCS-MLR model has been commonly applied for source apportionment (Guan et al., 2019; Jin et al., 2019; Wang et al., 2022). The PMF approach is challenging and time-consuming. In order for the PMF model to calculate correctly and produce a more precise outcome, it requires heavy metal(loid)s concentration data as well as uncertainty data (Diakite et al., 2021; Jain et al., 2018; Wang et al., 2022). Additionally, the model may produce disappointing results if the collected data have anomalies (Jin & Lv, 2020; Wang et al., 2022). The R^2 and P/M values of Co, Cr, and Sb are considerably lower in the PMF model, which may be due to this reason. The UNMIX model depends extensively on heavy metal(loid)s concentration data, hence the quality of the data clearly affects model performance. Additionally, earlier studies have shown that the UNMIX model may be used to identify sources with high contribution rates (Jain et al., 2018; Wang et al., 2022). However, the mechanism of the UNMIX model works similarly to that of the PMF model.

To evaluate the precision of the receptor models, R^2 and the predicted-to-measured (P/M) values are considered. Compared to all three models' R^2 values, which generally varied from 0.70 to 1, and their P/M values, which were very near to 1, indicated the models' strong accuracy and consistency. Although there are still some outliers and errors in the data, some metal(loid)s results showed lower R^2 values and lower accuracy of P/M values. The lower accuracy results were mostly reflected in PMF and UNMIX models. The result was completely significant for the APCS-MLR model, where R^2 values were greater than 0.90 and all heavy metal(loid)s had negligible errors in P/M values. As a result, it is challenging for the PMF model to reach optimal precision. Despite other factors affecting the contribution of some elements, UNMIX model can achieve a higher accuracy than that of PMF. In this study, APCS-MLR modelling was highly accurate, and previous studies using the approach showed that the data were consistent (Guan et al., 2019; Jin & Lv, 2020; Li et al., 2021a; Wang et al., 2022).

The cumulative contribution percentage of different contamination sources, as calculated by the three receptor models, to the total amount of heavy metal(loid)s concentration in soil was shown in Fig. 8.9. The highest source contribution occurred from past mining activities and industrialisation, followed by natural sources, groundwater, and human activities. The lowest source contribution percentage came from human activities, which revealed that soil contributed to a reasonable amount of anthropogenic contamination. Despite using different algorithms and mechanisms, the APCS-MLR, PMF, and UNMIX models produced consistent results in the study area, and both analyses were validated. Therefore, integrating multiple models can increase the accuracy and consistency of source apportionment results.

Conclusion

In this study, the source apportionment and intensity of eleven heavy metal(loid)s concentrations in the surface soil of a semi-arid region in Matehuala, Mexico was studied using comparisons of three receptor models (APCS-MLR, PMF, and UNMIX) in addition to contamination risk assessment. The concentration of As ranged from 13.14 to 591.31 mg/kg with a mean of 119.44 mg/kg and the concentration of Fe ranged from 1260.65 to 7241.70 mg/kg with a mean of 3700.52 mg/kg, and these two metals were the dominant metal(loid)s in the studied soils. The risk of contamination in this area is highest for As and Fe, according to the assessment of I_{geo}. The APCS-MLR model identified three source factors of contamination and four contamination sources were apportioned by PMF and UNMIX models. Factor 1 was individually dominated by contributions of As, reflecting the As-contaminated groundwater source for irrigation purposes. Factor 2 was dominated by contributions of all heavy metal(loid)s to various extents, reflecting the past mining and industrialisation activities. Factor 3 was governed by contributions of contaminants such as Cr, Fe, Mn, Ni, Pb, and Zn, reflecting the natural source of parental soil formation. Factor 4 had dominant contributions from Cu and Pb, reflecting anthropogenic factors since such effects were not identified by the APCS-MLR model. Overall, the contribution rate of source apportionment for the APCS-MLR model is as follows: past mining and industrialisation (57.33%) > natural source (24.51%) > groundwater (18.16%); for the PMF model: past mining and industrialisation (42.86%) > natural source (36.06%) > groundwater (15.57%) > human activities (5.51%); and for the UNMIX model: past mining and industrialisation (45%) > natural source (31.91%) > groundwater (14.73%) > human activities (8.36%). The heavy metal(loid)s concentrations in the study are a mainly came from past mining and industrialisation activities. The comparison of PMF, UNMIX, and APCS-MLR receptor models revealed that the analytical results of the three models were consistent and conformed to each other. In this study, the APCS-MLR model performed better because it can more correctly identify the source and estimate the proportion of contributions. A more accurate source diagnostic result could be obtained by combining elements of various models that are better suited for specific data sets. This study presented an efficient technique for quantifying the most significant contamination sources of concern for soil risk, which is critical for contamination prevention and risk mitigation with limited resources.

Acknowledgements Thanks to the Institute of Infrastructure and Environment, EGIS, Heriot-Watt University, Edinburgh. The authors are thankful to The School of Energy, Geoscience, Infrastructure and Society (EGIS), Heriot-Watt University, Edinburgh for providing a student bursary to the first author for doctoral research through the James Watt Scholarship. The author would also thanks to IPICyT, San Luis Potosi, Mexico for providing feedback and support.

Funding This study was financially supported by the Engineering and Physical Sciences Research Council (EPSRC-IAA), UKRI, and Heriot-Watt University, Edinburgh internal research grant (under award number D19R12025) through the Project 748502: Spreading The Chemical Free Arsenic Removal Technology To Various Communities In India, Bangladesh And Mexico.

Data Availability Data will be made available on request.

Declaration of Competing Interest The authors declare that they have no known competing financial interests or personal relationships that could have appeared to influence the work reported in this paper.

References

Abrahim, G. M. S., & Parker, R. J. (2008). Assessment of heavy metal enrichment factors and the degree of contamination in marine sediments from Tamaki Estuary, Auckland, New Zealand. *Environmental Monitoring and Assessment, 136*(1), 227–238. https://doi.org/10.1007/s10661-007-9678-2

Aguilera, A., Cortés, J. L., Delgado, C., Aguilar, Y., Aguilar, D., Cejudo, R., Quintana, P., Goguitchaichvili, A., & Bautista, F. (2022). Heavy metal contamination (Cu, Pb, Zn, Fe, and Mn) in urban dust and its possible ecological and human health risk in Mexican cities. *Frontiers in Environmental Science*, 195. https://doi.org/10.3389/fenvs.2022.854460

Arslan, H., & Ayyildiz Turan, N. (2015). Estimation of spatial distribution of heavy metals in groundwater using interpolation methods and multivariate statistical techniques; its suitability for drinking and irrigation purposes in the Middle Black Sea Region of Turkey. *Environmental Monitoring and Assessment, 187*(8), 1–13. https://doi.org/10.1007/s10661-015-4725-x

Ashraf, I., Ahmad, F., Sharif, A., Altaf, A. R., & Teng, H. (2021). Heavy metals assessment in water, soil, vegetables and their associated health risks via consumption of vegetables, District Kasur, Pakistan. *SN Applied Sciences, 3*(5), 1–16. https://doi.org/10.1007/s42452-021-04547-y

Bundschuh, J., Armienta, M. A., Morales-Simfors, N., Alam, M. A., López, D. L., Delgado Quezada, V., Dietrich, S., Schneider, J., Tapia, J., Sracek, O., Castillo, E., & Ahmad, A. (2021). Arsenic in Latin America: New findings on source, mobilization and mobility in human environments in 20 countries based on decadal research 2010–2020. *Critical Reviews in Environmental Science and Technology, 51*(16), 1727–1865. https://doi.org/10.1080/10643389.2020.1770527

Castillo-Nieto, F., & Carranza-Alvarado, M. (1996). Geological mining monograph of the state of San Luis Potosí. Consejo de Recursos Minerales, Pachuca, Mexico.
Castro-Larragoitia, J., Kramar, U., & Puchelt, H. (1997). 200 years of mining activities at La Paz/San Luis Potosí/Mexico—Consequences for environment and geochemical exploration. *Journal of Geochemical Exploration, 58*(1), 81–91. https://doi.org/10.1016/S0375-6742(96)00054-4
Chapa-Vargas, L., Mejia-Saavedra, J. J., Monzalvo-Santos, K., & Puebla-Olivares, F. (2010). Blood lead concentrations in wild birds from a polluted mining region at Villa de La Paz, San Luis Potosi, Mexico. *Journal of Environmental Science and Health, Part A, 45*(1), 90–98. https://doi.org/10.1080/10934520903389242
Chen, H., Teng, Y., Chen, R., Li, J., & Wang, J. (2016). Contamination characteristics and source apportionment of trace metals in soils around Miyun Reservoir. *Environmental Science and Pollution Research, 23*(15), 15331–15342. https://doi.org/10.1007/s11356-016-6694-1
De Koning, H. W. (1974). Lead and cadmium contamination in the area immediately surrounding a lead smelter. *Water, Air, and Soil Pollution, 3*(1), 63–70. https://doi.org/10.1007/BF00282727
Deng, M., Zhu, Y., Shao, K., Zhang, Q., Ye, G., & Shen, J. (2020). Metals source apportionment in farmland soil and the prediction of metal transfer in the soil-rice-human chain. *Journal of Environmental Management, 260*, 110092. https://doi.org/10.1016/j.jenvman.2020.110092
Diakite, M. L., Hu, Y., & Cheng, H. (2021). Source apportionment based on the comparative approach of two receptor models in a large-scale region in China. *Environmental Science and Pollution Research, 28*(40), 56696–56710. https://doi.org/10.1007/s11356-021-14602-1
Dong, B., Zhang, R., Gan, Y., Cai, L., Freidenreich, A., Wang, K., Guo, T., & Wang, H. (2019). Multiple methods for the identification of heavy metal sources in cropland soils from a resource-based region. *Science of the Total Environment, 651*, 3127–3138. https://doi.org/10.1016/j.scitotenv.2018.10.130
Đorđević, D., Petrović, S., Relić, D., & Mihajlidi-Zelić, A. (2013). Applying receptor models Unmix and PMF on real data set of elements in PM for sources evaluation of the sea coastal side region (Southeast Adriatic Sea). *Atmospheric Measurement Techniques Discussions, 6*(3), 4941–4969. https://doi.org/10.5194/amtd-6-4941-2013
Facchinelli, A., Sacchi, E., & Mallen, L. (2001). Multivariate statistical and GIS-based approach to identify heavy metal sources in soils. *Environmental Pollution, 114*(3), 313–324. https://doi.org/10.1016/S0269-7491(00)00243-8
Gholizadeh, M. H., Melesse, A. M., & Reddi, L. (2016). Water quality assessment and apportionment of pollution sources using APCS-MLR and PMF receptor modeling techniques in three major rivers of South Florida. *Science of the Total Environment, 566*, 1552–1567. https://doi.org/10.1016/j.scitotenv.2016.06.046
Gimeno-García, E., Andreu, V., & Boluda, R. (1996). Heavy metals incidence in the application of inorganic fertilizers and pesticides to rice farming soils. *Environmental Pollution, 92*(1), 19–25. https://doi.org/10.1016/0269-7491(95)00090-9
Gómez-Hernández, A., Rodríguez, R., Del Río, A. L., Ruiz-Huerta, E. A., Armienta, M. A., Dávila-Harris, P., Sen-Gupta, B., Delgado-Rodríguez, O., Ríos, A. D. A., & Martínez-Villegas, N. (2020). Alluvial and gypsum karst geological transition favors spreading arsenic contamination in Matehuala, Mexico. *Science of the Total Environment, 707*, 135340. https://doi.org/10.1016/j.scitotenv.2019.135340
Goswami, A. P., & Kalamdhad, A. S. (2022). Water quality variation and source apportionment using multivariate statistical analysis. *Environmental Forensics*, 1–23. https://doi.org/10.1080/15275922.2022.2125112
Guan, Q., Wang, F., Xu, C., Pan, N., Lin, J., Zhao, R., Yang, Y., & Luo, H. (2018). Source apportionment of heavy metals in agricultural soil based on PMF: A case study in Hexi Corridor, northwest China. *Chemosphere, 193*, 189–197. https://doi.org/10.1016/j.chemosphere.2017.10.151
Guan, Q., Zhao, R., Pan, N., Wang, F., Yang, Y., & Luo, H. (2019). Source apportionment of heavy metals in farmland soil of Wuwei, China: Comparison of three receptor models. *Journal of Cleaner Production, 237*, 117792. https://doi.org/10.1016/j.jclepro.2019.117792

Guo, X., Wu, Z., & He, M. (2009). Removal of antimony (V) and antimony (III) from drinking water by coagulation–flocculation–sedimentation (CFS). *Water Research, 43*(17), 4327–4335. https://doi.org/10.1016/j.watres.2009.06.033

He, S., Lu, Q., Li, W., Ren, Z., Zhou, Z., Feng, X., Zhang, Y., & Li, Y. (2017). Factors controlling cadmium and lead activities in different parent material-derived soils from the Pearl River Basin. *Chemosphere, 182*, 509–516. https://doi.org/10.1016/j.chemosphere.2017.05.007

He, B., Zhao, X., Li, P., Liang, J., Fan, Q., Ma, X., Zheng, G., & Qiu, J. (2019). Lead isotopic fingerprinting as a tracer to identify the pollution sources of heavy metals in the south eastern zone of Baiyin, China. *Science of the Total Environment, 660*, 348–357. https://doi.org/10.1016/j.scitotenv.2018.11.339

Henry, R. C. (2003). Multivariate receptor modeling by N-dimensional edge detection. *Chemometrics and Intelligent Laboratory Systems, 65*(2), 179–189. https://doi.org/10.1016/S0169-7439(02)00108-9

Hingorani, R., Jiménez-Relinque, E., Grande, M., Castillo, A., Nevshupa, R., & Castellote, M. J. C. E. J. (2020). From analysis to decision: Revision of a multifactorial model for the in situ assessment of NOx abatement effectiveness of photocatalytic pavements. *Chemical Engineering Journal, 402*, 126250. https://doi.org/10.1016/j.cej.2020.126250

Hong, N., Yang, B., Tsang, D. C., & Liu, A. (2020). Comparison of pollutant source tracking approaches: Heavy metals deposited on urban road surfaces as a case study. *Environmental Pollution, 266*, 115253. https://doi.org/10.1016/j.envpol.2020.115253

Hou, D., O'Connor, D., Nathanail, P., Tian, L., & Ma, Y. (2017). Integrated GIS and multivariate statistical analysis for regional scale assessment of heavy metal soil contamination: A critical review. *Environmental Pollution, 231*, 1188–1200. https://doi.org/10.1016/j.envpol.2017.07.021

Huang, Y., Li, T., Wu, C., He, Z., Japenga, J., Deng, M., & Yang, X. (2015). An integrated approach to assess heavy metal source apportionment in peri-urban agricultural soils. *Journal of Hazardous Materials, 299*, 540–549. https://doi.org/10.1016/j.jhazmat.2015.07.041

INEGI. (2009). Prontuario de Información Geográfica Municipal de los Estados Unidos Mexicanos. Matehuala, San Luis Potosí; Instituto Nacional de Estadística Geografía e Informática: Aguascalientes, México.

Iyama, W. A., Okpara, K., & Techato, K. (2021). Assessment of heavy metals in agricultural soils and plant (Vernonia amygdalina Delile) in Port Harcourt Metropolis, Nigeria. *Agriculture, 12*(1), 27. https://doi.org/10.3390/agriculture12010027

Jain, S., Sharma, S. K., Mandal, T. K., & Saxena, M. (2018). Source apportionment of PM10 in Delhi, India using PCA/APCS, UNMIX and PMF. *Particuology, 37*, 107–118. https://doi.org/10.1016/j.partic.2017.05.009

Jia, Z., Wang, J., Zhou, X., Zhou, Y., Li, Y., Li, B., & Zhou, S. (2020). Identification of the sources and influencing factors of potentially toxic elements accumulation in the soil from a typical karst region in Guangxi, Southwest China. *Environmental Pollution, 256*, 113505. https://doi.org/10.1016/j.envpol.2019.113505

Jin, G., Fang, W., Shafi, M., Wu, D., Li, Y., Zhong, B., Ma, J., & Liu, D. (2019). Source apportionment of heavy metals in farmland soil with application of APCS-MLR model: A pilot study for restoration of farmland in Shaoxing City Zhejiang, China. *Ecotoxicology and Environmental Safety, 184*, 109495. https://doi.org/10.1016/j.ecoenv.2019.109495

Jin, Z., & Lv, J. (2020). Integrated receptor models and multivariate geostatistical simulation for source apportionment of potentially toxic elements in soils. *CATENA, 194*, 104638. https://doi.org/10.1016/j.catena.2020.104638

Jose, J., & Srimuruganandam, B. (2021). Source apportionment of urban road dust using four multivariate receptor models. *Environmental Earth Sciences, 80*(19), 1–16. https://doi.org/10.1007/s12665-021-09960-5

Kosiorek, M., & Wyszkowski, M. (2020). Remediation of cobalt-contaminated soil using manure, clay, charcoal, zeolite, calcium oxide, main crop (Hordeum vulgare L.), and after-crop (Synapis alba L.). *Minerals, 10*(5), 429. https://doi.org/10.3390/min10050429

Lang, Y. H., Li, G. L., Wang, X. M., & Peng, P. (2015). Combination of Unmix and PMF receptor model to apportion the potential sources and contributions of PAHs in wetland soils from Jiaozhou Bay, China. *Marine Pollution Bulletin, 90*(1–2), 129–134. https://doi.org/10.1016/j.marpolbul.2014.11.009

Lang, Y. H., & Yang, W. (2014). Source apportionment of PAHs using Unmix model for Yantai costal surface sediments, China. *Bulletin of Environmental Contamination and Toxicology, 92*(1), 30–35. https://doi.org/10.1007/s00128-013-1164-7

Lee, D., Balachandran, S., Pachon, J., Shankaran, R., Lee, S., Mulholland, J. A., & Russell, A. G. (2009). Ensemble-trained PM2. 5 source apportionment approach for health studies. *Environmental Science & Technology, 43*(18), 7023–7031. https://doi.org/10.1021/es9004703

Lee, D. D., & Seung, H. S. (1999). Learning the parts of objects by non-negative matrix factorization. *Nature, 401*(6755), 788–791. https://doi.org/10.1038/44565

Li, F., Li, X., Hou, L., & Shao, A. (2018). Impact of the coal mining on the spatial distribution of potentially toxic metals in farmland tillage soil. *Scientific Reports, 8*(1), 1–10. https://doi.org/10.1038/s41598-018-33132-4

Li, F., Xiang, M., Yu, S., Xia, F., Li, Y., & Shi, Z. (2022). Source identification and apportionment of potential toxic elements in soils in an Eastern Industrial City, China. *International Journal of Environmental Research and Public Health, 19*(10), 6132. https://doi.org/10.3390/ijerph19106132

Li, P., Wu, T., Jiang, G., Pu, L., Li, Y., Zhang, J., Zhang, J., Xu, F., & Xie, X. (2021a). An integrated approach for source apportionment and health risk assessment of heavy metals in subtropical agricultural soils, Eastern China. *Land, 10*(10), 1016. https://doi.org/10.3390/land10101016

Li, Y., Kuang, H., Hu, C., & Ge, G. (2021b). Source apportionment of heavy metal pollution in agricultural soils around the Poyang Lake region using UNMIX model. *Sustainability, 13*(9), 5272. https://doi.org/10.3390/su13095272

Liao, S., Jin, G., Khan, M. A., Zhu, Y., Duan, L., Luo, W., Jia, J., Zhong, B., Ma, J., Ye, Z., & Liu, D. (2021). The quantitative source apportionment of heavy metals in peri-urban agricultural soils with UNMIX and input fluxes analysis. *Environmental Technology & Innovation, 21*, 101232. https://doi.org/10.1016/j.eti.2020.101232

Lu, A., Li, B., Li, J., Chen, W., & Xu, L. (2018). Heavy metals in paddy soil-rice systems of industrial and township areas from subtropical China: Levels, transfer and health risks. *Journal of Geochemical Exploration, 194*, 210–217. https://doi.org/10.1016/j.gexplo.2018.08.003

Lv, J. (2019). Multivariate receptor models and robust geostatistics to estimate source apportionment of heavy metals in soils. *Environmental Pollution, 244*, 72–83. https://doi.org/10.1016/j.envpol.2018.09.147

Manz, M., & Castro, L. J. (1997). The environmental hazard caused by smelter slags from the Sta. Maria de la Paz mining district in Mexico. *Environmental Pollution, 98*(1), 7–13. https://doi.org/10.1016/S0269-7491(97)00107-3

Martínez-Villegas, N., Briones-Gallardo, R., Ramos-Leal, J. A., Avalos-Borja, M., Castañón-Sandoval, A. D., Razo-Flores, E., & Villalobos, M. (2013). Arsenic mobility controlled by solid calcium arsenates: A case study in Mexico showcasing a potentially widespread environmental problem. *Environmental Pollution, 176*, 114–122. https://doi.org/10.1016/j.envpol.2012.12.025

Martínez-Villegas, N., Hernández, A., Meza-Figueroa, D., & Sen Gupta, B. (2018). Distribution of arsenic and risk assessment of activities on soccer pitches irrigated with arsenic-contaminated water. *International Journal of Environmental Research and Public Health, 15*(6), 1060. https://doi.org/10.3390/ijerph15061060

Massoud, R., Hadiani, M. R., Hamzehlou, P., & Khosravi-Darani, K. (2019). Bioremediation of heavy metals in food industry: Application of Saccharomyces cerevisiae. *Electronic Journal of Biotechnology, 37*, 56–60. https://doi.org/10.1016/j.ejbt.2018.11.003

Mendoza-Chávez, Y. J., Uc-Castillo, J. L., Cervantes-Martínez, A., Gutiérrez-Aguirre, M. A., Castillo-Michel, H., Loredo-Portales, R., SenGupta, B., & Martínez-Villegas, N. (2021). Paracyclops chiltoni inhabiting water highly contaminated with arsenic: Water chemistry, population structure, and arsenic distribution within the organism. *Environmental Pollution, 284*, 117155. https://doi.org/10.1016/j.envpol.2021.117155

Meng, L., Zuo, R., Wang, J. S., Yang, J., Teng, Y. G., Shi, R. T., & Zhai, Y. Z. (2018). Apportionment and evolution of pollution sources in a typical riverside groundwater resource area using PCA-APCS-MLR model. *Journal of Contaminant Hydrology, 218*, 70–83. https://doi.org/10.1016/j.jconhyd.2018.10.005

Ming-Kai, Q. U., Wei-Dong, L. I., ZHANG, C. R., Shan-Qin, W. A. N. G., Yong, Y. A. N. G., & Li-Yuan, H. E. (2013). Source apportionment of heavy metals in soils using multivariate statistics and geostatistics. *Pedosphere, 23*(4), 437–444. https://doi.org/10.1016/S1002-0160(13)60036-3

Muller, G. M. M. G. M. G. M. G. P. (1969). Index of geo accumulation in sediments of the Rhine River. *GeoJournal, 2*, 108–118.

Ogundele, D. T., Adio, A. A., & Oludele, O. E. (2015). Heavy metal concentrations in plants and soil along heavy traffic roads in North Central Nigeria. *Journal of Environmental & Analytical Toxicology, 5*(6), 1. https://doi.org/10.4172/2161-0525.1000334

Paatero, P. (1997). Least squares formulation of robust non-negative factor analysis. *Chemometrics and Intelligent Laboratory Systems, 37*(1), 23–35. https://doi.org/10.1016/S0169-7439(96)00044-5

Paatero, P., & Tapper, U. (1994). Positive matrix factorization: A non-negative factor model with optimal utilization of error estimates of data values. *Environmetrics, 5*(2), 111–126. https://doi.org/10.1002/env.3170050203

Puchelt, H. (1992). Environmental inorganic geochemistry of the continental crust. In The Natural environment and the biogeochemical cycles (pp. 27–63). Springer. https://doi.org/10.1007/978-3-540-46995-7_2

Radomirović, M., Ćirović, Ž, Maksin, D., Bakić, T., Lukić, J., Stanković, S., & Onjia, A. (2020). Ecological risk assessment of heavy metals in the soil at a former painting industry facility. *Frontiers in Environmental Science, 8*, 560415. https://doi.org/10.3389/fenvs.2020.560415

Ran, H., Guo, Z., Yi, L., Xiao, X., Zhang, L., Hu, Z., Li, C., & Zhang, Y. (2021). Pollution characteristics and source identification of soil metal (loid) s at an abandoned arsenic-containing mine, China. *Journal of Hazardous Materials, 413*, 125382. https://doi.org/10.1016/j.jhazmat.2021.125382

Razo, I., Carrizales, L., Castro, J., Díaz-Barriga, F., & Monroy, M. (2004). Arsenic and heavy metal pollution of soil, water and sediments in a semi-arid climate mining area in Mexico. *Water, Air, and Soil Pollution, 152*(1), 129–152. https://doi.org/10.1023/B:WATE.0000015350.14520.c1

Ruíz-Huerta, E. A., De la Garza Varela, A., Gómez-Bernal, J. M., Castillo, F., Avalos-Borja, M., SenGupta, B., & Martínez-Villegas, N. (2017). Arsenic contamination in irrigation water, agricultural soil and maize crop from an abandoned smelter site in Matehuala, Mexico. *Journal of Hazardous Materials, 339*, 330–339. https://doi.org/10.1016/j.jhazmat.2017.06.041

Saha, A., Gupta, B. S., Patidar, S., & Martínez-Villegas, N. (2022a). Spatial distribution and source identification of metal contaminants in the surface soil of Matehuala, Mexico based on positive matrix factorization model and GIS techniques. *Frontiers in Soil Science, 2*, 1041377. https://doi.org/10.3389/fsoil.2022.1041377

Saha, A., Gupta, B. S., Patidar, S., & Martínez-Villegas, N. (2022b). Spatial distribution based on optimal interpolation techniques and assessment of contamination risk for toxic metals in the surface soil. *Journal of South American Earth Sciences, 115*, 103763. https://doi.org/10.1016/j.jsames.2022.103763

Saha, A., Gupta, B. S., Patidar, S., & Martínez-Villegas, N. (2022c). Evaluation of potential ecological risk index of toxic metals contamination in the soils. *Chemistry Proceedings, 2022*(10), 59. https://doi.org/10.3390/IOCAG2022-12214

Saha, A., Gupta, B. S., Patidar, S., & Martínez-Villegas, N. (2022d) Estimating source apportionment of heavy metals contamination in surface soil based on positive matrix factorization (PMF) model around Cerrito Blanco in San Luis Potosi, Mexico. In *Proceedings of the 4th international electronic conference on geosciences, 1–15 December 2022*. MDPI, Basel, Switzerland. https://doi.org/10.3390/IECG2022-13746

Secretaria de Economía. (2006). NMX-AA-132-SCFI-2001—Muestreo de Suelos Para la Identificación y la Cuantificación de Metales y Metaloides, y Manejo de la Muestra (p. 32). Secretariade Economía, México City, México.

Secretaría de Medio Ambiente y Recursos Naturales. (2007). Que establececriterios para determinar las concentraciones deremediación de sueloscontaminadosporarsénico, bario, berilio, cadmio, cromohexavalente, mercurio, níquel, plata, plomo, selenio, talio y/o vanadio (pp. 35–96). https://www.gob.mx/cms/uploads/attachment/file/135331/48.-_NORMA_OFICIAL_MEXICANA_NOM-147-SEMARNAT-SSA1-2004.pdf

Shi, G. L., Liu, G. R., Peng, X., Wang, Y. N., Tian, Y. Z., Wang, W., & Feng, Y. C. (2014). A comparison of multiple combined models for source apportionment, including the PCA/MLR-CMB, Unmix-CMB and PMF-CMB models. *Aerosol and Air Quality Research, 14*(7), 2040–2050. https://doi.org/10.4209/aaqr.2014.01.0024

Shi, G. L., Tian, Y. Z., Guo, C. S., Feng, Y. C., Xu, J., & Zhang, Y. (2012). Sediment-pore water partition of PAH source contributions to the Yellow River using two receptor models. *Journal of Soils and Sediments, 12*(7), 1154–1163. https://doi.org/10.1007/s11368-012-0540-y

Shi, W., Li, T., Feng, Y., Su, H., & Yang, Q. (2022). Source apportionment and risk assessment for available occurrence forms of heavy metals in Dongdahe Wetland sediments, southwest of China. *Science of the Total Environment, 815*, 152837. https://doi.org/10.1016/j.scitotenv.2021.152837

Stoffers, P., Glasby, G. P., Wilson, C. J., Davis, K. R., & Walter, P. (1986). Heavy metal pollution in Wellington Harbour. *New Zealand Journal of Marine and Freshwater Research, 20*(3), 495–512. https://doi.org/10.1080/00288330.1986.9516169

Thurston, G. D., & Spengler, J. D. (1985). A quantitative assessment of source contributions to inhalable particulate matter pollution in metropolitan Boston. *Atmospheric Environment* (1967), *19*(1), 9–25. https://doi.org/10.1016/0004-6981(85)90132-5

United States Environmental Protection Agency (USEPA). (1994). Method 200.7: Revision 4.4, determination of metals and trace elements in water and wastes by inductively coupled plasma-atomic emission spectrometry. United States Environmental Protection Agency, Cincinnati, OH, USA.

United States Environmental Protection Agency. (2007). EPA Unmix 6.0 fundamentals & user guide. US Environmental Protection Agency, Office of Research and Development, Washington, DC.

United States Environmental Protection Agency. (2014). EPA positive matrix factorization (PMF) 5.0 fundamentals and user guide, Washington, DC.

Vega, E., Mugica, V., Carmona, R., & Valencia, E. (2000). Hydrocarbon source apportionment in Mexico City using the chemical mass balance receptor model. *Atmospheric Environment, 34*(24), 4121–4129. https://doi.org/10.1016/S1352-2310(99)00496-3

Vercoutere, K., Fortunati, U., Muntau, H., Griepink, B., & Maier, E. A. (1995). The certified reference materials CRM 142 R light sandy soil, CRM 143 R sewage sludge amended soil and CRM 145 R sewage sludge for quality control in monitoring environmental and soil pollution. *Fresenius' Journal of Analytical Chemistry, 352*(1), 197–202. https://doi.org/10.1007/BF00322326

Wang, H., Li, X., Chen, Y., Li, Z., Hedding, D. W., Nel, W., Ji, J., & Chen, J. (2020a). Geochemical behavior and potential health risk of heavy metals in basalt-derived agricultural soil and crops: A case study from Xuyi County, eastern China. *Science of the Total Environment, 729*, 139058. https://doi.org/10.1016/j.scitotenv.2020.139058

Wang, Z., Shen, Q., Hua, P., Jiang, S., Li, R., Li, Y., Fan, G., Zhang, J., & Krebs, P. (2020b). Characterizing the anthropogenic-induced trace elements in an urban aquatic environment: A source apportionment and risk assessment with uncertainty consideration. *Journal of Environmental Management, 275*, 111288. https://doi.org/10.1016/j.jenvman.2020.111288

Wang, J., Yang, J., & Chen, T. (2022). Source appointment of potentially toxic elements (PTEs) at an abandoned realgar mine: Combination of multivariate statistical analysis and three common receptor models. *Chemosphere, 307*, 135923. https://doi.org/10.1016/j.chemosphere.2022.135923

Wang, Z., Chen, J., Tian, F., Yang, P., Qiao, X., & Yao, Z. (2010). Application of factor analysis with nonnegative constraints for source apportionment of soil polycyclic aromatic hydrocarbons (PAHs) in Liaoning, China. *Environmental Forensics, 11*(1–2), 161–167. https://doi.org/10.1080/15275920903558968

Wei, W., Ling, S., Wu, X., & Li, X. (2023). Geochemical accumulation and source tracing of heavy metals in arable soils from a black shale catchment, southwestern China. *Science of the Total Environment, 857*, 159467. https://doi.org/10.1016/j.scitotenv.2022.159467

Wu, Q., Leung, J. Y., Geng, X., Chen, S., Huang, X., Li, H., Huang, Z., Zhu, L., Chen, J., & Lu, Y. (2015). Heavy metal contamination of soil and water in the vicinity of an abandoned e-waste recycling site: Implications for dissemination of heavy metals. *Science of the Total Environment, 506*, 217–225. https://doi.org/10.1016/j.scitotenv.2014.10.121

Wu, J., Margenot, A. J., Wei, X., Fan, M., Zhang, H., Best, J. L., Wu, P., Chen, F., & Gao, C. (2020). Source apportionment of soil heavy metals in fluvial islands, Anhui section of the lower Yangtze River: Comparison of APCS–MLR and PMF. *Journal of Soils and Sediments, 20*(9), 3380–3393. https://doi.org/10.1007/s11368-020-02639-7

Xia, F., Qu, L., Wang, T., Luo, L., Chen, H., Dahlgren, R. A., Zhang, M., Mei, K., & Huang, H. (2018). Distribution and source analysis of heavy metal pollutants in sediments of a rapid developing urban river system. *Chemosphere, 207*, 218–228. https://doi.org/10.1016/j.chemosphere.2018.05.090

Xiang, M., Li, Y., Yang, J., Lei, K., Li, Y., Li, F., Zheng, D., Fang, X., & Cao, Y. (2021). Heavy metal contamination risk assessment and correlation analysis of heavy metal contents in soil and crops. *Environmental Pollution, 278*, 116911. https://doi.org/10.1016/j.envpol.2021.116911

Xu, Q., Wang, J., & Shi, W. (2022). Source apportionment and potential ecological risk assessment of heavy metals in soils on a large scale in China. *Environmental Geochemistry and Health*, 1–15. https://doi.org/10.1007/s10653-022-01266-0

Xue, J. L., Zhi, Y. Y., Yang, L. P., Shi, J. C., Zeng, L. Z., & Wu, L. S. (2014). Positive matrix factorization as source apportionment of soil lead and cadmium around a battery plant (Changxing County, China). *Environmental Science and Pollution Research, 21*(12), 7698–7707. https://doi.org/10.1007/s11356-014-2726-x

Yan, G., Mao, L., Liu, S., Mao, Y., Ye, H., Huang, T., Li, F., & Chen, L. (2018). Enrichment and sources of trace metals in roadside soils in Shanghai, China: A case study of two urban/rural roads. *Science of the Total Environment, 631*, 942–950. https://doi.org/10.1016/j.scitotenv.2018.02.340

Yáñez, L., García-Nieto, E., Rojas, E., Carrizales, L., Mejía, J., Calderón, J., Razo, I., & Díaz-Barriga, F. (2003). DNA damage in blood cells from children exposed to arsenic and lead in a mining area. *Environmental Research, 93*(3), 231–240. https://doi.org/10.1016/j.envres.2003.07.005

Yang, P., Yang, M., Mao, R., & Shao, H. (2014). Multivariate-statistical assessment of heavy metals for agricultural soils in northern China. *The Scientific World Journal, 2014*. https://doi.org/10.1155/2014/517020

Yang, J., Sun, Y., Wang, Z., Gong, J., Gao, J., Tang, S., Ma, S., & Duan, Z. (2022). Heavy metal pollution in agricultural soils of a typical volcanic area: Risk assessment and source appointment. *Chemosphere, 304*, 135340. https://doi.org/10.1016/j.chemosphere.2022.135340

Zang, H., Zhang, Y., Yao, J., & Ma, H. (2022). Source analysis of heavy metal pollution using UNMIX and PMF models in soils along the Shuimo River in Urumqi, China. *International Journal of Environmental Research and Public Health, 19*(22), 14794. https://doi.org/10.3390/ijerph192214794

Zhang, M., Wang, X., Liu, C., Lu, J., Qin, Y., Mo, Y., Xiao, P., & Liu, Y. (2020). Identification of the heavy metal pollution sources in the rhizosphere soil of farmland irrigated by the Yellow River using PMF analysis combined with multiple analysis methods—Using Zhongwei city, Ningxia, as an example. *Environmental Science and Pollution Research, 27*(14), 16203–16214. https://doi.org/10.1007/s11356-020-07986-z

Zhang, H., Cai, A., Wang, X., Wang, L., Wang, Q., Wu, X., & Ma, Y. (2021). Risk assessment and source apportionment of heavy metals in soils from Handan city. *Applied Sciences, 11*(20), 9615. https://doi.org/10.3390/app11209615

Zhang, T., Wang, M., Bai, G., Liu, J., Li, P., Zhang, Y., & Xia, S. (2022). Distribution characteristics, risk assessment, and source analysis of heavy metals in surface sediments and near-lakeshore soils of a plateau lake in China. *Gondwana Research, 115*, 191–200. https://doi.org/10.1016/j.gr.2022.10.005

Zhang, Y., Guo, C. S., Xu, J., Tian, Y. Z., Shi, G. L., & Feng, Y. C. (2012). Potential source contributions and risk assessment of PAHs in sediments from Taihu Lake, China: Comparison of three receptor models. *Water Research, 46*(9), 3065–3073. https://doi.org/10.1016/j.watres.2012.03.006

Zhao, R., Guan, Q., Luo, H., Lin, J., Yang, L., Wang, F., Pan, N., & Yang, Y. (2019). Fuzzy synthetic evaluation and health risk assessment quantification of heavy metals in Zhangye agricultural soil from the perspective of sources. *Science of the Total Environment*, *697*, 134126. https://doi.org/10.1016/j.scitotenv.2019.134126

Zhao, H., Lan, X., Yu, F., Li, Z., Yang, J., & Du, L. (2022). Comprehensive assessment of heavy metals in soil-crop system based on PMF and evolutionary game theory. *Science of the Total Environment, 849*, 157549. https://doi.org/10.1016/j.scitotenv.2022.157549

Zhao, W., Hopke, P. K., & Karl, T. (2004). Source identification of volatile organic compounds in Houston, Texas. *Environmental Science & Technology, 38*(5), 1338–1347. https://doi.org/10.1021/es034999c

Zhou, X., Wang, H., Ding, B., Peng, W., & Wang, R. (2019). Multi-objective evolutionary computation for topology coverage assessment problem. *Knowledge-Based Systems, 177*, 1–10. https://doi.org/10.1016/j.knosys.2019.03.033

Part II
Process Development for Arsenic Removal from Groundwater

Chapter 9
Polymer Nanofilm Composite Membranes for Ionic and Molecular Separation: History, Challenges and Future Perspectives

Pulak Sarkar and Santanu Karan

Abstract The membrane-based separation process is one of the energy-efficient technology to overcome the demand for a large quantity of fresh water and is also known for its use in water treatment, desalination, solvent recycling, and environmental remediation. Most conventional membranes used for separation-based applications are mainly polymer-derived thin-film composite (TFC) membranes, where the top selective layer of the composite membrane is made by using the interfacial polymerization (IP) process at the aqueous-organic interface. Different fabrication strategies are introduced for their applications in the aqueous feed to produce a high-specification selective layer of membranes with high liquid permeance that are selective to specific ions or molecules. It is particularly interesting to develop ultrathin polymer nanofilm as a selective layer of composite membranes, which provides ultrahigh liquid permeance as the liquid permeance is inversely proportional to their thickness. Additionally, to achieve the required high ions/molecular selectivity and make them stable in organic solvents, highly crosslinked polymer nanofilms (e.g., polyamide, polyester, polyimine, etc.) are developed. Here, we discussed the fabrication of conventional high-specification polymer membranes made from ultrathin polymer nanofilm and their ionic or molecular separation performance. The possible strategy to improve membrane fabrication techniques, applications in seawater desalination, and aqueous nanofiltration performance are discussed. A summary of the overview, challenges and future needs in the membrane-based separation process are discussed.

The original version of the chapter has been revised: The PRIS number have been corrected in the acknowledgement section. The correction to this chapter can be found at
https://doi.org/10.1007/978-981-97-4764-1_23

P. Sarkar · S. Karan (✉)
Membrane Science and Separation Technology Division, CSIR-Central Salt and Marine Chemicals Research Institute, G. B. Marg, Bhavnagar, Gujarat 364002, India
e-mail: santanuk@csmcri.res.in

Academy of Scientific and Innovative Research, (AcSIR), Ghaziabad, Uttar Pradesh 201002, India

B. Sen Gupta and N. Martínez-Villegas (eds.), *Arsenic Remediation of Food and Water*,
https://doi.org/10.1007/978-981-97-4764-1_9

Keywords Interfacial polymerization · Thin-film composite membrane · Nanofiltration · Nanofilm composite membrane · Polyamide · Reverse osmosis · Seawater desalination

Abbreviations

TFC	Thin-film composite
IP	Interfacial polymerization
NF	Nanofiltration
RO	Reverse osmosis
UF	Ultrafiltration
PIP	Piperazine
MPD	*m*-Phenylene diamine
PEI	Polyethyleneimine
TMC	Trimesoyl chloride
PAN	Polyacrylonitrile
PES	Polyethersulfone
EDA	Ethylenediamine
PAA	Poly(acrylic acid)

Introduction

The deficiency of freshwater is increasing day by day due to population growth, industrialization, and climate change all over the world. An energy-efficient purification process is needed to overcome this deficiency to cope with the massive freshwater demand. Over the past few decades, different water purification techniques have been developed to produce potable water, where evaporation and distillation techniques are widely used (Curcio, 2005). However, these techniques involve high capital investment and energy consumption. The emerging technology of membrane-based separation in desalination is the most attractive process to remove mineral components from saline water by membrane filtration due to the low cost of operation and low energy consumption. An alternative to the distillation process, the membrane-based separation process has established its enormous demand in water treatment. Separation membranes work at a low operational cost as it does not require phase change of the working media and can operate near room temperature under a transmembrane pressure difference. These membranes are widely applied for food (Vandezande, 2008), chemical (Marchetti, 2014), seawater desalination (Misdan, 2012), wastewater treatment (Kim, 2010), pharmaceutical (Nghiem, 2005), and biological (Kruschitz, 2020) industries. For cost-effective freshwater production using a membrane, there is a significant requirement to develop high-specification membranes (Epsztein,

2020). For that, different types of water purification techniques were developed to purify polluted water and make them drinkable. The widely known membranes are thin-film composite (TFC) based reverse osmosis (RO), seawater RO (SWRO), and nanofiltration (NF), used for specific separation applications. TFC membranes represent the benchmark for the RO, SWRO, and NF membranes and are mainly formed via IP on the top of the porous ultrafiltration (UF) substrate.

NF or RO membranes are mainly composed of a very thin polyamide film on the top of a porous substrate, produced through an IP reaction between amine and acyl chloride monomers at the aqueous-organic interface. Different fabrication approaches and new materials were introduced to increase the permeation through the membrane by maintaining similar separation performance. High-performance composite membranes were developed by manipulating the interfacial kinetics (Liang, 2020; Sarkar, 2021a), decreasing the thickness of the selective layer (Sarkar, 2021b), post-surface modification (Huang, 2020; Peng, 2020), introducing different sacrificial layers (Karan, 2015; Wang, 2018), and different nanostructures (Sarkar, 2020) within the selective layer. Recent developments in the membrane-based separation process have some limitations because of the intrinsic property of the materials used, including the permeance–selectivity trade-off, which restricts the liquid permeance and ions/molecular selectivity. Therefore, to design high-performance TFC membranes, separation properties beyond the permeance-selectivity (ions/molecular selectivity) trade-off upper bound of the existing polyamide-based membranes (Geise, 2011; Park, 2017a, 2017b) are of primary need. This will enable us to achieve high liquid permeance and selectivity between ions and molecules. In addition, high-specification membranes with stability in different harsh solvents will be advantageous for their applications in various complex separation processes.

This article discussed the advantage of high-specification polymer membranes used for ionic and molecular separation. The membrane's designs, fabrication methods, and separation performance are discussed. We addressed the fabrication strategies (e.g., thickness reduction of the separation layer, incorporation of different porous materials into the separation layer, etc.) to get high liquid permeance and high salts or dye rejection. In summary, we discussed their development on a large scale and future needs in the membrane-based separation process.

Overview of Membrane Technologies

Membrane-based separation technology is one of the most energy-efficient and extensively used techniques for water purification. RO membrane was first developed by *Loeb and Sourirajan* from cellulose triacetate in the 1960s (LOEB, 1963) and commercialized in the early 1970s. Later, an aromatic polyamide TFC membrane made via IP was introduced (Francis, 1966; Lt, 1970). A detailed description of the IP is provided in Sect. Membrane Fabrication Techniques. They comprise three layers: non-woven polyester (or polypropylene) support at the bottom surface, an ultrafiltration porous interlayer, and a top thin polyamide separation layer. Cadotte developed the first TFC membranes in the 1970s; since the 1980s, they have been widely utilized. The dense polyamide separation layer selectively allows water to pass through molecular voids (free-volume) while rejecting all monovalent and divalent salts as well as neutral organic solutes (molecular weight > 100 Da). These membranes are widely used as spiral-wound modules, producing drinking water from brackish and seawater in a single-pass or two-pass configuration. Compared to the energy-intensive traditional distillation process, the membrane needs only 25% extra energy than the fundamental limits to purify saltwater to pure water under applied pressure (Sholl, 2016). While the TFC RO membranes perform well in a wide range of water matrices and produce potable water in bulk (in a few million liters per day), they suffer from corrosion, biofilm formation, scaling, biofouling, and aging (Al-Juboori, 2012; Pandey, 2012; Zhao, 2015a, 2015b). Several operational precautions are followed to overcome these limitations, such as pretreatment of feedwater, periodic cleaning of membrane modules, use of antiscalants, and backwashing (Kavitha, 2019; Wolf, 2005; Zhao, 2015a, 2015b). However, during prolonged use, the membrane modules are replaced at regular intervals upon continuous aging (permeance loss over time). Therefore, developing a new class of membranes that are more productive and have antifouling properties would decrease the operational and capital costs in many separation processes.

Along with developing RO membranes, *Loeb and Sourirajan* introduced ultrafiltration (UF) membranes (Baker, 2012), which showed separation performance between RO and microfiltration (MF) membranes in terms of their solute rejection (Ahmed, 2017). UF membranes were developed via phase inversion of polymer solution in a non-solvent, which produces integrally skinned asymmetric-type structures with relatively dense and thick (a few microns) layers on the top (separation layer). They often show a higher molecular weight cut-off (MWCO) ranging from 10,000–300,000 Da. Thus, separating small molecules with a molecular weight of 200–10,000 Da with improved liquid permeance was not efficiently achieved. A new porous membrane was sought for this MWCO range, and subsequently, in 1984, the 'nanofiltration' membranes were introduced and named by FilmTec, USA. They were fabricated from semi-aromatic polyamide and used for the selective removal of ions or molecules with desired MWCO range. The separation property was primarily determined by the molecular (solutes) sizes while also partly dependent on the surface charge.

The separation layer of TFC NF and RO membranes is a thin amorphous polyamide layer of ~50–200 nm thickness. It is produced via IP on the top of porous ultrafiltration support (substrate), where the polymerization process is dominated by the mass transfer of the monomers at the interface. Typically, a multifunctional amine aqueous solution is allowed to react with an immiscible organic solution of multifunctional acyl chloride at the aqueous-organic interface. A thin polyamide film (the separation layer) is thus formed on the top of the porous support. Besides their application in water treatment, they are also used for molecular separation in organic solvents (Marchetti, 2014). This demands special attention on the stability of the composite structure and performance deterioration in organic solvents as they suffer from structural deformation.

Strategies to Enhance the Composite Membrane's Separation Performance

Many fabrication strategies are reported in the literature to enhance the membrane separation performance in respect of permeance, and ions/molecular selectivity (Fig. 9.1) (Lu, 2021). These new strategies aim to provide high water permeance, but most of the process suffers from poor ions/molecular selectivity. Here, we have classified these emergent fabrication strategies into five categories and discussed limitations for developing high-performance TFC membranes.

Reduction of the Selected Layer's Thickness in the Composite Membrane

Recently, the 'thickness reduction' method of the top selective layer has been growing as an emerging technology to achieve significantly higher liquid permeance. As for a porous membrane, water permeability is defined by the following Hagen–Poiseuille equation (Werber, 2016a);

$$A = \frac{\varepsilon r_p^2}{8\mu\delta_m}$$

where μ is the solvent viscosity of the feed solution, r_p is the pore radius, ε is the porosity of the selective layer, and δ_m is the active layer thickness.

Hence, it is essential to reduce the selective layer thickness and make it as thin as possible to obtain a high permeance membrane. The tortuosity of a thin film is expected to be lower, considering the marginal effect on the porosity. Extensive research has been done to make high-performance membranes by reducing the active layer thickness without affecting their separation properties. This would

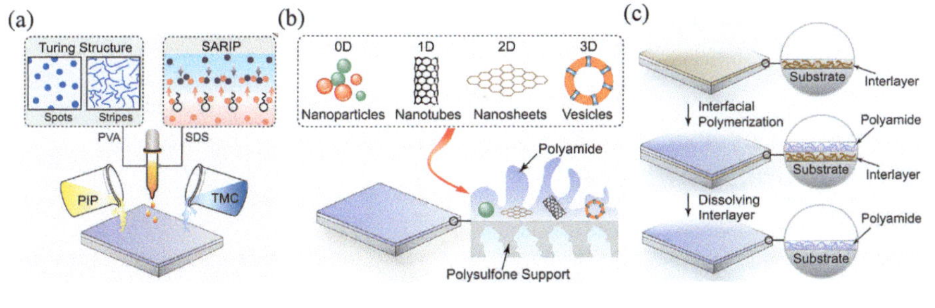

Fig. 9.1 **a–c** Schematic presentation of different strategies to fabricate high-performance TFC membranes. Reproduced with permission (Lu, 2021). Copyright 2021, the Royal Society of Chemistry

positively impact separation processes with reduced device footprint for their industrial applications. Many fabrication protocols were introduced, such as designing a selective layer with thickness down to ~10 nm (Gao, 2019; Jiang, 2018; Karan, 2015), 2D lamellar nano membranes (Yang, 2019a, 2019b; Zhang, 2019a, 2019b), electrostatic-modulated IP-based membranes (You, 2021), and atomic thin nanoporous membranes (Prozorovska, 2018; Yang, 2019a, 2019b). We recently reported a post-solvent washing method to decrease the polyamide selective layer thickness below ~ 5 nm to enhance the water permeance of the membranes (Sarkar, 2021b).

Enhancing Surface Area of the Selective Layer of the Composite Membrane

Developing the separation layer of a composite membrane with an increased surface area is a direct approach to achieving high separation performance. This strategy has been a topic of attention for over a decade. Enormous efforts were made to design wrinkled or crumpled structures of polyamide separation layer in RO (Karan, 2015) and NF (Tan, 2018) membranes to increase their separation performance. It is believed that wrinkled or crumpled type morphology can increase effective membrane surface area, giving a much higher permeability than smooth membranes (Karan, 2015). Due to the diffusion-based nature of the IP process, a controlled diffusivity of the aqueous phase monomers into the organic phase can be used to increase the membrane's effective surface area. Several new fabrication strategies have been made to tune the diffusivity of monomers in the IP process and improve the membrane separation performance. Notably, nano-foaming (Ma, 2018b), electrospraying (Chowdhury, 2018), 3D honeycomb-like spatial polyamide structure (Gui, 2020), and macromolecule incorporation (Tan, 2018; Zhu, 2019) are several examples of them (Fig. 9.1) (Lu, 2021). These membranes formed with a crumpled morphology of the separation layer with increased effective surface area and thus higher permeable area, leading to increased water permeance. Despite the high efficacy of these methods, the complexity involved in the preparation steps severely limits the development of such high-permeance membranes.

Separation Layer of the Membrane With the Incorporation of Microporous Structures

Incorporating non-porous/porous nanostructures within the top selective layer is a novel approach to increasing liquid permeance for their use in reverse osmosis and nanofiltration. Pre-customized nanostructures were mixed with the amine monomers (e.g., *m*-phenylene diamine (MPD), piperazine (PIP), polyethyleneimine (PEI), etc.) in the aqueous solution or the organic phase (with TMC), creating a thin film nanocomposite structure of the top separation layer via the IP method. As an example, zeolite imidazole framework (ZIF-8) (Zhai, 2019), covalent organic framework (COF), TiO_2 nanoparticles (Wang, 2017a, 2017b, 2017c), CeO_2 nanoparticles (Wang, 2019), carbon dots (Li, 2017), graphene oxide (Ali, 2016; Lai, 2016), silica nanoparticles (Shen, 2018), and multi-walled carbon nanotubes (MWCNTs) (Zhao, 2015a, 2015b), have been widely used to increase the separation performance (Fig. 9.1) (Lu, 2021). The polyamide selective layer's membrane surface properties are enhanced by incorporating nanostructures and creating extra molecular channels at the interface of the nanostructures and the polyamide matrix (Lu, 2021). Thus, incorporating nanostructures with clearly defined free volumes within the polyamide layer can significantly enhance the overall water permeance by maintaining the separation performance of the membranes (Lu, 2021).

Introducing Interlayers Beneath the Separation Layer

The introduction of an interlayer on top of UF porous substrates is another method for enhancing the separation performance of TFC membranes. These interlayer structures deliver uniformity of substrate pores with enhanced water-wetting properties compared to the conventional UF porous substrates. This interlayer structure enables a more uniform monomer distribution with more effective monomer diffusion during the IP process. As a result, a defect-free and highly crosslinked polyamide film is formed with significantly high liquid permeance and high ion/molecular selectivity. Examples of new materials to fabricate interlayer structures on the top of the porous substrate are sacrificial interlayer of $Cd(OH)_2$ nanowires (Karan, 2012, 2015), nanoscaffold (Yang, 2018), cellulose nanocrystals (Wang, 2017a, 2017b, 2017c), carbon nanotube (Wang, 2018; Zhou, 2018; Zhu, 2016), etc. (Fig. 9.1) (Lu, 2021). These interlayer-based TFC membranes surpass the limitations of conventional membranes and thus provide a new approach to developing high-performance TFC membranes.

Post-functionalization and Post-treatment Process

Surface modification of the TFC membranes is an ingenious strategy to increase membrane separation performance. The surface of the freshly prepared polyamide membranes via the IP process contains unreacted acyl chloride groups, which can be further modified using different functional groups. For example, microwave-assisted surface grafting (Huang, 2020; Peng, 2020), such as betaine, zwitterionic polymer (Marré Tirado, 2016), and phosphonium (Peng, 2020) modification, were used to improve the membrane separation performance. Zwitterion-modified membranes efficiently improve the water permeance and antifouling property by forming a strongly hydrated layer (Ji, 2017). Additionally, to increase the performance of the membrane, researchers have also developed various new fabrication strategies, which include additives (Liang, 2020; Sarkar, 2021a), supporting layer modification (Landaburu-Aguirre, 2016), and post-treatment with different solvents (Guo, 2020), etc. Among them, post-treatment with organic solvents is an easy and efficient approach to improve the membrane separation performance.

Membrane Fabrication Techniques

Interfacial Polymerization (IP) Process

Conventional Interfacial Polymerization (IP) Method to Fabricate Composite Membranes

The conventional IP process is a widely used technique to fabricate TFC membranes with a thin crosslinked polymer thin film on the top of a UF support from the reaction of two monomers taken separately in an aqueous and organic phase (Sarkar, 2021a, 2021b) (Fig. 9.2). Composite polyamide membranes are made between a diamine monomer (e.g., MPD for RO and PIP for NF) and an acyl chloride monomer (TMC). Recently, Tian et al. reported a direct synthesis method of ultrathin polyamide film using the *in-situ* IP (ISIP) method on the commercial porous substrate (Tian, 2021b). Ultrathin polyamide film with an 8.5 nm thickness was formed with a very low amount of PIP (0.05 wt%) and a very low amount of TMC (0.1 wt%). The prepared membranes exhibit excellent pure water permeance of 46.6 L m^{-2} h^{-1} bar^{-1} and Na_2SO_4 rejection (up to 98.1%) (Tian, 2021b). Sarkar et al. have reported polyamide nanofilms and their composite membranes using the conventional IP method on top of HPAN support (Sarkar, 2021a, 2021b). The prepared membranes showed excellent pure water permeance of up to 68.0 L m^{-2} h^{-1} bar^{-1} and Na_2SO_4 rejection (up to 99.99%) (Sarkar, 2021a; 2021b).

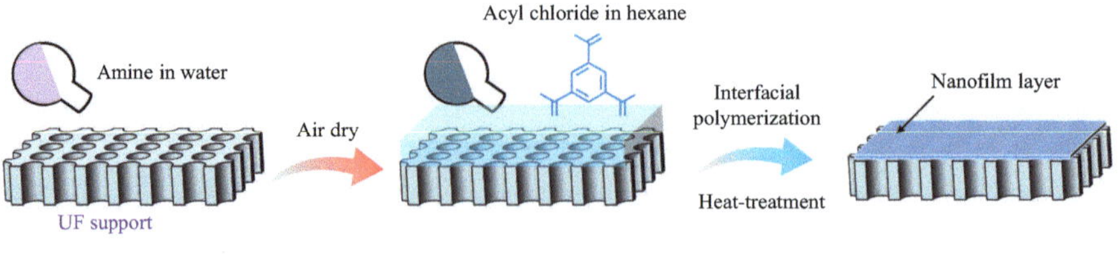

Fig. 9.2 Schematic presentation of the membrane fabrication process via conventional interfacial polymerization (IP)

Support-Free Interfacial Polymerization to Fabricate Composite Membranes

The morphology of the porous substrate during IP has a substantial impact on the creation of the polyamide layer. The structural integrity of the composite membranes is also dependent on the interfacial adherence between the UF support and the polyamide layer. Support-free interfacial polymerization (SFIP) is a process where the reaction occurs at the aqueous-organic interface without any external transfer process (Fig. 9.3) (Jiang, 2018). This SFIP method overcomes the limits of the traditional IP method. It allows a simple yet controllable fabrication of an ultrathin polyamide film (nanofilm) and its composite structure for NF or RO applications. For example, in 2017, Park et al. used the SFIP technique to fabricate polyamide RO membranes (Park, 2017a, 2017b). The prepared membranes via this method show excellent perm-selectivity, which is superior to conventional IP-based membranes (Park, 2017a, 2017b).

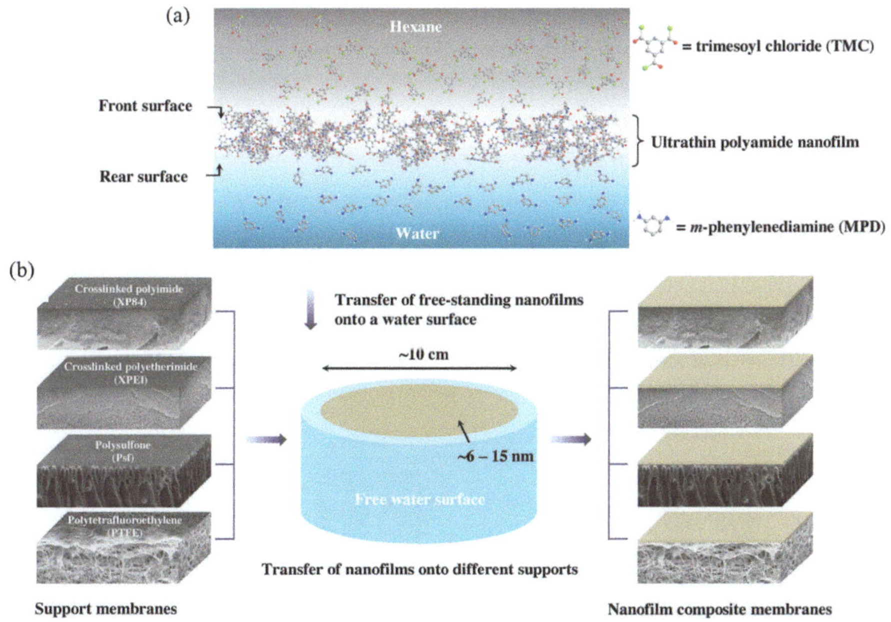

Fig. 9.3 **a**, **b** Schematic presentation of the polyamide nanofilms via support-free interfacial polymerization (SFIP) process and their composite membranes. Reproduced with permission (Jiang, 2018). Copyright 2018, the WILEY–VCH

Membrane Fabrication via Diffusion-Controlled Interfacial Polymerization

The thickness of the separating layer can also be tuned using a diffusion-controlled IP process. However, creating a defect-free ultrathin polyamide separation layer is quite a big challenge. In 2017, Shan et al. developed an ultrathin polyamide nanofilm (thickness ~ 25 nm) using poly(ethyleneimine) (PEI) and TMC via a diffusion-controlled IP process. The prepared membrane exhibits high water permeance (up to 1246 kg m^{-2} h^{-1} MPa^{-1}), which is 23 times greater than the reported conventional IP-based membranes. (Shan, 2017). In the literature, different additives (e.g., surfactant) were used to regulate the diffusivity of the monomers (Liang, 2020; Sarkar, 2021a). We recently reported the fabrication of diffusion-controlled NF membranes using a surfactant (e.g., sodium lauryl sulfate or SLS) in the aqueous solution to achieve a stoichiometric equilibrium at the interface (Fig. 9.4) (Sarkar, 2021a). Membranes prepared with this method show very high Na_2SO_4 rejection of up to 99.99% with a high monovalent to divalent ion selectivity value of greater than 4000, which is 30–275 times higher than the literature-reported TFC NF membranes (Sarkar, 2021a).

Role of the Sacrificial Layer Underneath the Separation Layer on the Membrane Separation Performance

Introducing an inter-layer as a sacrificial layer is crucial to fabricating high-performance polyamide membranes. They are deposited on the UF support before the fabrication of the separation layer. The continuous, smooth, and hydrophilic sacrificial layer generates a smaller and uniform pore size than the UF porous substrate. This sacrificial layer increases the storage of amine monomers and their diffusion

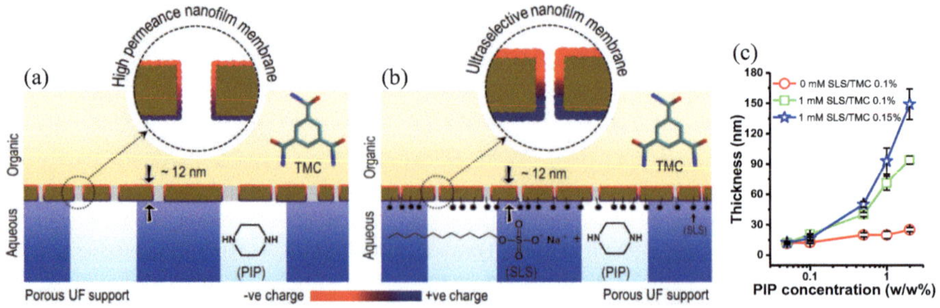

Fig. 9.4 **a**, **b** Schematic presentation of the fabrication of polyamide nanofilm and their composite membranes on top of UF support without (**a**) and with (**b**) diffusion-controlled IP process. **c** Thickness variation of the nanofilms prepared under different IP conditions. Reproduced with permission (Sarkar, 2021a). Copyright 2020, the WILEY–VCH

into the organic (or hexane) phase and keeps the interface flat to grow the ultrathin top polyamide film. For example, a cadmium hydroxide ($Cd(OH)_2$) nanostrand interlayer was vacuum filtered on the top of the UF substrates. The free standing ultrathin polyamide nanofilm was fabricated via controlled IP between MPD and TMC (Fig. 9.5) (Karan, 2015). $Cd(OH)_2$ nano strand interlayer was removed by hydrochloric acid or the acid formed during the IP reaction and produced defect-free freestanding nanofilm with a thickness of less than 10 nm. Controlling the interfacial reaction conditions, the morphology of the nanofilms was tailored to have a range of textures, from smooth to crumple. The composite membranes with a crumpled nanofilm prepared on alumina support showed high acetonitrile permeances of 112 L m^{-2} h^{-1} bar^{-1} with outstanding separation behaviour of organic solutes (Karan, 2015). The permeance value was two orders higher than the traditional membranes. Later, Wang et al. reported a polyamide NF membrane comprising a crumpled structure via interfacial polymerization on top of SWCNT/PES composite substrate loaded with zeolitic imidazolate framework (ZIF-8) nanoparticles (Wang, 2018). The nanoparticle's interlayer was then removed by water dissolution to create a rough polyamide surface with a wrinkled morphology. The prepared membrane shows very high water permeance (up to 53.5 L m^{-2} h^{-1} bar^{-1}) and Na_2SO_4 rejection (>95%), which is higher than the conventional NF membranes (Wang, 2018). Therefore, introducing a sacrificial interlayer on top of a porous substrate is a useful technique for making high-performance NF membranes.

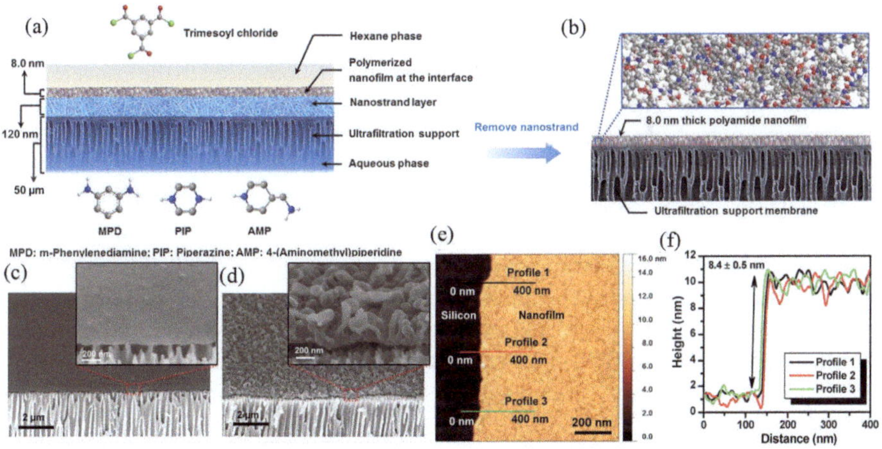

Fig. 9.5 Schematic presentation of the polyamide nanofilms **a, b** prepared on sacrificial $Cd(OH)_2$ nanostrand layer and removed the nanostrand layer by acid dissolution from the UF support membrane. **c, d** Cross-sectional view of smooth nanofilm and crumpled nanofilm on top of alumina support using SEM. **e, f** The thickness profile of a nanofilm using AFM. Reproduced with permission (Karan, 2015). Copyright 2015, the American Association for the Advancement of Science

Role of Wrinkled and Crumpled Structure on the Membrane Separation Performance

High-permeance TFC membranes can be developed by increasing the effective surface area of the top selective layer, which is now a growing interest in the membrane separation process (Tan, 2018). Tan et al. developed a novel method to produce Turing-type polyamide membranes via a diffusion-controlled IP process on top of porous polysulfone (PSf) support (Tan, 2018). They introduced a macromolecule (polyvinyl alcohol or PVA) in the aqueous phase containing an activator (PIP monomer) which interacts with the PIP monomers via hydrogen bonding and increases the solution viscosity and the diffusivity of the PIP monomer leading to a diffusion-driven instability at the interface and generates crumpled structures. Membranes showed high root-mean-square roughness up to 32.2 nm and increased surface area up to 35.5%. Membranes with increased surface roughness and surface area enabled outstanding separation performance in water permeance and selectivity (Tan, 2018).

Role of Interlayer on the Membrane Separation Performance

Introducing interlayers such as nanomaterials (nanowires or nanotubes) deposited on ultrafiltration (UF) support to prepare TFC membranes has attracted much attention from researchers (Wang, 2017a, 2017b, 2017c; Zhou, 2018; Zhu, 2016). These TFC membranes showed very good separation performance compared to the conventional fabrication process. The crosslinked top separation layer was made on the interlayer support, which shows uniform pore size distribution, as well as higher surface porosity compared to the UF, supports (Gao, 2019; Gong, 2019; Shi, 2013; Wang, 2017a, 2017b, 2017c; Zhang, 2019a, 2019b; Zhou, 2018; Zhu, 2016) (Fig. 9.6). In 2016, Zhu et al. synthesized ultrathin polyamide NF membrane on the top of a single-walled carbon nanotube coated with polydopamine (PD/SWCNTs) interlayer prepared on PES porous support (Zhu, 2016). Due to the increased porosity, smoother surface morphology, and polar nature of the PD/SWCNTs interlayer, an ultrathin polyamide thin film of 12 nm thickness was achieved. High water permeance of 32 $L\ m^{-2}\ h^{-1}\ bar^{-1}$ and moderate separation of divalent ions (~95.9%) was achieved (Zhu, 2016).

Recently, Zhang et al. introduced a unique approach for developing high-permeance NF membranes using covalent organic framework (COF) nanofiber as an interlayer during an IP reaction between PIP and TMC on PES substrates (Zhang, 2019a, 2019b). This hydrophilic COF nanofiber interlayer increases the diffusivity of PIP monomers into the hexane phase containing TMC, forming ~20 nm thick polyamide film. The membranes exhibit good water permeance (up to 31.1 $L\ m^{-2}\ h^{-1}\ bar^{-1}$) and moderate rejection of Na_2SO_4 (up to 95%) (Zhang, 2019a,

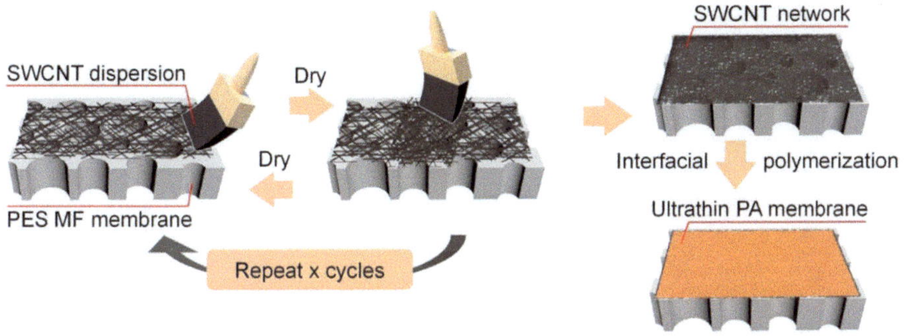

Fig. 9.6 Schematic presentation of the fabrication process of SWCNT-based composite membranes via IP method. Reproduced with permission (Gao, 2019). Copyright 2019, the American Chemical Society

2019b). This research shows the importance of developing and modifying support layers to produce high-performance NF membranes for efficient desalination.

Surface Modification of Composite Membranes via Post Functionalization

One of the promising ways to improve membrane performance is surface modification or post-functionalization. Tirado et al. developed surface-modified RO membranes to minimize membrane biofouling using zwitterionic polymer (Marré Tirado, 2016). Recently, Peng et al. reported phosphonium-modified composite membranes, which showed high water permeance and antibacterial properties (Peng, 2020).

They used different monomers (e.g., tetrakis(hydroxymethyl)phosphonium chloride (THPC), tetrakis (2-hydroxyethyl) ammonium bromide (THAB), 2-amino-*N*,*N*,*N*-triethylethane ammonium bromide (ATAB), ethylenediamine (EDA), and PIP) to modify the membrane surface. Membrane modified with THPC showed excellent water permeance of 50.5 L m^{-2} h^{-1} bar^{-1}, good rejection of Na_2SO_4 (greater than 98%), and good antibacterial properties (Peng, 2020). Huang et al. reported zwitterionic-modified NF membranes using betaine monomer (Huang, 2020) (Fig. 9.7). The prepared membrane showed high water permeance of 40.8 L m^{-2} h^{-1} bar^{-1}, which is 2.2 times higher than the conventional membranes, and the rejection of Na_2SO_4 was maintained at 97.0% (Huang, 2020).

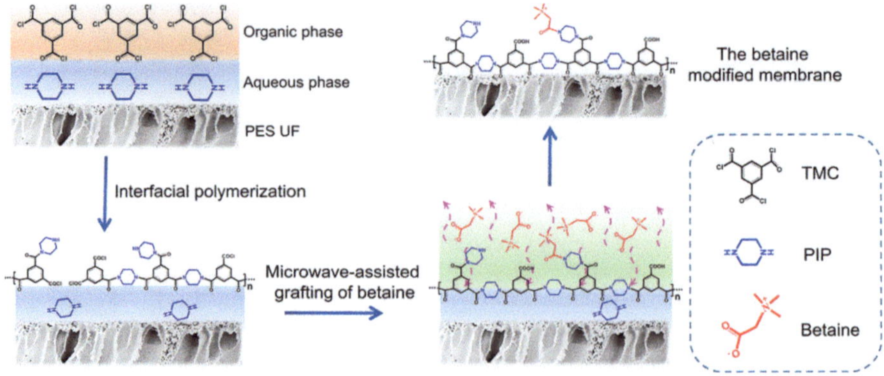

Fig. 9.7 Schematic presentation of the fabrication of post-modified polyamide membrane via interfacial polymerization and microwave-assisted grafting of betaine. Reproduced with permission (Huang, 2020). Copyright 2020, American Chemical Society

Fabrication of Composite Membranes Using Electrospray Process

Electrospray is widely used to form nanofibers, nanocoating, and nanoparticles (Jaworek, 2008). In this process, the fluid at the outlet of the nozzle is subjected to electric shear stress by keeping the nozzle at a high electrical potential. The advantage of this method is that the droplets can be made be very small, down to a few nanometers, and the size and charge of the droplets can be tuned by controlling electrical potential, that is, by regulating the flow rate and applied voltage. Several groups have employed this electrospray method to form a polyamide layer to fabricate TFC membranes (Chen, 2020; Chowdhury, 2018; Ma, 2018a; Yang, 2020). Recently, Ma et al. developed ultrathin polyamide RO membranes using electrospray-based IP (Fig. 9.8) (Ma, 2018a). The resulting membranes showed three times higher pure water permeance than those prepared with the conventional IP technique. The electrospray-assisted IP occurs at the interface of fine microdroplets of the MPD and TMC monomers. It is observed that the membrane formed in the electrospray-based IP process shows a smoother structure, whereas a ridge-and-valley structure formed in the conventional IP process (Ma, 2018a). The thickness of the polyamide layers showed a linear increment rate of 1 nm/min with electrospraying time and resulted in a controlled layer thickness from 10 to 100 nm range. Overall, this method provides practical uses of monomers without wasting the monomer solution, allowing greener production (Ma, 2018a). Chowdhury et al. reported a polyamide RO membrane with a controlled growth of ~5 nm per layer using a similar method (Chowdhury, 2018). The prepared membrane in this technique also showed a smoother surface structure with RMS roughness from 7 to 40 nm, which is lower than half of the RMS roughness observed in commercial SW30XLE membrane (RMS: >80 nm) (Chowdhury, 2018). The prepared membranes exhibited very good perm-selectivity compared

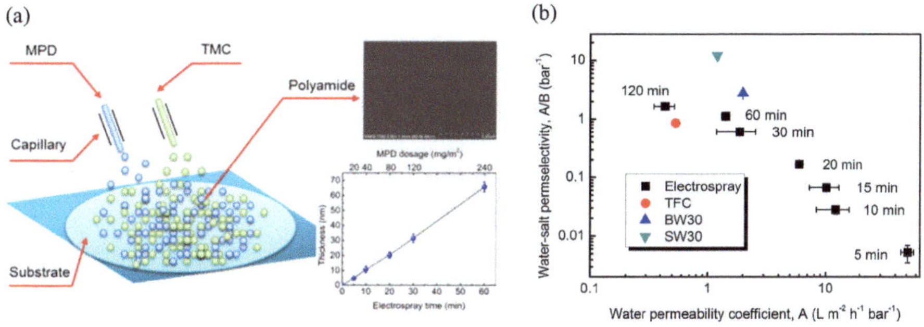

Fig. 9.8 **a** Schematic illustration of the fabrication of polyamide membrane via the electrospray method and its surface morphology, and thickness variation at different electrospray time. **b** Separation performance of the electrosprayed membranes in terms of water-salt permselectivity (A/B_{NaCl}) and water permeability (A), in comparison with the conventional polyamide-based TFC membranes. Reproduced with permission (Ma, 2018a). Copyright 2018, the American Chemical Society

to the commercial benchmarking membrane (Chowdhury, 2018). Recently, Chen et al. and co-workers developed polyamide membranes with a hierarchical structure using a dual-needle electrostatic atomization-assisted IP process (Chen, 2020). This technique showed very good control of membrane nanostructure; the amount of monomer controls the thickness and compressibility depending on the monomer ratio. These electrospray-based TFC membranes show good water permeance (up to 23.7 L m^{-2} h^{-1} bar^{-1}) and high retention of acid yellow dye (~99.2%) (Chen, 2020). The dual polyamide layer membrane showed good filtration performance compared to only one layer of loose or dense membranes. Notably, the double-layered membrane exhibited outstanding acid stability up to pH 4, even after 30 days of the experiment (Chen, 2020). Overall, the controllable advantage of the electrospray process gives a good reason for the design of new materials, opening a new method for fabricating the next-generation high-performance TFC membranes.

Fabrication of Composite Membranes via Chemical Vapor Deposition (CVD) Process

Diamond-like carbon (DLC) membranes are now a growing research interest in fabricating inorganic-based membranes. DLC membranes show incredibly high mechanical strength similar to diamond due to the highly crosslinked sp^3 carbon networks. These DLC membranes are fabricated via the CVD technique (Pan, 1990; Robertson, 2002). Additionally, these DLC membranes show high solvent stability in different organic solvents. For example, Karan et al., in 2012, reported ultrathin freestanding DLC membranes prepared from a parallel-plate plasma-enhanced CVD reactor (Fig. 9.9) (Karan, 2012).

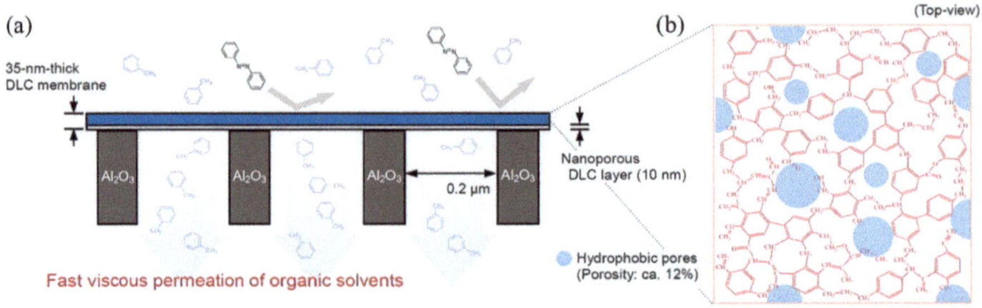

Fig. 9.9 **a**, **b** Schematic presentation of the fabrication of ultrathin DLC membrane via CVD process. Reproduced with permission (Karan, 2012). Copyright 2012, the American Association for the Advancement of Science

The membranes showed high mechanical strength with Young's moduli in a range of 90–170 GPa and were also stable in the organic solvents. The DLC membrane, with ~1 nm hydrophobic pores, permits the ultrahigh filtration performance in different organic solvents. It was seen that the membrane separation performance highly depends on the monomers used during CVD. The 35 nm-thick acetylene-based DLC membranes showed 67.3 L m^{-1} h^{-1} of ethanol flux and 94% rejection of azobenzene which is about three orders of magnitude higher than the current OSN membranes (Karan, 2012). This CVD-based approach provides a new way of developing high-permeance membranes for organic solvents.

Fabrication of Separation Membranes via Molecular Layer-by-Layer (mLbL) Method

Layer-by-layer (LbL) technique is the most extensively used coating technology in various fields, including catalysis, separation, optics, energy, and biomedicine (Richardson, 2015). Layer-by-layer (LbL) based membranes are now a growing interest for advancing high-specification TFC membranes. This Layer-by-layer (LbL) fabrication method was first developed in 1997 by Decher to fabricate polymer multi-composites (Decher, 1997). Most of the research has studied the LbL method by using adsorption. It is done by immersing the membrane within a specific solution and rinsing it in designated electrolyte solutions (El-Hashani, 2008; Hong, 2006). Several research groups developed TFC membranes using the molecular LbL-deposition (MLD) method (Choi, 2015; Gu, 2013, 2015; Qian, 2012; Shui, 2021) (Fig. 9.10). For the first time in 2012, Qian et al. developed an organic networks-based NF membrane on top of UF support with the MLD method (Qian, 2012). In this process, with precise control at the single molecular level, polyamide structure was formed via crosslinking between tetrakis (4-aminophenyl) methane diamine and tetraphenylmethane-4,4′,4″,4‴-tetraacyl chloride monomers with a 1.6 nm per MLD cycle. Ultrathin MLD NF membranes exhibit good water permeance (up to

2.7 L m^{-2} h^{-1} bar^{-1}), and good rejection of divalent salts (e.g., Na_2SO_4) was up to 96.2%. Gu et al. reported the molecular layer-by-layer (mLbL) method to fabricate RO membranes by using MPD and TMC on the top of porous support (Gu, 2013). They have successfully fabricated polyamide selective films with precise, controlled film thickness, negligible surface roughness, and selective chemical configurations. Membranes prepared with 15 cycles of mLbL assembly showed an excellent rejection of NaCl (up to 98%), and the water permeance was 75% higher than the conventional IP-based polyamide membrane. This mLbL process decreases the thickness of the top polyamide layer (~25 nm) and increases water permeance by reducing the hydraulic resistance and the length of the water's diffusive route. (Gu, 2013). Shui et al. recently reported TFC membranes with ultrathin polyamide top layers using the mLbL method for organic solvent nanofiltration (OSN) (Shui, 2021). They fabricated defect-free crosslinked polyamide layers through a layer-by-layer reaction between 1,3,5-tris (4-amino-phenyl) benzene with TMC. The prepared membrane showed excellent solvent permeance as well as high rejection of methyl orange (MW: 327 Da) (~93.2%), which is better than the performance of existing TFC membranes (Shui, 2021). Overall, this mLbL deposition method established broad applicability in developing high-performance TFC membranes with controllable top layer thickness, roughness, and separation performance.

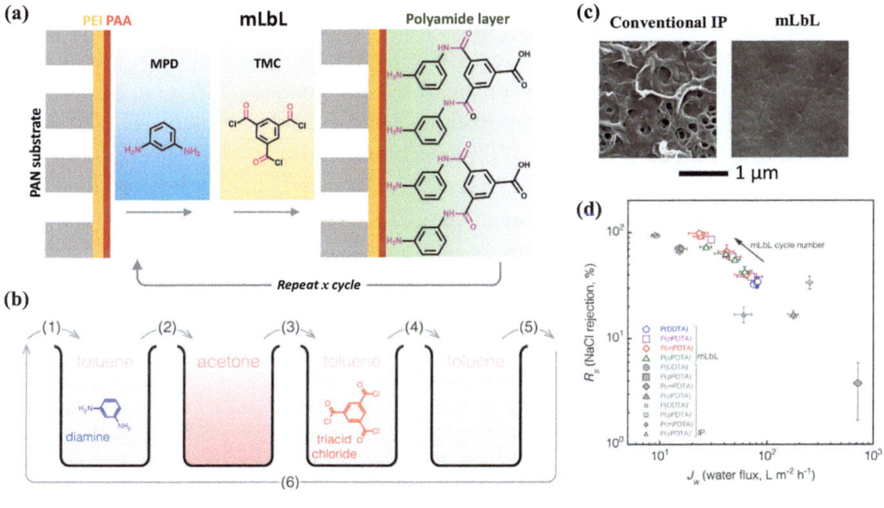

Fig. 9.10 **a** Schematic presentation of the fabrication of polyamide TFC membranes via molecular layer-by-layer (mLbL) method. **b** Schematic presentation of the fabrication of polyamide TFC membranes via the mLbL method by dipping in various monomer solutions. **c** The surface morphology of the crumple and smooth polyamide film fabricated via conventional IP and mLbL assembly method. **d** The separation performance in terms of NaCl rejection and water flux of the prepared TFC membranes. Reproduced with permission (Choi, 2015). Copyright 2015, American Chemical Society

Reverse Osmosis (RO) and Nanofiltration (NF) Applications

Reverse Osmosis (RO) Applications

Seawater Desalination

Seawater desalination is a rapidly growing interest to increase the potable water demand all over the world. Among all the traditional technologies (e.g., thermal desalination, distillation, etc.), the TFC RO membrane-based desalination process is more effective due to its cost and energy effectiveness. In this process, highly concentrated salt water is forced against a RO membrane in a cross-flow condition at a high operating pressure of ~60–70 bar. Desalination membranes can reject most of the dissolved salt (>99%) and allow water to pass through them in seawater RO (SWRO) as well as brackish water RO (BWRO) conditions. Many new strategies were adopted to enhance water permeance and the ion/molecular selectivity of the membrane. In addition, the control of the membrane surface morphology (Chowdhury, 2018) or the introduction of different nanofillers (Jeong, 2007) has been widely used to enhance the water permeance, leading to lower selectivity due to the creation of defects. Some biological aquaporin-based membranes exhibited high water permeance and salt rejection (Agre, 2004; Eriksson, 2013). Incorporating graphene (Mi, 2014), carbon nanotubes (Yang, 2019a, 2019b), or other nanostructures also produced increased water permeance with a compromised salt rejection during seawater desalination. Recently, Vincenzo et al. developed artificial water channels (AWCs) in membranes using biological proteins for their applications in seawater desalination (Di Vincenzo, 2021). The prepared biomimetic AWCs membranes showed high rejection of NaCl (up to 99.5%) with high water flux of 75 L m^{-2} h^{-1} under 65 bar applied pressure with 35 g L^{-1} NaCl feed solution (Di Vincenzo, 2021). Very recently, Lee et al. reported aquaporins (AQP)-based biomimetic membranes using AQPs and graphene oxide (GO) (Lee, 2021) producing high water permeance of up to 7.8 L m^{-2} h^{-1} bar^{-1} and high rejection of NaCl (up to 99.1%), which is high compared to the existing SWRO membranes (Lee, 2021).

Applications in Boron Removal

Boron is a critical component for both animals and plants and a valuable component for many industries. For both human consumption and agricultural use, low boron water is essential. For decades, membrane technologies for boron removal have drawn more attention (Tu, 2010). SWRO membranes are widely accepted for boron removal from saline or seawater. Most RO membranes could not achieve high boron removal efficiency (Tagliabue, 2014). Recent research has focused on enhancing the boron removal efficiency of RO processes using membrane design, optimization of operating parameters, and system design. As an example, commercial SWRO membranes exhibit high rejection (up to 90%), whereas commercial BWRO membranes show extremely poor rejection (65%) of neutral boron (Tagliabue, 2014; Tu, 2010). Under pH 8 and high-pressure conditions, Toray TM800K SWRO membranes exhibit excellent boron rejection of up to 96% (Tu, 2010). Li et al. reported fluorinated SWRO membranes fabricated via fluorine-containing monomers (Li, 2022). The resulting fluorinated membranes showed high salt rejection with good boron rejection of ~88.0%, which is good compared to the control and commercial SWRO membranes. Recently Wang et al. reported the performance of polyamide SWRO membranes via surface engineering using different functional groups (Wang, 2022). The resulting membrane displays a very good boron removal of up to 95.3% while maintaining water flux (Wang, 2022).

Nanofiltration (NF) Applications

Nanofiltration is a current advance in the membrane-based separation process, which lies between UF and RO-based membranes. The main exciting properties of the NF membranes are their selective separation of ions or molecules, such as monovalent and divalent ions. NF membranes are mainly used in ion separation (Marchetti, 2014), heavy metal removal (Qasem, 2021), wastewater treatment (Kim, 2010), the food industry (Vandezande, 2008), the textile industry (Lau, 2009), and the pharmaceutical industry (Nghiem, 2005). The details of nanofiltration applications are described in the following.

Selective Separation of Monovalent and Divalent Salts

Selective separation of ions using membrane technology is very important to produce high-quality water for its use in sensor technology, water reclaiming, recycling, and conservation, as well as environmental pollution control-based applications. There are different techniques, such as solvent extraction, chemical precipitation, and adsorption. Still, the membrane-based separation process is more effective due to low energy consumption. NF membranes are more useful for the selective separation of monovalent to divalent salts or ions from a single or mixed feed system. The design of membranes with sharp pore-size distribution and an increase in cross-linking degree using different fabrication methods enables a more selective separation of ions or molecules. Due to this sharp pore size distribution, the membrane can more precisely differentiate two ions or molecules, enabling higher solute–solute selectivity. Recently, we have reported a new fabrication method using surfactant in the aqueous solution to control the IP process. Our membranes show a selectivity of 4310 between NaCl to Na_2SO_4 salts, two orders of magnitude higher than the existing TFC NF membranes (Sarkar, 2021a).

Applications in Heavy Metal Removal

Contamination of heavy metals (e.g., cadmium, zinc, lead, nickel, copper, etc.) in water due to industrialization is a severe issue for our health because they are highly toxic and non-biodegradable. Therefore, removing heavy metals from wastewater has become essential for our health. To address this, there are many processes, such as ion exchange, precipitation using chemicals, adsorption, coagulation, etc. Among them, NF membranes are the most promising and feasible technique to remove heavy metals from contaminated water. Bera et al. developed PIP-based NF membranes to remove heavy metals from water (Bera, 2018). The prepared membranes showed greater than 99% rejection of lead ions, up to 97% rejection of cadmium ions, up to 96% rejection of cobalt ions, and up to 99% rejection of copper ions with water permeance 8 L m^{-2} h^{-1} bar^{-1}. Recently, Ibrahim et al. developed poly (homopiperazine–amide) based NF membranes by reacting homopiperazine and TMC. The prepared membranes showed high retention of lead and cadmium ions, up to 98.1% and 96.3%, respectively (Ibrahim, 2020).

Applications in Wastewater Treatment

Membrane-based separation is a widely used process for removing contaminants from wastewater. NF membranes have a pore size of around ~1 nm, so small uncharged solutes and monovalent ions get separated where monovalent ions are permeated due to the surface electrostatic properties. For these features, NF membranes are very advantageous for the selective removal of ions or salts from the single or mixed feed streams. NF membranes also remove sulfate from the seawater and concentrate NaCl selectively.

Arsenic Exposure and Their Remediation Using Membrane Technology

Due to the lack of surface water and rainfall, groundwater is frequently used for home, commercial, and industrial applications, particularly in desert regions. Groundwater quality has suffered due to global population growth, industrialization, increased agriculture productivity, and climate variability (Podgorski, 2020). The WHO limit for some chemical species is exceeded in groundwaters, including mercury (Hg), zinc (Zn), chromium (Cr), lead (Pb), cadmium (Cd), fluoride (F), and metal elements, including selenium (Se) and arsenic (As). The majority of them seriously harm people's health.

Due to arsenic's extreme toxicity and carcinogenicity, it is now a frequent pollutant found in both surface and groundwater and has developed into a severe environmental problem (Mohan, 2007). As arsenic is exceedingly harmful to both humans and animals, its presence in groundwater affects drinking water quality. The two most frequent oxidation forms of arsenic discovered in drinking water are oxyanions of trivalent arsenic (As[III]) and pentavalent arsenic (As[V]), even though that arsenic can exist in four different oxidation states (−3, 0, +3, +5) (Sharma, 2009).

Because of its toxicity to humans, arsenic (As) pollution has recently drawn significant attention. Arsenic pollutants can also impact the condition of the reservoirs in groundwater. Human activities that cause arsenic contamination have rendered it one of the most severe environmental hazards. The toxicity of arsenic and its treatment is currently the focus of many organizations, including industries, environmental organizations, and the general public. Around 200 million people worldwide are contaminated with arsenic at harmful levels, and significant resources were put into research and potential relief techniques.

Around the world, numerous technologies have been developed to reduce the amount of arsenic in groundwater to below the permissible level. As an example, adsorption, precipitation, ion exchange, and membrane technology e.g., reverse osmosis (RO) and nanofiltration (NF), etc. are widely used methods to arsenic from contaminated water (Ghosh, 2019; Regis, 2022). During recovery, the first three processes either produce significant amounts of sediment or release hazardous chemical reagents into the environment. In contrast with other remediation processes, the membrane-based technique doesn't generate any sediment or accumulate arsenic, making disposal challenging.

Human Health Effects of Arsenic

The upper skin layer of the earth naturally contains arsenic, which is widely dispersed in the air, water, and land. The inorganic form is extremely poisonous. By consuming contaminated food, drinking polluted water, and using polluted water to prepare food and water plants, people are exposed to larger concentrations of inorganic arsenic (Maharjan, 2005). Chronic arsenic poisoning can result from long-term exposure to inorganic arsenic, primarily from food and beverages. The main risk to human health from arsenic contamination is from contaminated groundwater (Podgorski, 2020). As, inorganic arsenic is naturally present at a more significant percentage in the groundwater in most countries like India, Bangladesh, China, Pakistan, the United States of America, Argentina, etc. The main source of exposure to arsenic is from drinking contaminated water, and food prepared with polluted water. Another way of arsenic exposure is by consuming food with dietary sources of arsenic (e.g., fish, meat, etc.) (Maharjan, 2005).

For example, humans' arsenic exposure can result in immediate symptoms with significant health issues such as vomiting, diarrhea, cardiovascular illness, and harm to the nervous system through ingestion, inhalation, or skin absorption (Maharjan, 2005). In severe conditions, they suffered muscle pain, numbness and tingling in the legs, and mortality. Skin blemishes and skin cancer are among the most noticeable effects of long-term exposure to high concentrations of inorganic arsenic (for example, by drinking water and food). In addition, to skin problems, the long-term consumption of arsenic can also cause cancers of the bladder and lungs (Maharjan,

Table 9.1 The global standard for arsenic content in drinking water (Sreeram, 2017)

Country	Drinking water standard (μg/L)
World Health Organization	10
Australia	7
China	50
European Union	10
India	50
Taiwan	10
US	10

2005). Regardless of the source of exposure, arsenic and its derivatives are incredibly carcinogenic to humans, according to the International Agency for Research on Cancer (IARC). The WHO recommended 10–50 μg/L as the maximum permissible limit for arsenic consumption in drinking water (Smith, 2002). Table 9.1 presents the maximum contamination level of arsenic in different countries of the world.

Arsenic has been associated with a variety of adverse health effects due to long-term exposure, such as abnormalities in several body organ systems, including the integumentary, neurological, respiration, cardiovascular, hematological, immunological, endocrine, hepatic, renal, and reproductive functions, and their growth (Fig. 9.11) (Singh, 2021). Arsenic exposure over a long period can harm the unborn child's health and, therefore, lead to infant mortality. Numerous studies suggested that exposure to arsenic significantly negatively impacts young adults' intellectual development, IQ, and memory. Different people, community groups, and areas experience different diseases and indications due to their long-term, high-level consumption of inorganic arsenic (Smith, 2009). As a result, there is no specific definition for the diseases that arsenic produces. This makes it more challenging to evaluate the impact of arsenic on human health.

Sources of Arsenic

As a naturally occurring substance, arsenic is always present in the soil, air, groundwater, animals, plants, food, etc. Arsenic contamination primarily results from earthquakes, geological weathering, volcanic eruptions, and wildfires. Surprisingly, over 200 various types of minerals contain arsenic in significant quantities. In actuality, stones and rocks have been found to contain 99% of arsenic. In particular, arsenic is primarily found in sulfidic minerals like arsenopyrite, which primarily develops in anaerobic environments, and other stone components like silicate, phosphate, sulfide, and oxides (Chen, 2021). Sediments with significant concentrations of iron oxide and other reducing agents may also contain arsenic with concentrations between 3 and 10 mg/kg. These arsenic-containing minerals can easily dissolve and contaminate the groundwater. Arsenic in the soil, which remains as either As(III) or As(V) under

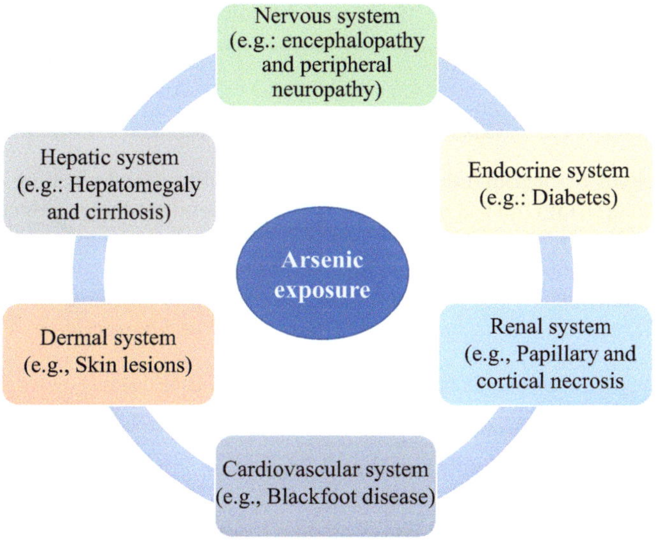

Fig. 9.11 Some examples of the health effects of arsenic on humans (Singh, 2021)

oxidizing or reducing conditions, respectively, is the primary source of arsenic pollution of the surface water and groundwater. The approximate concentration of arsenic present in the soil is around 3–4 mg/kg. Aquifer sediments, primarily due to the release of arsenic from Fe oxides, are also one of the leading causes of groundwater contamination.

To distinguish the effects of exposure to arsenic in drinking water, extensive research has been conducted. In seawater, arsenic can also be found in concentrations ranging from 1 to 8 μg/L, but it does not stay in the solution since it can easily bind to certain minerals or be consumed by a wide range of marine organisms, including algae, fish, mollusks, phytoplankton, etc. (Khosravi-Darani, 2022). For example, compared to other food types (e.g., rice), fish might contain 100 times more arsenic. Additionally, it has been shown that seaweed, an excellent organic soil fertilizer used in dairy farming, also contains a significant amount of arsenic, with a concentration of up to 100 mg/kg as compared to grains (Khosravi-Darani, 2022). Other significant human-caused sources of arsenic include mining, fossil fuel combustion, pesticides that contain arsenic, and pharmaceutical compounds. Some examples of naturally occurring and man-made sources of arsenic are presented in Fig. 9.12. However, there is still a need for further research regarding the source of arsenic, other arsenic compounds, and their hazardous consequences on human health.

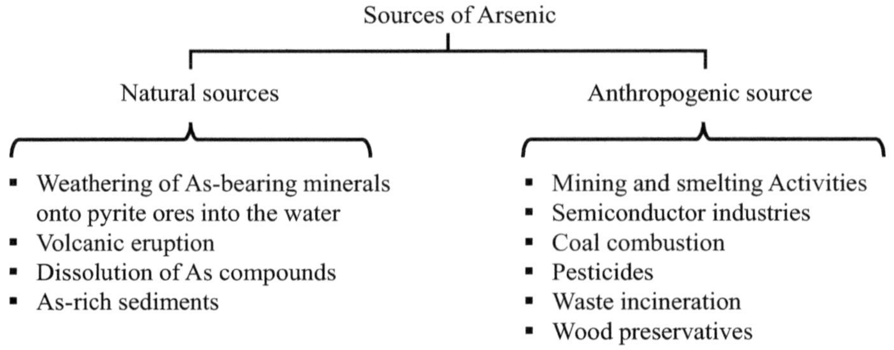

Fig. 9.12 Examples of arsenic sources are both naturally occurring and man-made (Singh, 2021)

Arsenic Removal Using Membrane-Based Separation Process

Everywhere in the world, the removal of arsenic from drinking water is a big problem. Adsorption, precipitation, coagulation, ion exchange, and membrane filtration techniques are commercially available technologies to remove arsenic (Shih, 2005; Yoon, 2009). More research and improvement is required to understand these separation processes better to provide high-quality drinking water. Among them, the most effective method to remove arsenic from the ground/contaminated water is membrane technology, which can effectively remove 96% of the arsenic from contaminated groundwater. Arsenic from contaminated water is removed without accumulating, preventing bacteria from entering the membrane. The most efficient and widespread methods for removing arsenic from contaminated water among all the cutting-edge methods are RO and NF because of their advantages, such as their straightforward operational procedures, high removal performance, and minimal sludge development during operation (Shih, 2005; Yoon, 2009). The RO membrane has a lower water permeance than NF membranes and can separate or desalinate all dissolved ions and compounds. NF has largely supplanted RO in many applications, due to its low cost, low applied pressure, low energy consumption, high water permeance rate, and excellent arsenic removal effectiveness. Details of the nanofiltration-based separation of arsenic are discussed in the following.

Nanofiltration Process

NF membranes are widely used to separate monovalent and divalent ions and exhibit good arsenate removal effectiveness from contaminated water. It is known that the performance of NF membranes follows a complex relationship depending on their surface properties, roughness, active layer thickness, and charge rather than being entirely determined by membrane pore size. It is observed that membranes with higher negative surface charges are more suitable for arsenic removal at neutral pH.

So, in the NF membrane-based arsenic separation process, charge interaction is very crucial.

After being introduced for the first time in the late 1980s, NF membranes have undergone remarkable progress. Around 1993, they started to be utilized for the removal of arsenic as well (Chang, 1994; Thompson, 1993). The separation efficiency of some of the commercial and modified NF membranes are listed in Table 9.2. In these early studies, arsenic removed by NF membranes was up to 97%. Since then, a wide variety of commercial and modified NF membranes have been used by researchers to enhance separation performance and address the limitations. To develop improved membranes for As removal, extensive research has been done.

For example, Table 9.2 presents some commercial state-of-the-art NF membranes to remove arsenate and arsenite from contaminated water, which includes NF-1, NF-2, NF-20, NF-45, NF-300, ES-10, NTR-7450, UTC-70, ESNA-1-LF, HODRA-CORE, NF90-4040, Dow/FilmTec (e.g., NF45, NF70, NF90, NF270), etc. (Siddique, 2020). Since then, researchers have also introduced a range of NF membranes made with various manufacturing and modification techniques to improve the removal efficiency of arsenate and arsenite from polluted/contaminated water. As an example, Richards et al. reported renewable energy-powered UF–NF/RO membrane for arsenic removal (Richards, 2011). Membranes under optimal conditions show 85% rejection of As (Richards, 2011). Harisha et al. investigated arsenic removal using commercial NF (NF-300) membranes and reported water permeance of 3.2 L m^{-2} h^{-1} bar^{-1} with a high rejection efficiency rate of up to 99% (Harisha, 2010). He et al. fabricated a TFN membrane by incorporating polyhedral oligomeric silsesquioxane (POSS) and removing arsenic from water while showing water permeance of 5.4 L m^{-2} h^{-1} bar^{-1} with rejection efficiency greater than 93% (He, 2016). A year later, they reported TFN membranes by incorporating UiO-66 nanoparticles to efficiently separate arsenic (He, 2017). The resulting TFN membrane shows high pure water permeance up to 11.5 L m^{-2} h^{-1} bar^{-1} and high rejections of $HAsO_4^{2-}$ up to 98.6% (He, 2017). Later, He et al. reported TFN membranes incorporated with Na^+ functionalized carbon quantum dots (Na-CQD) for arsenic removal from the contaminated ground and surface water (He, 2018a). Their fabricated Na-CQD modified TFN membranes show excellent pure water permeance (PWP) and high rejection due to their high surface negative charge and reduced pore sizes. It is seen that TFN membranes after incorporating 0.05 wt% of Na-CQDs (TFN0.05) show good pure water permeance up to 10.4 L m^{-2} h^{-1} bar^{-1} and excellent rejection of $HAsO_4^{2-}$ up to 99.5% (He, 2018a). The Na-CQD modified TFN membranes also show excellent antifouling qualities as well as long-term stability (He, 2018a). In the same year, He et al. reported TFC membranes by incorporating zwitterionic co-polymer via IP method to remove arsenic (As) ions from water (He, 2018b). They employ P[MPC-co-AEMA], a water-soluble zwitterionic co-polymer produced by free-radical polymerization of 2-methacryloyloxyethyl phosphorylcholine (MPC) and 2-aminoethyl methacrylate hydrochloride (AEMA) (He, 2018b). The fabricated TFC membrane by incorporating 50 wt% of P[MPC-co-AEMA] shows good pure water permeance up to 8.5 L m^{-2} h^{-1} bar^{-1} and excellent rejections of $HAsO_4^{2-}$ up to 99.8% (He, 2018b). Polyamide core–shell biofunctionalized (PA-CSBF) based matrix membranes for

Table 9.2 Performances of nanofiltration membranes are used for the removal of arsenic (Siddique, 2020)

Membrane	Materials	Arsenic rejection (%)		References
		Arsenate	Arsenite	
ES-10	TFC polyamide membranes	87–93	50–89	Urase (1998)
NTR-7450	Sulfonated polyethersulfone	80.5–84.5	–	Ahmed (2010)
UTC-70	TFC polyamide membranes	>95	–	
NF-300	TFC polyamide membranes	60–99	–	Padilla (2010)
NF-1, NF-2 and NF-20	TFC polyamide membranes	50–100	–	Sen (2010)
NF-300	TFC polyamide membranes	>95	–	Saitua (2011)
ESNA-1-LF	TFC polyamide membranes	>94	–	Yu (2013)
HODRA-CORE	Sulfonated polyethersulfone	<47	–	
NF-1	TFC polyamide membranes	93–98	–	Pal (2014)
NF and LPRO	TFC polyamide membranes	40 and 90	–	Chang (2014)
NF	TFC polyamide membranes	92–94	–	Xi (2014)
DL/DK	TFC polyamide membranes	55.8–76.2	–	Fang (2014)
NF90-4040	TFC polyamide membranes	94	90	Harfoush (2018)
Dow/FilmTec NF90	TFC polyamide membranes	98	–	Jadhav (2016)
Dow/FilmTec NF270	TFC polyamide membranes	94	–	
FilmTec NF45	TFC polyamide membranes	60–90	–	Vrijenhoek (2000)
FilmTec NF70	TFC polyamide membranes	97	–	Waypa (1997)
TFN-30—0.15 wt%	TFN membranes incorporated with 30 nm UiO-66 (MOF) particles	98.6	–	He (2017)

(continued)

Table 9.2 (continued)

Membrane	Materials	Arsenic rejection (%)		References
		Arsenate	Arsenite	
TFC-50	TFC membranes incorporated with 50 wt% P[MPC-co-AEMA] water soluble zwitterionic co-polymer	99.8	–	He (2018b)
PMIA	PMIA based asymmetric TFC membranes	>90	–	Zhao (2012)
1.5 wt% SPEEK	Sulfonated poly(ether ether ketone) coated composite NF membrane	>95	–	Song (2015)
PA-CSBF4	TFC nanofiltration membrane modified with core shell biofunctionalized (C-SBF) nanoparticles	–	99	Zeeshan (2020)

the effective removal of arsenic and selenium from wastewater were reported by Zeeshan et al. (Zeeshan, 2020). They fabricated membranes by incorporating core–shell biofunctionalized (C-SBF) nanoparticles at various concentrations ranging from 10 to 50 mg. The modified PA-CSBF4 membrane with 40 mg C–S BF content gives very good water permeance of up to 444 L m^{-2} h^{-1} and excellent separation rate of As(III) and Se ions up to 99% and 98%, respectively (Zeeshan, 2020). However, the reported membrane's main issue is its inability to achieve excellent separation performance in terms of high water permeance as well as rejection of different ions/molecules simultaneously. As a result, superior performance membranes are urgently needed for As removal from water.

Applications in Fluoride Removal

The crisis has grown worldwide due to fluoride contamination in drinking water. As a result of geochemical reactions, geological changes, or human activity, it is necessary for small amounts. Still, it becomes toxic to humans with intakes of more than 1.5 mg L^{-1} in mostly contaminated drinking water. Fluoride accumulation above the permissible limits causes severe health problems in infants, children, and adults. Many physical, chemical and biological strategies were developed to reduce the level of fluoride in drinking water. Among them, highly selective NF membranes are widely used and exhibit superior performance compared to the other membranes for removing fluoride (Tahaikt, 2007). For example, commercial NF90 membrane shows 98% rejection of fluoride and pure water permeance of up to 5.9 L m^{-2} h^{-1} bar^{-1} at pH 7.2 (Tian, 2021a). Diawara et al. reported NF and Low-Pressure Reverse Osmosis (LPRO) to remove fluoride from brackish water (Diawara, 2011). They demonstrate that the fluorine retention rate of the NF membrane reaches up to 71%, whereas the fluorine rejection rate of the LPRO membrane can reach 97–98.9% (Diawara, 2011).

Challenges and Future Perspectives

The past few decades have seen a significant amount of interest, and research will continue in the future into the specific design of RO and NF membranes with desired water permeance and solute separation properties for specific applications. Among them, water permeance and ion/molecular selectivity of the membranes is the main key factor in determining a membrane's overall performance. This property of membranes highly depends on steric hindrance, the Donnan effect, and the dielectric exclusion principle (Zhou, 2020). Many fabrication methods and new materials were introduced to increase the membrane ions/molecular selectivity by tuning the membrane properties, i.e., all three separation phenomena. Whereas the properties of an ideal membrane should have both high values of water permeance or permeability and ions/molecular selectivity. Recent studies show that membranes with high ion/molecular selectivity are more critical than high water permeance or permeability in designing high-performance membranes (Cohen-Tanugi, 2014; Werber, 2016b). Based on the modeling of the desalination process, it is observed that the water permeability of a particular RO membrane has a marginal impact on energy consumption. For example, in seawater desalination for a TFC membrane, if water permeance increases up to fivefold, the energy consumption decreases to 3.7% (Werber, 2016b). Therefore, increasing the water permeance and ions/molecular selectivity of the membranes beyond the trade-off of the existing membranes is a more efficient strategy for designing high-performance membranes to improve the system products and water quality. Over the past several decades, significant progress has been made primarily in designing controllable RO and NF membranes. Still, little effort has gone into advancing NF theories, a deeper and more thorough understanding of the relationships between synthesis, property, and performance, or the ongoing development of controllably easy membrane fabrication methods. Therefore, substantial work must be done by academia and industry to develop and advance high-specification RO and NF membranes for real-scale applications.

Conclusions

This article delivers a timely update on high-precision polymer-based membranes for their separations in the aqueous system. We have summarized recent development on polymer-based membranes with improved liquid permeance and separation properties. Tuning of the top selective layer has been widely discovered using new materials, introducing non-porous/porous nanostructures, decreasing the top layer thickness, modification with different functional groups, and controlling the interfacial polymerization process. These fabrication approaches can impact the membrane separation performance in terms of liquid permeance and ion/molecular selectivity. Thus, it is promising to tune the top selective layer of the composite structure and increase the membranes' performance. A new fabrication protocol or the use of additional novel materials may be employed in future research to increase the performance of the existing TFC RO or NF membranes, where both the water permeance and ions/molecular selectivity can be improved. Studies on scaling up polymer-based membranes and making membrane modules for industrial applications may be another interesting topic to be explored.

Acknowledgements P. S. and S. K. acknowledge the financial support from the Science and Engineering Research Board, Department of Science and Technology, Government of India (project grant number: SB/S2/RJN-030/2016). P. S. and S. K. acknowledge the technical support from the Centralized Instrumental Facility, CSIR-CSMCRI, Bhavnagar. A PRIS number (CSIR-CSMCRI 117/2024) is assigned for this manuscript.

References

Agre, P. (2004). Aquaporin water channels (Nobel lecture). *Angewandte Chemie International Edition, 43*, 4278–4290.

Ahmed, I. (2017). Importance and significance of UF/MF membrane systems in desalination water treatment. *Desalination, 187*.

Ahmed, S. (2010). Performance of nanofiltration membrane in a vibrating module (VSEP-NF) for arsenic removal. *Desalination, 252*, 127–134.

Ali, M. E. A. (2016). Thin film composite membranes embedded with graphene oxide for water desalination. *Desalination, 386*, 67–76.

Al-Juboori, R. A. (2012). Biofouling in RO system: Mechanisms, monitoring and controlling. *Desalination, 302*, 1–23.

Baker, R. W. (2012). Membrane technology and applications (3rd ed.). John Wiley & Sons, Inc.

Bera, A. (2018). Anti-organic fouling and anti-biofouling poly(piperazineamide) thin film nanocomposite membranes for low-pressure removal of heavy metal ions. *Journal of Hazardous Materials, 343*, 86–97.

Chang, S. (1994). Removal of arsenic by enhanced coagulation and membrane technology, critical issue in water and wastewater treatment. In *Proceedings of the national conference on environmental engineering, Boulder, CO, USA* (pp. 11–13).

Chang, F.-F. (2014). Comparison of polyamide nanofiltration and low-pressure reverse osmosis membranes on as (III) rejection under various operational conditions. *Desalination, 334*, 10–16.

Chen, J. (2020). Accurately controlling the hierarchical nanostructure of polyamide membranes: Via electrostatic atomization-assisted interfacial polymerization. *Journal of Materials Chemistry A, 8*, 9160–9167.

Chen, Q. Y. (2021). Arsenic: A global environmental challenge. *Annual Review of Pharmacology and Toxicology, 61*, 47–63.

Choi, W. (2015). Tailor-made polyamide membranes for water desalination. *ACS Nano, 9*, 345–355.

Chowdhury, M. R. (2018). 3D printed polyamide membranes for desalination. *Science, 361*, 682–686.

Cohen-Tanugi, D. (2014). Quantifying the potential of ultra-permeable membranes for water desalination. *Energy & Environmental Science, 7*, 1134–1141.
Curcio, E. (2005). Membrane distillation and related operations—A review. *Separation and Purification Reviews, 34*, 35–86.
Decher, G. (1997). Fuzzy nanoassemblies: Toward layered polymeric multicomposites. *Science, 277*, 1232–1237.
Di Vincenzo, M. (2021). Biomimetic artificial water channel membranes for enhanced desalination. *Nature Nanotechnology, 16*, 190–196.
Diawara, C. K. (2011). Performance of nanofiltration (NF) and low pressure reverse osmosis (LPRO) membranes in the removal of fluorine and salinity from brackish drinking water. *Journal of Water Resource and Protection, 3*, 912–917.
El-Hashani, A. (2008). Use of layer-by-layer assembled ultrathin membranes of dicopper-[18]azacrown-N_6 complex and polyvinylsulfate for water desalination under nanofiltration conditions. *Journal of Membrane Science, 318*, 65–70.
Epsztein, R. (2020). Towards single-species selectivity of membranes with subnanometre pores. *Nature Nanotechnology, 15*, 426–436.
Eriksson, U. K. (2013). Subangstrom resolution X-ray structure details aquaporin-water interactions. *Science, 340*, 1346–1349.
Fang, J. (2014). Rejection and modeling of arsenate by nanofiltration: Contributions of convection, diffusion and electromigration to arsenic transport. *Journal of Membrane Science, 453*, 42–51.
Francis, P. S. (1966). *Fabrication and evaluation of new ultrathin reverse osmosis membranes*. U.S. Department of the Interior.
Gao, S. (2019). Ultrathin polyamide nanofiltration membrane fabricated on brush-painted single-walled carbon nanotube network support for ion sieving. *ACS Nano, 13*, 5278–5290.
Geise, G. M. (2011). Water permeability and water/salt selectivity tradeoff in polymers for desalination. *Journal of Membrane Science, 369*, 130–138.
Ghosh, S. (2019). Technology alternatives for decontamination of arsenic-rich groundwater—A critical review. *Environmental Technology and Innovation, 13*, 277–303.
Gong, G. (2019). New insights into the role of an interlayer for the fabrication of highly selective and permeable thin-film composite nanofiltration membrane. *ACS Applied Materials & Interfaces, 11*, 7349–7356.
Gu, J. E. (2013). Molecular layer-by-layer assembled thin-film composite membranes for water desalination. *Advanced Materials, 25*, 4778–4782.
Gu, J. E. (2015). Tailoring interlayer structure of molecular layer-by-layer assembled polyamide membranes for high separation performance. *Applied Surface Science, 356*, 659–667.
Gui, L. (2020). Ultrafast ion sieving from honeycomb-like polyamide membranes formed using porous protein assemblies. *Nano Letters, 20*, 5821–5829.
Guo, S. (2020). Custom-tailoring loose nanofiltration membrane for precise biomolecule fractionation: New insight into post-treatment mechanisms. *ACS Applied Materials & Interfaces, 12*, 13327–13337.
Harfoush, M. (2018). Arsenic removal from drinking water using low-pressure nanofiltration under various operating conditions. *Water Practice Technology, 13*, 295–302.
Harisha, R. S. (2010). Arsenic removal from drinking water using thin film composite nanofiltration membrane. *Desalination, 252*, 75–80.
He, Y. (2016). Concurrent removal of selenium and arsenic from water using polyhedral oligomeric silsesquioxane (POSS)-polyamide thin-film nanocomposite nanofiltration membranes. *Industrial and Engineering Chemistry Research, 55*, 12929–12938.
He, Y. (2017). UiO-66 incorporated thin-film nanocomposite membranes for efficient selenium and arsenic removal. *Journal of Membrane Science, 541*, 262–270.
He, Y. (2018a). Na^+ functionalized carbon quantum dot incorporated thin-film nanocomposite membranes for selenium and arsenic removal. *Journal of Membrane Science, 564*, 483–491.
He, Y. (2018b). Novel thin-film composite nanofiltration membranes consisting of a zwitterionic co-polymer for selenium and arsenic removal. *Journal of Membrane Science, 555*, 299–306.

Hong, S. U. (2006). Optimization of flux and selectivity in Cl^-/SO_4^{2-} separations with multilayer polyelectrolyte membranes. *Journal of Membrane Science, 283*, 366–372.
Huang, B. Q. (2020). High-performance zwitterionic nanofiltration membranes fabricated via microwave-assisted grafting of betaine. *ACS Applied Materials & Interfaces, 12*, 35523–35531.
Ibrahim, S. (2020). Poly(homopiperazine-amide) thin-film composite membrane for nanofiltration of heavy metal ions. *ACS Omega, 5*, 28749–28759.
Jadhav, S. V. (2016). A pilot scale concurrent removal of fluoride, arsenic, sulfate and nitrate by using nanofiltration: Competing ion interaction and modelling approach. *Journal of Water Process Engineering, 13*, 153–167.
Jaworek, A. (2008). Electrospraying route to nanotechnology: An overview. *Journal of Electrostatics, 66*, 197–219.
Jeong, B. H. (2007). Interfacial polymerization of thin film nanocomposites: A new concept for reverse osmosis membranes. *Journal of Membrane Science, 294*, 1–7.
Ji, Y. L. (2017). Recent advances in the fabrication of membranes containing "ion pairs" for nanofiltration processes. *Polymers, 9*(12), 715.
Jiang, Z. (2018). Water transport through ultrathin polyamide nanofilms used for reverse osmosis. *Advanced Materials, 30*, 1705973.
Karan, S. (2012). Ultrafast viscous permeation of organic solvents through diamond-like carbon nanosheets. *Science, 335*, 444–447.
Karan, S. (2015). Sub-10 nm polyamide nanofilms with ultrafast solvent transport for molecular separation. *Science, 348*, 1347–1351.
Kavitha, J. (2019). Pretreatment processes for seawater reverse osmosis desalination systems—A review. *Journal of Water Process Engineering, 32*, 100926.
Khosravi-Darani, K. (2022). Arsenic exposure via contaminated water and food sources. *Water, 14*, 1884.
Kim, J. (2010). The use of nanoparticles in polymeric and ceramic membrane structures: Review of manufacturing procedures and performance improvement for water treatment. *Environmental Pollution, 158*, 2335–2349.
Kruschitz, A. (2020). Downstream processing technologies in the biocatalytic production of oligosaccharides. *Biotechnology Advances, 43*, 107568.
Lai, G. S. (2016). Graphene oxide incorporated thin film nanocomposite nanofiltration membrane for enhanced salt removal performance. *Desalination, 387*, 14–24.
Landaburu-Aguirre, J. (2016). Preparation of loose polypiperazine amide membranes. Effect of the nanocomposite sublayer on the NF process performance. *Chemical Engineering Journal, 294*, 431–438.
Lau, W. J. (2009). Polymeric nanofiltration membranes for textile dye wastewater treatment: Preparation, performance evaluation, transport modelling, and fouling control—A review. *Desalination, 245*, 321–348.
Lee, C. S. (2021). Aquaporin-incorporated graphene-oxide membrane for pressurized desalination with superior integrity enabled by molecular recognition. *Advancement of Science, 8*, 2101882.
Li, C. (2022). Fabrication of fluorinated polyamide seawater reverse osmosis membrane with enhanced boron removal. *Journal of Membrane Science, 662*, 121009.
Li, Y. (2017). Influence of hydrophilic carbon dots on polyamide thin film nanocomposite reverse osmosis membranes. *Journal of Membrane Science, 537*, 42–53.
Liang, Y. (2020). Polyamide nanofiltration membrane with highly uniform sub-nanometre pores for sub-1 Å precision separation. *Nature Communications, 11*.
Loeb, S. (1963). Sea water demineralization by means of an osmotic membrane. *Advances in Chemistry, 38*, 117–132.
Lt, R. (1970). Development of new reverse osmosis membranes for desalination. U S, Off. Saline Water, Res Dev. Progr Rep 531.
Lu, X. (2021). Fabrication of desalination membranes by interfacial polymerization: History, current efforts, and future directions. *Chemical Society Reviews, 50*, 6290–6307.

Ma, X. H. (2018a). Interfacial polymerization with electrosprayed microdroplets: Toward controllable and ultrathin polyamide membranes. *Environmental Science & Technology Letters, 5*, 117–122.
Ma, X. H. (2018b). Nanofoaming of polyamide desalination membranes to tune permeability and selectivity. *Environmental Science & Technology Letters, 5*, 123–130.
Maharjan, M. (2005). Arsenic contamination in drinking water and skin manifestations in lowland Nepal: The first community-based survey. *American Journal of Tropical Medicine and Hygiene, 73*, 477–479.
Marchetti, P. (2014). Molecular separation with organic solvent nanofiltration: A critical review. *Chemical Reviews, 114*, 10735–10806.
Marré Tirado, M. L. (2016). Assessing biofouling resistance of a polyamide reverse osmosis membrane surface-modified with a zwitterionic polymer. *Journal of Membrane Science, 520*, 490–498.
Mi, B. (2014). Graphene oxide membranes for ionic and molecular sieving. *Science, 343*, 740–742.
Misdan, N. (2012). Seawater reverse osmosis (SWRO) desalination by thin-film composite membrane—Current development, challenges and future prospects. *Desalination, 287*, 228–237.
Mohan, D. (2007). Arsenic removal from water/wastewater using adsorbents—A critical review. *Journal of Hazardous Materials, 142*, 1–53.
Nghiem, L. D. (2005). Pharmaceutical retention mechanisms by nanofiltration membranes. *Environmental Science and Technology, 39*, 7698–7705.
Padilla, A. P. (2010). Performance of simultaneous arsenic, fluoride and alkalinity (bicarbonate) rejection by pilot-scale nanofiltration. *Desalination, 257*, 16–21.
Pal, P. (2014). A nanofiltration–coagulation integrated system for separation and stabilization of arsenic from groundwater. *Science of the Total Environment, 476*, 601–610.
Pan, X. D. (1990). Diamond-like carbon films prepared by rf plasma deposition. *Vacuum, 41*, 1360–1363.
Pandey, S. R. (2012). Fouling in reverse osmosis (RO) membrane in water recovery from secondary effluent: A review. *Reviews in Environmental Science & Biotechnology, 11*, 125–145.
Park, H. B. (2017a). Maximizing the right stuff: The trade-off between membrane permeability and selectivity. *Science, 356*, eaab0530.
Park, S. J. (2017b). Fabrication of polyamide thin film composite reverse osmosis membranes via support-free interfacial polymerization. *Journal of Membrane Science, 526*, 52–59.
Peng, H. (2020). Phosphonium modification leads to ultrapermeable antibacterial polyamide composite membranes with unreduced thickness. *Advanced Materials, 32*, 2001383.
Podgorski, J. (2020). Global threat of arsenic in groundwater. *Science, 368*, 845–850.
Prozorovska, L. (2018). State-of-the-Art and future prospects for atomically thin membranes from 2D materials. *Advanced Materials, 30*, 1801179.
Qasem, N. A. A. (2021). Removal of heavy metal ions from wastewater: A comprehensive and critical review. *Npj Clean Water, 4*, 36.
Qian, H. (2012). Ultrathin films of organic networks as nanofiltration membranes via solution-based molecular layer deposition. *Langmuir, 28*, 17803–17810.
Regis, A. O. (2022). Pressure-driven membrane processes for boron and arsenic removal: pH and synergistic effects. *Desalination, 522*, 115441.
Richards, L. A. (2011). Renewable energy powered membrane technology: Salt and inorganic contaminant removal by nanofiltration/reverse osmosis. *Journal of Membrane Science, 369*, 188–195.
Richardson, J. J. (2015). Technology-driven layer-by-layer assembly of nanofilms. *Science, 348*(6233), aaa2491.
Robertson, J. (2002). Diamond-like amorphous carbon. *Materials Science and Engineering R: Reports, 37*, 129–281.
Saitua, H. (2011). Experimental investigation on arsenic removal with a nanofiltration pilot plant from naturally contaminated groundwater. *Desalination, 274*, 1–6.

Sarkar, P. (2020). Effect of porous and nonporous nanostructures on the permeance of positively charged nanofilm composite membranes. *Advanced Materials Interfaces, 7*, 2000251.

Sarkar, P. (2021a). Ultraselective and highly permeable polyamide nanofilms for ionic and molecular nanofiltration. *Advanced Functional Materials, 31*, 2007054.

Sarkar, P. (2021b). Fast water transport through sub-5 nm polyamide nanofilms: The new upper-bound of the permeance-selectivity trade-off in nanofiltration. *Journal of Materials Chemistry A, 9*, 20714–20724.

Sen, M. (2010). Removal of arsenic from contaminated groundwater by membrane-integrated hybrid treatment system. *Journal of Membrane Science, 354*, 108–113.

Shan, L. (2017). Microphase diffusion-controlled interfacial polymerization for an ultrahigh permeability nanofiltration membrane. *ACS Applied Materials & Interfaces, 9*, 44820–44827.

Sharma, V. K. (2009). Aquatic arsenic: Toxicity, speciation, transformations, and remediation. *Environment International, 35*(4), 743–759.

Shen, H. (2018). Preparation of polyamide thin film nanocomposite membranes containing silica nanoparticles via an in-situ polymerization of $SiCl_4$ in organic solution. *Journal of Membrane Science, 565*, 145–156.

Shi, Z. (2013). Ultrafast separation of emulsified oil/water mixtures by ultrathin freestanding single-walled carbon nanotube network films. *Advanced Materials, 25*, 2422–2427.

Shih, M. C. (2005). An overview of arsenic removal by pressure-driven membrane processes. *Desalination, 172*, 85–97.

Sholl, D. S. (2016). Seven chemical separations to change the world. *Nature, 532*, 435–437.

Shui, X. (2021). Tailoring ultrathin microporous polyamide films with rapid solvent transport by molecular layer-by-layer deposition. *Journal of Membrane Science, 628*, 119249.

Siddique, T. A. (2020). Nanofiltration for arsenic removal: Challenges, recent developments, and perspectives. *Nanomaterials, 10*(7), 1323.

Singh, G. (2021). Occurrence, fate, and remediation of arsenic. In *Pollutants and water management: Resources, strategies and scarcity* (pp. 349–376).

Smith, A. H. (2002). Arsenic epidemiology and drinking water standards. *Science, 296*, 2145–2146.

Smith, A. H. (2009). Health effects of arsenic and chromium in drinking water: Recent human findings. *Annual Review of Public Health, 30*, 107–122.

Song, J. (2015). Arsenic removal using a sulfonated poly(ether ether ketone) coated hollow fiber nanofiltration membrane. *Environmental Science: Water Research & Technology, 1*, 839–845.

Sreeram, A. (2017). Optimisation of the removal of arsenate from water using nanochitosan. *Desalination and Water Treatment, 70*, 235–243.

Tagliabue, M. (2014). Boron removal from water: Needs, challenges and perspectives. *Journal of Cleaner Production, 77*, 56–64.

Tahaikt, M. (2007). Fluoride removal from groundwater by nanofiltration. *Desalination, 212*, 46–53.

Tan, Z. (2018). Polyamide membranes with nanoscale Turing structures for water purification. *Science, 360*, 518–521.

Thompson, M. (1993). Evaluating arsenic removal technologies. In *Proceedings of the 1993 AWWA annual conference, San Antonio, TX, USA* (pp. 6–10).

Tian, J. (2021a). Progress in research and application of nanofiltration (Nf) technology for brackish water treatment. *Membranes, 11*, 662.

Tian, J. (2021b). Direct generation of an ultrathin (8.5 nm) polyamide film with ultrahigh water permeance via in-situ interfacial polymerization on commercial substrate membrane. *Journal of Membrane Science, 634*, 119450.

Tu, K. L. (2010). Boron removal by reverse osmosis membranes in seawater desalination applications. *Separation and Purification Technology, 75*, 87–101.

Urase, T. (1998). Effect of pH on rejection of different species of arsenic by nanofiltration. *Desalination, 117*, 11–18.

Vandezande, P. (2008). Solvent resistant nanofiltration: Separating on a molecular level. *Chemical Society Reviews, 37*, 365–405.

Vrijenhoek, E. M. (2000). Arsenic removal from drinking water by a "loose" nanofiltration membrane. *Desalination, 130*, 265–277.

Wang, J. (2017a). Construction of TiO_2@graphene oxide incorporated antifouling nanofiltration membrane with elevated filtration performance. *Journal of Membrane Science, 533*, 279–288.

Wang, J. J. (2017b). Nanofiltration membranes with cellulose nanocrystals as an interlayer for unprecedented performance. *Journal of Materials Chemistry A, 5*, 16289–16295.

Wang, S. (2017c). Aggregation suppressed thin film nanocomposite (TFN) membranes prepared with an: In situ generation of TiO_2 nanoadditives. *RSC Advances, 7*, 26136–26144.

Wang, S. (2022). Surface engineering design of polyamide membranes for enhanced boron removal in seawater desalination. *Journal of Membrane Science, 651*, 120425.

Wang, Y. (2019). Cerium oxide doped nanocomposite membranes for reverse osmosis desalination. *Chemosphere, 218*, 974–983.

Wang, Z. (2018). Nanoparticle-templated nanofiltration membranes for ultrahigh performance desalination. *Nature Communications, 9*, 2004.

Waypa, J. J. (1997). Arsenic removal by RO and NF membranes. *Journal American Water Works Association, 89*, 102–114.

Werber, J. R. (2016a). Materials for next-generation desalination and water purification membranes. *Nature Reviews Materials, 1*, 16018.

Werber, J. R. (2016b). The critical need for increased selectivity, not increased water permeability, for desalination membranes. *Environmental Science & Technology Letters, 3*, 112–120.

Wolf, P. H. (2005). UF membranes for RO desalination pretreatment. *Desalination, 182*, 293–300.

Xi, B. (2014). Fluoride and arsenic removal by nanofiltration technology from groundwater in rural areas of China: Performances with membrane optimization. *Separation Science and Technology, 49*, 2642–2649.

Yang, H. (2019a). Covalent organic framework membranes through a mixed-dimensional assembly for molecular separations. *Nature Communications, 10*, 2101.

Yang, Y. (2019b). Large-area graphene-nanomesh/carbon-nanotube hybrid membranes for ionic and molecular nanofiltration. *Science, 364*, 1057–1062.

Yang, S. (2020). Electrosprayed polyamide nanofiltration membrane with intercalated structure for controllable structure manipulation and enhanced separation performance. *Journal of Membrane Science, 602*, 117971.

Yang, Z. (2018). Tannic acid/Fe^{3+} nanoscaffold for interfacial polymerization: Toward enhanced nanofiltration performance. *Environmental Science and Technology, 52*, 9341–9349.

Yoon, J. (2009). Removal of toxic ions (chromate, arsenate, and perchlorate) using reverse osmosis, nanofiltration, and ultrafiltration membranes. *Chemosphere, 77*, 228–235.

You, X. (2021). Electrostatic-modulated interfacial polymerization toward ultra-permselective nanofiltration membranes. *iScience, 24*, 102369.

Yu, Y. (2013). Effects of ion concentration and natural organic matter on arsenic (V) removal by nanofiltration under different transmembrane pressures. *Journal of Environmental Sciences, 25*, 302–307.

Zeeshan, M. H. (2020). Polyamide intercalated nanofiltration membrane modified with biofunctionalized core shell composite for efficient removal of arsenic and selenium from wastewater. *Journal of Water Process Engineering, 34*, 101175.

Zhai, Z. (2019). In situ assembly of a zeolite imidazolate framework hybrid thin-film nanocomposite membrane with enhanced desalination performance induced by Noria-polyethyleneimine codeposition. *ACS Applied Materials & Interfaces, 11*, 12871–12879.

Zhang, M. (2019a). Controllable ion transport by surface-charged graphene oxide membrane. *Nature Communications, 10*, 1253.

Zhang, Z. (2019b). Ultra-permeable polyamide membranes harvested by covalent organic framework nanofiber scaffolds: A two-in-one strategy. *Chemical Science, 10*, 9077–9083.

Zhao, C. (2012). Arsenic removal from drinking water by self-made PMIA nanofiltration membrane. *Advances in Chemical Engineering and Science, 2*, 366–371.

Zhao, D. (2015a). A review of recent advance in fouling mitigation of NF/RO membranes in water treatment: Pretreatment, membrane modification, and chemical cleaning. *Desalination and Water Treatment, 55*, 870–891.

Zhao, F. Y. (2015b). A novel type of polyelectrolyte complex/MWCNT hybrid nanofiltration membranes for water softening. *Journal of Membrane Science, 492*, 412–421.

Zhou, X. (2020). Intrapore energy barriers govern ion transport and selectivity of desalination membranes. *Science Advances, 6*, eabd9045.

Zhou, Z. (2018). High-performance thin-film composite membrane with an ultrathin spray-coated carbon nanotube interlayer. *Environmental Science & Technology Letters, 5*, 243–248.

Zhu, J. (2019). MOF-positioned polyamide membranes with a fishnet-like structure for elevated nanofiltration performance. *Journal of Materials Chemistry A, 7*, 16313–16322.

Zhu, Y. (2016). Single-walled carbon nanotube film supported nanofiltration membrane with a nearly 10 nm thick polyamide selective layer for high-flux and high-rejection desalination. *Small, 12*, 5034–5041.

Chapter 10
Novel Cellulose-Based Hectocycle Nanopolymers for Arsenic Removal from Groundwater

Bayan Khalaf, Shehdeh Jodeh, and Subhi Samhan

Abstract Cellulose-based nano polymers with heterocyclic moieties were prepared by reacting cellulose with 2-amino pyridine (Cell-X), pyridine-2-carbonyl chloride (Cell-Y), or furan-2-carbonyl chloride (Cell-Z). The novel synthesized derivatives were characterized and evaluated as adsorbents for Arsenic removal from an aqueous solution. In order to maximize the adsorption efficiency of Arsenic, the optimum solution conditions of time, pH, temperature, adsorbent dosage, and Arsenic initial concentration were determined. The highest removal percentages of Arsenic from water were 91.65%, 96.73% and 95.54% for Cell-X, Cell-Y and Cell-Z, respectively. The best equilibrium isotherm of each process was investigated based on the experimental and calculated R^2 values of Freundlich and Langmuir adsorption models. In addition, the adsorption kinetics were investigated using pseudo-first-order and pseudo-second-order adsorption kinetic models. Van't Hoff plot was also determined for each process in order to detect the changes in adsorption enthalpy (ΔH), Gibbs free energy (ΔG) and Entropy (ΔS). The obtained results showed that all adsorption processes fitted with both Freundlich and Langmuir adsorption isotherm, and all mechanisms followed pseudo second-order kinetic model. Negative values of the thermodynamic parameter ΔG° indicate spontaneous adsorption processes. The results indicate that the synthesized nano polymers could be promising absorbents on a commercial scale for Arsenic removal from water to a drinkable for safe consumption.

B. Khalaf (✉)
Faculty of Science, Department of Industrial Chemistry, Arab American University, Jenin, Palestine
e-mail: bayan.kh107@hotmail.com

S. Jodeh
Faculty of Science, Department of Chemistry, An-Najah National University, Nablus, Palestine
e-mail: sjodeh@najah.edu

S. Samhan
Research and Development Center, Palestinian Water Authority, Ramallah, Palestine

B. Sen Gupta and N. Martínez-Villegas (eds.), *Arsenic Remediation of Food and Water*,
https://doi.org/10.1007/978-981-97-4764-1_10

Keywords Cellulose · Arsenic · Furan-2-carbonyl chloride · Pyridine-2,6-dicarbonyl dichloride · 2-Amino pyridine · Water purification

Introduction

Arsenic found in groundwater in measurable concentrations, either as a result of natural or anthropogenic resources, represents a serious concern in several regions around the world. Many people in rural and developing countries are exposed to high levels of arsenic through consuming arsenic contaminated drinking water (World Health Organization (WHO), 2011; Shankar et al., 2014; Abdul et al., 2015). The main concerns related to arsenic presence in groundwater are its high toxicity and carcinogenicity. Pollution of water with arsenic can cause many adverse effects on human health, including but not limited to skin lesions, cardiovascular disease diabetes, and cancer (skin, urinary tract, kidney, lungs, and liver). The World Health Organization (WHO) recommended a maximum allowable concentration of arsenic in drinking water of 10 ppb in order to reduce and minimize its adverse health effects. (World Health Organization (WHO), 2011; Nicomel et al., 2016; Pathan & Bose, 2018; Medina-Ramirez et al., 2019; Mukherjeea et al., 2016; Poguberović et al., 2016; Mohan & Pittman, 2007; Kabir & Chowdhury, 2017). Trivalent arsenic As (III) is approximately 60 times more toxic than the oxidized pentavalent. Arsenic As (IV), as well as, inorganic arsenic compounds are about 100 times more toxic than organic ones. Organic forms of arsenic compounds are quantitatively insignificant and are found mostly in surface water or in areas that are usually affected by industrial pollution. The relative amounts of As (III) to As (V) vary widely. The two important factors controlling the speciation of arsenic and its solubility are pH and redox potential in a geographic region. Techniques for arsenic removal from water can include ion exchange, membrane technology, adsorption, or chemical precipitation. Among these techniques, adsorption offers many advantages such as: simplicity and stable operation, low costs, easy handling of waste, and absence of added reagents (Asere et al., 2019; Attia et al., 2012; Chandra et al., 2010; Chen et al., 2013; Habuda-Stanić & Nujić, 2015; Idswater, 2010; Jain & Ali, 2000; Reynolds & Richards, 2000; Siddiqui & Chaudhry, 2017; Smedley & Kinniburg, 2002; Wong et al., 2017). Cellulose is one of the most abundant renewable and biological materials in the world. It has been an industrial feedstock of a large number of derivatives with an unlimited number of commercial applications. Surface-modified cellulose can also be of great interest due to a wide range of applications (Ayrilmis & Buyuksari, 2010; Azbar et al., 2010; Balamurugan et al., 2011; Hamed et al., 2012; Khalaf et al., 2021a, 2021b; Medronho & Lindman, 2014; Ramos et al., 2011; Salameh, 2009). This study aims to purify water from arsenic using novel synthesized cellulosic nano polymers modified with 2-amino pyridine (Cell-X), furan-2-carbonyl chloride (Cell-Z) or pyridine-2,6-dicarbonyl dichloride (Cell-Y). The efficiency of the prepared cellulose adsorbents toward arsenic was investigated as a function of adsorbent dose, temperature, pH and time. In addition, isotherms, kinetics and, thermodynamic parameters are investigated.

Materials and Methods

Materials

All Chemicals used in this work were of analytical grade, they purchased from Sigma Aldrich Chemical Company (Munich, Germany) and used without any further purification unless otherwise specified. Chemicals used in this study include: cellulose, ethanol, triethylamine, anhydrous Dimethyl Acetamide (DMAc), dimethyl sulfoxide, isopropyl alcohol, 2-amino pyridine, furan-2-carbonyl chloride, pyridine-2,6-dicarbonyl dichloride, methanol, anhydrous lithium chloride (LiCl), nitric acid, acetonitrile.

Methods

^{1}H NMR spectra were recorded at Forschungszentrum Jülich, Germany, using a Bruker 600 MHz spectrometer, equipped with a 5 mm broadband Cryo Probe Prodigy. The acquisition parameters are: 90° pulse calibrated at 12 μs, 1.3 s acquisition time, 2 s relaxation time, no spinning, 300 K, 2048 scans. Ultraviolet-Visible Spectroscopy was recorded using UV-1601 (SHIMADZU), IR Spectrometer (Nicolet iS5, iD3 ATR, Thermo Scientific, Japan) was used in this work. Thermogravimetric analysis on Q50 V20.10 TGA, Build 36 instrument at a heating rate of 10 °C/min in N_2 gaseous atmosphere). To study the nano adsorbents surface morphology, scanning electron microscopy (SEM) was used. Such that, each sample was fixed onto a double-sided C tape and mounted onto a sample holder. The sample was subjected to a sputter coating of a 10 nm-thick gold film onto the sample in order to minimize charging effects for taking high magnification images. The sample was placed into the vacuum sample chamber and micrographs were obtained using an emission scanning electron microscope (JEOL 7400F Oxford Instruments Inca, Tubney Woods, Abingdon, Oxon OX13 5QX, UK).

Cellulose Dissolution

The dissolution of cellulose was carried out by cellulose activation (mass = 1 g) first in (100 mL) water at room temperature, the suspension was stirred for two hours, followed by suction filtration and suspending the collected cellulose for one hour in 100 mL methanol. The procedure was repeated twice followed with two successive exchanges of (25 mL) volume of the anhydrous *N*,*N*-Dimethyl Acetamide (DMAc). The first DMAc exchange was done for an hour, while the second one was carried out for 24 h. After each solvent exchange, DMAc was removed using filtration under vacuum by a glass centered funnel. The activated cellulose was then transferred into a two-necked-round bottomed flask equipped with a condenser and magnet stir bar. After that, the flask was connected into a poplar via the condenser. A solution of (mass = 6.5 g) of anhydrous LiCl in 100 mL anhydrous (DMAc) was added to the flask. The mixture was then stirred for two hours until a clear gel was obtained.

Cellulose Functionalization with 2-Amino Pyridine

To the cellulose solution in LiCl/DMAc prepared before, trimethylamine (0.5 mL) was added dropwise under N_2, followed by adding a solution of 2-amino pyridine (mass = 1.741 g) in 10 mL DMAc. The produced reaction mixture was stirred for three hours at 70 °C. It was then transferred to a beaker containing 500 mL distilled water. After that, the produced precipitate was collected by suction filtration and washed several times with distilled water and isopropyl alcohol, then dried at room temperature.

Cellulose Crosslinking with Pyridine 2,6-Dicarbonyl Dichloride

The above procedure was repeated exactly except that, 2,6-pyridine dicarbonyl dichloride (mass = 3.775 g) in 10 mL DMAc was added to the solution of cellulose in the solvent mixture LiCl/DMAc. A precipitate of crosslinked cellulose started to appear after about 30 min of heating the reaction mixture at 70 °C. After that, the produced precipitate was collected by suction filtration and washed several times with distilled water and ethanol, then dried at room temperature.

Cellulose Acylation with Furan-2-Carbonyl Chloride

To the cellulose solution in LiCl/DMAc prepared before, trimethylamine (0.5 mL) was added dropwise under N_2, followed by a solution of Furan-2-carbonyl chloride (2.415 g, 0.0185 mol) in 10 mL DMAc. The produced reaction mixture was stirred for about three hours at 70 °C. After that, the mixture was transferred to a beaker containing 500 mL distilled water, the produced precipitate was then collected by suction filtration and washed several times with distilled water, and with isopropyl alcohol, then dried at room temperature.

Adsorption Study of Arsenic

Several samples of groundwater from Jenin and Nablus Cities in Palestine were taken in order to measure arsenic ions concentrations. The results indicated the absence of arsenic in the subjected groundwater samples. All experiments were performed in 50 mL plastic vials, containing groundwater samples that were polluted with a specific amount of Arsenic, then placed in a shaker immersed in a water bath equipped with a thermostat and a shaker. The effect of various variables on each of the nano polymers efficiency, such as initial concertation (C_0), pH, adsorbent dosage, time, and temperature were evaluated. After each run, each sample was collected from the mixture using a 10 mL plastic syringe, filtered through a 0.45 μm filter and subjected to analysis by flame Atomic Absorption Spectrophotometer. All adsorption experiments were performed in triplicate, and the mean of the three runs was reported. The amount of arsenic adsorbed by each adsorbent was determined according to Eqs. (10.1) and (10.2), respectively.

$$R(\%) = \frac{C_0 - C_e}{C_0} \tag{10.1}$$

$$Q_e = \frac{C_0 - C_e}{m} V \tag{10.2}$$

where, C_0 and C_e are the initial and the equilibrium concentrations (ppm) of Arsenic in solution, respectively, Q_e is the equilibrium adsorption capacity (ppm), m is the mass of Cell-X, Cell-Y and Cell-Z (mg), and V is solution volume (L).

Results and Discussion

Polymer Analysis

Grafting of cellulose with furan-2-carbonyl group (Cell-Z), 2-amino pyridine (Cell-X) and crosslinking with 2,6-pyridine dicarbonyl dichloride (Cell-Y) were carried in DMAc/LiCl as shown in Schemes 1, 2 and 3. The reagents were chosen for the functional groups they have, since the presence of the aromatic heterocyclic, the carbonyl group and the hydroxyl groups of the cellulose make them novel adsorbents for Arsenic (Foo & Hameed, 2010).

The prepared cellulose derivatives Cell-X, Cell-Y and Cell-Z were characterized by SEM, FT-IR, Proton NMR and TGA. They showed good chemical and thermal

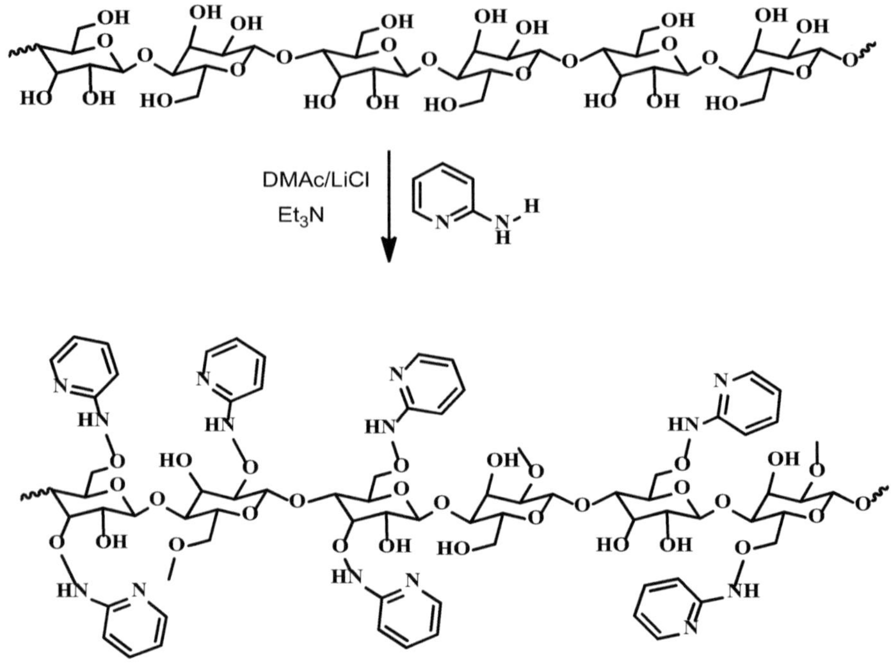

Scheme 1. Cellulose functionalized with 2-amino pyridine

stabilities, as well as high porosity. Characterization results for Cell-Y and Cell-Z nano-polymers are showed in our previous publications (Khalaf et al., 2021a, 2021b). Grafting of cellulose with 2-amino pyridine reagent was carried in DMAc/LiCl as shown in Scheme 1. This nanopolymer was characterized by proton NMR spectroscopy. The ^{1}H NMR of Cellulose functionalized with 2-amino pyridine is shown in Fig. 10.1. Cellulose solution in DMAc-d_6/LiCl was obtained by solvent exchanged as shown in the experimental part. The four peaks showed between δ 2.04 and 3.10 are from the four C2, C3, C4 and C5 protons, the peak at δ 3.0 is consistent with the chemical shift of H2 (Fig. 10.2).

FT-IR was done for Cellulose functionalized with 2-amino pyridine. The functional groups for the synthesized polymer are appeared in FT-IR characterization.

The obtained SEM image of Cellulose functionalized with 2-amino pyridine is shown in Fig. 10.3. As shown below, the surface morphology for this cellulose derivative detects a high surface to volume ratio, hence suggesting a good adsorption efficiency for Arsenic removal.

The TGA was performed on Cellulose functionalized with 2-amino pyridine, the sample mass is measured over time with temperature variation, in order to determine its thermal stability, hence knowing the decomposition behavior for each compound with increasing temperature. As shown in Fig. 10.4, the synthesized nano-polymer shows good thermal stability.

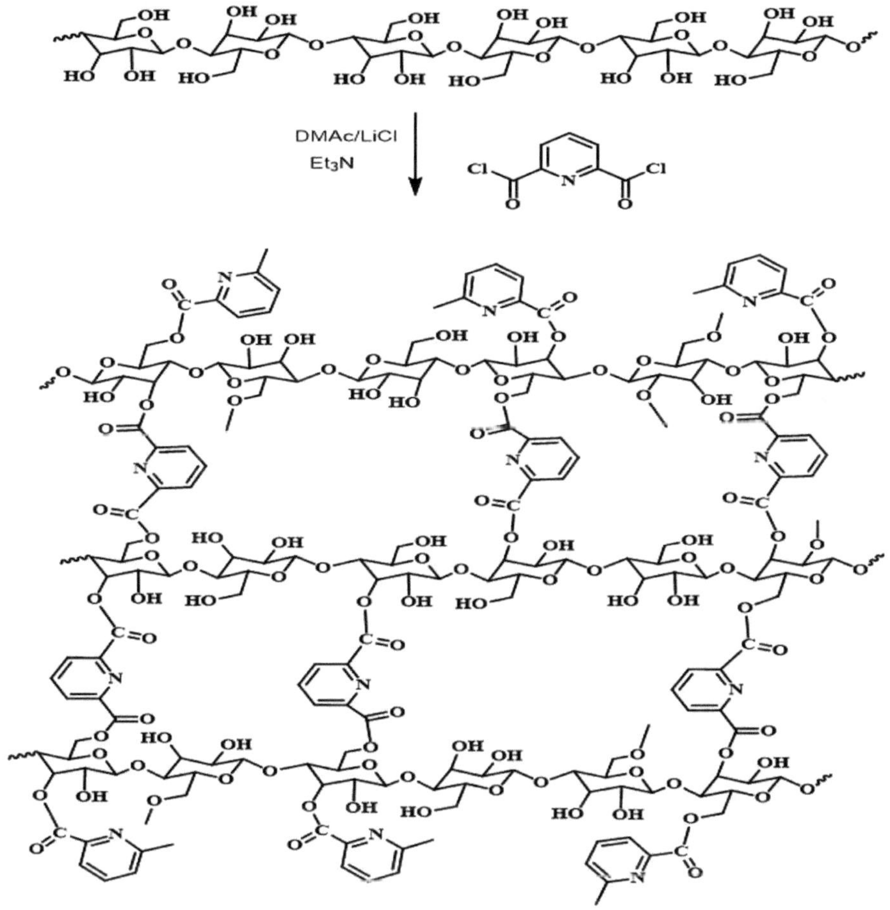

Scheme 2. Cellulose crosslinking with 2,6-pyridine dicarbonyl dichloride

Adsorption Results

Effect of Contact Time

Arsenic removal from water was carried out on Cell-X, Cell-Y and Cell-Z nano adsorbents. The adsorption was evaluated as a function of time at a constant pH 7, with adsorbent dose = 10 mg, solution volume = 10 mL, initial concertation = 10 ppm, and temperature of 20 °C. As shown in Fig. 10.5, the adsorption started rapidly and almost reached the plateau after few minutes, due to the availability of the binding sites at the beginning. After reaching the equilibrium time for each experiment, all adsorption sites will be occupied.

Figure 10.5 shows that the percentages of removal reached 82.97%, 74.38% and 79.72% for Cell-X, Cell-Y and Cell-Z, respectively, with contact time of 50 min,

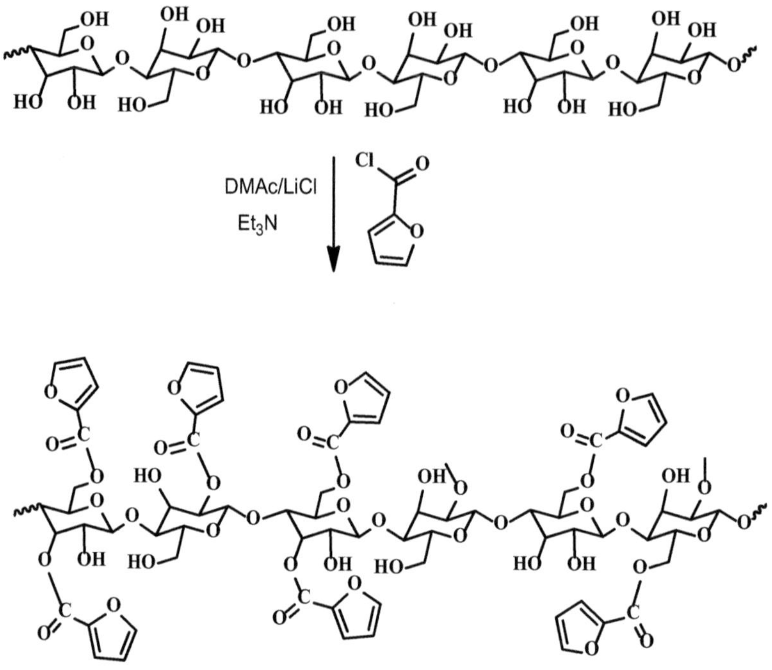

Scheme 3. Cellulose acylation with furan-2-carbonyl chloride

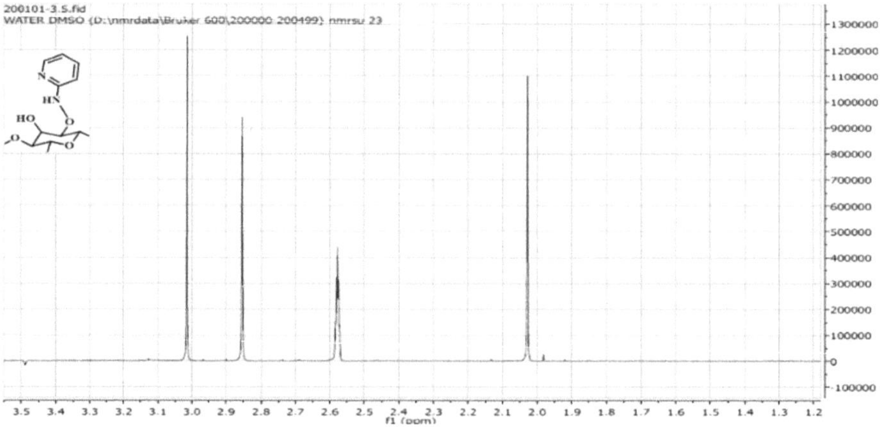

Fig. 10.1 ^{1}H NMR spectra of cellulose functionalized with 2-amino pyridine

10 min and 5 min respectively, which make these cellulose functionalized derivatives behave as unique adsorbents with the ability to trap Arsenic strongly.

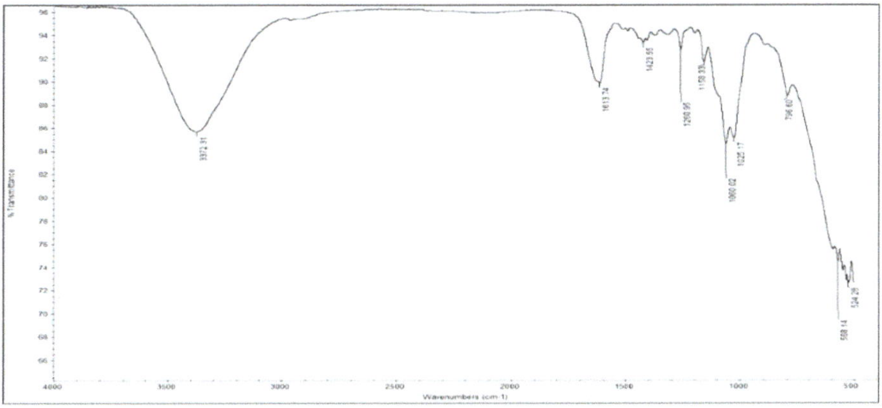

Fig. 10.2 FT-IR characterization of cellulose functionalized with 2-amino pyridine

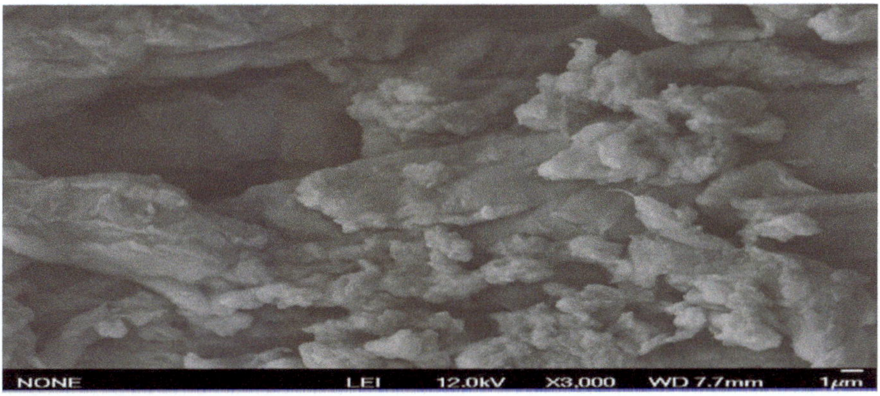

Fig. 10.3 SEM image of cellulose functionalized with 2-amino pyridine at 1 µm

Fig. 10.4 TGA analysis of cellulose functionalized with 2-amino pyridine

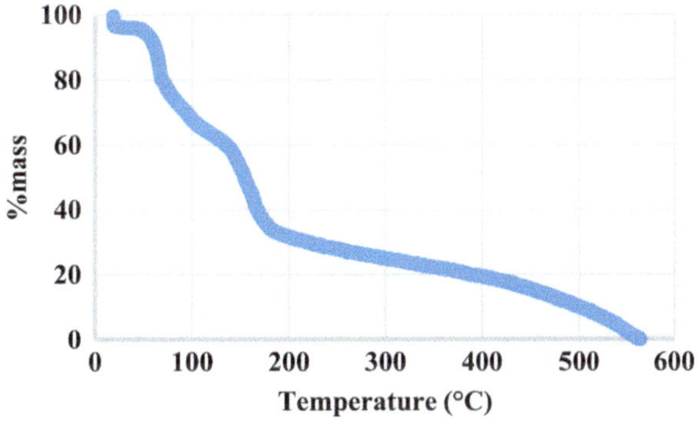

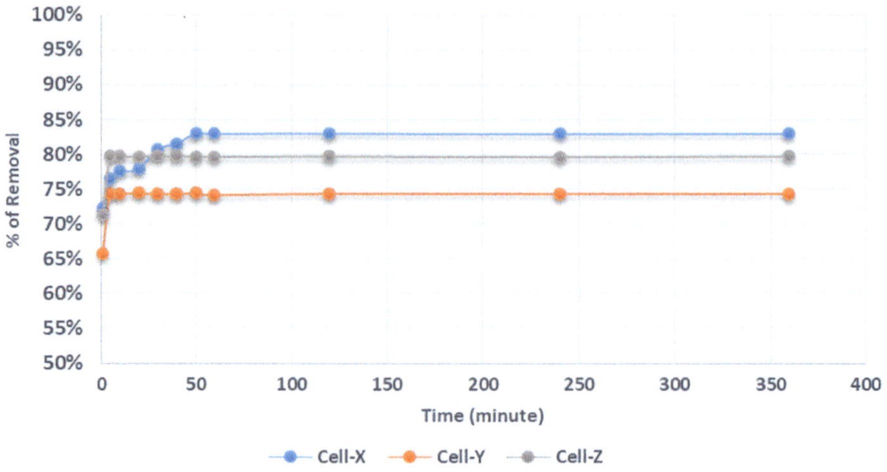

Fig. 10.5 Contact time effect for arsenic removal by Cell-X, Cell-Y and Cell-Z. Initial concertation (C_0) = 10 ppm, adsorbent dose = 10 mg, solution volume = 10 mL, pH = 7, temperature = 20 °C)

Effect of pH

The pH value is a critical factor in adsorption, because pH value could control the surface charge between the nano-adsorbent and Arsenic. The adsorption efficiency as a function of the pH is shown in Fig. 10.6, while the other solution variables were kept constant. At pH lower than 6, adsorption efficiency was very low (less than 50%) since the amines were in the ammonium form ($-NR_2H_2^+$), also the carboxyl and hydroxyl groups were in protonated form (COOH and OH), so the adsorption efficiency was low. Whereas, at pH values higher than 6, the amines were in the basic form, such that, a lope pair of electrons are available on the N for binding. The highest adsorption efficiency of Arsenic was observed at pH value in the range from 6 to 8. The highest removal percentages were 87.89%, 88.45% and 89.65% for Cell-X, Cell-Y and Cell-Z, respectively.

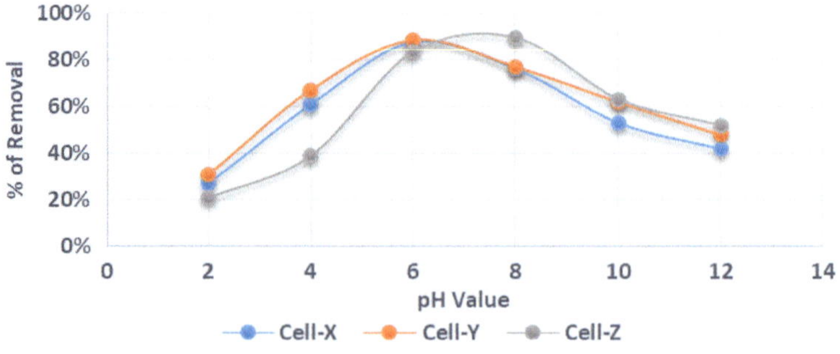

Fig. 10.6 pH effect of arsenic removal by Cell-X, Cell-Y and Cell-Z (CI = 10 ppm, adsorbent dose = 10 mg, solution volume = 10 mL, temperature = 20 °C, time = 30 min)

Fig. 10.7 Arsenic concentration effect by Cell-X, Cell-Y and Cell-Z (adsorbent dose = 10 mg, solution volume = 10 mL, temperature = 20 °C)

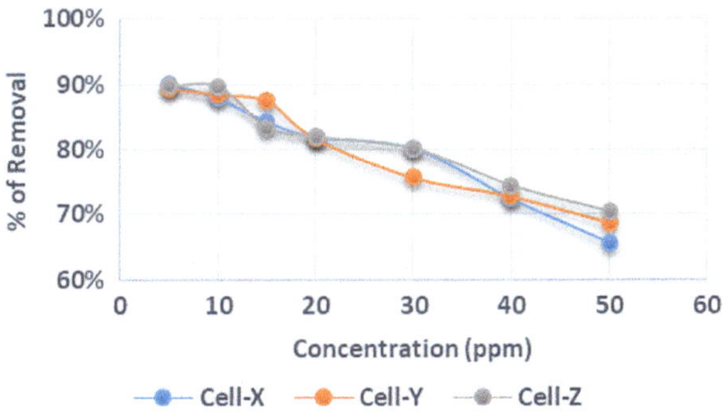

Effect of Arsenic Concentration

Arsenic initial concentration effect on the percentage removal using Cell-X, Cell-Y and Cell-Z was detected. The maximum percent of Arsenic removal was at 5 ppm Arsenic concentration (Fig. 10.7). As shown in the figure, as the Arsenic concentration increases, the percentage of removal decreases. The Arsenic adsorption and the sufficient binding sites in each nano-polymer are controlled by ion diffusion process. Such that, the binding sites will be more saturated with increasing arsenic concentration in the aqueous solution.

Effect of Temperature

The temperature effect on Arsenic adsorption by Cell-X, Cell-Y and Cell-Z was evaluated; the other variables were kept constant. As shown in Fig. 10.8, the maximum adsorption efficiency was around room temperature. At a temperature over 30 °C, the percentage of Arsenic removal started to decrease in which heating process will adversely affect the thermal stability of each nano-polymer.

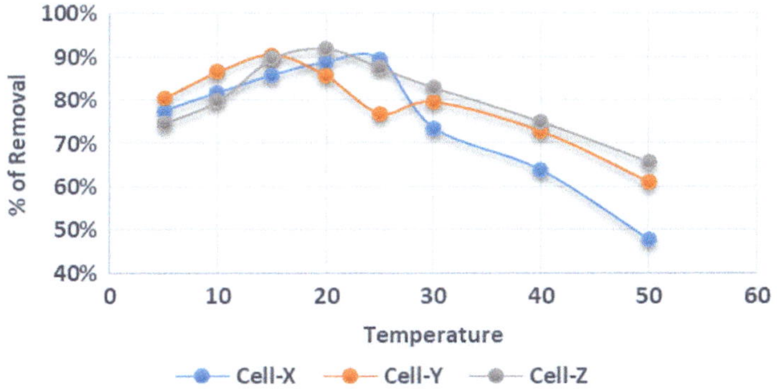

Fig. 10.8 Temperature effect of arsenic removal by Cell-X, Cell-Y and Cell-Z (adsorbent dose = 10 mg, solution volume = 10 mL, pH = 7)

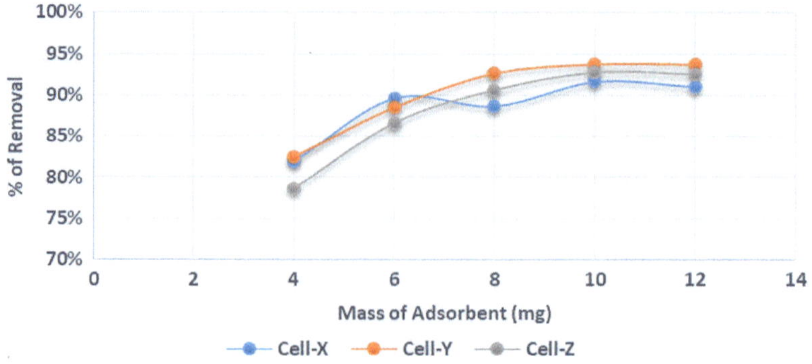

Fig. 10.9 Adsorbent dose effect on arsenic removal by Cell-X, Cell-Y and Cell-Z (solution volume = 10 mL)

Effect of Adsorbent Dose

The rate of Arsenic removal increased as the polymers dosage including Cell-X, Cell-Y or Cell-Z increased. The results indicated that adsorption could be controlled by diffusion and surface coordination mechanisms. An increase in the polymer dosage resulted in an increase in the number of available adsorption sites. When all adsorption sites get occupied, the diffusion process begins, that is controlled by osmosis. As the concentration of the adsorbate in the nano-polymer is equal to that in the solution, the adsorption reaches the plateau. As shown in Fig. 10.9. The optimum dose of Cell-X, Cell-Y, or Cell-Z was 12 mg due to increasing the surface to volume ratio (Foo & Hameed, 2010).

Adsorption Analysis

The best adsorption isotherm was investigated depending on the closeness of the correlation coefficient to one, for both Freundlich and Langmuir models. Adsorption kinetic models were also determined using pseudo-first-order and pseudo-second-order adsorption models. Van't Hoff graph was investigated in order to determine if the adsorption is spontaneous or not, and if each process is exothermic or endothermic one, this was done by detecting changes in Entropy, Enthalpy and Gibbs free energy for each process.

Isotherm Models

Analysis of isotherm data is vital since the results could be used to accurately explain the adsorption mechanism. The most common isotherms that are applied in solid/

liquid extraction systems the Langmuir and Freundlich isotherms (Foo & Hameed, 2010; Qiu et al., 2009).

Langmuir Adsorption Isotherm

Langmuir represents the chemisorption process. This adsorption is limited to a monolayer coverage, such that, adsorbed molecule cannot migrate across the adsorbent surface or interact with neighboring molecules. In addition, the adsorption sites are all have the same amount of energy (Qiu et al., 2009). Langmuir model studies adsorbate coverage on the adsorbent solid surface to the medium concentration above the solid surface in which the process occurs at fixed temperature value. The Langmuir model is expressed in Eq. (10.3).

$$\frac{\mathbf{C_e}}{\mathbf{q_e}} = \frac{1}{\mathbf{bQ_o}} + \frac{1}{\mathbf{Q_o}}\mathbf{C_e} \tag{10.3}$$

where C_e: equilibrium concentration of Arsenic (mg L^{-1}), b: Langmuir affinity constant (L/mg), Q_o: adsorption capacity at equilibrium (mg/g), q_e: amount of Arsenic divided by the adsorbent mass (mg/g). A graph of (C_e/q_e) versus C_e is used for finding out the Langmuir parameters. Which are ($1/Q_o$) as graph slope and ($1/bQ_o$) as the plot y-intercept (Foo & Hameed, 2010; Qiu et al., 2009).

Freundlich Adsorption Isotherm

Freundlich isotherm detects the adsorption to surfaces supporting sites of different affinities or to adsorption on heterogeneous adsorbent surfaces. This isotherm assumed that stronger adsorbents binding sites are occupied in the beginning in which the binding strength decays with increasing site occupation degree. According to Freundlich model, the adsorbed mass per mass of adsorbent can be written by Eq. (10.4) (Adejo, 2014).

$$\mathbf{log q_e} = \mathbf{log K_F} + \frac{1}{\mathbf{n}}\mathbf{log C_e} \tag{10.4}$$

where;

K_F is constant for Freundlich Isotherm detects the adsorption capacity (mg/g), and n represents the Heterogeneity coefficient, that used to detect if an adsorption is favorable or not (g/L). A plot of log q_e versus log C_e is used to calculate the Freundlich parameters, which are, log K_F as graph y-intercept and (1/n) as the plot slope. The adsorption of Arsenic on Cell-Z, Cell-Y and Cell-Z was fitted to Langmuir and Freundlich isotherms. The adsorption parameters were investigated by plotting C_e/q_e versus C_e for Langmuir adsorption isotherm, and log q_e versus log C_e

for Freundlich adsorption isotherm. These adsorption processes are fitted with both adsorption isotherms with R^2 that very close to one.

Adsorption Kinetic Models

The kinetic models give information about the adsorption system behavior and the rate at which specific constituent is removed by the adsorbents. In addition, the kinetic models can investigate the rate-determining-step, as well as, determining the type of adsorption. Several adsorption kinetic models have been established. Examples are including, external mass transfer model, first-order reversible reaction model, pseudo-first-order, pseudo-second-order rate model intra-particle diffusion model and many others (Ferrari et al., 2010; He et al., 2010; Sridev & Rajendran, 2009).

Pseudo First-Order Kinetics

The equation for this kinetic model is shown below:

$$\mathbf{log(q_e - q_t) = logq_e - \frac{K_1}{2.303}t} \tag{10.5}$$

where,

- q_e and q_t are the masses of Arsenic per unit mass of adsorbent at equilibrium, at time t respectively (mg/g).
- k_1: Rate constant for pseudo-first-order kinetic adsorption (mg g^{-1} min^{-1}).

A plot of log ($q_e - q_t$) versus t will give a straight line for the pseudo-first-order model in which $logq_e$ is the plot y-intercept and ($-k_1/2.303$) represents the graph slope (Lin & Wang, 2009).

Pseudo-Second-Order Kinetic Model

This adsorption model of kinetics detects that the rate-determining step can be fitting with chemical adsorption involving valence bonds in which the electrons are shared or exchanged between the adsorbent and the adsorbate.

Pseudo-second-order model can be written by Eq. (10.6):

$$\frac{\mathbf{t}}{\mathbf{q_t}} = \frac{1}{\mathbf{k_2 q_e^2}} + \frac{1}{\mathbf{q_e}}t \tag{10.6}$$

where;

- k_2: Rate constant of pseudo-second-order kinetic adsorption at equilibrium (g mg^{-1} min^{-1}). The graph of t/q_t versus t gives a linear relationship that allows

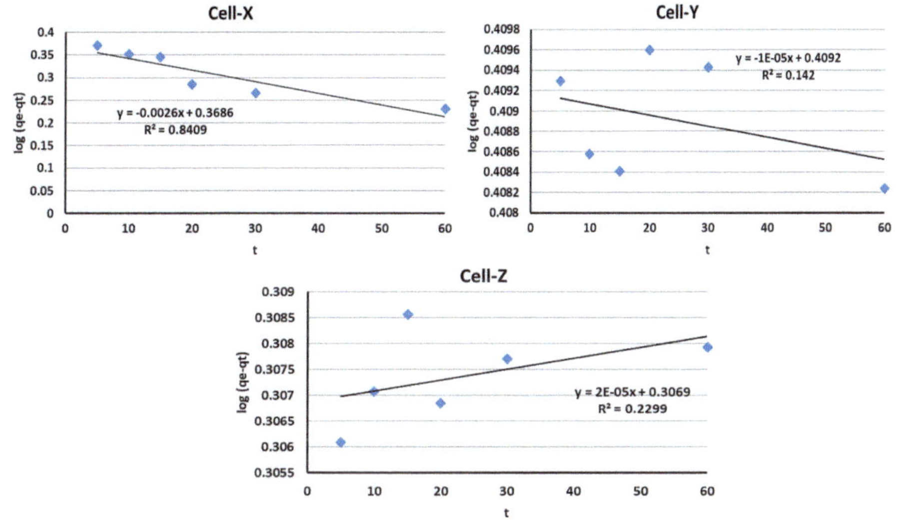

Fig. 10.10 Pseudo-first-order kinetic model results for arsenic adsorption on Cell-X, Cell-Y and Cell-Z adsorbents, respectively

computation of the rate constants in Pseudo-second-order kinetic, which are k_2 and q_e values (Agrawal & Sahu, 2009).

The kinetic experimental data of Arsenic pesticide on Cell-X, Cell-Y and Cell-Z fitted with pseudo-first-order and pseudo-second-order adsorption models for investigating the adsorption mechanism for each process. The kinetics parameters and values of R^2 were calculated from the linear plots of log $(q_e - q_t)$ versus t for pseudo-first-order adsorption model (Fig. 10.10), (t/q_t) versus t for pseudo-second-order model. According to the correlation coefficient values, the adsorptions of Arsenic on all three nano polymers Cell-X, Cell-Y and Cell-Z followed pseudo-second-order kinetic model (Fig. 10.11). These results indicate strong binding between each adsorbent and Arsenic.

Adsorption Thermodynamics

Thermodynamic studies are vital in order to determine if an adsorption process is a favorable or not. The thermodynamic adsorption behavior is detected depending on calculating the thermodynamic parameters: Change in Gibbs free energy (ΔG), Entropy change (ΔS) and Enthalpy change (ΔH) in which; (J) is the unit of ΔG and ΔH, whereas; the unit of ΔS is (J/K) (Agrawal & Sahu, 2009; Li et al., 2003; Tan et al., 2007; Ugurlu et al., 2008).

Equation (10.8) represents the general relation between the thermodynamic parameters:

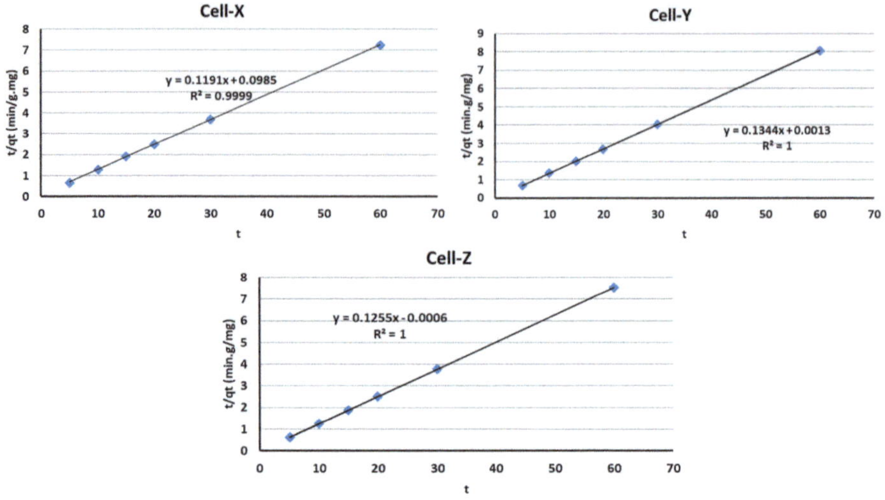

Fig. 10.11 Pseudo-second-order kinetic model results for arsenic adsorption on Cell-X, Cell-Y and Cell-Z adsorbents, respectively

$$\Delta G = \Delta H - T\Delta S \tag{10.7}$$

where, T: Absolute temperature (K).

Additionally, Gibbs free energy change can be expressed using Eq. (10.9):

$$\Delta G = -RT \, \mathrm{In}\, K_d \tag{10.8}$$

where.

- R: Universal gas constant (8.314 J mol^{-1} K^{-1}).
- K_d: Thermodynamic equilibrium constant which equals (q_e/C_e) with a unit of mol or L g^{-1}.

The combination of Eqs. (10.7) and (10.8) will result in Eq. (10.9).

$$lnK_d = \frac{\Delta S}{R} - \frac{\Delta H}{RT} \tag{10.9}$$

A graph of $\mathrm{In}K_d$ versus (1/T) results in a line with a plot slope (−ΔH/R) and y-intercept (ΔS/R). The graph is called Van't Hoff plot (Tan et al., 2007). Van't Hoff graph was used in order to calculate the thermodynamic parameters (ΔH, ΔG and ΔS values) for Arsenic adsorption by Cell-X, Cell-Y and Cell-Z. According to negative ΔG° and ΔH values. This detects that the adsorption processes of Arsenic on by Cell-X, Cell-Y and Cell-Z are exothermic and spontaneous at the tested temperatures. In addition, negative ΔS indicates a presence of an associative mechanism, such that,

Fig. 10.12 Effect of adsorbent recovery on arsenic adsorption by Cell-X, Cell-Y and Cell-Z

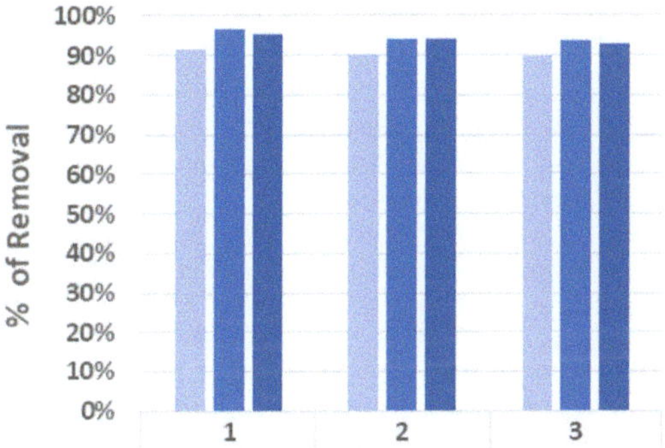

	1	2	3
1st use	91.65%	90.24%	89.65%
1st regeneration	96.73%	94.24%	93.76%
2nd regeneration	95.54%	94.11%	92.87%

no significant change will occur in the internal structure of Cellulose modified nano polymers during the adsorption process.

Adsorbents Regeneration

Each adsorbent was rinsed with diluted HNO_3 (0.1 N), water, and ethanol in order to ensure the complete removal of Arsenic. Then was used for three adsorption cycle, results are shown in Fig. 10.12. In the three-cycle, small drop in the adsorption efficiency was detected. These results indicated that, Cell-X, Cell-Y and Cell-Z could be promising materials on a commercial scale for removing Arsenic from water.

Conclusion

Cellulose based derivatives with heterocyclic rings were synthesized by reacting Cellulose modified with 2-amino pyridine, furan-2-carbonyl chloride or pyridine-2,6-dicarbonyl dichloride. The prepared nano-polymers were characterized by FT-IR, SEM, TGA and ^{1}H NMR instruments and were evaluated as novel adsorbents for Arsenic removal from groundwater. The optimum adsorption conditions including contact time, pH, temperature, adsorbent dose and Arsenic initial concentration were determined. The removal efficiency of Arsenic from water was promising for the three adsorbents. The results concluded that the adsorption process followed both Freundlich and Langmuir models. The kinetics study showed that, each adsorption is fitted with the pseudo second-order kinetic model. The obtained thermodynamic values (−negative $\Delta G°$) indicate spontaneous adsorptions.

References

Abdul, M. K. S., Jayasingheb, S. S., Chandanaa, E. P. S., Jayasumanac, C., Mangala, P., & De Silva, C. S. (2015). Arsenic and human health effects: A review. *Environmental Toxicology and Pharmacology, 40*, 828–846.

Adejo, S. (2014). Proposing a new empirical adsorption isotherm known as Adejo-Ekwenchi isotherm. *Journal of Applied Chemistry, 6*, 66–71.https://doi.org/10.9790/5736-0656671

Agrawal, A., & Sahu, K. K. (2009). Kinetics and isotherm studies of cadmium adsorption on manganese nodule residue. *Journal of Hazardous Materials, 137*(2), 915–924.

Asere, T. G., Stevens, C. V., & Laing, G. D. (2019). Use of (modified) natural adsorbents for arsenic remediation: A review. *Science of the Total Environment, 676*, 706–720.

Attia, T. M. S., Hu, X. L., & Yin, D. Q. (2012). Synthesized magnetic nanoparticles coated zeolite (MNCZ) for the removal of arsenic (As) from aqueous solution. *Journal of Experimental Nanoscience*, 1–10.

Ayrilmis, N., & Buyuksari, U. (2010). Utilization of olive mill sludge in the manufacture of fiberboard. *BioResources, 5*(3), 1859–1867.

Azbar, N., Bayram, A., Filibeli, A., Muezzinoglu, A., Sengul, F., & Ozer, A. (2010) A review of waste management options in olive oil production. *Critical Reviews in Environmental Science and Technology, 34*(3), 209–247.

Balamurugan, R., Sundarrajan, S., & Ramakrishna, S. (2011). Recent trends in nanofibrous membranes and their suitability for air and water filtrations. *Membranes, 1*(3), 232–248. https://doi.org/10.3390/membranes1030232

Chandra, V., Park, J., Chun, Y., Lee, J. W., Hwang, I., & Kim, K. S. (2010). Water dispersible magnetite reduced graphene oxide composites for arsenic removal. *ACS Nano, 4*(7), 3979–3986.

Chen, B., Zhu, Z., Ma, J., Qiu, Y., & Chen, J. (2013). Surfactant assisted Ce–Fe mixed oxide decorated multiwalled carbon nanotubes and their arsenic adsorption performance. *Journal of Materials Chemistry A, 1*, 11355–11367.

Ferrari, L., Kaufmann, J., Winnefeld, F., & Plank, J. (2010). Interaction of cement model systems with super plasticizers investigated by atomic force microscopy, zeta potential, and adsorption measurements. *Journal of Colloid and Interface Science, 347*(1), 15–24.

Foo, K., & Hameed, B. (2010). Insights into the modeling of adsorption isotherm systems. *Chemical Engineering Journal, 156*(1), 2–10.

Habuda-Stanić, M., & Nujić, M. (2015). Arsenic removal by nanoparticles: A review. *Environmental Science and Pollution Research, 22*, 8094–8123.

Hamed, O., Foad, Y., Hamed, E., & Al-Hajj, N. (2012). Cellulose powder from olive industry solid waste. *BioResources, 7*(3), 4190–4201.

He, J., Hong, S., Zhang, L., Gan, F., & Ho, Y. (2010). Equilibrium and thermodynamic parameters of adsorption of methylene blue onto rectorite. *Fresenius Environmental Bulletin, 19*(11), 2651–2656.

Idswater. (2010). *ActiGuard AAFS50 activated alumina for the removal of arsenic from potable water*. www.idswater.com/water/us/activated_alumina/457/products.html

Jain, C. K., & Ali, I. (2000). Arsenic: Occurrence, toxicity and speciation techniques. *Water Research, 34*(17), 4304–4312.

Kabir, F., & Chowdhury, S. (2017). Arsenic removal methods for drinking water in the developing countries: Technological developments and research needs. *Environmental Science and Pollution Research., 24*(31), 24102–24120.

Khalaf, B., Hamed, O., Jodeh, S., Bol, R., Hanbali, G., Safi, Z., & Samhan, S. (2021a). Cellulose-based hectocycle nanopolymers: Synthesis, molecular docking and adsorption of difenoconazole from aqueous medium. *International Journal of Molecular Sciences, 22*(11), 6090.

Khalaf, B., Hamed, O., Jodeh, S., Hanbali, G., Bol, R., Dagdag, O., & Samhan, S. (2021b). Novel, environment-friendly cellulos based derivatives for tetraconazole removal from aqueous solution. *Polymers, 13*, 450.

Li, Y., Wang, S., Luan, Z., Ding, J., Xu, C., & Wu, D. (2003). Adsorption of cadmium (II) from aqueous solution by surface oxidized carbon nanotubes. *Carbon, 41*(5), 1057–1062.

Lin, J., & Wang, L. (2009). Comparison between linear and non-linear forms of pseudo-first-order and pseudo-second-order adsorption kinetic models for the removal of methylene blue by activated carbon. *Frontiers of Environmental Science & Engineering*, 320–324.

Medina-Ramirez, A., Gamero-Melo, P., Ruiz-Camacho, B., Minchaca-Mojica, J. I., Romero-Toledo, R., & Gamero-Vega, K. Y. (2019). Adsorption of aqueous As(III) in presence of coexisting ions by a green Fe-modified W zeolite. *Water, 11*, 281.

Medronho, B., & Lindman, B. (2014). Competing forces during cellulose dissolution: From solvents to mechanisms. *Current Opinion in Colloid & Interface Science, 19*(1), 32–40.

Mohan, D., & Pittman, C. U., Jr. (2007). Arsenic removal from water/wastewater using adsorbents, a critical review. *Journal of Hazardous Materials, 142*(1–2), 1–53.

Mukherjeea, D., Ghosh, S., Majumdar, S., & Annapurnaa, K. (2016). Green synthesis of Fe_2O_3 nanoparticles for arsenic (V) remediation with a novel aspect for sludge management. *Journal of Environmental Chemical Engineering, 4*, 639–650.

Nicomel, N. N., Leus, K., Folens, K., De Voort, P. V., & Laing, G. D. (2016). Technologies for arsenic removal from water: Current status and future perspectives. *International Journal of Environmental Research and Public Health, 13*, 62.

Pathan, S., & Bose, S. (2018). Arsenic removal using "Green" renewable feedstock-based hydrogels: Current and future perspectives. *ACS Omega, 3*, 5910–5917.

Poguberović, S. S., Krčmar, D. M., Maletić, S. P., Kónya, Z., Pilipović, D. D. T., Kerkez, D. V., & Rončević, S. D. (2016). Removal of As(III) and Cr(VI) from aqueous solutions using "green" zero-valent iron nanoparticles produced by oak, mulberry and cherryleaf extracts. *Ecological Engineering, 90*, 42–49.

Qiu, H., Lv, L., Pan, B., Zhang, Q., & Zhang, W. (2009). Critical review in adsorption kinetic models. *Journal of Zhejiang University, Science, A, 10*(5), 716–724.

Ramos, L., Morgado, D. L., Gessner, F., Frollini, E., & El Seoud, O. A. (2011). A physical organic chemistry approach to dissolution of cellulose: Effects of cellulose mercerization on its properties and on the kinetics of its decrystallization. *Arkivoc, 7*, 416–425.

Reynolds, T. D., & Richards, P. E. (2000). *Operational and process units in environmental engineering*. PWS Publishing Company.

Salameh, Y. (2009). Methods for extracting cellulose material from olive pulp [MSc thesis, An-Najah National University].

Shankar, S., Shanker, U., Shikha. (2014). Arsenic contamination of groundwater: A review of sources, prevalence, health risks, and strategies for mitigation. *The Scientific World Journal*, 304524.

Siddiqui, S. I., & Chaudhry, S. A. (2017). Arsenic removal from water using nano-composites: A review. *Current Environmental Engineering, 4*, 81–102.

Smedley, P. L., & Kinniburg, D. G. (2002). A review of the source, behavior and distribution of arsenic in natural waters. *Applied Geochemistry, 17*, 517–568.

Sridev, D., & Rajendran, K. V. (2009). Synthesis and optical characteristics of ZnO nanocrystals. *Bulletin of Materials Science, 32*(2), 165–168.

Tan, I., Hameed, B., & Ahmad, A. (2007). Equilibrium and Kinetic studies on basic dye adsorption by oil palm fiber activated carbon. *Journal of Chemical Engineering, 127*(1–3), 111–119.

Ugurlu, M., Gurses, A., & Acıkyıldız, M. (2008). Comparison of textile dyeing effluent adsorption on commercial activated carbon and activated carbon prepared from olive stone by $ZnCl_2$ activation. *Microporous and Mesoporous Materials, 111*(1–3), 228–235.

Wong, W. W., Wong, H. Y., Borhan, A., Badruzzaman, M., Goh, H. H., & Zaman, M. (2017). Recent advances in exploitation of nanomaterial for arsenic removal from water: A review. *Nanotechnology, 28*, 1–31.

World Health Organization. (WHO). (2011). *Guidelines drinking-water quality* (4th ed.). WHO.

Chapter 11
Investigation of Physicochemical Characteristics for Alumina Selection for Fluoride and Arsenic Removal

Rajendra S. Thakur, Payal A. Kaneria, Anil R. Gupta, and Saroj Sharma

Abstract This study attempts a correlation of physicochemical characteristics of four commercially available activated alumina samples with their efficacy towards harmful ions such as fluoride and arsenic removal to enable identification and possible preparation of suitable varieties. The equilibrium adsorption capacities of alumina samples for fluoride and arsenic were measured by batch experiments under mechanical stirring, wherein two samples labeled AA-1 and AA-2 have shown high fluoride (~90%) and arsenic (57–72%) removal. Surface area, phase composition by powder XRD, infrared spectroscopy, and weight loss with temperature (TGA), point of zero charge of alumina varieties were characterized. Alumina varieties better suited for fluoride and arsenic removal, AA-1 and AA-2, were found to have good surface area (~300 m^2/g), positively charged surface (pH_{pzc} ~ 9) and consist mainly of amorphous phases.

Keywords Activated alumina · Defluoridation · Dearsenification · Physicochemical properties

Electronic supplementary material The online version contains supplementary material available at https://doi.org/10.1007/978-981-97-4764-1_11.

R. S. Thakur (✉) · P. A. Kaneria
Analytical and Environmental Science Discipline and Centralized Instruments Facility, CSIR-Central Salt and Marine Chemicals Research Institute (CSIR-CSMCRI), Council of Scientific and Industrial Research (CSIR), Gijubhai Badheka Marg, Bhavnagar, Gujarat 364002, India
e-mail: rthakur@csmcri.res.in

P. A. Kaneria
Department of Science, Marwadi University, Rajkot-Morbi Road, Rajkot, Gujarat 360003, India

A. R. Gupta · S. Sharma
Membrane Science and Separation Technology Division, CSIR-Central Salt and Marine Chemicals Research Institute, G. B. Marg, Bhavnagar, Gujarat 364002, India

A. R. Gupta
Academy of Scientific and Innovative Research (AcSIR), Ghaziabad, Uttar Pradesh 201002, India

B. Sen Gupta and N. Martínez-Villegas (eds.), *Arsenic Remediation of Food and Water*,
https://doi.org/10.1007/978-981-97-4764-1_11

Introduction

Arsenic and fluoride are "two of the chemicals of greatest health concern in some natural waters" and are "responsible for large-scale health effects through drinking water exposure" (WHO, 2011). Elevated fluoride levels in drinking water cause deformity of bones of the human body from dental carries to crippling fluorosis (WHO, 2011). Ingestion of arsenic-contaminated drinking-water causes dermal lesions such as hyperpigmentation and hypopigmentation, peripheral neuropathy, skin cancer, bladder and lung cancers (WHO, 2011).

Few remediation techniques have been developed and studied to maintain the permissible level of both these ions in water. Among them, adsorption by activated alumina and separation by membrane-based technique i.e., reverse osmosis are considered Best Available Technologies (BAT) by EPA for the removal of both these ions (EPA, 2007). Activated alumina system is low cost, easy to operate, and requires minimum attention, whereas reverse osmosis systems are costly and require advanced operator skill level (EPA, 2007). Activated alumina is used for many applications and has widespread availability of known specifications produced as an industrial chemical with worldwide production of over 40 million tonnes per annum in 1996 and 1997 (Hudson et al., 2012).

The first US patent of Fluoride removal using Activated alumina was granted in the year 1936 wherein acid was added to water and a high rate of fluoride removal, 40 L of treated water per hour per kg of alumina, was obtained (Churchill, 1936). Good fluoride removal was obtained with water at a pH of 5.0–6.5 by acid addition, and regeneration was carried out using dilute sodium/potassium hydroxide solutions, followed by 1% hydrochloric acid (HCl) (5). Sulphuric acid (H_2SO_4) replaced hydrochloric acid (HCl) in large scale defluoridation plant, probably to protect operators and equipment from acid fumes (Rubel, 1984). Locally produced activated alumina was attempted in many studies in different countries for defluoridation (Eswar & Devaraj, 2011; Jagtap et al., 2012). Few attempts were made for activated alumina based defluoridation without the acid addition to water, however a low rate of fluoride removal was obtained, 3 L of treated water per hour per kg of alumina, in a fixed column setup (Chauhan et al., 2007; UNICEF). The addition of acid to water was probably needed to counter competitively adsorbing species. Hydroxide and bicarbonate are known to strongly interfere with defluoridation using alumina (Jagtap et al., 2012). Typical groundwater has a pH of around 8 and bicarbonate/carbonate species are about 300 mg/L (White et al., 1963), and thus they can possess strong interference to ~10 mg/L of Fluoride present in water which may explain the slow rate of fluoride adsorption onto alumina. Recently, a high rate of fluoride removal, 40 L of treated water per hour per kg of alumina, was obtained in a suspended bed column by pre-treatment of activated alumina using 1% aqueous solution of solid acid (sodium or potassium hydrogen sulphate) wherein the same step also regenerated the alumina for four cycles (Bavda et al., 2016) and the mechanism presented in another study (Bavda et al., 2018). This method relies on the adsorption of sulphate/bisulphate ions onto alumina due to adsorption thereby effectively neutralizing bicarbonate and hydroxide ions on the alumina surface and also present in fluoride-containing water. Affinity for sulphate allows the use of a safer choice in place of highly corrosive sulphuric acid, and using the same approach, highly corrosive sodium hydroxide could be replaced with a safer choice. Thus, faster defluoridation (contact time ~ 1.5 min) was realized with the elimination of chemicals associated with storage and handling hazards. No acid or its substitute is added to water, thereby requiring a much lesser chemical for the pre-treatment step and treated water obtained with a pH of around 8.

The effectiveness of adsorption by alumina for fluoride has attracted the attention of researchers for better materials. Recently aluminium oxide/hydroxide nanoparticles prepared using electrolysis wherein 300 min of contact time was prescribed for maximum removal capacity of ca. 1 and 3 mg/g for arsenic and fluoride respectively (Rathore & Mondal, 2017). Mesoporous alumina was used for arsenic delivering capacities of 121 mg/g of AS(V) however time period is mentioned of few hours and a significant observation was that the surface area does not greatly influence the adsorption capacity (Kim et al., 2004). Composite adsorbent of titanium and lanthanum impregnated on activated carbon has shown much higher fluoride and arsenic removal capacity than same weight of aluminium based (alumina) and iron based (E33p) adsorbents (Jing et al., 2012). Locally available sawdust was mixed with ferric hydroxide and activated alumina in aqueous media to form a paste which upon heating produced particles of different sizes removed arsenic and fluoride from groundwater in a column (Dhanasekaran & Sahu, 2021). Metallopolymer containing zirconium ions chemically bound with polymer matrix has produced synchronous removal of arsenic and fluoride with maximum removal capacity of 7 and 116 mg/g for arsenic and fluoride respectively (Gupta et al., 2022). The metallopolymer acts as an effective adsorbent wherein the metal ion is uniformly arranged throughout the polymer matrix (Gupta et al., 2022).

Arsenic and fluoride removal has been performed using reverse osmosis or adsorption. The main adsorbent used for arsenic and fluoride removal are alumina covering nano-alumina, iron-based E33p, composite adsorbent of titanium and lanthanum impregnated on activated carbon, mixture of sawdust, ferric hydroxide, and activated alumina and metallopolymer containing zirconium ions chemically bound in a polymer matrix. Often adsorption study is performed in small batches and time is reported in hours for arsenic and fluoride removal. Aluminium oxide/hydroxide nanoparticles prepared using electrolysis followed by crushing, shaped particles unlikely, were fixed into bricks after it was spent, probably after a single use (Rathore & Mondal, 2017). Composite adsorbent of titanium and lanthanum impregnated on activated carbon appears a costly preparation however the study does not mention reuse of the adsorbent (Jing et al., 2012). Sawdust mixed with ferric hydroxide and aluminium oxide study again talks about degradation and disposal (Dhanasekaran & Sahu, 2021). Adsorbents having metal ions deposited on saw dust and activated carbon are unlikely to either possess material strength or produce shaped particles (Dhanasekaran & Sahu, 2021; Jing et al., 2012). Reproducible synthesis and large-scale fluoride or arsenic removal with customized adsorbents is yet to be reported. Metallopolymers have been studied for the effectiveness of regeneration reagents (Gupta et al., 2022). Authors have worked with activated alumina and metallopolymers spherical particles with mechanical strength, needed respectively for column design and reusability.

A survey of literature reveals that many alumina preparations have been reported with different levels of performance probably arising from their physicochemical characteristics. Many alumina preparations are discussed in literature along with the observation that during heating in the air of aluminium hydroxides or oxide hydroxides, materials of high surface area are formed with little change in the habit of the primary crystals and but different structural forms which are thermodynamically not stable, however "are well reproducible and characteristic for a given temperature range and starting material" (Hudson et al., 2012). Further "The highest surface area and lowest crystalline order of the solid is obtained around 675 K (400 °C)" (Hudson et al., 2012). The accurate description of activated alumina is transition alumina, and few of them were characterised by X-ray diffraction and electron microscopy (Hudson et al., 2012; Santos et al., 2000). Activated alumina specification of IS: 9700: 1991 Grade 1 with a minimum surface area of 350 m^2/g, pore volume of 0.40 mg/L, bulk density of 0.75 g/mL was recommended for defluoridation (UNICEF). Alumina specification from UNICEF does not discuss surface charge, phase composition or other physicochemical properties. Gong et al. has prepared few different varieties of alumina for a combination of solution pH and calcination temperature and found that alumina prepared at pH of 5 and 7 and calcined at 60 °C were having higher fluoride adsorption capacity than those prepared at higher pH of 9 and higher calcination temperatures of 60–150 °C (Gong et al., 2012). It was also found that good alumina preparations were essentially amorphous while inferior varieties were more crystalline (Gong et al., 2012). Parida et al. prepared γ-alumina using controlled precipitation and presented a fluoride removal capacity of around 80% from 30 mg/L aqueous Fluoride solution (Parida et al., 2009). Ion adsorption was correlated with zeta potential for γ-alumina membrane (de Lint et al., 2003). Fluoride adsorption was correlated with zeta potential for a few α-Al_2O_3 preparations (Reyes Behena et al., 2002). Alumina powders are commercially available as acidic, neutral, and basic varieties for column chromatography (Sigma, 2020). Acid strength of synthetic zeolite was correlated with catalytic activity (Otouma et al., 1969), which suggests a possible role of point of zero charge towards defluoridation capacity however a range of point of zero charges has been reported in different alumina varieties (Gulicovski et al., 2008). Characterization and preparation of various alumina varieties were attempted (Assih et al., 1988; Gong et al., 2012; Martens et al., 2001; Morterra et al., 1992; Ramos et al., 2008). Activated alumina is often locally prepared and utilized for a low-cost and easy to setup solution in developing countries, a specification can be useful to handle performance. To the best of our knowledge, a systematic study to correlate the physicochemical characteristics of commercial activated alumina to the defluoridation capacity is not available.

In the present study, the removal efficiency of four commercial activated alumina samples for the arsenic and fluoride has been determined. Two samples have been earlier employed to demonstrate faster and safer methods (Bavda et al., 2016) and two others were chosen for study. The physicochemical characterization of activated alumina is carried out by point of zero charge (pH_{pzc}), surface texture by scanning electron microscope (SEM), molecular spectra by Fourier Transform infra-red spectroscopy (FT-IR), elemental composition by energy dispersive X-ray analysis (EDX), and Brunauer–Emmett–Teller (BET) surface area analysis, and phase composition by powder X-ray diffraction (PXRD) and thermo gravimetric analysis (TGA). Correlation of physicochemical properties with defluoridation/dearsenification capacities is discussed. Such a study can pave the way for production of alumina or alumina coated materials for fluoride and arsenic removal.

Material and Methods

Four types of commercially available activated alumina varieties were employed for the defluoridation and dearsenification study. AA-1 refers to DF-101 grade from Siddhartha Chem. Industries, Vadodara, AA-2 to DF-101 grade from Aquaplus Technologies, Salem, AA-3, and AA-4 refers to gamma and DF-101 grades from Sorbead India. AA-1 and AA-2 were used to demonstrate fast and safe defluoridation (Bavda et al., 2016). Following chemicals are used without additional purification; NaF (Spectrochem Pvt. Ltd.—99%), Sodium Hydrogen Arsenate (Na_2HAsO_4), (Fisher Scientific, India), HCl (Himedia—35%), NaOH (Finar Limited—97%), KCl (S. D. Fine Chem. Ltd.—99.5%), $KHSO_4$ (Rankem—99%). All the alumina varieties were spherical beads of size ~1 mm for AA-1 and AA-2, ~4 mm for AA-3 and ~2 mm for AA-4.

Equilibrium fluoride and arsenic (As V) removal capacities of alumina varieties were determined in a batch adsorption experiment. Alumina weighing 10 g was added to 1 L of Milli-Q water (typical TDS in the range of 1–2 ppm) containing 8.2 ppm of Fluoride (Table 11.1) and stirred for 60 min in an overhead stirrer at 200 rpm. Arsenic removal experiment was performed in the same manner wherein 3 g alumina was used for 1 L of 110 ppb of Arsenate (As (V)) solution. Determination of point of zero charge (pH_{pzc}) of the alumina beads was performed by the batch equilibration method (Gulicovski et al., 2008), using KCl 0.01 M as the background electrolyte, whose initial pH values were adjusted by small additions of diluted (0.1 M) hydrochloric acid or sodium hydroxide. These solutions (25 mL) were equilibrated with alumina varieties (0.1 g) for 36 h with occasional stirring.

SEM images were recorded using a JEOL JSM-7100F instrument with OXFORD, X-MAX attachment for recording EDX spectra. The BET surface areas and BJH pore size distributions were measured by N_2 adsorption isotherm at 77 K using Micromeritics ASAP 2020 instrument. The sample was degassed at 120 °C for 4 h under vacuum before measurement of N_2 adsorption–desorption isotherms. Thermogravimetric analysis is performed using TG 2091 Libra instrument from Netzsch for a temperature range of 30–800 °C by maintaining a heating rate of 10 °C/min. Powder XRD diffractograms were recorded on a PANalytical Empyrean system operated with CuKα radiation in the range of 5–80° (2θ). Infrared (IR) spectra were recorded

Table 11.1 Fluoride and arsenic (V) removal using alumina. 10 versus 3 g of activated alumina was added to 1 L of 8.2 ppm fluoride versus 110 ppb arsenic containing water, stirred at 200 rpm for 1 h at room temperature

Alumina variety	Fluoride removal			Arsenic (V) removal		
	Initial level (ppm)	Final level (ppm)	% Fluoride removed	Initial level (ppb)	Final level (ppb)	% Arsenic removed
AA-1	8.2	0.9	89	110	12	89
AA-2	8.2	0.8	90	110	6	94
AA-3	8.2	7.9	4	110	39	64
AA-4	8.2	6.1	26	110	17	84

on a Perkin Elmer Spectrum GX FTIR system as KBr pellets in the range of 400–4000 cm^{-1}. Raman spectra were recorded using a LabRAM Evolution system from Horiba with a laser of 532 nm. Oakton pH 700 m was used for measuring pH and pH_{pzc}. Fluoride ion concentration was measured using an ion-selective electrode (9609BNWP) attached to Orion Versa Star make 4-channel pH/ion/conductivity meter provided by Thermo Scientific. The concentration of arsenic in water was measured by an Inductively Coupled Plasma-Mass Spectrometer (ICPMS) (Thermo Fisher Scientific iCAP RQ). All these instruments were routinely calibrated using standards.

Results and Discussion

Fluoride (F-) and Arsenic (As V) Removal Capacity of Activated Alumina

Fluoride and arsenic removal capacities were determined in a batch adsorption experiment. Alumina samples of weight 10 or 3 g were added to 1 L solution of 8.2 ppm Fluoride or 110 ppb Arsenic respectively. The mixture was stirred in an overhead stirrer at 200 rpm for an hour. Observations regarding removal capacity are summarized in Table 11.1, wherein it can be noticed that AA-1 and AA-2 are showing higher removal capacities for both Fluoride (~90%) and Arsenic (V) (89–94%). Removal efficiencies by AA-3 and AA-4 are for Fluoride (4–26%) for Fluoride and 64–84% for Arsenic (V). AA-1 and AA-2 are good variety of activated alumina for fluoride and arsenic removal while AA-3 and AA-4 are not suitable for the applications.

It is known that alumina exists in many forms and many varieties of activated alumina can be prepared by different combinations of starting materials and preparation methodologies. Physicochemical characterization of activated alumina is hereby presented to correlate with fluoride and arsenic removal capacities.

Elemental Composition of Alumina Varieties

The elemental composition of alumina varieties, obtained via EDX study shown in Fig. 11.1, suggests similar elemental composition in all the samples where Aluminium and Oxygen are major components. In all samples, Al content is around 54%, and O content is around 46%, wherein small variations are probably arising from limitation of detection method or varying content of H_2O groups. Elemental composition can be utilized as characterization method for alumina material.

Molecular Spectroscopy: Infrared (IR), Raman, and Powder XRD

IR spectra (in transmittance mode) of alumina varieties are presented in the top graph of Fig. 11.2, where all the spectra are similar to reported spectra. IR peak at 800 cm^{-1} correspond to Al–O bonds and falls in the fingerprint region. The peak around 3500 cm^{-1} corresponds to O–H from Al–O–H and C–O–H groups. Similar positions and appearance for major peaks, around 3500 for hydroxyl stretching, around 800 cm^{-1} for AlO_4 stretching, and broad peak from 500 to 700 cm^{-1} for AlO_6 stretching for all the samples suggesting similar molecular environments (Assih et al., 1988; Martens et al., 2001; Morterra et al., 1992). A low-intensity peak at 1390 cm^{-1} attributed to adsorbed carbonate species may be noted in AA-3 and AA-4 and is absent for other varieties, and the same peak appears/disappears in all samples when washed with base/acid (supplementary data). Peaks at around 1150 and 1070 cm^{-1}, for δ(OH) vibrations of bulk hydroxyls of boehmite (Assih et al., 1988; Martens et al., 2001; Morterra et al., 1992) are present as broad in AA-1 and AA-2 while AA-3 and AA-4 only display 1070 cm^{-1}. Differences in 1155 and 1070 cm^{-1} peaks between alumina samples remain present upon acid/base washing (supplementary data), and they may be considered characteristic of suitable alumina for fluoride and arsenic removal.

Raman absorption spectra of the alumina varieties are presented in the middle and bottom graph of Fig. 11.2. Raman spectra of AA-1 and AA-2 show intense peaks around 350 and 450 cm^{-1} while broad peaks with reduced intensity in the spectra of AA-3 and AA-4. These sets of peaks are reported in dry alumina gel and pseudoboehmite, broadness arising from low crystallinity (Assih et al., 1988; Martens et al., 2001). Powder X-ray diffraction studies were performed to further explore presence of pseudoboehmite and crystallinity.

Powder X-ray diffractograms of alumina varieties are shown in Fig. 11.3 which are primarily similar and containing few phases of aluminium oxide, namely aluminium oxide (JCPDS 00-010-0425, described as γ-alumina by Santos et al. (2000), boehmite (JCPDS 98-002-7865), and gibbsite (JCPDS 98-003-6233). Boehmite peaks are of negligible intensity in AA-1 and AA-2 in the X-ray diffractogram, while they are of higher intensity and broad in AA-3 and AA-4, indicating that it is present in

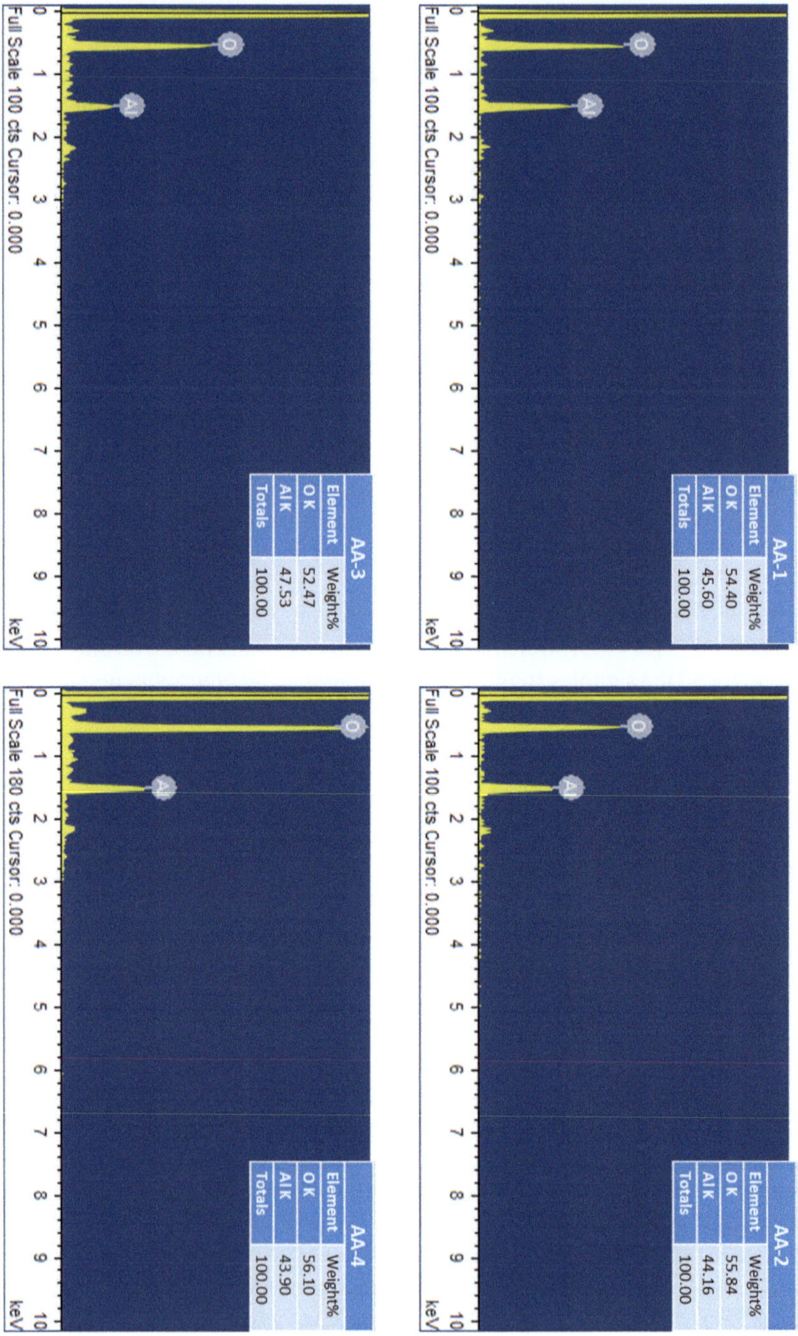

Fig. 11.1 EDX spectra showing elemental composition for AA-1 to AA-4

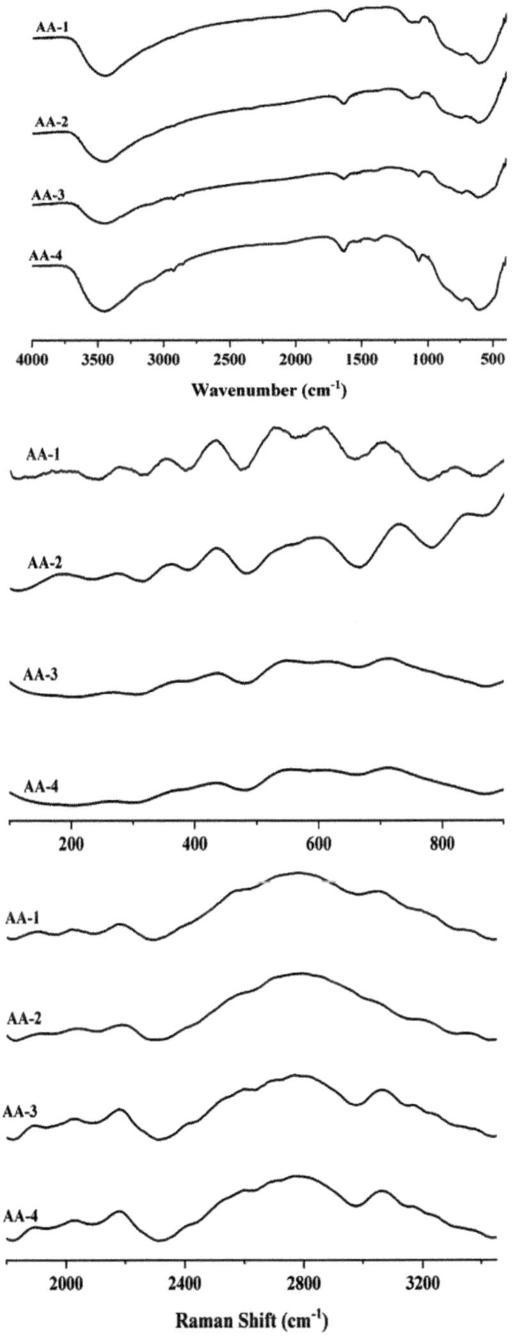

Fig. 11.2 Infrared spectra (IR) of AA varieties in the top graph. Raman spectra of AA varieties are shown in the middle (100–900 cm^{-1}) and bottom (1800–3300 cm^{-1}) graphs

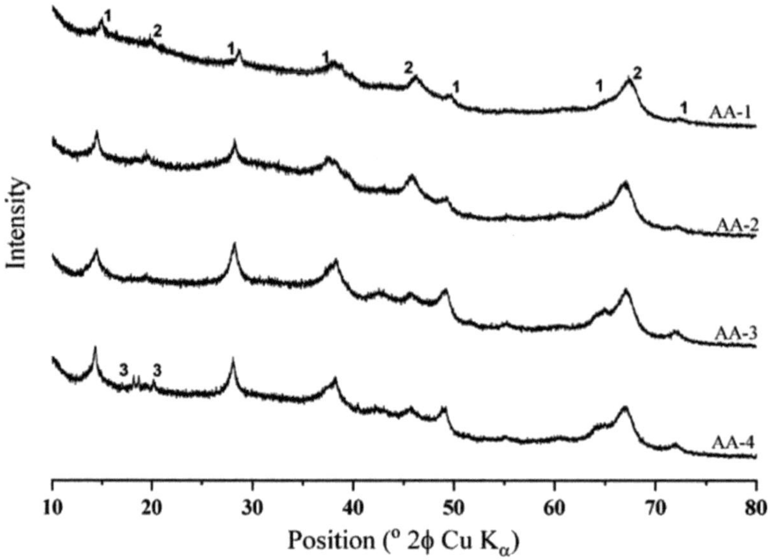

Fig. 11.3 Powder X-ray diffractograms showing phase composition for AA-1 to AA-4. (Pseudo) boehmite, aluminum oxide, gibbsite peaks are marked with 1, 2, and 3, respectively

the form of pseudoboehmite (Guzmán-Castillo et al., 2001). AA-1 and AA-2 have a higher ratio of (Aluminium oxide/pseudoboehmite) than AA-3 and AA-4. AA-4 also contains a small fraction of gibbsite. It appears that a higher ratio of (aluminium oxide/pseudoboehmite) of alumina variety is correlated to higher fluoride and arsenic removal capacity.

Morphology Study: SEM, BET Surface Area, and TGA

Surface morphology of activated alumina varieties, shown in Fig. 11.4, was observed using a scanning electron microscope at magnification factors of 1000 and 8000. Surfaces of AA-1 and AA-1 appear continuous while those of AA-3 and AA-4 have few chunks in the range of 10 μm at a lower magnification factor of 1000. Surfaces of AA-1, AA-2 and AA-4 can be observed to small lumps in the range of 1 μm at a higher magnification factor of 8000 with lumps of size around 1 μm while AA-3 appears porous owing to large sized lumps. Shapes like hexagonal plates for gamma-Al_2O_3, mentioned by Santos (Santos et al., 2000), was not observed in any SEM pictures, probably due to use of binder and other steps performed during production of spherical particles. SEM pictures composition can also be utilized as characterization method for alumina material.

The surface area of Activated alumina has been measured using Brunauer Emmett Teller (BET) method. The BET surface area, total pore volume and average pore

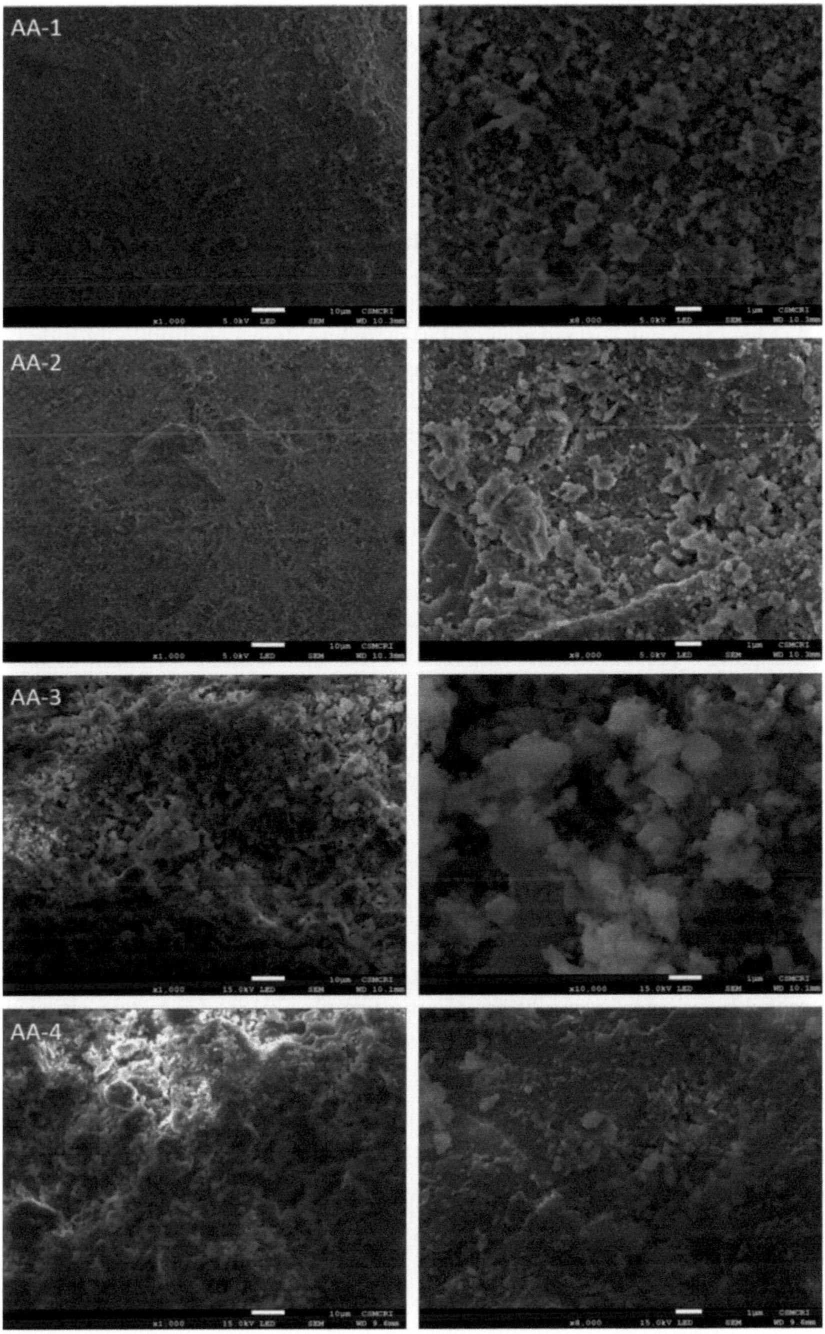

Fig. 11.4 SEM images showing surface morphology of activated alumina varieties, AA-1 (top row), AA-2 (second row), AA-3 (third row), and AA-4 (bottom row) at magnification of 1000 (left column) and 8000 (right column)

diameter are shown in Table 11.2. The BET surface area was 255 m^2/g for AA-3 and around 300 m^2/g for others. Average pore diameter and Pore volume are comparable for AA-1 and AA-2, around 47.1–49.8 Å and 0.37 cm^3/g, respectively. The values for average pore diameter and Pore volume are higher for AA-3 and AA-4, around 57.4–70.2 Å and 0.43–0.45 cm^3/g, respectively. Higher Pore diameter and volume for AA-3 and AA-4 compared to AA-1 and AA-2 could be due to presence of crystalline form of boehmite in the former and pseudoboehmite in the latter. During preparation of alumina, precipitation of aluminium hydroxide is followed by drying steps wherein formation of crystalline form may lead to tighter packing giving rise to larger pore diameter and volume and cause also cause drop in surface area.

AA-3 having a low surface area of 250 m^2/g has almost negligible fluoride (4%) and arsenic (14%) removal capacity among alumina varieties. AA-4 has lower fluoride (<1/3, 26% vs. 90%) and arsenic (~1/2, 33% vs. 57–72%) removal capacity than AA-1 and AA-2, although they have a similar surface area of 300 m^2/g. A low surface area may be related to low defluoridation capacity; however high surface area may not lead to high defluoridation capacity. This aspect needs to be carefully considered as alumina specification mentioned in WHO document lists high surface area as a requirement.

Weight loss obtained as a function of temperature during heating (Thermogravimetric Analysis TGA) is a characteristic of adsorbents delivering information about different phases present shown in Fig. 11.5 for all the alumina varieties display around 10% weight loss till 150 °C. Differential TG (DTG) graph suggests a slower rate of weight loss until 100 °C was observed for AA-1 and AA-2 than AA-3 and AA-4. It can be inferred that AA-1 and AA-2 have a higher affinity for water and carbon dioxide than AA-3 and AA-4. DTG graphs further show that AA-3 and AA-4 have a broad transition around 450 °C, likely arising from pseudoboehmite to transition alumina (Guzmán-Castillo et al., 2001). A transition around 500 °C was observed for AA-1 and AA-2 may arise from boehmite to transition alumina (Guzmán-Castillo et al., 2001), the broadness may arise from small quantities of boehmite is present in different crystallite sizes or multiple environments surrounded by amorphous aluminium oxide. A transition around 280 °C observed only in AA-4 is probably arising from gibbsite (Earnest et al., 2018). TGA data also shows presence of pseudoboehmite and low crystallinity in alumina varieties suitable for defluoridation (Fig. 11.5).

Table 11.2 Surface area (BET) values of alumina varieties

Sample name	BET surface area (m^2/g)	Average pore diameter (Å)	Pore volume (cm^3/g)
AA-1	292.7	49.8	0.37
AA-2	303.4	47.1	0.37
AA-3	255.1	70.2	0.45
AA-4	297.5	57.4	0.43

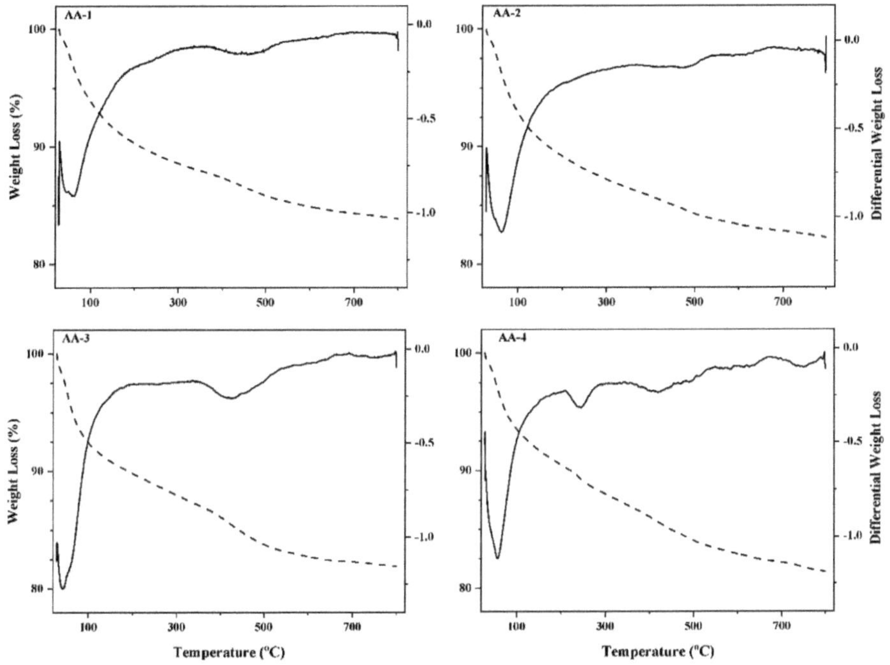

Fig. 11.5 TGA and DTG profile showing weight loss upon heating for AA-1 to AA-4

Surface Charge of Alumina

The graphical representation of final pH versus initial pH is shown in Fig. 11.6 where the centre of the plateau, having a similar final pH value for a range of initial pH values, is called the point of zero charge (pH_{pzc}). All the alumina varieties show absorption in acidic and basic ranges of pH and only larger plateau, having a larger difference between initial and final pH, may be considered.

AA-1 and AA-2 possess pH_{pzc} of 9 i.e., a plateau at a pH value of 9 while AA-3 and AA-4 possess a major plateau around a pH value of 5 and another minor plateau around 9. It suggests that AA-1 and AA-2 possess higher absorption capacity for anions like fluoride than those for AA-3 and AA-4. AA-3 and AA-4 possess higher absorption capacity for cations than that for negatively charged species suggesting their unsuitability for fluoride removal. The method used for measuring pH_{pzc} by zeta potential is known to produce doubtful results while the method used here is suitable for commercial alumina preparation (Analytic, 2016).

A graphical summary of the finding is shown in Fig. 11.7, depicting alumina variety with low crystallinity and positive charge is suitable for fluoride and arsenic removal from water.

Activated alumina with a surface area of $\geq$350 m^2/g (IS: 9700: 1991 Grade 1) was recommended for defluoridation (UNICEF). We have observed that alumina with a

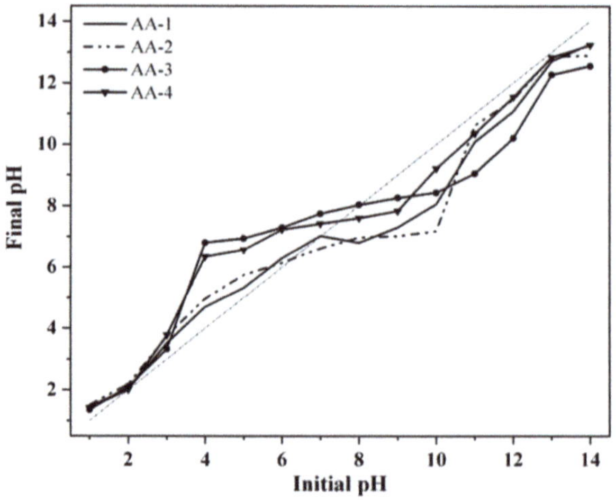

Fig. 11.6 Point of zero charge determination using batch equilibration method of alumina varieties. Graph showing pH_{pzc} determination using batch equilibrium method, consumption of H^+/OH^- for AA-1 to AA-4

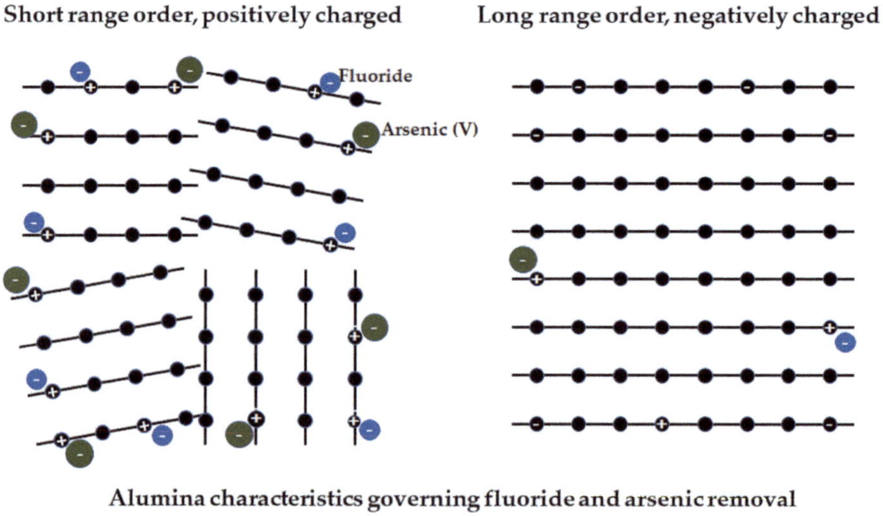

Fig. 11.7 Graphic representation of alumina varieties, having low crystallinity and positive charge, suitable for fluoride and arsenic removal

high surface area can deliver a low fluoride removal capacity however other conditions also need to be fulfilled. Our study suggests that alumina suitable for removing arsenic and fluoride consists mainly of aluminium oxide, an amorphous form having short-range order, which probably enables a larger number of unsaturated sites in the alumina. It can be inferred that such alumina, barring good binders suitable for aqueous operation, may release alumina particles. The leaching of aluminium ions (Al^{3+}) observed in our experiments was 0.1 ppm or lower, similar to those found in the lime-alum process (WHO, 2011) as, like many other metals, the aluminium oxide is known to form chains, cycles, and layers. We have observed a milky solution when water containing AA-1 and AA-2 were stirred by an overhead stirrer or water was passed through a suspended-bed column filled with AA-1 and AA-2, arising probably from alumina particles which settle quickly. For large operation, batch-wise drawing of treated water may be performed followed by decantation after settling to remove alumina sediments.

Alumina prepared by calcination of aluminium hydroxide can assume acidic or basic surface depending upon pH of the mother solution used for ageing (Gong et al., 2012), probably due to the amphoteric nature of aluminium ions. The surface of the alumina undergoes crystallisation to the extent determined by calcination temperature. Commercially available alumina varieties for column chromatography are stable at room temperature, and their activity can be restored upon drying. This physicochemical characterisation may help, or contribute to, specify grade suitable for removing fluoride and arsenic. It may be a worthwhile idea to prepare alumina for defluoridation having (1) a positive surface (pH_{pzc} around 9) by ageing in mother solution at pH of 5–7 and (2) a high proportion of aluminium oxide and low proportion of boehmite/pseudoboehmite (by PXRD, IR, and DTG), by calcination at suitable temperatures. Hudson et al. have described aluminium oxide PXRD phase, mentioned as gamma by Santos et al. (Santos et al., 2000), can be synthesised from boehmite, which can be synthesized, hydrothermally at 180–250 °C, from aluminium trihydroxide (Hudson et al., 2012). It may also be noted that washing of alumina surface by $KHSO_4$ may perform regeneration by neutralizing hydroxide and carbonate and desorption of fluoride by sulphate (SO_4^{2-}) ions. Alumina preparation recipe may be modified to prepare a negatively charged surface for uranium present in water in the form of uranyl (UO_2^{2+}/UO_2OH^+) ions. Parts of this study may be also useful for production of alumina or alumina coated materials for fluoride and arsenic removal.

Conclusions

A study to correlate physicochemical characteristics of alumina with fluoride and arsenic removal capacity was attempted. Four different alumina varieties with varying degrees of fluoride and removal capacities were chosen from our past experience. All the varieties have a similar elemental composition (EDX) and overall molecular structure (IR, Raman, XRD, TGA). The Different appearance of a peak around 1100 cm^{-1} in the IR spectra was found, doublet for AA-1 and AA-2 while singlet or AA-3 and AA-4. Phase composition from PXRD and DTG suggests AA-1 and AA-2 possess a higher ratio of aluminium oxide to psuedoboehmite phase than AA-3 and AA-4. AA-1 and AA-2 possess the point of zero charge (pH_{pzc}) around 9 while it is around 5 for AA-3 and AA-4. Surface area and weight loss upon heating suggest that these parameters may be needed to be above a threshold value for high defluoridation capacity; however, merely crossing the threshold they are not correlated. It may be inferred from the study that phase composition and pH_{pzc} should also be characterized for the use of alumina varieties to remove arsenic and fluoride.

Acknowledgements The work was done under DST project WTI-354-2K12 and CSIR projects OLP-0007 and MLP-0045. The authors are thankful to Analytical Discipline and Centralized Instruments Facility for providing instrumentation facilities. CSIR-CSMCRI Communication No. PRIS/182/2020.

Appendix

Statement of Significance

We have correlated analytical characteristics recorded using sophisticated instruments with the fluoride and arsenic removal of commercial alumina varieties. Alumina is an industrial preparation and is the material in a Best Available Technology for fluoride and arsenic removal. Recently we have patented a fast and safe process for defluoridation using alumina and also published a mechanism for the same. It was observed that only two varieties of alumina were suited for defluoridation. Thus, a specification is needed for alumina suitable for fluoride and arsenic removal. Data recorded shows that phase composition and surface charge govern removal capacity besides surface area. We have also proposed a graphical summary.

Material characteristic is expected to impact removal of harmful ions in a given process. Fluoride and arsenic removal capacity by alumina is correlated to (1) the surface area where large area can provide large number of sites, (2) phase composition as amorphous phase can have a larger number of surface unsaturated sites than that of crystalline phase, and (3) surface charge as positively charged surface can facilitate attachment of negative ions.

We are proposing that alumina of desired physicochemical characteristics can be used for arsenic and fluoride removal. This study can be used to produce alumina or other adsorbent for fluoride and arsenic removal. As activated alumina is locally prepared and utilized for a low-cost and easy to setup solution in developing countries, an specification can be useful to handle performance.

Graphical Abstract Synopsis

Fluoride and arsenic removal capacity by alumina is correlated to (1) the surface area where large area can provide large number of sites, (2) phase composition as amorphous phase can have a larger number of surface unsaturated sites than that of crystalline phase, and (3) surface charge as positively charged surface can facilitate attachment of negative ions.

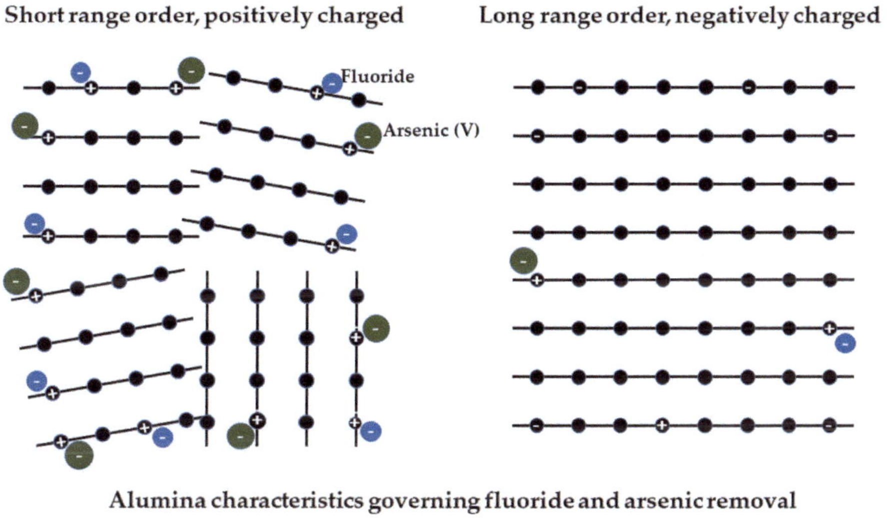

References

Analytic. (2016). *Colloidal alumina: Stability statement by zeta potential doubtful*. https://analytik.co.uk/wp-content/uploads/2016/12/application-note-colloidal-alumina-stability-statement-by-zeta-potential.pdf

Assih, T., Ayral, A., Abenoza, M., & Phalippou, J. (1988). Raman study of alumina gels. *Journal of Materials Science, 23*(9), 3326–3331.

Bavda, A. R., Sharma, S., Sharma, C. R., Bapat, P. S., & Thakur, R. S. (2016). *A process for the defluoridation of drinking water*. PCT patent 035101.

Bavda, A. R., Bapat, P. S., Joshi, V. B., & Thakur, R. S. (2018). Investigating the chemistry of alumina based defluoridation using safe H^+ source. *International Journal of Hydrogen Energy, 43*(47), 21699–21708.

Chauhan, V. S., Dwivedi, P. K., & Iyengar, L. (2007). Investigations on activated alumina based domestic defluoridation units. *Journal of Hazardous Materials, 139*(1), 103–107.

Churchill, H. V. (1936). *Removal of fluorine from water*. US patent US2059553.

de Lint, W. B. S., Benes, N. E., Lyklema, J., Bouwmeester, H. J. M., van der Linde, A. J., & Wessling, M. (2003). Ion adsorption parameters determined from zeta potential and titration data for a γ-alumina nanofiltration membrane. *Langmuir, 19*(14), 5861–5868.

Dhanasekaran, P., & Sahu, O. (2021). Arsenate and fluoride removal from groundwater by sawdust impregnated ferric hydroxide and activated alumina (SFAA). *Groundwater for Sustainable Development, 12*, 100490.

Earnest, C. M., Gann, K., & Stong, B. (2018) Improved quantification of gibbsite in bauxite ores by thermogravimetric methods (TGA and DTG). *Advances in Applied Chemistry and Biochemistry, 1*, 09–17.

Environmental Protection Agency. (2007). *Removing multiple contaminants from drinking water: Issues to consider*. https://www.epa.gov/sites/production/files/2015-04/documents/poster_treatment_technologies.pdf

Eswar, P., & Devaraj, C. G. (2011). Water defluoridation: Field studies in India. *Indian Journal of Dental Advancements, 3*, 526–533.

Gong, W.-X., Qu, J.-H., Liu, R.-P., & Lan, H.-C. (2012). Adsorption of fluoride onto different types of aluminas. *Chemical Engineering Journal, 189–190*, 126–133.

Gulicovski, J. J., Čerović, L. S., & Milonjić, S. K. (2008). Point of zero charge and isoelectric point of alumina. *Materials and Manufacturing Processes, 23*(6), 615–619.

Gupta, A. R., Joshi, V. C., Yadav, A., & Sharma, S. (2022). Synchronous removal of arsenic and fluoride from aqueous solution: A facile approach to fabricate novel functional metallopolymer microspheres. *ACS Omega, 7*(6), 4879–4891.

Guzmán-Castillo, M. L., Bokhimi, X., Toledo-Antonio, A., Salmones-Blásquez, J., & Hernández-Beltrán, F. (2001). Effect of boehmite crystallite size and steaming on alumina properties. *The Journal of Physical Chemistry B, 105*(11), 2099–2106.

Hudson, L. K., Misra, C., Perrotta, A. J., Wefers, K., & Williams, F. S. (2012). Aluminium oxide. In *Ullmann's encyclopedia of industrial chemistry* (Vol. 2, pp. 607–646). Wiley-VCH Verlag GmbH & Co KGaA.

Jagtap, S., Yenkie, M. K., Labhsetwar, N., & Rayalu, S. (2012). Fluoride in drinking water and defluoridation of water. *Chemical Reviews, 112*(4), 2454–2466.

Jing, C., Cui, J., Huang, Y., & Li, A. (2012). Fabrication, characterization, and application of a composite adsorbent for simultaneous removal of arsenic and fluoride. *ACS Applied Materials & Interfaces, 4*(2), 714–720.

Kim, Y., Kim, C., Choi, I., Rengaraj, S., & Yi, J. (2004). Arsenic removal using mesoporous alumina prepared via a templating method. *Environmental Science & Technology, 38*(3), 924–931.

Martens, W. N., Frost, R. L., Bartlett, J., & Kloprogge, J. T. (2001). The ageing of alumina hydrolysates synthesized from sec-butoxyaluminium (III). *Journal of Materials Chemistry, 11*(6), 1681–1686.

Morterra, C., Emanuel, C., Cerrato, G., & Magnacca, G. (1992). Infrared study of some surface properties of boehmite (γ-AlO_2H). *Journal of the Chemical Society, Faraday Transactions, 88*(3), 339–348.

Otouma, H., Arai, Y., & Ukihashi, H. (1969). The acidic nature of synthetic zeolites: Acid strength and its correlation with the catalytic activity. *Bulletin of the Chemical Society of Japan, 42*(9), 2449–2453.

Parida, K. M., Pradhan, A. C., Das, J., & Sahu, N. (2009). Synthesis and characterization of nano-sized porous gamma-alumina by control precipitation method. *Materials Chemistry and Physics, 113*(1), 244–248.

Ramos, R. L., Castillo, N. A. M., Azuara, A. J., Barron, J. M., Rodriguez, L. E. L., Rosales, J. M. M., & Piña, A. A. (2008). Fluoride removal from water solution by adsorption on activated alumina prepared from pseudo-boehmite. *Journal of Environmental Engineering and Management, 18*(5), 301–309.

Rathore, V. K., & Mondal, P. (2017). Competitive adsorption of arsenic and fluoride onto economically prepared aluminum oxide/hydroxide nanoparticles: Multicomponent isotherms and spent adsorbent management. *Industrial and Engineering Chemistry Research, 56*(28), 8081–8094.

Reyes Bahena, J. L., Robledo Cabrera, A., López Valdivieso, A., & Herrera Urbina, R. (2002). Fluoride adsorption onto α-Al_2O_3 and its effect on the zeta potential at the alumina–aqueous electrolyte interface. *Separation Science and Technology, 37*(8), 1973–1987.

Rubel Jr., F. (1984). *Design manual: Removal of fluoride from drinking water supplies by activated alumina* (EPA-600/2-84-134).

Santos, P. S., Santos, H. S., & Toledo, S. P. (2000). Standard transition aluminas electron microscopy studies. *Materials Research, 3*(4), 104–114.

Sigma. (2020). *Commercially available acidic/basic/neutral alumina.* https://www.sigmaaldrich.com/content/dam/sigma-aldrich/docs/Sigma-Aldrich/Product_Information_Sheet/a1522pis.pdf

UNICEF. *Draft specifications for domestic de-fluoridation filter unit and activated alumina for defluoridation*. http://www.unicefiec.org/document/draft-specifications-for-domestic-household-defluoridation-filter-unit-booklet-english

White, D. E., Hem, J. D., & Waring, G. A. (1963). Chemical composition of sub-surface waters. In *Data of geochemistry*.

WHO. (2011). *Guidelines for drinking-water quality* (4th ed.). WHO. ISBN: 978-92-4-154815-1.

Chapter 12
Arsenic Remediation from Water in Burkina Faso Using Local Materials as Adsorbents: Overview, Mitigation and Prospects

Yacouba Sanou

Abstract In Burkina Faso, arsenic is one of pollutants which affect the quality of water and food, mainly the vegetables. To mitigate this arsenic pollution, it is important to develop a cheap and efficient technology using a mixture of adsorbents, which is eco-friendly and cost-effective and based on laterite soils. The present study aimed to contribute to the development of an effective method for the treatment of arsenic contaminated water using the local materials. Batch experiments were carried out for arsenic removal using As(V) aqueous solutions. Experimental results indicated an improvement of removal capacity of new material and the yield of water treatment increased. Indeed, the increase of flow rate caused a decrease in the efficiency of the water treatment from 98.25 to 88.14% while the increase in the mass of doped laterite showed an increase in the service times. From column experiments, best efficiency (≈99%) in arsenic removal was obtained for pH values between 3 and 5 with a flow rate of 9 mL/min and 12 g of doped laterite. From batch studies, the mechanism of As(V) removal was described by a physico-chemical adsorption following pseudo-second order kinetic. The adsorption capacity of doped laterite (0.83 mg/g) was higher compared to the one of natural laterite (0.79 mg/g) indicating the better efficiency of doped laterite. Consequently, the doping of laterite increased the iron content which is responsible of high arsenic removal regarding the affinity iron-arsenic.

Keywords Adsorption · Arsenic · Characterization · Laterite · Remediation

Y. Sanou (✉)
Laboratory of Analytical, Environmental and Bio-Organic Chemistry, UFR/SEA, University Joseph KI-ZERBO, 03 BP 7021 Ouagadougou 03, Burkina Faso
e-mail: prosper_yacson@yahoo.fr; yacouba.sanou@ujkz.bf

B. Sen Gupta and N. Martínez-Villegas (eds.), *Arsenic Remediation of Food and Water*,
https://doi.org/10.1007/978-981-97-4764-1_12

Introduction

Arsenic contamination in groundwater for drinking purposes has become a worldwide problem and a serious environmental challenge in the world (Hughes, 2002; Jain & Ali, 2000; Smedley & Kinniburgh, 2002). Ingestion of arsenic-contaminated water can cause important threats to the human health. Long-term uptake of arsenic-contaminated drinking water causes liver, lung, kidney, bladder, skin and nerve tissue injuries as well as cardiovascular problems (Ansone et al., 2013; Dupont et al., 2007; Nemade et al., 2009). To mitigate this pollution, several technologies have been developed for the treatment of arsenic contaminated water including adsorption, precipitation, ion exchange, and electrocoagulation and filtering methods (Burggraaf & Cot, 1996; Mahvi et al., 2005; Sanou, 2019). Among them, adsorption is one of the most popular processes for arsenic removal from water because of its cost-effective, high affinity for dissolved arsenic and it doesn't produce sludge (Sanou et al., 2019; Tuna et al., 2013). So far, various adsorbents natural or synthetic have been investigated for arsenic removal. Development and implementation of cheap natural adsorbents such as activated carbons, clay, ferric hydroxide, ferrihydrite and active silica were successfully used to remove arsenic from water (Kundu & Gupta, 2005; Maji et al., 2007; Sanou et al., 2016). In general, commercial adsorbents including carbons, granular ferric hydroxide, and others are described as limited in their use due to high operational cost (Sanou et al., 2019). For this reason, other economic and eco-friendly materials for water purification were looked for. A preliminary study using laterite soils in Burkina Faso indicated their low adsorption capacity compared to granular ferric hydroxide, ferrihydrite and activated carbons (Sanou et al., 2020a, 2020b). However, the use of lateritic soils does not allow the easy production of drinking water in very large quantities due to their relatively slow absorption kinetics compared to other adsorbent materials such as ferrihydrite (Mälher & Persson, 2013). Thus, Dehou (Bossy, 2010) proposed a method for the doping of laterite with ferrihydrite which allows more fixation of ferrihydrite on laterite compared to the other methods. The application of this method requires a chemical pretreatment of the laterite in order, on the one hand, to open the pores of the laterite for better fixation of ferrihydrite, and, on the other hand, to remove organic elements responsible for inconsistent effects during batch adsorption tests. In order to increase the adsorption capacity of laterite soil, the study focused to prepare a new material by doping laterite with ferrihydrite and assess its adsorption capacity for As(V) in comparison to the one of natural laterite using the batch experiments.

Overview and Problem to Address

Burkina Faso is among a developing country which faces water safety problems, with some lack of appropriate system for water collection and treatment, or simply containing groundwater being used like drinking water in rural area without any prior treatment. Indeed, the presence of high arsenic concentration in the Northern part of Burkina Faso caused the pollution of water with some consequences such as melanosis, hyperkeratosis and skin changes on rural populations. Nowadays, almost tube wells and fountains in Mogtedo city are polluted by arsenic species and investigations showed an arsenic concentration up to 230 μg/L. In Burkina Faso, groundwater is the main source water supply in rural areas. However, the physico-chemical quality and microbiological status of these ground waters are altered by human or natural activities including the original contamination geochemical by fluorine and arsenic is frequently encountered (Bossy, 2010; Sanou, 2019). In Burkina Faso, besides cyanides and mercury, arsenic is the pollutant that caused more health problems with cases intoxication of populations, diseases dermal and pollution of aquifers (Bretzler et al., 2017). One of the first studies concerning the arsenic contamination in water in Burkina Faso revealed concentrations above the limit value in drinking water (10 μg/L) in the boreholes of Mogtédo area (De Jong & Kikietta, 1981). Previous studies revealed that melanosis and hyperkeratosis are the consequences of water pollution by arsenopyrite in Yatenga Province (Smedley et al., 2007; Somé et al., 2012). Arsenic concentration in water from tube wells was ranging between 1 and 124 μg/L while 87% of villagers use the water from tube wells. Among the first recognized consequences from chronic exposure to arsenic was melanosis, a skin disorder of hyperpigmentation or keratosis where the skin goes rough and dry with skin papules (Somé et al., 2012).

Mitigation

To mitigate this pollution, conventional methods of treatment of waters such as filtration, coagulation/precipitation, reverse osmosis, adsorption, and coupled techniques have been developed to remove pollutants affecting water quality and among which fixed bed percolation would be the more suitable in rural areas (Nicomel et al., 2016; Sanou et al., 2016). This work aims to study the As(V) removal in aqueous solutions using Granular Ferric Hydroxide (GFH) and Natural Laterite at ambient temperature. Column experiments were conducted to investigate the As(V) removal mechanism and effects of parameters affecting the adsorption were studied to follow the adsorption kinetics. Maximum removal of arsenic (99.99 and 99.5%) was achieved at 15 min of contact time with an initial concentration of 20 mg/L using 10 g of GFH and laterite in 50 mL volume of solution, respectively. Implementation of a treatment design using Granular Ferric Hydroxide (GFH) and activated Carbons in column and batch experiments have shown an efficiency in removing arsenic from Water. The cost of GFH, make it not affordable for developing countries such as Burkina Faso and materials for the preparation of activated carbons are too high to be applied in Burkina Faso. The suitable materials and applicable at the rural area in developing countries in such a process has to be cheap, easily available such as natural products to keep the total costs to an affordable level with a maintenance process not too complicated. Indeed, laterite soils showed a possible use and capacity for arsenic removal but they could not remove up to WHO admissible value (10 ppb). Using activated carbons, 61% of arsenic was removed with an adsorption capacity of 153 μg/g during 60 min of treatment (Sanou et al., 2019). High cost of those materials does not facilitate their use in Burkina Faso and laterite soils showed a low possibility to remove arsenic comparatively to GFH using natural groundwater. Analysis of pH value and initial arsenic concentration showed that laterite removed more than 95% of total arsenic from groundwater and that GFH removed 99% of the arsenic at pH 3 and 0.47 mg As/L. Using a molar sodium hydroxide solution, laterite was more regenable than GFH under the same experimental conditions. It has been demonstrated the high arsenic removal capacity of laterite collected in Vietnam for groundwater treatment using column experiments (Nguyen et al., 2020). An aqueous 1 M NaOH solution could regenerate more laterite (99%), compared to the GFH material (20%) after adsorption. This suggests that natural laterite can be used as an efficient and low-cost material to reduce arsenic contamination in groundwater.

However, this technology could not allow to achieve the WHO guideline value using one cycle of flow rate. The research of inexpensive adsorbent materials available a leads to favoring the choice of geomaterials natural materials such as laterites for the treatment of arsenic waters. Laterite soils have been used in Burkina Faso as fixe bed in continuous mode using column tests for Arsenic treatment experiments. Results showed that the adsorption capacity of laterite was influenced by pH, flow rate, initial arsenic concentration, and particle size requiring the optimization of those parameters. Indeed, the efficiency of the treatment decreased from 100 to 64% when flowrate increase from 1 to 12 mL/min. The increase of initial arsenic concentration caused an increase of adsorption capacity from 1.05 to 11.62 μg/g. The treatment becomes more efficient at low grains size which caused a turbidity of treated water higher than limit value (5 NTU). Better treatment (100%) in our operating conditions was obtained at low flow rate less than 8 mL/min with pH of water between 6 and 8, corresponding to an adsorption capacity of 3.25 μg/g (Sanou et al., 2020b). Laterite soils seem to be promising alternative adsorbents for the treatment of arsenic waters because of their accessibility, and their availability but their adsorption capacity does not allow to reduce the concentration of arsenic up to the admissible value (Sanou et al., 2020b). Another study in Burkina Faso using laterites thermally treated and used as adsorbents in dynamicmode showed it effectiveness for the removal of arsenic in water (Sanou et al., 2020c). Results revealed a decrease of arsenic removal efficiency from 100 to 79% with flow rate increasing. Maximum removal of 100% for arsenic was obtained for pH values between 3.5 and 6. The increase of initial arsenic concentration and phosphate amount caused an increase of arsenic adsorption up to 24 μg/g. NaOH solution could desorb 86.8% of arsenic and the reuse of regenerated laterite indicated its efficiency in same experimental conditions (Sanou et al., 2020c). It seems that mixedmaterials are efficient for arsenic removal. Therefore, GFH and laterite have been mixed to prepare a new adsorbent in two different ratios. Indeed, the increase of the contact time between 5 and 90 min involved an increase of adsorption capacity up to 49.47 μg/g while a change of initial pH caused a variation of adsorption capacity from 49 to 42.38 μg/g. An increase of initial arsenic concentration showed a proportional increase of adsorption capacity for both mixed material while this capacity decreased when the adsorbent amount increased (Sanou & Pare, 2021). It has been demonstrated that Ferrihydrite is effective as an adsorbent for arsenic removal with high adsorption capacity (Mälher & Persson, 2013). However, the cost of its preparation, the low regeneration after use and the column clogging constitute some disadvantage for its use in dynamic studies. To solve the ferrihydrite clogging problem, soils laterites could be used as support for ferrihydrite to make it stronger and increase the performance of laterite in arsenic removal using column experiments.

Prospects

For efficient removal of arsenic from water (ground water or surface water), it is suitable to combine two or three methods in order to improve the yield of the treatment. At this regard, filtration-adsorption noted FI-AD has been tested and results revealed an increase of 40% in treatment efficiency comparatively to filtration and adsorption applied only. Other mixtures are possible such as adsorption-electrocoagulation, filtration-electrocoagulation, electro-oxidation/filtration, etc. In addition, it is possible to test a mixture of local materials with iron oxides such as Laterite-Granular ferric hydroxide, Ferrihydrite-Laterite and Activated carbons-laterite soil. Finally, it is possible to use two layers of adsorbents (laterite and ferrihydrite) in column experiments by applying a multi-round or cycles of arsenic influent water using many values of flowrate.

In this work, we are testing the adsorbing capacity of ferrihydrite-laterite for arsenic remediation in water.

Materials and Methods

Preparation Doped Laterite (DL)

Laterite used in this work was collected in Balkuy district (12°17′23″ N, 1°27′47″ W), one of the peripheral districts of Ouagadougou (Burkina Faso). Raw laterite was sampled and sent to the SGS mineral laboratory for crushing and washing using the protocol described as follow. The natural laterite (NL) was crushed with hammer, washed thoroughly with distilled water and dried in an oven (Interlabs DP1-I Instruments) at 105 °C for 24 h. Then, it was crushed and sieved (AFNOR sieve) to retain particles of sizes comprised between 0.25 and 2 mm.

The doping of laterite with ferrihydrite was carried out according to the method described by Dehou (2011) whose principle requires a chemical pretreatment of the laterite before the doping process. In this process, the range of pH between 7 and 8 helps to prevent the degradation of ferrihydrite which precipitates at $pH > 8$ and the drying of coated laterite at 60 °C makes it possible to avoid transformation of the ferrihydrite into goethite at temperature above 60 °C (Dehou, 2011; Zhao, 2016).

Technical Characterization of Laterite

Determination of Bulk Density

It is a physical parameter used to estimate the mineralogical composition of raw materials. It was calculated according to the following equation previously established and used (Sanou, 2019; Sanou et al., 2020a):

$$\rho = \frac{m_1 - m_0}{V} \tag{12.1}$$

with m_0 is the mass of the empty conical flask, m_1: mass of the flask filled with laterite.

V: volume of the empty flask.

Determination of pH at Zero Point Charge

The pH at zero point charge noted pH_{ZPC} is the pH for which the overall electrical charge of the adsorbent is neutral and corresponds to isoelectric pH. Its value was determined according to the method described by Noh and Schwarz (1989).

$$pH_{ZPC} = \frac{1}{n}\left(\sum_{i=1}^{n} pH_i\right) \tag{12.2}$$

where pH_i represents the pH of each solution "i" and *n*, *the number of* solutions used.

Residual Porosity

The porosity of the adsorbent bed constitutes a very important parameter in column experiments. It is determined using the method described in the literature, the principle of which is to load the column with the adsorbent and set a flow rate of water at the inlet of the column (Btatkeu et al., 2015). Subsequently, the flow rate F_0 at the outlet of the column is determined 15 min later and the flow rate F_1 at the outlet of the column 1 h later. Porosity was calculated using the following formula:

$$Porosity = \frac{F_1}{F_0} \times 100 \tag{12.3}$$

Batch Experiments for Arsenic Remediation

Arsenic (V) adsorption tests were performed in batch mode with natural laterite and doped laterite in order to study the effect of doping with ferrihydrite on adsorption capacity of laterite. Arsenic (V) solutions were prepared by the dilution of an H_3AsO_4 standard solution (1000 mg/L) in distilled water. Solutions with concentrations ranging from 1 to 9 mg/L were used in the following experiments. Arsenic (V) standard solution was commercially obtained from chemical stores.

Kinetic Study of As(V) Removal

In order to study not only the adsorption capacities of natural laterite and that doped as a function of the contact time, but also the adsorption kinetics, solutions of 100 mL of As(V) at 5 mg/L were prepared and placed in Erlenmeyer flasks with a capacity of 250 mL. To each Erlenmeyer flask, we added 1 g of adsorbent material to obtain an adsorbent dose of 10 g/L. The content of each Erlenmeyer flask was stirred for an appropriate time. Stirring time was varied from 1 to 24 h to allow sufficient time to reach equilibrium. After stirring, the contents of each Erlenmeyer flask are recovered and centrifuged using a centrifuge (CENTAUR, 87/4655). After filtration, arsenic concentration was analyzed by induction coupled optical/plasma emission spectrometry (ICP-OES). The adsorption capacity Q_e (mg/g) was calculated as follows:

$$Q_e = \frac{(C_0 - C_e)}{m} * V \tag{12.4}$$

where C_0 and C_e representing initial and final concentrations of arsenic in solution (μg/L), respectively.

V represents the volume of the solution (L) and m: the weight or mass of adsorbent.

Isotherms Study

In order to better understand the mechanisms of arsenic removal using natural and doped laterites, adsorption experiments were carried out by putting 1 g of adsorbent into contact with 100 mL of arsenic (V) solutions in Erlenmeyer flasks at different initial arsenic concentrations ranged from 1 to 9 mg/L. The mixtures in each Erlenmeyer flask were stirred for 17 h. After stirring, the content of each Erlenmeyer

Table 12.1 Physicochemical parameters of NL and DL

Characteristic	Quantitative value	
	Natural laterite	Doped laterite
pH_{ZPC}	7.19	5.36
Density	2.6	3.1
Porosity (%)	91.62	86.24

flask was collected, centrifuged and filtered. After filtration, As(V) concentration in each solution was analysis using ICP-OES. The adsorption capacity Q_e (mg/g) was calculated using the relation (12.4).

Results and Discussion

Physico-chemical Characteristics of Absorbents

Physicochemical parameters such as the zero point charge (pH_{ZPC}) or isoelectric point, the bulk density and the porosity of both two laterites were determined and reported in Table 12.1.

Data (Table 12.1) showed that the pH at zero charge point of doped laterite (acid pH) was lower than the one of natural laterite (neutral pH). This low pH value of DL could be explained by the binding of iron and the release of hydroxyl ions during chemical treatment. Moreover, there was an increase in the bulk density of DL comparatively to NL and that would be the result of the fixation of the iron during the doping process and the reduction of the content of some less heavy elements in NL. The low porosity of doped laterite could be explained by its higher iron composition and its high density which would form a more compact and less porous bed adsorbent.

Kinetic Modeling

The effect of contact time on adsorption capacity of both adsorbents was studied between 60 and 1440 min (Fig. 12.1). We noted an increase of adsorption capacity of each material with the contact time following two phases. A rapid step between 60 and 540 min would be due to the availability of active sites on the surface of materials at the beginning to fix arsenate ions. The constant step after 540 min up to 1440 min which could be explained by the saturation of active sites on each adsorbent. The increase in adsorption capacity with contact time would be due to the fact that the driving force gradually decreases over time, while the agitation remains constant leading the system to equilibrium (1020 min) where no additional shrinkage is only possible with increasing time. The increasing phase would be the result of the formation of stronger complex or the migration of arsenates onto the surface of the adsorbent (Kundu & Gupta, 2005; Sanou et al., 2019). The similar result was obtained by Mahmood et al. (2011) which concluded that both to-steps process in arsenic sorption may be due to the involvement of a fast ion exchange followed by chemisorption. Consequently, the equilibrium time was reached after 1020 min. It was observed a greater adsorption capacity of doped laterite at any time as indicated by its curve above that obtained with natural laterite. This performance of the doped laterite would be due to its strong chemical composition rich in iron oxide. Maximum capacity of DL was 0.49 mg/g while NL has a capacity of 0.47 mg/g over the contact time.

In order to evaluate the kinetics of arsenic (V) adsorption process on two laterites, two kinetic equations given by Lagergren and Svenska (1898), and Ho and Mckay (1999) were used respectively to describe the pseudo-first order model and pseudo-second order model. The graphical representation $(Q_e - Q_t) = f(t)$ allowed to determine the constants of the pseudo-first order model ($Q_{\exp}$ and k_1) (Fig. 2a). The plot $t/Q_t = f(t)$ allowed to determine the constants of the pseudo-second order model ($Q_{\exp}$ and k_2) (Fig. 2b). Values of all those constants were regrouped in Table 12.2.

We notice that the experimental ($Q_{\exp}$) and theoretical (Q_{theo}) values are of the same order of magnitude for the pseudo-second order kinetic model for the two laterites with a higher adsorption capacity with the doped laterite. Likewise, the constant k_2 which expresses the rate constant of the adsorption of As(V) on the doped laterite is greater than the constant k_2 with the natural laterite. In addition, we find higher values of the correlation coefficients R^2 for the pseudo-second order model with the two laterites. These results indicate that arsenic (V) adsorption onto two laterites could be explained by the pseudo-second order kinetic model and chemisorption could be the rate-limiting step (Ho & McKay, 2000; Maiti et al., 2007).

A more detailed mechanism for As(V) sorption on the surface of hydrous ferric oxides was proposed based on the published spectroscopic data as indicated in following equations (Katsoyiannis & Zouboulis, 2002).

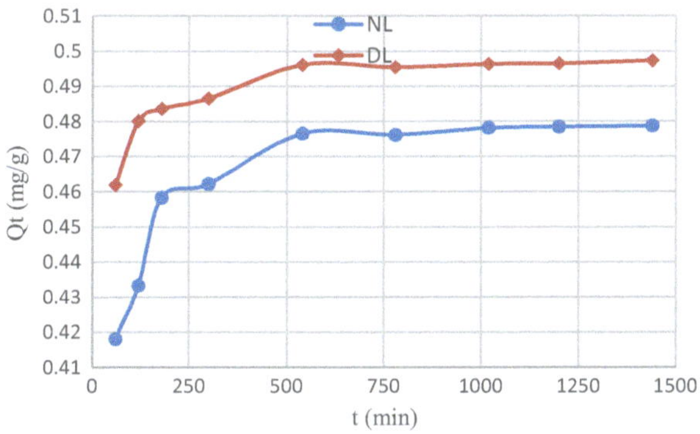

Fig. 12.1 Effect of the contact time on the adsorption capacity using NL and DL

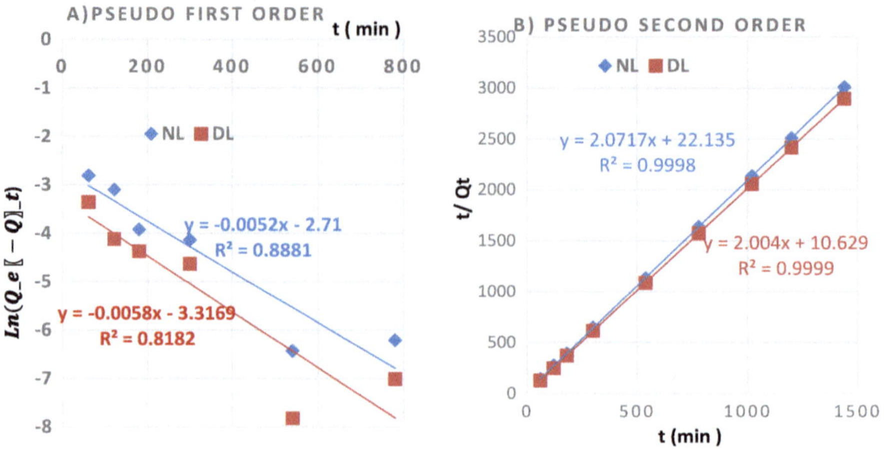

Fig. 12.2 Curves of pseudo first order and pseudo-second order models in arsenic (V) removal

Table 12.2 Kinetic constants in the adsorption of As(V) on two laterites

Pseudo-first order model							
Natural laterite (NL)				Dopted laterite (DL)			
k_1 (min^{-1})	Q_{theo} (mg/g)	Q_{exp} (mg/g)	R^2	k_1 (min^{-1})	Q_{theo} (mg/g)	Q_{exp} (mg/g)	R^2
0.005	0.06	0.47	0.89	0.006	0.03	0.49	0.82
Pseudo-second order model							
Natural laterite (NL)				Dopted laterite (DL)			
k_2 (min^{-1})	Q_{theo} (mg/g)	Q_{exp} (mg/g)	R^2	k_2 (min^{-1})	Q_{theo} (mg/g)	Q_{exp} (mg/g)	R^2
0.19	0.48	0.47	0.99	0.38	0.49	0.49	0.99

$$FeOH^{-1/2} + H^+ + AsO^{3-}_{4(aq)} \leftrightarrow FeAsO_3 + H_2Ok_1 \text{ (monodentate)} \quad (12.5)$$

$$2FeOH^{-1/2} + 2H^+ + AsO^{3-}_{4(aq)} \leftrightarrow Fe_2O_2AsO_2 + 2H_2Ok_2 \text{ (bidentate)} \quad (12.6)$$

$$2FeOH^{-1/2} + 3H^+ + AsO^{3-}_{4(aq)} \leftrightarrow Fe_2O_2AsOOH + 2H_2Ok_3 \text{ (protonated bidentate)} \quad (12.7)$$

From equations above (12.5), (12.6) and (12.7), literature suggests that arsenic adsorption onto iron oxides/hydroxides materials (laterite, ferrihydrite, ferric hydroxides, etc.) is occurred through a chemisorption mechanism where arsenic is directly bounded with the surface functional groups (Katsoyiannis & Zouboulis, 2002). This assumption is in agreement with a previous study which suggests that the physico-chemical properties of iron oxides after coating on the surface of polymeric materials do not change after the coating procedure (Katsoyiannis & Zouboulis, 2002). In our study, the sorption mechanism could be similar to that described in the literature (Ansone et al., 2013; Katsoyiannis & Zouboulis, 2002). The rate of pseudo-second-order reaction may be dependent on the amount of solute sorbed and the amount sorbed at equilibrium time. To confirm this assumption on chemisorption mechanism, some isotherms models available in literature have been applied in order to help in mechanism explanation.

Isotherms Models

The effect of the initial arsenic concentration was studied in the range of 1 to 9 mg/L with 1 g of each adsorbent. Results showed that the amount of adsorbed As(V) on each laterite increases from 0.1 to 0.8 mg/g when the initial concentration increased between 1 and 9 mg/L (Fig. 12.3). This result would be due to the occupation of active sites inaccessible at low arsenic concentrations (Sanou et al., 2019, 2020b). In addition, adsorption capacity of doped laterite was similar to the one of natural laterite at As(V) concentration between 1 and 5 mg/L. But, a low difference was observed in adsorption capacities of both adsorbents at high arsenic concentration as indicated by the curve $Q = f(C_0)$ of DL above that of NL. For a more detailed understanding in the adsorption capacity of As(V) on two laterites, three isotherm models viz. Langmuir, Freundlich and Dubinin–Radushkevich using 10 g/L as dose of each adsorbents were studied (Freundlich, 1906; Langmuir, 1918; Singh & Pant, 2004).

The line $C_e/Q_e = f(C_e)$ represents the Langmuir model (Langmuir, 1918) whose the constants (Q_m and b) are obtained from the slope and the intercept of this straight line. The plot $\ln Q_e = f(\ln C_e)$ represents Freundlich's model (Freundlich, 1906) and the constants were calculated from the intercept and the slope of the line and values of constants were reported in Table 12.3.

Fig. 12.3 Effect of initial As(V) concentration on adsorption capacity

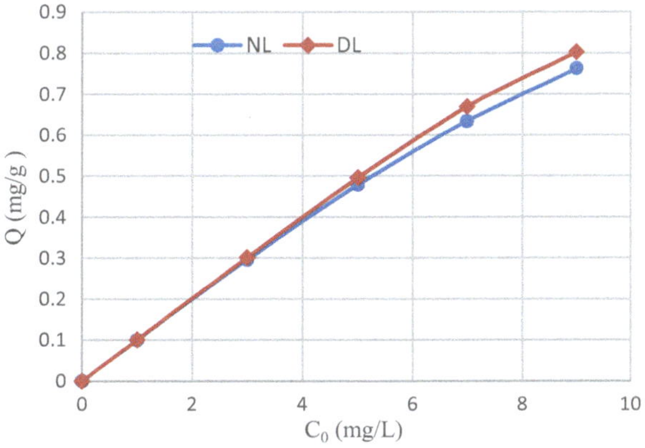

Table 12.3 Constants of Langmuir and Freundlich isotherms

Adsorbent	Langmuir constants				Freundlich constants		
	Q_m (mg/g)	b (L/mg)	R^2	ΔG^0 (kJ/mol)	n	K_F (mg/g)	R^2
NL	0.79	10.05	0.99	−2.52	2.64	0.76	0.94
DL	0.82	32.69	0.99	−3.75	3.11	0.98	0.86

The value of Q_m using DL (0.82 mg/g) was greater than the one obtained with NL (0.79 mg/g). In addition, the values of the correlation coefficients (R^2) were close to unity (1) using two laterites. Consequently, this result showed that As(V) adsorption was fitted to Langmuir isotherm indicating a monolayer adsorption process occurred on identical adsorption sites (Langmuir, 1918). Moreover, Gibbs free energy (ΔG^0) was calculated and its negative values (−3.75 and −2.52 kJ/mole) suggests the adsorption of arsenic (V) was produced according to a spontaneous process.

It was noted that the values of K_F and n for the doped laterite were higher than those obtained with the natural laterite. However, the correlation coefficients using Freundlich isotherm were lower than those obtained with Langmuir isotherm indicating the best application of Langmuir model. The removal of As(V) would be produced by monolayer adsorption onto the localized sites. Since Langmuir and Freundlich isotherms do not provide a clear information about the type of the adsorption, the isotherm of Dubinin–Radushkevich (D–R) was applied and the value of adsorption energy can indicate an idea on the adsorption type.

The graph $\ln Q_e = f(\varepsilon^2)$ representing the D–R model (Singh & Pant, 2004) gives the values of D–R's constants from the slope and the intercept of this line and were reported in Table 12.4.

It was noted that the values of capacity Q_m and the absorption energy E using DL were higher than those obtained with NL. In addition, the fact that values of E were less than 8 kJ/mole, indicates that the adsorption of arsenic (V) on the surface of two laterites would be physicochemical adsorption (Han et al., 2013; Sanou et al., 2019). Thus, it would coexist both two physisorption and chemisorption. Indeed, there are the existence of a physical adsorption at the start of stirring due to the diffusion of As(V) species in the pores of adsorbents, following by a chemisorption due to the establishment of chemical bonds with the increase of contact time. This result is in agreement with those available in literature on arsenic (V) removal mechanism using laterite soils and their based materials (Messeaouda, 2016; Roy et al., 2013).

Table 12.4 Constants of Dubinin–Radushkevich isotherm

Adsorbent	K (kJ^2/mole^2)	Q_m (mg/g)	E (kJ/mole)	R^2
NL	0.013	0.69	6.08	0.98
DL	0.009	0.83	7.41	0.97

Conclusions

The doping of laterite with ferrihydrite increased the adsorption capacity of laterite for an effective adsorption of As(V) but the process was influenced by the contact time and initial arsenic concentration requiring their optimization for a best As(V) adsorption. Indeed, the kinetic study concluded that the adsorption of arsenic (V) using two laterites (NL and DL) followed a kinetic model of pseudo-second order. The study of adsorption layer showed that the process was more adapted to Langmuir isotherm comparatively to Freundlich model indicating a monolayer adsorption. The application of Dubinin–Radushkevich (DR) isotherm concluded on a physico-chemical adsorption with a maximum adsorption capacity of 0.83 mg/g and 0.79 mg/g for DL and NL, respectively. In this regard, doped laterite can be a hopeful adsorbent for treatment of arsenic contaminated water using continuous experiments.

Acknowledgements Author would like to thank the NAM S&T Centre, for opportunity to publish this study. TWAS and UNESCO are acknowledged for supporting this study.

References

Ansone, L., Klavins, M., Viksna, A. (2013). Arsenic removal using natural biomaterial-based sorbents. *Environmental Geochemistry and Health* (Springer). https://doi.org/10.1007/s10653-013-9546-7

Bossy, A. (2010). *Origines de l'arsenic dans les eaux, sols et sédiments du district aurifère de St-Yrieix-la-Perche (Limousin, France): Contribution du lessivage des phases porteuses d'arsenic.* Thèse de l'Université de limoges, France.

Bretzler, A., Lalanne, F., Nikiema, J., Podgorski, J., Pfenninger, N., Berg, M., Schirmer, M. (2017). Groundwater arsenic contamination in Burkina Faso, West Africa: Predicting and verifying regions at risk. *Science of the Total Environment, 584*(585), 958–970.

Btatkeu, K. B. D., Tchatchueng, J. B., Noubactep, C., & Car, S. (2015). Designing metallic iron based water filters: Light from methylene blue discoloration. *Journal of Environmental Management, 166*, 567–573.

Burggraaf, A. J., & Cot, L. (1996). Fundamentals of inorganic membrane science and technology. *Membrane Science and Technology* (Elsevier series) *4*, 67–118.

De Jong, S. J., & Kikietta, A. (1981). *Une particularité bien localisée, heureusement présence d'arsenic en concentration toxique dans un village près de Mogtédo (Haute-Volta)* (p. 44). Bulletin de Liaison du Comité Interafricain d'Etudes Hydrauliques.

Dehou, S.-C. (2011). *Etude des propriétés d'adsorption des oxyhydroxydes de fer déposés sur un support naturel (la brique): Application à l'élimination du fer dans les eaux de forages en République Centrafricaine* (p. 157). Thèse, Université Lille 1, France.

Dupont, L., Jolly, G., & Aplincourt, M. (2007). Arsenic adsorption on lignocellulosic substrate loaded with ferric ion. *Environmental Chemistry Letters, 5*(3), 125–129.

Freundlich, H. M. F. (1906). Über die Adsorption in Lösungen. *Zeitschrift für Physikalische Chemie - Leipzig, 57A*, 385–470.

Han, C., Li, H., Pu, H., Yu, H., Deng, L., Huang, S., & Luo, Y. (2013). Synthesis and characterization of mesoporous alumina and their performances for removing arsenic (V). *Chemical Engineering Journal, 217*, 1–9.

Ho, Y. S., & McKay, G. (1999). Pseudo second order model for sorption process. *Process Biochemistry, 34*, 451–465.

Ho, Y., & McKay, G. (2000). The kinetics of sorption of divalent metal ions onto sphagnum moss peat. *Water Research, 34*(3), 735–742.

Hughes, M. F. (2002). Arsenic toxicity and potential mechanisms of action. *Toxicology Letters, 133*, 1–16.

Jain, C. K., & Ali, I. (2000). Arsenic: Occurrence, toxicity and speciation techniques. *Water Research, 8*, 4304–4312.

Katsoyiannis, I. A., & Zouboulis, A. I. (2002). Removal of arsenic from contaminated water sources by sorption onto iron-oxide-coated polymeric materials. *Water Research, 36*, 5141–5155.

Kundu, S., & Gupta, A. K. (2005). Analysis and modeling of fixed bed column operations on As(V) adsorption onto iron oxide-coated cement (IOCC). *Journal of Colloid and Interface Science, 290*, 52–60.

Lagergren, S., & Svenska, B. K. (1898). About the theory of so-called adsorption of soluble substances. *Kungliga Svenska Vetenskapsakademiens. Handlingar, 24*(4), 1–39.

Langmuir, I. (1918). The adsorption of gases on plane surfaces of glass, mica and platinum. *Journal of the American Chemical Society, 40*, 1361–1403.

Mahmood, T., Saddique, T., Naeem, A., Aslam, M., & Ali, R. (2011). Kinetic and thermodynamic study of Cd(II), Co(II) and Zn(II) adsorption from aqueous solution by NiO. *Chemical Engineering Journal, 171*, 935–940.

Mahvi, A. H., Naghipour, D., Vaezi, F., & Nazmara, S. (2005). Tea waste as an adsorbent for heavy metal removal from industrial wastewaters. *American Journal of Applied Sciences, 2*, 372–375.

Maiti, A., Dasgupta, S., Basu, J. K., & De, S. (2007). Adsorption of arsenite using natural laterite as adsorbent. *Separation and Purification Technology, 55*, 350–359.

Maji, S. K., Pal, A., Pal, T., & Adak, A. (2007). Modeling and fixed bed column adsorption of As(V) on laterite soil. *Journal of Environmental Science and Health, Part A, 42*, 1–95.

Mälher, J., & Persson, I. (2013). Rapid adsorption of arsenic form aqueous solutions by ferrihydrite-coated sand and granular ferric hydroxide. *Applied Geochemistry, 37*, 179–189.

Messeaouda, S. (2016). *Etude de la capacité de rétention et d'élimination des cations métalliques par des adsorbants naturels*. Université de Mustapha Stambouli, Mascara, Algérie.

Nemade, P. D., Kadam, A. M., & Shankar, H. S. (2009). Adsorption of arsenic from aqueous solution on naturally available red soil. *Journal of Environmental Biology, 30*(4), 499–504.

Nguyen, P. T. T., Sanou, Y., Pare, S., & Ha M. B. (2020). Removal of arsenic from groundwater using Lamdong laterite as a natural adsorbent. *Polish Journal of Environmental Studies, 29*(2), 1305–1314. https://doi.org/10.15244/pjoes/103028

Nicomel, N., Leus, K., Folens, K., Van Der Voort, P., & Du Laing, G. (2016). Technologies for arsenic removal from water: Current status and future perspectives. *International Journal of Environmental Research and Public Health, 13*(1), 62.

Noh, J. S., & Schwarz, J. A. (1989). Estimation of the point of zero charge of simple oxides by mass titration. *Journal of Colloid and Interface Science, 130*, 157–164.

Roy, P., Mondal, N. K., Bhattacharya, S., Das, B., & Das, K. (2013). Removal of arsenic (III) and arsenic (V) on chemically modified low-cost adsorbent: Batch and column operations. *Journal of Applied Water Engineering, 3*(1), 293–309.

Sanou, Y. (2019). *Traitement des eaux avec des charbons actifs*, GFH et latérite. Editions Universitaires Européennes, tvisrasco (pp. 1–157).

Sanou, Y., Balogoun, C. K., Tiendrebeogo, R., Kabore, R., Tchakala, I., & Pare, S. (2020a). Physicochemical and spectroscopic properties of two laterite soils for applications in arsenic water treatment. *International Journal of Multidisciplinary Research and Development, 7*(5), 12–17. http://www.allsubjectjournal.com/search/7-4-27

Sanou, Y., Kabore, R., & Pare, S. (2020c). Adsorption of arsenic and phosphate from groundwater onto a calcined laterite as fixed bed in column experiments. *French-Ukrainian Journal of Chemistry, 8*(2), 227–242. https://doi.org/10.17721/fujcV8I2P227-243

Sanou, Y., Nguyen, P. T. T., Pare, S., & Nguyen, P. V. (2019). The removal of As(V) from aqueous solutions using ferromagnetic activated carbon: Equilibrium and kinetic studies. *Revue des Sciences de l'Eau, 32*(2), 179–192. https://doi.org/10.7202/1065206ar

Sanou, Y., & Pare, S. (2021). Comparative study of adsorption capacity of two mixed materials for arsenic remediation. *Journal of Environmental Treatment Techniques, 9*(3), 559–565. https://doi.org/10.47277/JETT/9(3)565

Sanou, Y., Pare, S., Nguyen, P. T. T., & Nguyen, P. V. (2016). Experimental and kinetic modeling of As(V) adsorption on granular ferric hydroxide and laterite. *Journal of Environmental Treatment Techniques, 4*(3), 62.

Sanou, Y., Tiendrebeogo, R., & Pare, S. (2020b). Développement d'un pilote de traitement des eaux de forage contaminées par l'arsenic pour une application en zones rurales au Burkina Faso. *Journal de la Société Ouest-Africaine de Chimie, 49*, 22–30.

Singh, T. S., & Pant, K. K. (2004). Equilibrium, kinetics and thermodynamic studies for adsorption of As (III) on activated alumina. *Separation and Purification Technology, 36*, 139–147.

Smedley, P. L., & Kinniburgh, D. G. (2002). A review of the source, behavior and distribution of arsenic in natural water. *Applied Geochemistry, 17*, 517–568.

Smedley, P. L., Knudsen, J., & Maiga, D. (2007). Arsenic in groundwater from mineralized Proterozoic basement rocks of Burkina Faso. *Applied Geochemistry, 22*, 1074–1092.

Somé, I. T., Sakira, A. K., Ouédraogo, M., Ouédraogo, T. Z., Traoré, A., Sondo, B., & Guissou, P. I. (2012). Arsenic levels in tube-wells water, food, residents' urine and the prevalence of skin lesions in Yatenga province, Burkina Faso. *Interdisciplinary Toxicology, 5*(1), 38–41.

Tuna, A. O. A., Ozdemir, E., Simsek, B. E., & Beker, U. (2013). Removal of As(V) from aqueous solution by activated carbon-based hybrid adsorbents: Impact of experimental conditions. *Chemical Engineering Journal, 223*, 116–128.

Zhao, L. (2016). *Développement et mise en œuvre de nouveaux matériaux adsorbants d'anions à base de ferrihydrite ou d'Hydroxydes Doubles Lamellaires intégrés dans un gel d'alginate.* Université de Limoges, France.

Part III
Technological Interventions

Chapter 13
Arsenic and Iron Removal from Groundwater Using Indigenously Developed Ceramic Membranes

Swachchha Majumdar, Subhendu Sarkar, Ganesh Chandra Sahoo, Sudhendu Sensarma, and Sibdas Bandyopadhyay

Abstract CSIR-Central Glass and Ceramic Research Institute, Kolkata, India had developed a new technology for simultaneous removal of Arsenic and Iron from contaminated groundwater using indigenously developed ceramic membranes. This technology is based on adsorption of Arsenic on suspended colloidal and subsequent separation of Arsenic and Iron to produce quality drinking water. The technology had been verified in pilot scale and then demonstration scale before commercialisation took place. Till date, 250 plants have been installed out of which 150 plants are for simultaneous removal of Arsenic and Iron from groundwater and rest are only Iron removal plants as the groundwater at the locations of these plants is devoid of Arsenic but Iron contaminations are found.

Keywords CSIR-CGCRI · Technology · Ceramic membranes · Arsenic and iron removal · Groundwater · Groundwater · Ceramic membrane · Iron · Arsenic

Introduction

Currently in the world, 108 countries have issues with Arsenic contamination in groundwater. It is estimated that 300 million people worldwide do consume Arsenic contaminated groundwater. Sources of Arsenic is either geogenic or anthropogenic. In general, rocks containing Arsenic bearing mineral salts gradually break down and these Arsenic containing salts are leached out through runoff. There could be a few

S. Majumdar (✉) · S. Sarkar · G. C. Sahoo · S. Sensarma · S. Bandyopadhyay
CSIR-Central Glass and Ceramic Research Institute, Kolkata, India
e-mail: swachchha@cgcri.res.in

S. Sarkar
e-mail: subhendu@cgcri.res.in

G. C. Sahoo
e-mail: gcsahoo@cgcri.res.in

B. Sen Gupta and N. Martínez-Villegas (eds.), *Arsenic Remediation of Food and Water*,
https://doi.org/10.1007/978-981-97-4764-1_13

other sources which emit arsenic in the atmosphere which finally results in the entry of Arsenic in the water cycle. Coal-fired Power plants are one such example (Guo et al., 2014). The problems associated with arsenic contamination are extremely serious in large parts of India particularly the state of West Bengal and Bihar and also the adjoining countries like Bangladesh and Nepal. Arsenic contamination is found in the United States of America also (Bissen & Frimmel, 2003). Arsenic exposure in Indo Gangetic plains of Bihar is causing increased cancer risk (Cullen & Reimer, 1989). Arsenic in groundwater is found as oxy anions, which predominantly occurs in two states: arsenic (+III)/arsenite and arsenic (+V)/arsenate. Several factors including pH condition and redox potential govern the fact that out of these two states which is predominant in the groundwater (Kumar et al., 2021).

Details of the Technological Process

CSIR-Central Glass & Ceramic Research Institute (CGCRI) has been pioneer in the field of research and development of ceramic membranes in India. Ceramic membranes have unique advantages like thermal and chemical stability, mechanical strength, better resistance to alkali, acid and microbial attack, durability, abrasion resistance etc. The ceramic membranes developed by CSIR-CGCRI are manufactured using a mixed matrix of Alumina and Clay, and in tubular configuration having several channels. Though generally ceramic membranes are categorised as microfiltration, ultrafiltration and nanofiltration based on the mean pore size and porosity range, for the purpose of Arsenic and Iron removal from the groundwater, microfiltration membranes are being used in CSIR-CGCRI technology based plants. Microfiltration is a pressure driven process in which particulate maters having less than 0.1 micron is passed through the membrane due to application of a trans-membrane pressure in the range of 0.1–3 bar whereas particulates having larger than 0.1 micron size are arrested by the membrane surface. The separated stream containing the less than 0.1 micron size particles is called permeate and the stream containing more than 0.1 micron size particulates is called retentate.

The technology is a combination process of oxidation-adsorption and subsequent filtration through ceramic membranes. Iron is found in underground water in soluble ferrous state (Fe^{2+}). For separation of Iron, conversion from soluble ferrous state to insoluble Ferric state (Fe^{3+}) is required. The oxidation reaction is as followings:

$4\,Fe^{2+} + O_2 + 4H^{+} 4\,Fe^{3+} + 2H_2O$ (Meng et al., 2002).

In groundwater, Inorganic Arsenic is found in negatively charged arsenate [As(V), $HAsO_4^{2-}$] and neutral arsenite [As(III), H_3AsO_3]. Arsenite can be upto 50–90% of the total Arsenic under high reducing conditions (Munter et al., 1999). Arsenite shows lower affinity towards any adsorbent material (Lin et al., 2018). Therefore, Arsenite needs to be oxidised to Arsenate so that Arsenic removal can take place through adsorption.

Therefore, CSIR-CGCRI technology for Iron and Arsenic removal is a hybrid process in which aerial oxidation is imparted for conversion of Ferrous ions to insoluble Ferric Ions and Arsenite to Arsenate. The insoluble Ferric is easily filterable through ceramic membranes. However, CSIR-CGCRI technology has a unique feature of using a proprietary Iron based colloidal adsorbent media which adsorbs the Arsenates. As such the Arsenates in water have very small particle size and cannot be retained in usual microfiltration membrane surface as such membranes have larger pore diameter which allows the arsenates to pass through. But, when colloidal adsorbent media is charged in the water and specific residence time is allowed, the Arsenates find the adsorbent sites for attachment. Once the Arsenates are attached with the adsorbent media, the particulate size of the media attached with Arsenates become larger and thus become filterable through the membrane surface. The technology uses cross flow microfiltration technique.

CSIR-CGCRI process for Iron and Arsenic removal has been patented in Bangladesh (BD5912002, dated 28.03.2002), Chile (CL605-2002 dated 28.03.2002) and Taiwan (TW91106287 dated 29.3.2002), and the US Patent no. 7,014,771, issued on March 21, 2006 (Figs. 13.1 and 13.2).

Special features of the process:

i. A combined process having two steps in which one is adsorption of Arsenic using colloidal Iron based media and subsequent membrane based separation for solid–liquid separation through ceramic micro-filtration membrane based plants.
ii. Higher removal efficiency as there is contacts and attachment of Arsenic species with colloidal adsorbent media particles and then total separation of these attached fine particulates of iron and media through the surface of microfiltration membranes.

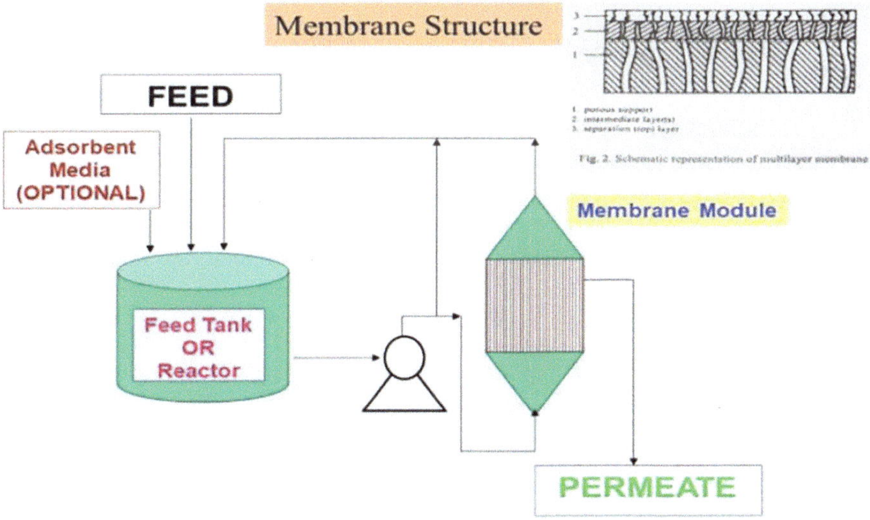

Fig. 13.1 Schematic of membrane based technology Arsenic and Iron removal

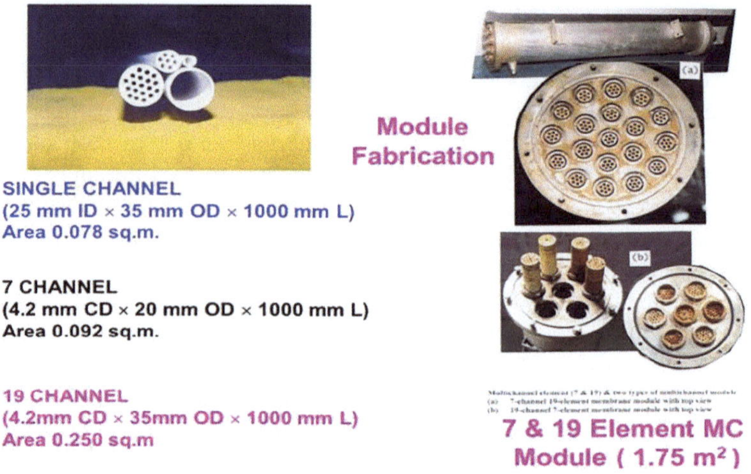

Fig. 13.2 Ceramic membranes and modules

iii. Use of colloidal adsorbent media whereas other technologies use granular media loaded in packed bed.
iv. Use of colloidal media results into a much better adsorption of Arsenic particles.
v. The level of purification achieved is less than 10 ppb for Arsenic.
vi. In case of presence of Iron only (absence of arsenic), colloidal media is not required.
vii. Ease of operation and maintenance particularly electrically operated semi-automatic plant with minimum human intervention, ease of periodical cleaning (once in 3–6 months) of the modules, back washing with minimum amount of filtered water, no loss of water, smaller space requirement for plant installation, lower sludge volume, etc. make this process ideal for rural area (Fig. 13.3).

Typical community level or small plants are generally made of food grade stainless steel with SS piping, valves, SS impeller for pump and food grade PVC storage tanks. The use of stainless steel makes these plants costly and capacity enhancement of the same system is not wise due to higher price of the stainless steel. Higher capacity plants are developed with the innovative idea of using PVC modules and piping and also MS modules which has brought down the cost of the plants.

The salient features of the technology:

(a) Elimination of arsenic and iron from contaminated ground water to produce Arsenic and Iron free drinking water (having Arsenic concentration less than 10 ppb, which is WHO specified) and iron content (Fe around 0.1 ppm, which is less than BIS specified 0.3 ppm).
(b) Adsorption capacity of the media is 7–8 times higher than conventionally used granular Ferric Hydroxide in other technologies.
(c) Production capacity ranging from 500 to 20000 LPD based on 12 hourly operation per day.

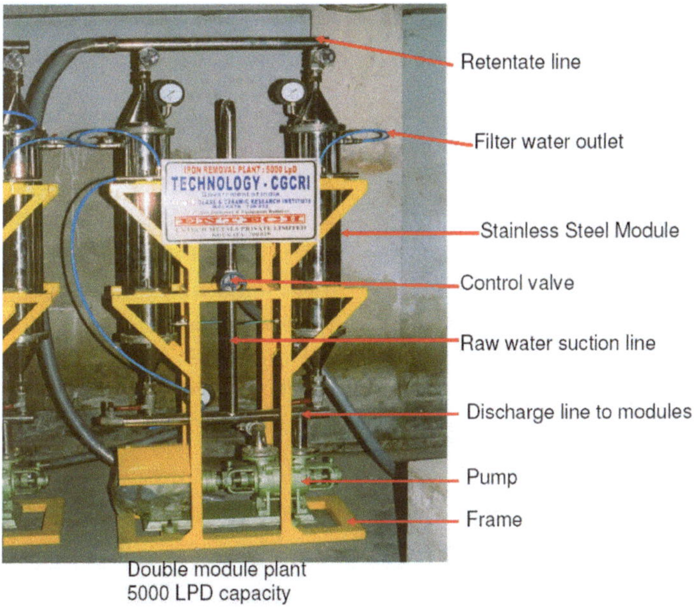

Fig. 13.3 A typical small capacity modular plant of 5000 Litre per day capacity

(d) Simple and user-friendly operational protocol.
(e) Media requirement is only once in the period of 3–6 months which is sufficient in case of 0.5–1 ppm Arsenic concentration in feed water.
(f) Cost of production is INR 0.10–0.20 per litre depending upon the capacity of the plant.
(g) Low sludge volume.

The technologies developed by CSIR-CGCRI have been transferred to various organisations as given in the following Table 13.1.

The propagation of the above technologies has taken place through installation of more than 250 water purification plants of different capacities. DST, Govt. of India, PHED, Govt. of West Bengal and PHED, Govt. of Bihar and private initiatives are the major sources of finance for the propagation of these technologies. This has been possible due to CSIR-CGCRI's expertise in process design, and apt support from the licensees. Followings are some details of the process development and technology implementation:

a. Multi-element modules up to 55 element modules were fabricated using the ceramic membrane elements for removal of water contaminants.
b. Modular and compact design with easy operation and maintenance, easy scale up options.
c. High level of purification efficiency achieved
d. Sustainable and cost effective technology for removal of suspended particulate matters, Iron and Arsenic

Table 13.1 Technologies developed by CSIR-CGCRI transferred to various organisations

Title	Recipient
Ceramic membrane based high capacity modules for pretreatment of turbid water for polishing of iron and arsenic contaminated water using micro filtration technique	Porel Dass Water and Effluent Control Pvt. Ltd., Kolkata on 11.05.2011
	Sarvo Technologies Limited, Faridabad on 30.8.2017
Ceramic membrane based technology for arsenic and iron removal from groundwater	Entech Metals Pvt. Ltd., Kolkata on 23/04/03
	Entech Metals Pvt. Ltd. on 22/07/13
	Victor Engineering Co., Assam on 7.10.2015
	Porel Dass Water and Effluent Control Pvt. Ltd., Kolkata on 24.04.2015
	Zeolite (India) Pvt. Ltd. On 11.9.2017
	Rollabss Hi Tech Industries, 27.03.2018

e. One ceramic membrane based plant of 80,000 LPD capacity coupled with reverse osmosis plant of 30,000 LPD capacity has been handed over to the Chairman, Taki Municipality on 27th August, 2013
f. One 10,000 LPD capacity plant coupled with Brackish water reverse osmosis unit has been installed at Bally, Howrah West Bengal
g. One 8000 LPH capacity Iron and Arsenic Removal plant at Ramchandrapur, Malda has been handed over to PHED, Govt. of West Bengal in March, 2017
h. 34 Iron removal plants installed at North Eastern States under DST support
i. 23 S.S.I based plants installed for private entrepreneurship development.
j. 99 Plants installed at different border outposts of Border Security Force at Nadia and North 24 Parganas districts of West Bengal (Figs. 13.4 and 13.5).
k. 50 Iron and Arsenic removal plants installed at Bihar

Fig. 13.4. 5000 LPD entrepreneurship plant at Dakshin Jagaddal, Near Kolkata

Fig. 13.5 A 5000 LPD capacity plant at BSF outpost, Nadia District, West Bengal

Conclusion

The Arsenic and Iron removal technology developed by CSIR-CGCRI could demonstrate that an indigenously developed technology can cater to the needs of the country. The Arsenic menace is very alarming in the country and this technology can be very attractive for decentralised and community level plants which can offer a sustainable solution to the people suffering from Arsenic contamination.

References

https://www.who.int/india/health-topics/arsenic

Bissen, M., & Frimmel, F. H. (2003). Arsenic- a review; part II: Oxidation of arsenic and its removal in water treatment. *Acta Hydrochimica et Hydrobiologica, 31*(2), 97–107.

Cullen, W. R., & Reimer, K. J. (1989). Arsenic speciation in the environment. *Chemical Reviews, 89*, 713

Guo, H., Wen, D., Liu, Z., Jia, Y., & Guo, Q. (2014). A review of high arsenic groundwater in Mainland and Taiwan, China: Distribution, characteristics and geochemical processes. *Applied Geochemistry, 41*, 196–217; CODEN: APPGEY; ISSN:0883-2927.

Kumar, A., Ali, M., Kumar, R., Kumar M., Sagar, P., Pandey, R. K., Akhouri, V., Kumar, V., Anand, G., Niraj, P. K., Rani, R., Kumar, S., Kumar, D., Bishwapriya, A., & Ghosh, A. K. (2021). *Scientific Reports, 11*, 2376. https://doi.org/10.1038/s41598-021-81579-9

Lin, F., Zhang, S., Ma, G., Qiu, L., & Sun, H. (2018). Application of ceramic membrane in water and wastewater treatment. In E3S Web of Conferences, ICAEER 2018 (p. 53, Vol. 04032). https://doi.org/10.1051/e3sconf/20185304032

Meng, X., Korfiatis, G. P., Bang, S., & Bang, K. W. (2002). Combined effects of anions on arsenic removal by iron hydroxides. *Toxicology Letters, 133*(1), 103–111; CODEN: TOLED5; ISSN:0378-4274.

Munter, R., Trapido, M., Veressinina, Y., & Kallas, J. (1999). Chemical oxidation of ferrous iron in aqueous solutions and groundwater samples. *Proceedings of the Estonian Academy of Sciences. Chemistry, 48*(4), 174–181. https://doi.org/10.3176/chem.1999.4.04)

Chapter 14
An Overview of the Strategies to Mitigate Arsenic in Deltaic Aquifers

Surajit Chakraborty, Utsab Ghosal, and Srijita Ghosh

Abstract The knowledge of arsenic (As) contamination has expanded enormously in the past two to three decades. From the very beginning of scientific research in India, the Bengal Basin attracted researchers working in the different fields. Therefore, plenty of information regarding Bengal Basin is available. Many research works have been carried out on the quantity and quality of groundwater in various parts of Bengal Basin. The ones relevant to the present study have been presented in this chapter. Here information has been gathered under broad heads according to the field of study. A few works, not exactly carried out within the present study area, but carried out in the surrounding areas and relevant to the present study, have been also incorporated here.

Introduction

The knowledge of arsenic (As) contamination has expanded enormously in the past two to three decades. From the very beginning of scientific research in India, the Bengal Basin attracted researchers working in the different fields. Therefore, plenty of information regarding Bengal Basin is available. Many research works have been carried out on the quantity and quality of groundwater in various parts of Bengal Basin. The ones relevant to the present study have been presented in this chapter. Here information has been gathered under broad heads according to the field of study. A few works, not exactly carried out within the present study area, but carried out in the surrounding areas and relevant to the present study, have been also incorporated here.

S. Chakraborty (✉)
Department of Environment Management, Indian Institute of Social Welfare and Business Management, College Square (West), Kolkata 700073, India
e-mail: surajitc@iiswbm.edu

U. Ghosal · S. Ghosh
Department of Geology, Presidency University, 86/1, College Street, Kolkata 700073, India

B. Sen Gupta and N. Martínez-Villegas (eds.), *Arsenic Remediation of Food and Water*,
https://doi.org/10.1007/978-981-97-4764-1_14

Rivers are the natural resources that create a linkage between the ecosystem and human civilization. Rivers transport sandy and porous sediments and deposit those in their floodplains and in deltas where the sediments eventually develop important underground reservoirs of freshwater named deltaic or alluvial aquifers. These aquifers supposedly yield wholesome groundwater which further led to the promotion of the use of groundwater in the Bengal Basin as a safer alternative to conventional surface water for domestic supply (Royal Society of Chemistry, Environmental Chemistry Group, Bulletin-January 2009). However, the groundwater in these alluvial aquifers was soon proven to be polluted by naturally occurring dissolved Arsenic (As) and the case was severe and widespread in Bengal Basin both in case of West Bengal and Bangladesh (McArthur et al., 2010) and also in the southernmost Nepal and Pakistan. The World Health Organization (WHO) described the situation in Bangladesh as the *'largest poisoning of a population in history'* (Smith et al., 2000). In Bangladesh and West Bengal (the Bengal Basin) at least 85% of the combined population of 240 million, use groundwater for domestic water supply, including for drinking (Biswas et al., 2012). It is estimated that in 1998–99 around 27 million people were drinking water containing more than the national standard of 50 μg/L of As. Worse, the WHO and many countries now consider 50 μg/L unsafe, and recommend a limit of only 10 μg/L. At this level, around 50 million people in Bangladesh, about 40% of the total population, and about 12 million people in West Bengal, are consuming dangerous concentrations of As. For many alluvial aquifers, Arsenic concentrations of several hundred milligrams per litre in groundwater are a very common scenario. Dissolved arsenic is odourless, tasteless, poisonous and carcinogenic. Its carcinogenic risk is measured in larger scales of years to decades (Royal Society of Chemistry, Environmental Chemistry Group, Bulletin-January 2009). The adverse health effects of As in drinking water are acute in southern Asian countries where the degree and extent of As-pollution in groundwater is also severe. In Bangladesh only, the effect of natural As-pollution of groundwater was termed "the worst mass poisoning of a population in history" by Smith et al. (2000).

The vertical spread of As-pollution in shallow groundwater of deltaic aquifers can clearly be explained and understood in regard to sea-level change (McArthur et al., 2004; Umitsu, 1993) and flushing (McArthur et al., 2004; Van Geen et al., 2008) although the logic behind the lateral diversity of As-pollution of absolutely zero to >1000 μg/L in deltaic aquifers over distances from tens of meters to kilometres is still fuzzy. In case of southern West Bengal, the As-pollution is confined to abandoned palaeo-channels only (Nath et al., 2005) which can be a probable reasoning for its lateral diversity. This is supported by *palaeosol* model by McArthur et al. (2008) that in southern West Bengal, groundwater from palaeo-channel aquifers is As-polluted (defined as >50 μg/L As) and that from palaeo-interfluvial aquifers capped by a palaeosol is As-free (defined as <10 μg/L As).

Scale of the Pollution

The four countries (namely: India, Bangladesh, Pakistan and Sri Lanka) that are highly affected by As-pollution (Fig. 14.1) have dominance in rural population and their economy is broadly dependent on agricultural activities. As concentration in groundwater in a large number of wells typically exceeds 50 μg/L in these countries. As per statistics from 2016, around 330 people among the dwellers of Bangladesh live in the 'average' village (DHS, 2016) and around 65% of the inhabitants live in rural areas (http://data.worldbank.org/indicator/SP.RUR.TOTL.ZS). The vast rural population is largely dependent on private or communal hand pumps where water is sourced from local groundwater that is widely As-polluted. As concentrations as high as 4000 μg/L have been recorded in southern West Bengal where the alluvial aquifers are mainly comprised of fluvial and deltaic sands composed of quartz, mica, and feldspar.

In Nepal, the Terai (the lowland alluvial plain) is well populated by about 12.5 million residing there and around 90% of them depend on groundwater for their daily needs. A study by Shrestha and Shrestha (2004) reveals that 24% of groundwater wells in the Terai region yielded water with >10 μg/L As while 7% exceeded 50 μg/L with wells having <30 m deep being most contaminated (32% > 10 μg/L As). A

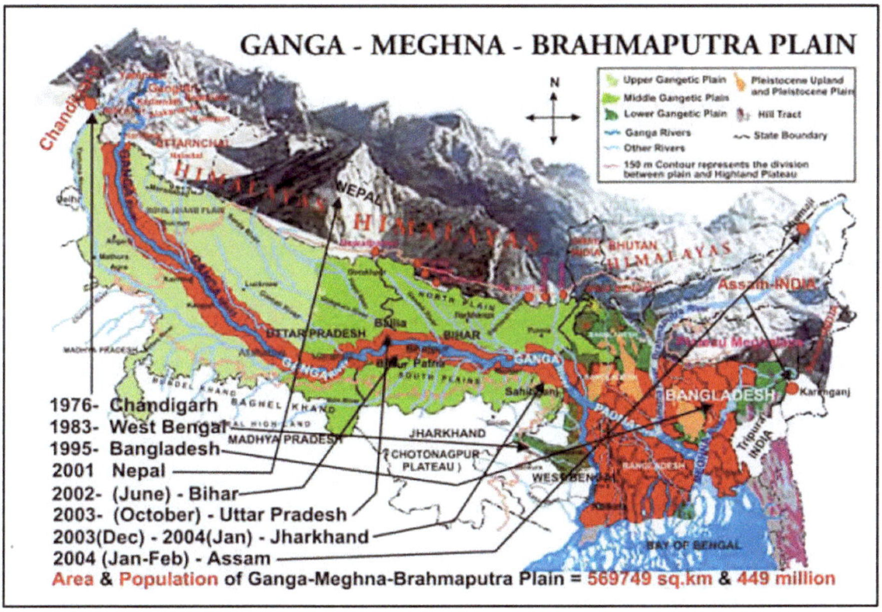

Fig. 14.1 Showing groundwater As-pollution status in different countries and different states of India in Ganga-Meghna-Brahmaputra Plain (*Source* Bhu-Jal news CGWB, Volume-24, Apr-Sept 2009)

recent estimate by UNICEF suggests that around 1 in 10 (1.4 million) are at risk from the As-pollution of groundwater (https://www.unicef.org/wash/nepal_35975.html).

A report of 2004 states that, around 36% of the population of Sindh Province in southern Pakistan was declared to be at risk from As in groundwater (Ahmed et al., 2004). Ramay et al. (2006) reported that out of 34 districts in Punjab province, 18 have As polluted groundwater with 40% of wells yielding groundwater with >10 μg/L As and 9% with >50 μg/L As.

Mechanism of As-Pollution in Groundwater

Anthropogenic Mechanism

Among the many potential sources of human-induced As-pollution, the vitals are leaching from ash deposited from the smoke plumes of brick-kilns, leaching of residues of arsenical pesticides and herbicides that are applied to crops; leaching of arsenical tick-control solutions from ponds used for dipping livestock; release of As from soils by application of phosphate fertilizer which promotes the competitive exchange of PO_4 with As, the oxidation of gangue sulphides in mine-waste, in which rates of oxidation and weathering exceed than those that occur naturally in situ, and the poor disposal practices at factories that manufacture As-pesticides (McArthur et al., 2018). These effects are not significant as sources of As-pollution in groundwater in the Bengal Basin except in Kolkata (formerly Calcutta) where the source of As-pollution is dominantly arsenical waste-products from the manufacture of Paris Green, an As-rich pesticide (Chatterjee et al., 1993).

Natural Mechanism

There are five different mechanisms as described by McArthur (2019) through which As can naturally be released into groundwater. These are:

i. Hydrothermal activity e.g. hot springs and geysers in volcanic areas (e.g. Yellowstone National Park, USA)
ii. Oxidation of arsenic-bearing pyrite in fractured crystalline rocks and in sediments (rarely)
iii. Microbially-mediated reductive dissolution of sedimentary iron oxyhydroxides, which strongly sorb As and PO_4 and liberate both to the solution when destroyed
iv. Desorption from mineral surfaces in response to rising pH as much as pH above 8, the rise being driven by processes unconnected to As
v. Evaporative concentration.

The microbially-mediated reductive dissolution of sedimentary iron oxyhydroxides is considered the major mechanism process in the Bengal Basin. Reduction of iron oxyhydroxides is profound in low-lying, flat terrains like deltas and alluvial plains of large rivers which can hold stagnant or slowly moving groundwater that approaches anoxia for lacking both dissolved oxygen and dissolved nitrate (Gulens et al., 1979; Nickson et al., 2000). Liberation of As in groundwater is usually associated with the release of Fe and with PO_4 which also strongly sorbs to sedimentary iron oxides. In groundwater where reductive dissolution occurs, As, Fe, and PO_4 show a positive correlation which is often weak owing to several secondary processes. In the early stages of reduction, the free As is resorbed to unreacted Fe-oxyhydroxides, as demonstrated by modelling (Welch et al., 2000) and visualised by McArthur et al. (2004). In anoxic groundwater, SO_4 can potentially be reduced to H_2S, inducing a reaction with dissolved Fe and As (and undissolved Fe-oxides) which further forms pyrite containing hundreds to thousands of ppm of As thereby removing both Fe and As from groundwater in a ratio different from that in groundwater (Kresse & Fazio, 2003; Kirk et al., 2004). Both Fe and As are then removed by sorption onto freshly precipitated Fe-oxyhydroxides as a result of anoxic oxidation of Fe(II) by NO_3 (Smith et al., 2017).

Figure 14.2 represents a schematic diagram of how As is transported from its source and deposited in the alluvial sediments. The glaciers in the Himalayas break down the rocks into finer particles which are then transported rapidly to the delta and end up in being deposited there (McArthur, 2019). All the way along, the particles undergo chemical weathering causing them to gain a thin, patchy coating of iron-oxyhydroxide from the weathering of Fe-bearing minerals. Occurrence of the thin and patchy coating is substantiated by a slower rate of weathering in low temperatures and the absence of vegetation in the source regions (McArthur, 2019). The oxyhydroxide coatings strongly sorb As and PO_4 from river water on their way to delta. In addition to As-bearing Fe-oxyhydroxides, the sediments also do bury organic matter present in the vegetation in delta regions along with them which serve as the main source of food for the bacteria present in the deltaic sediments. They grow unhindered through oxidation by using the only available oxygen from sedimentary Fe-oxyhydroxides thereby leading to the conversion of insoluble Fe(III) to soluble Fe(II) and its release in the groundwater along with As and PO_4.

Distribution of Arsenic Pollution

As pollution occurs in diverse geological and climatic conditions. It occurs most commonly in sands deposited by large rivers, and most of the worst cases are found in the tropical river basins of Asia. However, As-contaminated groundwater is found in unconsolidated sediments and sedimentary, igneous, and metamorphic rocks ranging in age from a few thousand to more than a billion years old. As pollution is found in climates ranging from hot and humid tropics to Arctic Alaska and hyper-arid deserts. Despite this diversity, in any given location, contamination usually has a

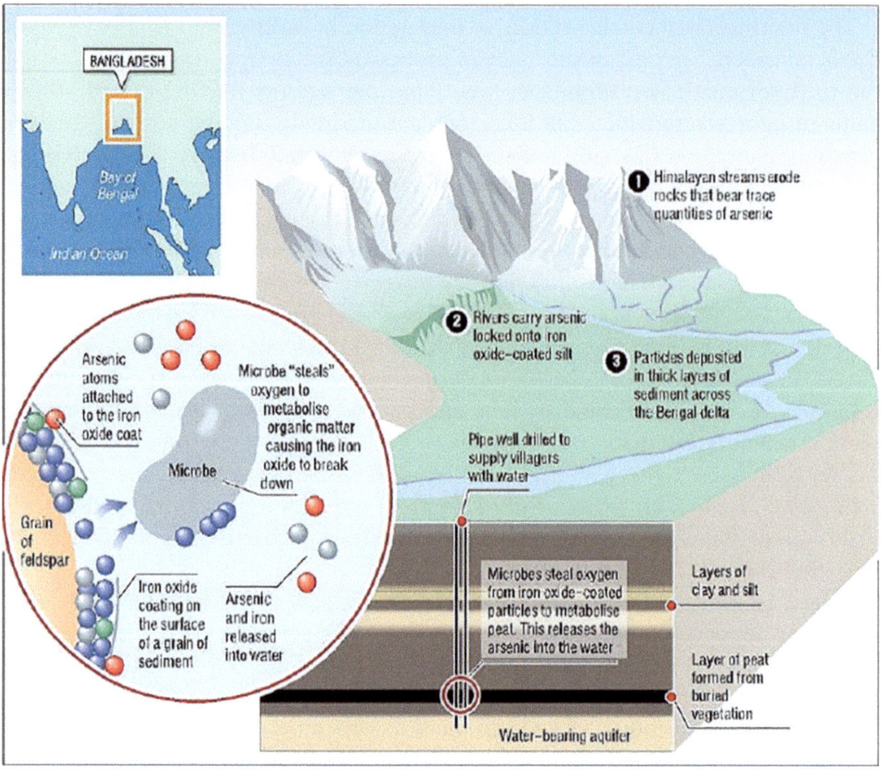

Fig. 14.2 Showing the transportation of As from its source Himalayan rocks to its sink Bengal basin (*Source* Arsenic speciation in the environment by John McArthur, ECG Bulletin January, 2006)

well-defined relationship to particular strata, or to particular depths of wells. In many areas where groundwater contains high levels of As, so too do the soils. Although the quantities vary greatly, most plants take up As through their roots and into the edible parts. Where As-rich groundwater is used for irrigation, the As content of soils gradually builds up, and leads to more As being taken-up by plants. Thus, the effects of As in food and water are both additive and cumulative. The worst conditions occur in the subsistence rice economies of Asia, where rice is irrigated with As-contaminated water. The diet of the rural poor typically comprises locally grown rice with little fruit, vegetables or meat, and so a deficiency of vitamins, minerals and protein reduces their ability to resist the toxic effects of As. If their food is cooked in, and washed down with, polluted well-water, the daily intake of As can be ten times the recommended maximum. Thus, poverty and environmental hazards combine to exacerbate the suffering of poor, rural populations.

The extent of human exposure due to naturally occurring dissolved As in the Holocene sedimentary aquifers of Bengal Basin has been recognized as world's largest mass poisoning in human history (Smith et al., 2000). The presence of elevated

As (>10 μg/L) in groundwater was first reported from West Bengal, India in early 1980s (Saha, 1984). In the subsequent years, such enrichment was also accounted from the aquifers of Bangladesh (Dhar et al., 1997; Roy et al., 1999). Currently about 60 million of the people from these regions are at risk of chronic As toxicity due to exposure from drinking water as well as other dietary components (Chakraborti et al., 2008; Chatterjee et al., 2010; Halder et al., 2012). Since after first reporting of As in groundwater, significant progresses have been made in the context of its source characterization, identification of mobilization and immobilization processes, spatial and vertical distributions, etc. (Bhattacharya et al., 1997; BGS and DPHE, 1999, 2001; Biswas et al., 2012; Harvey et al., 2002; Islam et al., 2004; Mukherjee et al., 2011; Nath et al., 2009; Nickson et al., 1998; Polizzotto et al., 2008; Polya & Charlet, 2009; Ravenscroft et al., 2005; Van Geen et al., 2003a, 2003b). These developments have led to the Governments of West Bengal, India, and Bangladesh to introduce various mitigation strategies together with international aid agencies to ensure a safe drinking water supply among the affected population. The undertaken strategies include treatment of surface water (pond as well as river water), rainwater harvesting, installation of dug wells, tubewell switching (changing of drinking water source from high As tube-wells to nearest low As TWs), installation of deep TWs (depth >150 m) and distribution of households and community filters (Jakariya et al., 2003, 2007). Nevertheless, the community acceptance of most of these strategies is not very promising because of technical, economic and social constraints (Ahmed et al., 2006; Jakariya et al., 2003, 2007).

Groundwater Composition

Drilling information into the contaminated zone in the subsurface layer of Bengal basin reveals a surficial layer of organic-poor weathered clay, light brown silty clay, and sandy silt, followed at depth by three different sedimentary sequences: (a) shallow palaeochannel (SPC) of Holocene- to Recent grey sand at depths below 35 m (mbgl); (b) palaeointerfluve (PI)), with late Pleistocene brown sands beneath a red clay Last Glacial Maximum Palaeosol (LGMP) cap, at depths >37 ± 3 mbgl; and (c) mostly Holocene deep palaeo-channel (DPC) of grey sands at depths >35 mbgl (Fig. 14.3). Screening of aquifer types for developing wells was done on the basis of compositional differences between PI and PC groundwater away from the drill sites. For the DPC aquifers, limiting concentrations were > 10 μg/L for As, >0.9 mg/L for Fe, and 1 μg/L for Uranium (U) and Vanadium (V).The groundwater chemistry of gray PC and brown PI aquifers are quite different (Davies, 1995; DPHE, 1999, 2001; Van Geen et al., 2003a, 2003b; Von Bromssen et al., 2007, 2008 et seq; Zheng et al., 2005). The PC aquifers typically host groundwater that contains >10 μg/L of As, > 0.9 mg/L of Fe, 0.2 mg/L of Mn, and low concentrations of V and U in ppb range (Hoque et al., 2012, 2014; McArthur et al., 2012). Groundwater in the pre-LGM gray sands mainly contains As < 10 μg/L (DPHE, 1999, 2001; Hug et al., 2011; Ravenscroft et al., 2013). Few instances caused due to the mixing of PI and PC groundwater with

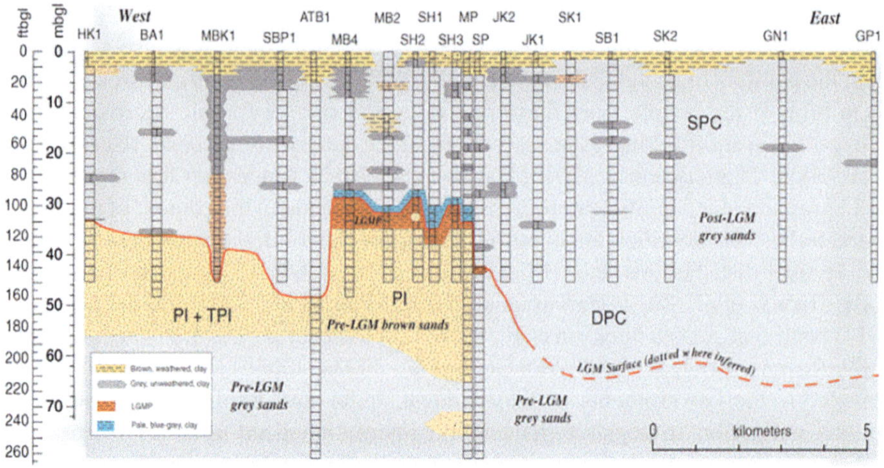

Fig. 14.3 Showing Sedimentary sequences across the study area demarcating palaeo-channel and palaeo-interfluvial zones (*Source* Ghosal et al., 2015)

wastewater and at PI margins were recorded that were least significant to make a remark in the observations.

Mitigation Strategies

To provide low-As groundwater for As mitigation, a detailed hydrogeological study was carried out across 235 km^2 of central West Bengal, in the western part of the Bengal Basin (Ghosal et al., 2015). By well-completion colour survey, drilling, and chemical analysis of 535 well waters up to a depth of 70 m bgl, PI aquifers with groundwater <10 μg/L As have been identified which covers one-third of the study area. The PI aquifers form two, north–south trending, subsurface features that are separated laterally by a DPC aquifer of grey sands, and are flanked on the east by another DPC aquifer, both of which host As-polluted groundwater. The PI aquifers are overlain by SPC aquifers of grey sand in which groundwater is usually As-polluted and is highly exploited by the local people. The LGMP now protects the PI aquifers from downward migration of As-polluted groundwater from overlying SPC aquifers. The SPC and DPC groundwater are also widely contaminated by wastewater derived from pit latrines, septic tanks, and other methods of sanitary disposal. Groundwater from wells in the PI setting is not affected by waste water and As unless sited within tens of metres of the margins of the PI, where the PI aquifers are invaded by adjacent contaminated PC groundwater.

In order to implement proper mitigation strategies for this As-pollution, the PI aquifers underlying the LGMP must be exploited successfully and a preferable way to do this is to emplace the tops of the screens in the PI sands beneath the LGMP

(Fig. 14.3) and also some meters below it. The attainable drilling depth in West Bengal is 60 m where the typical screen length is 3.7 m (12 feet). In such a scenario, the screen tops would be placed at a depth of 46 m, which is even below the top of the PI aquifers in all localities in both areas. It is an easy initiative to convey the fact to the villagers that a PI aquifer, yielding low As groundwater, is present at an accessible depth under the protective LGMP or other aquiclude as this fact is already known to local communities like local drillers who are well aware of the stratigraphy of their areas. If it becomes possible to make these people to learn the implication of the stratigraphy with respect to As pollution, it would be a good step ahead towards mitigation of As pollution across the Bengal Basin in cost effective way.

A quite feasible physical As-pollution mitigation strategy is exploitation of aquifers of brown sand which is now recognized as the upper part of the PI aquifers. With a proper understanding of the distribution of buried PC and PI, the exploitation of deep Pleistocene aquifers would be beneficiated at a greater scale.

Across most of the areas in Bengal basin, SPC aquifers at depths <34 mbgl overlying both PI and DPC aquifers and hosting As-polluted groundwater are easy and cost-effective to be exploited with respect to slightly deeper PI aquifers. This has inflicted an adverse health effect on the rural population (Ghosal et al., 2015). Therefore, mitigation of As-pollution in groundwater can be easily attained by simply drilling wells a little deeper in order to tap the PI aquifers.

A systematic monitoring programme for water quantity and water quality of selected drinking water wells would be great both for water-quality characterization in the aquifers and estimation of a prior indication for the impending arrival of contaminants in water supply wells. Since motor-fitted wells make the movement of As-rich water easier through a preferential flow path from the PC aquifers, proper caution should be done for the development of new motor-fitted wells in the deep aquifer below PI or PC aquifers (Ghosal et al., 2015). Wells fitted with hand pump can be constructed at any depth below the LGMP in a PI setting but for deeper aquifers not below the before attaining the depth of 225 mbgl to provide safe drinking water.

Capacity Building

In order to secure As-free groundwater for domestic supply, capacity building is a vital step to be followed. The knowledge that wells tapping PC aquifers should not be used for domestic consumption, not only because of the hazards posed by As-pollution and waste water, but also because a local, cheap, and easily accessed source of good water is available by drilling wells deeper to exploit the PI aquifers below 37 m depth which will provide a long-term source of low-As groundwater should be conveyed to the people living over the affected areas. Well-completion and domestic utensils colour can also serve as useful tools for indication of As-pollution in a way like a red stain on well-completions and domestic utensils developed from precipitation of iron oxide indicate that the wells tapping groundwater certainly have a high concentration of Fe and possess a high chance of As-pollution whereas black stain on well-completions and domestic utensils developed from precipitation of manganese oxide indicates a higher probability of the wells being As-free.

Testing for As concentration of the collected water samples from the households and primary schools in Haringhata block of Nadia district, West Bengal were done using *Wagtech Digital Arsenator* in the field (Fig. 14.4) or in the laboratory and the test results were conveyed to the well-owners instantly or over the telephone. During the ongoing As-testing spare time needs to be utilized by answering the queries or issues regarding the usage of water by any family member or any local individual with clarity. To encourage good practices for corrective action during capacity-building programmes, two sets of photos should be given to the participants in capacity development programmes to encourage best remediation practises. While the first set displays the normal PI and PC sediment cores which could be easily used to identify the colour variations in the two settings and learn more about the relationship of sand color to As. The second image, which shows the red and black stains on the well completions, might be utilised by local people to understand the most likely quality of the water they are consuming. As a whole, it is best to have a practice of comparing standard photographs of well-platform and sediment colours to promote the best remedy during capacity-building programs.

The proposed layouts of the capacity building programme at district and community levels are

Sl no	Training/ orientation	Key contents	Whom
District Level			

(continued)

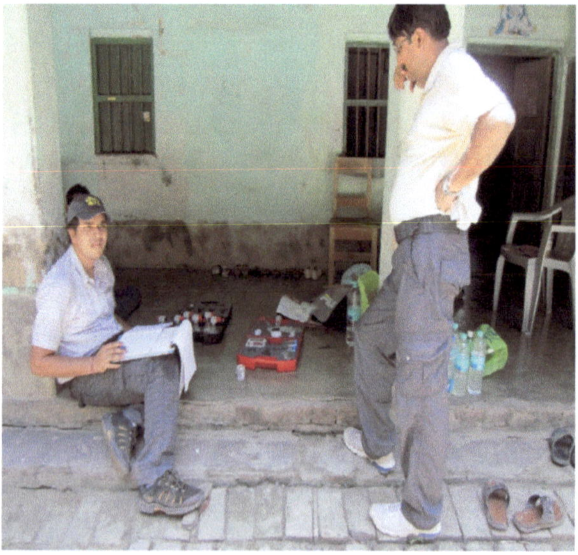

Fig. 14.4 Showing testing of arsenic in the field with the help Wagtech Digital Arsenator (*Source* Ghosal and Sikdar (Bhu-vidya, 2022)

(continued)

Sl no	Training/ orientation	Key contents	Whom
1	Orientation on arsenic, and Program activities	• Overview of arsenic, sources and chemistry of arsenic; • Arsenic in Bengal • Health effects due to arsenic consumption; • Arsenic mitigation options; • Existing Policies and guidelines; • Coordination mechanism; • Current program activities on arsenic mitigation	District level stakeholders such as representatives from government agencies, heads of NGOs and CBOs
2	District level Drillers	• Overview of arsenic, sources and chemistry of arsenic; • Health effects due to arsenic consumption; • Arsenic in Bengal • Understanding the generic sedimentology in As-polluted areas and their implications to provide safe water	Drillers from BDO office and PHED
3	Health officials training on arsenic	• Overview of arsenic, sources and chemistry of arsenic; • Health effects due to arsenic consumption; • Arsenic in Bengal • Arsenic mitigation options; • Case detection and management • Levels of case referral	Health officials from DPHO; Medical doctors
4	Local Journalists training on arsenic	• Overview of arsenic, sources and chemistry of arsenic; • Health effects due to arsenic consumption; • Arsenic in Bengal • Arsenic mitigation options; • Arsenic awareness through mass media	Local journalists

Sl. no	Training/ orientation	Key contents	Whom
Community Level			

(continued)

(continued)

Sl. no	Training/ orientation	Key contents	Whom
1	Training on arsenic to local drillers	• Overview of arsenic, sources, and chemistry of arsenic • Understanding the generic sedimentology • Importance of hand flap drilling for shallow domestic wells • Health effects due to arsenic consumption • Arsenic mitigation options; • Current program activities on arsenic mitigation	Local drillers
2	School teachers training on arsenic	• Overview of arsenic, sources and chemistry of arsenic; • Understanding the generic sedimentology • Health effects due to arsenic consumption; • Arsenic mitigation options; • Current program activities on arsenic mitigation	School teachers & selected school students of STD VIII and above
3	Training to village leaders and social workers	• Overview of arsenic, sources and chemistry of arsenic • Understanding the generic sedimentology • Health effects due to arsenic consumption • Arsenic mitigation options; • Current program activities on arsenic mitigation • Tube well and dug well parts, O&M procedures	VDC secretary, social workers and leaders from arsenic affected VDCs and communities
4	Orientation on filter O&M	• Key filter components • O & M procedures • Simple Troubleshooting • Safe water storage and hygenic handling of filter	Filter users

Thus, to summarize, while it is extremely critical to exploit this knowledge for mitigation exists among the local drillers in the communities, who know the stratigraphy of their areas. What drillers are not aware of is the implication of that stratigraphy with respect to As pollution. Teaching them the implication of the stratigraphy would do more to mitigate As pollution across the Bengal Basin than most other mitigation strategies, and at a fraction of the cost.

References

Ahmed, M. F., Ahuja, S., Alauddin, M., Hug, S. J., Lloyd, J. R., Pfaff, A., Pichler, T., Saltikov, C., Stute, M., & van Geen, A. (2006). Ensuring safe drinking water in Bangladesh. *Science, 314*, 1687–1688.

Ahmed, K. M., Bhattacharya, P., Hasan, M. A., Akhter, S. H., Alam, S., Bhuyian, M. A. H. (2004). Arsenic enrichment in groundwater of the alluvial aquifers in Bangladesh. An overview. *Applied Geochemistry, 19*, 181–200.

Arsenic pollution in groundwater: another piece of the puzzle falls into place. Royal Society of Chemistry, Environmental Chemistry Group, Bulletin-January 2009, ISSN 1758-6224

Bhattacharya, P., Chatterjee, D., & Jacks, G. (1997). Occurrence of arsenic–contaminated groundwater in alluvial aquifers from Delta Plains, Eastern India: Options for safe drinking water supply. *International Journal of Water Resources Development, 13*, 79–92.

Biswas, A., Nath, B., Bhattacharya, P., Halder, D., Kundu, A. K., Mandal, U., et al. (2012). Testing tubewell platform color as a rapid screening tool for arsenic and manganese in drinking water wells. *Environmental Science & Technology, 46*(1), 434–440.

Chakraborti, D., Singh, E. J., Das, B., et al. (2008). Groundwater arsenic contamination in Manipur, one of the seven North-Eastern Hill states in India: A future danger. *Environmental Geology, 56*(2), 381–390: https://doi.org/10.1007/s00254-007-1176-x.

Chatterjee, A., Das, D., & Chakraborti, D. (1993). A study of ground water contamination by arsenic in the residential area of behala, calcutta due to industrial pollution. *Environmental Pollution, 80*(1), 57–65.

Chatterjee, D., Halder, D., Majumder, S., Biswas, A., Nath, B., & Bhattacharya, P. (2010). Assessment of arsenic exposure from groundwater and rice in Bengal Delta Region, West Bengal, India. *Water Research, 44*, 5803–5812.

Davies, J. (1995). The hydrogeochemistry of alluvial aquifers in central Bangladesh. In: H. Nash, & G. J. H. McCall (Eds.), *Groundwater quality, chapman and hall, London.*

Dhar, R. K., Biswas, B. K., Samanta, G., Mandal, B. K., Chakraborti, D., Roy, S., et al. (1997). Groundwater arsenic calamity in Bangladesh. *Current Science, 73*, 48–59.

DHS. (2016). Bangladesh demographic and health survey 2014. Ministry of health and family welfare, Dhaka, Bangladesh. Mitra and Associates, Dhaka, Bangladesh; The DHS Program, Rockville, Maryland, U.S.A.

DPHE. (1999). Groundwater studies for arsenic contamination in Bangladesh. Phase I: Rapid investigation., department of public health engineering (DPHE) of government of Bangladesh, British geological survey (BGS) and Mott MacDonald Ltd (MML) UK.

DPHE. (2001). Arsenic contamination of groundwater in Bangladesh. In: D. G. Kinniburgh, & P. L. Smedley (Eds.), Department of public health engineering (DPHE) of government of Bangladesh and British geological survey (BGS) Keyworth (p. 267)

Ghosal, U., Sikdar, P. K., & McArthur, J. M. (2015). Palaeosol control of arsenic pollution: The Bengal Basin in West Bengal, India. *Groundwater, 53*(4), 588–599. https://doi.org/10.1111/gwat.12251.

Gulens, J., Champ, D. R., & Jackson, R. E. (1979). Influence of redox environments on the mobility of arsenic in groundwater. In: E.A. Jenne (Ed.), *Chemical Modeling in Aqueous Systems; Speciation, Sorption, Solubility, and Kinetics, American Chemical Society Symposium Series* (Vol. 93, pp. 81–95).

Halder, D., Bhowmick, S., Biswas, A., Mandal, U., Nriagu, J., Guha Mazumdar, D. N., et al. (2012). Consumption of brown rice: A potential pathway for arsenic exposure in rural Bengal. *Environmental Science and Technology, 46*, 4142–4148.

Harvey, C. F., Swartz, C. H., Badruzzaman, A. B. M., Keon-Blute, N., Yu, W., Ali, M. A., Jay, J., Beckie, R., Niedan, V., Brabander, D., Oates, P. M., Ashfaque, K. N., Islam, S., Hemond, H. F., & Ahmed, M. F. (2002). Arsenic mobility and groundwater extraction in Bangladesh. *Science, 298*(5598), 1602–1606.

Hoque, M. A., McArthur, J. M., & Sikdar, P. K. (2014). Sources of low-arsenic groundwater in the Bengal Basin: Investigating the influence of the last glacial maximum palaeosol using a 115-km traverse across Bangladesh. *Hydrogeology Journal, 22*(7), 1535–1547. https://doi.org/10.1007/s10040-014-1139-8

Hoque, M. A., McArthur, J. M., & Sikdar, P. K. (2012). The palaeosol model of arsenic pollution of groundwater tested along a 32 km traverse across West Bengal, India. *Science of the Total Environment, 431*, 157–165.

Hug, S. J., Gaertner, D., Roberts, L. C., Schirmer, M., Ruettimann, T., Rosenberg, T. M., et al. (2011). Avoiding high concentrations of arsenic, manganese and salinity in deep tubewells in Munshiganj District, Bangladesh. *Applied Geochemistry, 26*, 1077–1085.

Islam, F. S., Gault, A. G., Boothman, C., Polya, D. A., Chatterjee, D., & Lloyd, J. (2004). Direct evidence of arsenic release from Bengal sediments due to cooperatives, local government division, Banmetal-reducing bacteria. *Nature, 430*, 68–71.

Jakariya, M., Vahter, M., Rahman, M., Wahed, M. A., Hore, S. K., Bhattacharya, P., Jacks, G., & Persson, L. î (2007). Screening of arsenic in tubewell water with field test kits: Evaluation of the method from public health perspective. *Science of the Total Environment, 379*(2–3), 167–175.

Jakariya, M., Chowdhury, A. M. R., Hossain, Z., Rahman, M., Sarker, Q., Khan, R. I., et al. (2003). Sustainable community-based safe water options to mitigate the Bangladesh arsenic catastrophe—an experience from two upazilas. *Current Science, 52*(2), 141–146.

Kirk, M. F., Holm, T. R., Park, J., Jin, Q. S., Sanford, R. A., Fouke, B. W., & Bethke, C. M. (2004). Bacterial sulfate reduction limits natural arsenic contamination in groundwater. *Geology, 32*, 953–956.

Kresse, T., & Fazio, J. (2003). Occurrence of arsenic in ground waters of Arkansas and implications for source and release mechanisms. Arkansas ambient ground-water monitoring program, Arkansas department of environmental quality, Water Quality Report WQ03-03-01 March, 2003.

McArthur, J. M. (2019). Arsenic in groundwater. In *Groundwater development and management: Issues and challenges in South Asia* (279–308 p.). Springer International Publishing, Cham, Switzerland, with Capital Publishing Company, New Delhi, India.

McArthur, J. M., Banerjee, D. M., Hudson-Edwards, K. A., Mishra, R., Purohit, R., Ravenscroft, P., Cronin, A., Howarth, R. J., Chatterjee, A., Talukder, T., Lowry, D., Houghton, S., & Chadha, D. K. (2004). Natural organic matter in sedimentary basins and its relation arsenic in anoxic ground water: The example of West Bengal and its worldwide implications. *Applied Geochemistry, 19*, 1255–1293.

McArthur, J. M., Banerjee, D. M., Sengupta, S., Ravenscroft, P., Klump, S., Sarkar, A., et al. (2010). Migration of As, and 3H/3He ages, in groundwater from West Bengal: Implications for monitoring. *Water Research, 44*, 4171–4185.

McArthur, J. M., Sikdar, P. K., Nath, B., Grassineau, N., Marshall, J. D., & Banerjee, D. M. (2012). Sedimentological control on Mn, and other trace elements in groundwater of the Bengal delta. *Environmental Science and Technology, 46*(2), 669–676.

McArthur, J. M., Sikdar, P. K., Leng, M. J., Ghosal, U., & Sen, I. (2018). Groundwater quality beneath an Asian megacity on a Delta: Kolkata's (Calcutta's) disappearing Arsenic and present manganese. *Environmental Science & Technology, 52*(9), 5161–5172.

Mukherjee, A., Fryar, A. E., Scanlon, B. R., Bhattacharya, P., & Bhattacharya, A. (2011). Elevated arsenic in deeper groundwater of the western Bengal basin, India: Extent and controls from regional to local scale. *Applied Geochemistry, 26*, 600–613.

Nath, B., Chakraborty, S., Burnol, A., Stüben, D., Chatterjee, D., & Charlet, L. (2009). Mobility of arsenic in the sub-surface environment: An integrated hydrogeochemical study and sorption model of the sandy aquifer materials. *Journal of Hydrology, 364*, 236–248.

Nath, B., Berner, Z., Basu, M. S., Chatterjee, D., Charlet, L., & Stueben, D. (2005). Characterization of aquifers conducting groundwaters with low and high arsenic concentrations: A comparative case study from West Bengal, India. *Mineralogical Magazine, 69*(05), 841–854.

Nickson, R., McArthur, J. M., Burgess, W. G., Ahmed, K. M., Ravenscroft, P., & Rahman, M. (1998). Arsenic poisoning in Bangladesh groundwater. *Nature, 395*, 338.

Nickson, R., McArthur, J. M., Ravenscroft, P., Burgess, W. G., & Ahmed, K. M. (2000). Mechanism of arsenic release to groundwater, Bangladesh and West Bengal. *Applied Geochemistry, 15*, 403–413.

Polizzotto, M. L., Kocar, B. D., Benner, S. G., Sampson, M., & Fendorf, S. (2008). Near-surface wetland sediments as a source of arsenic release to ground water in Asia. *Nature, 454*, 505–509.

Polya, D., & Charlet, L. (2009). Rising arsenic risk? *Nature Geoscience, 2*, 383–384.

Ramay, M. I., Ahmad, T., Shipin, O. V., Jezeph, D., & Kadushkin, A. (2006). Arsenic contamination of groundwater and its mitigation in the province of Punjab (Pakistan) in the light of the situation in South Asia.

Ravenscroft, P., McArthur, J. M., & Hoque, M. (2013). Stable groundwater quality in deep aquifers of southern Bangladesh: The case against sustainable abstraction. *Science of the Total Environment, 454–455C*, 627–638.

Ravenscroft, P., Burgess, W. G., Ahmed, K. M., Burren, M., & Perrin, J. (2005). Arsenic in groundwater of the Bengal Basin, Bangladesh: Distribution, field relations, and hydrogeological setting. *Hydrogeology Journal, 4*, 727–751.

Roy, C. T., Basu, G. K., Mandal, B. K., et al. (1999). Arsenic poisoning of Bangladesh groundwater. *Nature, 401*, 545–546.

Saha, K. C. (1984). Melanokeratosis from arsenical contamination of tubewell water. *Indian Journal of Dermatology, 29*, 37–46.

Smith, A. H., Marshall, G., Roh, T., Ferreccio, C., Liaw, J., & Steinmaus, C. (2017). Lung, bladder, and kidney cancer mortality 40 years after arsenic exposure reduction. *JNCI: Journal of the National Cancer Institute, 110*(3), 241–249.

Smith, A. H., Lingas, E. O., & Rahman, M. (2000). Contamination of drinking water by arsenic in Bangladesh: A public health emergency. *Bulletin of the World Health Organization, 78*, 1093–1103.

Shrestha, B. R., & Shrestha, K. B. (2004). Spatial distribution of arsenic concentration in groundwater in the Terai, Nepal. In D. R. Kansakar (Ed.), *Summary project report. Department of Irrigation, Lalitpur, Nepal, HMG/ Nepal* (pp. 85–96).

Umitsu, M. (1993). Late Quaternary sedimentary environments and landforms in the Ganges Delta. *Sedimentary Geology, 83*(3–4), 177–186.

Van Geen, A., Zhang, Y., Goodbred, S., Horneman, A., Aziz, Z., Cheng, Z., et al. (2008). Flushing history as a hydrogeological control on the regional distribution of arsenic in shallow groundwater of the Bengal basin. *Environmental Science and Technology, 42*, 2283–2288.

Van Geen, A., Ahmed, K. M., Seddique, A. A., Shamsudduha, M. (2003a). Community wells to mitigate the arsenic crisis in Bangladesh. *Bulletin of the World Health Organization, 81*(9), 632–638.

Van Geen, A., Zheng, Y., Versteeg, R., Stute, M., Horneman, A., Dhar, R. K., Steckler, R., Gelman, M., Small, C., Ahsan, H., et al. (2003b) Spatial variability of arsenic in 6000 tubewells in a 25 km^2 area of Bangladesh. *Water Resources Research, 39*(5), 1140.

Von Brömssen, M., Jakariya, M., Bhattacharya, P., Ahmed, K. M., Hasan, M. A., Sracek, O., Jonsson, L., Lundell, L., & Jacks, G. (2007). Targeting low-arsenic aquifers in Matlab Upazila, South eastern Bangladesh. *Science of the Total Environment, 379*(2–3), 121–132.

Von Brömssen, M., Larsson, S. H., Bhattacharya, P., Hasan, M. A., Ahmed, K. M., Jakariya, M., Sikder, M. M. A., Sracek, O., Bivén, A., Doušová, B., Patriarca, C., Thunvik, R., & Jacks, G. (2008). Geochemical characterisation of shallow aquifer sediments of MatlabUpazila, Southeastern Bangladesh—Implications for targeting low-As aquifers. *Journal of Contaminant Hydrology, 99*(1–4), 137–149.

Welch, A. H., Westjohn, D. B., Helsel, D. R., & Wanty, R. B. (2000). Arsenic in ground water of the United States: Occurrence and geochemistry. *Ground Water, 38*(4), 589–604.

Zheng, Y., Datta, S., Stute, M., Dhar, R., Hoque, M. A., Rahman, M. W., Ahmed, K. M., Schlosser, P., & van Geen, A. (2005). Stable isotopes (18O, 2H) and arsenic distribution in the shallow aquifers in Araihazar, Bangladesh. Eos Trans. AGU, vol. 86, no. 52, Fall Meeting Supplement, Abstract H31B-1305.

Chapter 15
A Chemical and Waste-Free Community Plant for Treating Arsenic-Contaminated Water in Tepul Village, North 24 Parganas, West Bengal

Bhaskar Sen Gupta, Isita Sen Gupta, Soumyadeep Mukhopadhyay, Joey Sen Gupta, Nibedita Chatterjee, and Chanchal Majumder

Abstract Subterranean arsenic removal (SAR) is a sustainable water treatment process to remove arsenic and iron from groundwater without any chemical use or waste generation. The SAR plant creates an aerobic high Eh bed in the aquifer by recharging oxygenated groundwater, thereby supporting the growth of arsenic and iron oxidizing aerobic bacteria. The arsenic and iron are immobilized in the aquifer sand in the forms of As (V) and Fe(III), respectively. This work reports the installation and operation of a SAR plant installed at Tepul village in West Bengal, India. The groundwater was sampled, PHREEQ modelling was done to understand the chemical states of arsenic and iron, and a lithological study was done to identify various layers of the aquifer before selecting the Tepul site for installation of the SAR plant. The naturally occurring second aquifer water at Tepul had an arsenic concentration of 254 ppb. Once the plant was installed and the operation started, the arsenic level decreased to below the WHO limit of 10 ppb for 3000 L per day (LPD) water within 50 days of its operation.

Keywords Arsenic · Iron · Community water supply · Sustainable in-situ groundwater treatment · Chemical free · Waste free · Subterranean arsenic removal (SAR)

B. S. Gupta (✉)
School of Energy, Geoscience, Infrastructure and Environment, Heriot-Watt University, Edinburgh EH14 4AS, UK
e-mail: b.sengupta@hw.ac.uk

I. S. Gupta · S. Mukhopadhyay · J. S. Gupta · N. Chatterjee
Scientific Event Limited, Belfast, Northern Ireland, UK

C. Majumder
Indian Institute of Engineering Science and Technology (IIEST), Shibpur, West Bengal, India

B. Sen Gupta and N. Martínez-Villegas (eds.), *Arsenic Remediation of Food and Water*,
https://doi.org/10.1007/978-981-97-4764-1_15

Introduction

Arsenic contamination of groundwater affects more than 50 million people in Eastern India, exposing them to a high risk of developing various forms of cancer. According to the WHO, arsenic contamination in West Bengal and Bangladesh is the severest form of mass poisoning in the history of mankind leading to arsenicosis and cancer. There is a shortage of scalable low-cost processes that can be used for various communities suffering from arsenicosis without necessary access to medical treatment. It is well documented that a large number of people, irrespective of the age group, fail to engage in gainful employment at the advanced stage of arsenicosis. Symptoms include internal organ problems or failures, painful skin lesions, and gangrene of the foot among various other serious ailments. This initiative was aimed to plug the gap between academic research and large-scale implementation on the ground, creating economic, social, and health impacts.

For this purpose, a commercial-scale zero waste and chemical-free arsenic treatment plant was commissioned in Tepul village in West Bengal, close to the Bangladesh border using Subterranean Arsenic Removal (SAR) Technology (Sen Gupta et al., 2009). The process involves the abstraction of groundwater, aerating it to increase dissolved oxygen (DO) level and returning a part of it into the abstraction zone under gravity. This increases the DO level in the subterranean water, supporting aerobic bacteria, which enzymatically oxidise dissolved As(III) and Fe(II) to insoluble As(V) and Fe(III), trapping them in the aquifer sand. This process can reduce arsenic concentration from 200 + $\mu g.L^{-1}$down to < 10 $\mu g.L^{-1}$(WHO limit = 10 $\mu g.L^{-1}$).

The in-situ subterranean process (SAR Technology; www.insituarsenic.org) would be inexpensive, to the tune of INR 100 or 1.5 USD daily operating cost for producing 4000 L of potable water for community supply, which is more affordable than all the other available technologies in rural India. This would make clean and safe water accessible to local communities for whom maintenance of household filters by backwashing or recharging filter candles is not an attractive option. This project revealed that setting up SAR plants for commercial use by local entrepreneurs would be a highly profitable micro-business as the daily profit could be 3000–5000 INR/ day (approx 30–50 GBP/day) for an initial investment of 5000 GBP and monthly operating cost of 30.00 GBP (approx).

Unlike other arsenic remediation processes, the Subterranean Arsenic Removal (SAR) or 'In-situ treatment' method neither uses any chemicals nor produces a waste stream. Their installation is similar to that of an ordinary tubewell and can be installed by village technicians (www.insituarsenic.org).

SAR technology has been effective for the Ganga and Mekong delta where the arsenic is of arsenopyrite origin. This landmass covers the arsenic-affected zones of eastern India, Bangladesh, Cambodia, Vietnam, and Thailand. An estimated 70 million people are affected in India and Bangladesh by arsenic exposure and another 30 million in other ASEAN countries. The technology holds the promise of achieving the following goals.

1. Providing safe drinking and irrigation water to millions at an affordable price in arsenic-affected areas.
2. Prevention of arsenic-related diseases such as cancer of the liver, lungs, and bladder as well as skin lesions.
3. Reducing arsenic exposure through the food chain; the arsenic level in rice has reached an alarming level noting that the Food Standard Agency (UK) may restrict the import of rice from the arsenic-affected regions to the UK.
4. Using the huge water resource of shallow aquifers in South and Southeast Asia, which may be unsuitable for drinking and irrigation purposes due to high arsenic content.

Description of the Process

Subterranean groundwater treatment is based on the principle of oxidation and filtration used in conventional water treatment plants for the removal of Fe and Mn from water (Sen Gupta, 2010). However, the process is carried out in the aquifer zone by controlled oxidation of iron and arsenic. The key challenge lies in converting the soluble As(III) to sparingly soluble As(V), which is accelerated by autotrophic bacteria under favourable oxic conditions. It is believed that chemoautotrophic arsenite-oxidising bacteria or CAOs, that exist naturally in the aquifer are responsible for oxidising [$H_2AsO_3^-$] to [$H_2AsO_4^{2-}$], turning As(III) to As(V), which binds on iron oxyhydroxide precipitate (FeOOH) under oxic conditions. The oxidation of soluble iron, Fe(II) to its insoluble form, Fe(III) in the presence of dissolved oxygen can be explained by the following equations.

$$Fe^{2+} \rightarrow Fe^{3+} + e^- \tag{15.1}$$

$$O_2 + 4H^+ + 4\,e^- \rightarrow 2\,H_2O \tag{15.2}$$

The insoluble ferric iron, Fe(III) that forms when aerated water is introduced into the aquifer precipitates in the form of highly compact goethite or lepidocrocite according to the following equation.

$$Fe^{3+} + 2\,H_2O \rightarrow FeOOH + 3H^+ \tag{15.3}$$

Owing to their compact structure, goethite, and lepidocrocite do not block the aquifer and the permeability of the aquifer is maintained. This is very important for the long-term term productivity of the aquifer. The oxic condition helps the iron oxidising bacteria accelerate the reactions, 1–3, enzymatically. Furthermore, cyclic injection of oxygenated water and withdrawal of reduced water maintains a balance between dissolved oxygen (DO) and other nutrients for the growth of iron and arsenic oxidising bacteria.

In the SAR process, reduced groundwater in the aquifer is changed to an oxidized state by periodically charging aerated water. This transforms the microbial flora in the aquifer into a rich body of Fe, Mn, and As oxidising bacteria that enzymatically removes As along with Fe and Mn.

A schematic diagram of the plant is illustrated in Fig. 15.1. Commonly available showerheads couldbe used to increase the DO level in the water to up to 6 mg/L by spraying it in a plastic tank before returning it to the aquifer in a predefined operating sequence consisting of delivery, intermission, and infiltration (Fig. 15.2). Water was pumped from the underground aquifer using a submersible pump and sprayed in the oxygenation chamber by the showerheads. About 25–50% of this water was returned to the aquifer at the same depth under gravity after aeration while the remainder is supplied as arsenic-free water to the consumers. It takes about 10–12 weeks to prepare an aquifer as a fully functional and biochemically mediated As removal unit. The plant components were sourced from local shops and were installed by local technicians. The wellhead was sealed and pipes for delivery, and infiltration as well as a device for aeration and de-aeration of the well went through the wellhead to allow gravity infiltration. The device for aeration and de-aeration was constructed in a way that no water flows out during the infiltration process.

During the groundwater delivery, primarily Fe (II) is adsorbed onto the surface of the soil grains as an interlayer. However, during recharge, oxygen-rich water oxidises Fe(II) deposits into insoluble ferric hydroxide that remove arsenic as a co-precipitation product. The arsenic and iron removal processes are accelerated by the autocatalytic effect of the oxidation products and by the autotrophic microorganisms. The process is illustrated in Fig. 15.3. The method is very highly cost-effective in reducing the concentration of As below 10 $\mu g.L^{-1}$. The other advantage of this process is that there is no sludge handling cost. The flood proof borewell installation, the water tank and the spray system based on shower heads are illustrated in Figs. 15.4 and 15.5. Figure 15.6 depicts the arsenic profile in the drinking water and the plant at Tepul village over the first 54 days. Beyond this period, there was no significant change in the arsenic concentration was noticed and the water was fully compliant with the WHO standard for arsenic.

Site Information

The following sections describe the process of site selection and provide some snapshots of plant installation at Tepul village in the North 24 Parganas district of West Bengal.

In West Bengal, 79 blocks in 8 districts have Arsenic beyond the permissible limit of 0.05 mg/l. The most affected areas are on the eastern side of the Bhagirathi River in the districts of Malda, Murshidabad, Nadia, North 24 Parganas, and South 24 Parganas and the western side of the districts of Howrah, Hugli, and Bardhaman. Arsenic in groundwater is confined mainly in the aquifers up to 100 m depth. The

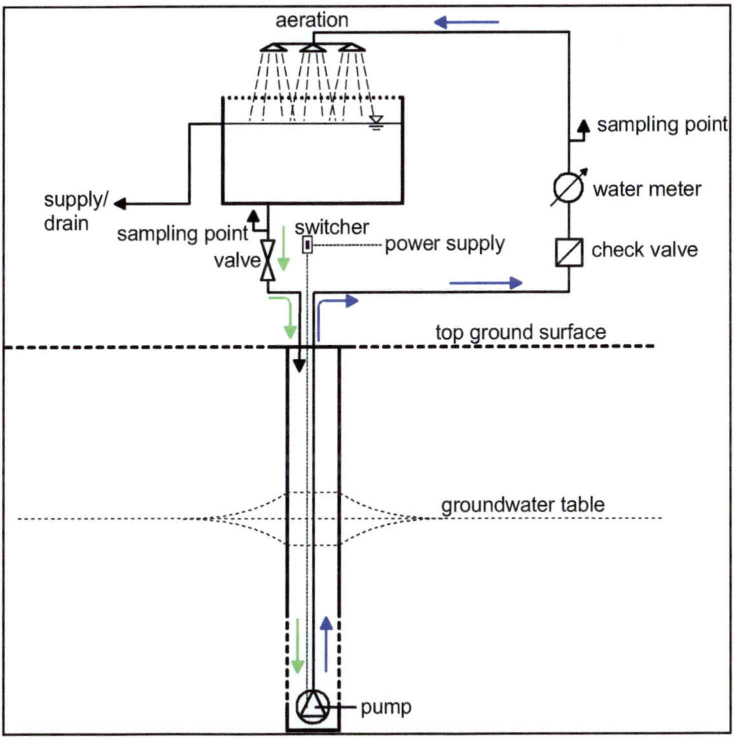

Fig. 15.1 Schematic diagram of the process *Source* Sen Gupta et al., 2009, https://doi.org/10.1016/j.envpol.2009.09.014

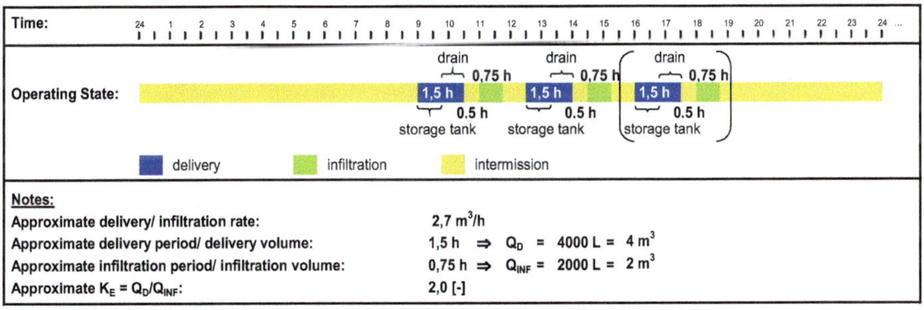

Fig. 15.2 Sequence of operation in terms of delivery, infiltration, and transmission (each cycle would produce 2000 L; daily output = 6000 L (2000 × 3); this includes 2000 l/cycle for consumption and the same amount for recharge. *Source* Sen Gupta et al., 2009, https://doi.org/10.1016/j.envpol.2009.09.014

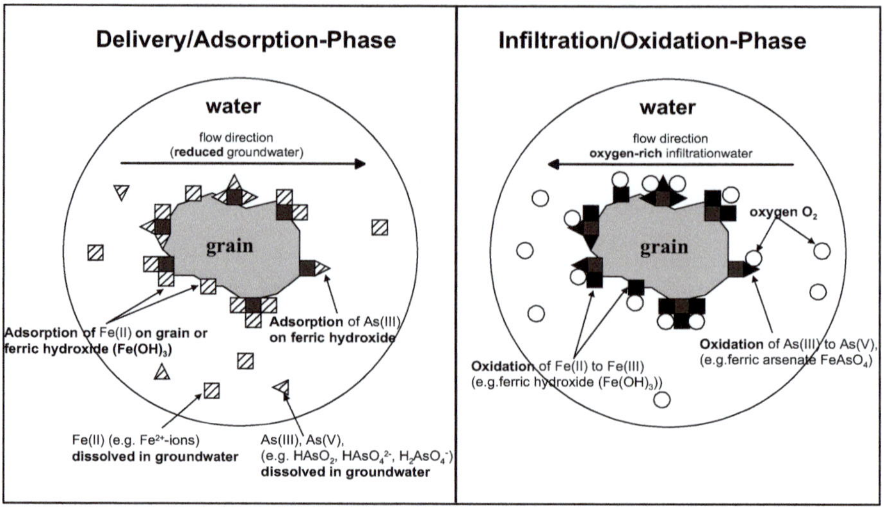

Fig. 15.3 Oxidation of Fe(II) and As(III) during Delivery and Infiltration *Source* Sen Gupta et al., 2009, https://doi.org/10.1016/j .envpol.2009.09.014

Fig. 15.4 Flood proof ground installation (raised vent pipe to ensure floodwater cannot enter the system. *Source* Sen Gupta et al., 2009, https://doi.org/10.1016/j.envpol.2009.09.014

deeper aquifers are free from Arsenic contamination. Groundwater arsenic contamination in West Bengal in the last 19 years was analysed by SOES (Jadavpur University, Kolkata). Out of 1, 40, 150 samples analysed for arsenic to date by SOES (JU), 48.1% had arsenic above 10 μg/L (the WHO guideline value). Importantly, 3.3% of the analyzed borewells had arsenic concentrations above 300 μg/L (the concentration predicting overt arsenical skin lesions (56). A total of 187 (0.13%) hand-operated tube wells were highly contaminated (>1000 μg/L) and were classified into three categories: Severely affected, mildly affected, and arsenic-safe. Nine districts (Maldah, Murshidabad, Nadia, North-24-Parganas, South-24-Parganas, Bardhaman, Haora, Hugli, and Kolkata), where more than 300 μg/L arsenic concentrations were

Fig. 15.5 Pictures from an earlier trial system, commissioned by the author which was used to collect plant data: **a** installation on the ground and storage tanks on roof space; **b** showerheads inside the tank to increase the DO level up to 6 mg/L. *Source* Sen Gupta, 2010 TCE, Feb 2010, IChemE (UK)

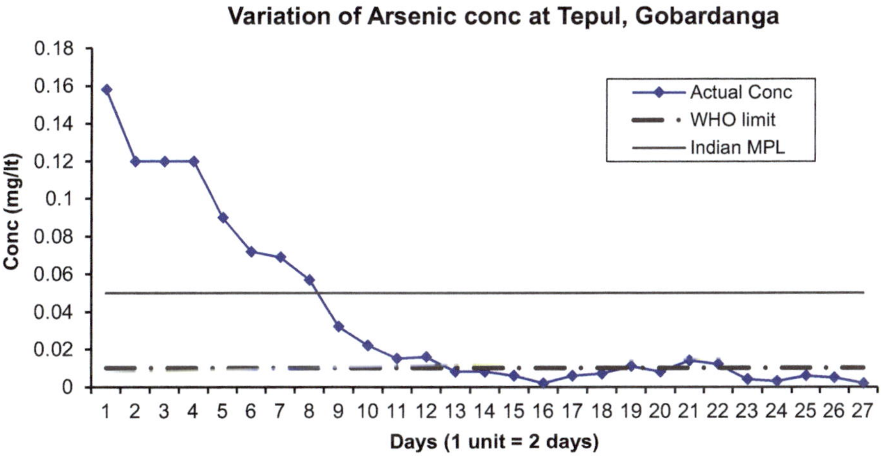

Fig. 15.6 The arsenic and iron concentration of water over the first 50 days of operation; steady drinking water quality achieved after 25 days as well as beyond 50 days (maximum permissible limit of As in drinking water; WHO- 0.01 $mg.L^{-1}$; operating DO 4 mg/L.)

found in borewells are categorized as severely affected. Out of 1,35,555 samples analysed from these districts, 67,306 (49.7%) had arsenic concentrations above 10 μg/L. Central Ground Water Board (CGWB) of India has studied the arsenic concentration in 79 administrative blocks of West Bengal. Different Government/ Non-government Organizations/Institutions have come forward to identify the arsenic-affected tube wells yielding arsenic-contaminated water, and also trying to monitor the magnitude and the extent of the pollution to mitigate the problems. The arsenic scenario in north 24 Pargana district is shown in Fig. 15.7. Figure 15.8 shows the location of the study area, the Tepul Village.

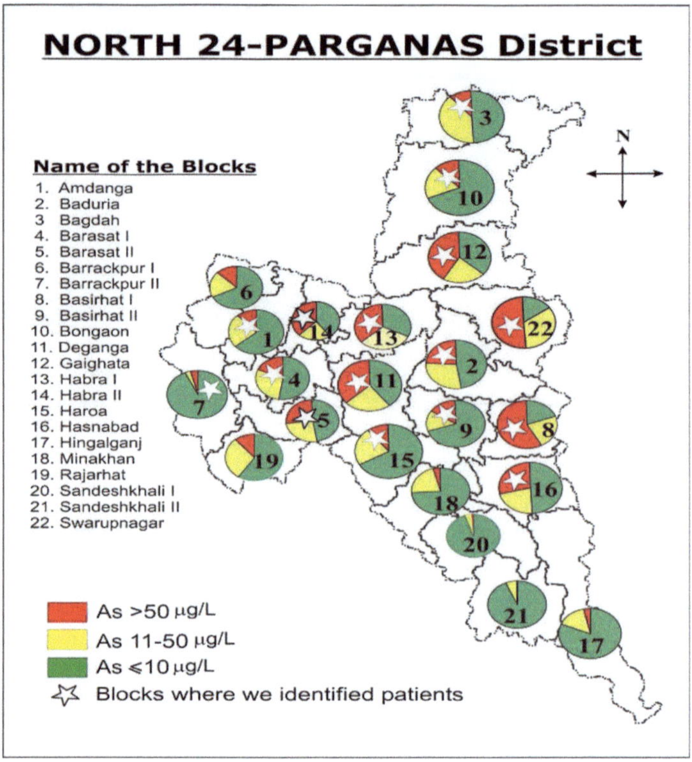

Fig. 15.7 Groundwater arsenic contamination situation in each block of North 24 Parganas district shown by pie-diagram (http://www.soesju.org/arsenic/wb2.htm)

Hydrogeology of the Study Area

Geologically, the North 24 Parganas district is mainly within the upper delta plain of the Ganga-Bhagirathi river systems. Arsenic in groundwater is mainly restricted in shallow aquifers (depth 15–70, meters below ground level or mbgl) which are mainly built-up of sediments deposited by meandering streams and levees. The migration of the meander belts and levee back swamps with a fining upward sequence in a vertical section depends upon the density of the channel network the composition of sediments changes laterally across the delta plain even within a few meters to several hundreds of metres. This may be responsible for the lateral variation of arsenic in groundwater.

In the present study area, the main water-bearing formations are Quaternary formations mainly recent and Pleistocene alluvial deposits, and the aquifer materials comprising gravels and sand of varying grades. Groundwater occurs within the water table and in semi-confined to confined conditions. It is also noticed that the

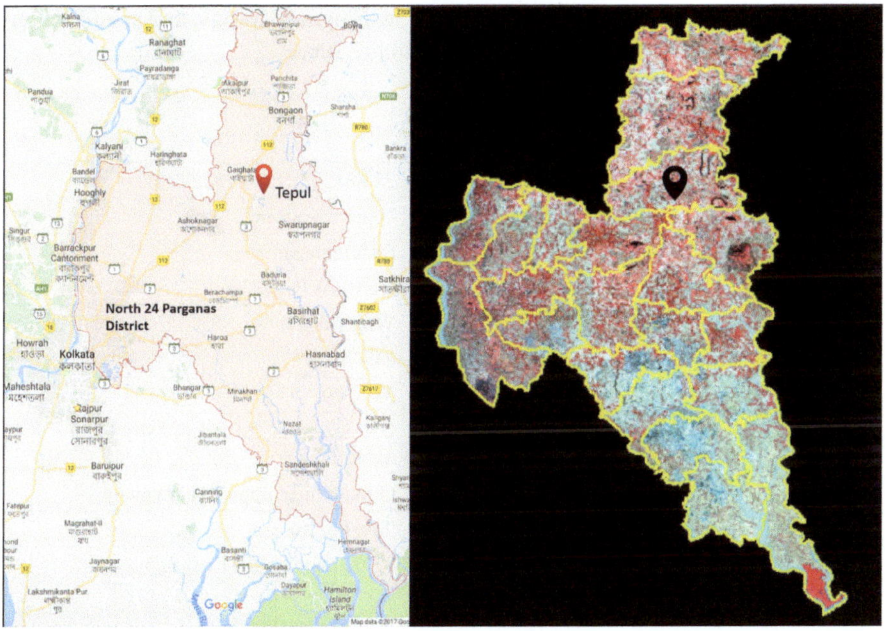

Fig. 15.8 North 24 Parganas with block boundaries (The pin shows the location of Tepul village), red colour shows a high level of arsenic

nature of aquifer materials in horizontal extent is not uniform, and it changes from place to place. This is due to the facies variation during the sedimentation.

The variation of arsenic concentration in space, from the same aquifer, can be explained as: (a) during the movement of arsenic-bearing water through the aquifer, local clay lenses may adsorb arsenic from the arsenic-rich water; (b) the nature of the release of arsenic in groundwater from the geological horizons may depend on local factors like presence of organic matter in the geological horizons, local geochemical conditions like redox potential (Eh), hydrothermal conditions and physicochemical characters etc.

Based on the available data, the Habra II block of North 24 Parganas was found to be a suitable spot for a commercial plant installation. The Tepul village in Gobardanga, which is close to the Bangladesh border, is well known for its high Arsenic concentration in groundwater of shallow aquifers. Documents of major water projects being implemented by the Public Health Engineering Directorate of the West Bengal Government indicated that no major water supply scheme is being implemented in this area.

The actions taken during the site visit are noted below:

After discussion with the local people and observation from the local ponds and wells, it was determined that the 1st layer of water is at 30 ft depth, the 2nd layer is below 200 ft and the 3rd layer is at 600 ft depth. The hand borewells are between 40–50 ft in depth.

The target is to find out the market demand, willingness-to-pay, and arsenic status in the adjacent areas by employing a questionnaire survey.

SPECIFIC INFORMATION ON TEPUL VILLAGE	
GPS coordinates of the site:	22.881525 N; 88.785227 E
Gram Panchayat:	Tepur Mirzapur
Block / Tehsil:	Swarupnagar
District:	North 24 Parganas
State:	West Bengal
PIN:	743,247
Village code:	322,779
Location:	20.2 km away from sub-district HQ Swarupnagar
Nearest town:	Gobordanga, approx 2 km away
Geographical area:	284.9 hectares
Total population:	3,762 people (2011 census)

PHREEQC: Geochemical Speciation Programming

Kinetic rate expressions can be defined in a completely general way in PHREEQC using Basic statements in the rates data block. The rate expressions can be used in batch-reaction or transport calculations by using the kinetics data block. Equilibrium is calculated before a kinetic calculation is initiated and again when a kinetic reaction increment is added. Equilibrium is calculated for all solution species, and for all exchange, equilibrium-phase, solid solutions, surface assemblages, and gas phases that have been defined. Kinetic reactions between solids and the aqueous phase can be calculated without any modification of the database. PHREEQC can also calculate kinetic reactions among aqueous species that are normally assumed to be in equilibrium, but this requires that the database be redefined.

A web-based version of the USGS model PHREEQC has been used for this study from http://www.ndsu.nodak.edu/webphreeq/.

This program was used to determine the chemical speciation of the water samples taken from the 1st aquifer, 2nd aquifer, and 3rd aquifer at the Tepul site. The water samples were analysed in a certified laboratory and were found to have the following parameters in Table 15.1, which were then used in the PHREEQC software.

After analysing the results, it was decided that the best results for oxidizing the Fe and Mn in the groundwater of the first aquifer will be obtained on recharging aerated water with the DO level of 4–5 mg/L and Redox Potential of + 150 mV and above.

Table 15.1 Groundwater parameters (average) at Tepul Site

	1st Aquifer	2nd Aquifer	3rd Aquifer
Depth (ft)	30	220	600
pH	7.49	7.87	8.2
Temp (°C)	24	23	22
Pe (volt)	−1.63	−2.445	−4.075
Density	1	1	1
Ca (mg/L)	35	43	56
C as HCO3 (mg/L)	156	163	112
Cl (mg/L)	45	48	48
Arsenic (mg/L)	0.4	2.54	0.03
Fe (mg/L)	0.551	0.67	0.41
Mn (mg/L)	0.08	0.1	0.12
N as N03- (mg/L)	1.4	1.45	1.20
O (mg/L)	2	2	1.5
K (mg/L)	1.2	1.8	2.3
Na (mg/L)	4	4.8	5.8

Study of Arsenic-Rich Lithology Sample from Tepul

Determination of arsenic content and other chemical constituents in lithological samples has been done from the borehole samples down to the depth of 81 m drilled at Tepul village. The lithology of the borehole is shown in Table 15.2.

In general, the concentration of arsenic is high in clay rather than in sand samples. This value is very high compared with the values found in the aquifer sands. The organic-rich layer of 1.35 m thickness is of immense importance for arsenic release in Aquifer B. Concentrations of extractable-arsenic in the sediment profile are moderate in Aquifer B and show minima related to organic matter concentrations that confirm the operation of solubilization of arsenic via reduction of iron oxides. The concentration of arsenic is high in clay rather than in sand samples. Concentrations of Zn,

Table 15.2 Lithology of the borehole at Tepul Village

Depth range (m)	Material
0.0–2.19	Topsoil
2.20–6.47	Aquitard, silty grey clay
6.48–14.6	Aquitard, silty grey clay
14.61–33.40	Aquifer A; grey sand (Arsenic-rich
33.41–34.76	Organic rich clay
34.77–56.48	Aquifer B; grey sand (Arsenic-rich
56.49–78.65	Aquiclude; sandy clay
78.66–81.00	Grey sandy clay

Cu, and Ni, correlate well with each other and show an association with traces of sedimentary iron sulfide, which is authigenic in origin in a stable reducing environment that is maintained as anoxic by the high concentrations of organic matter in the aquitards and organic-rich layer. As expected, concentrations of Arsenic show associations with both FeOOH and iron sulfides. These influences can be separated using measurements of TOS (total sulfur), which resides primarily in iron sulfides. A similar organic-rich horizon has been identified in arsenic-affected parts of North 24 Parganas district within the depth of 30 m and it seems to play an important role in the release of arsenic in groundwater.

Arsenic-rich aquifers constituted of grey-coloured medium to fine sand, sub-angular to sub-rounded in shape with mineral assemblages of biotite, garnet, lignite, and opaques, indicating a provenance of dominantly metamorphic origin. Sand grains in the arsenic-rich aquifer are coated with iron and arsenic-rich material. The presence of clay with enriched organic matter in the sediments having high iron and arsenic sulphides deposited in the reducing environment is the principal source of high arsenic concentration in groundwater. Arsenic is released to solution by reductive dissolution of FeOOH and release of its sorbed arsenic to groundwater.

The target of SAR technology in this project is to stabilize the iron and arsenic-coated aquifer layer by creating an oxidized environment, bypassing the high oxygen-depleting ability of the organic-rich layer.

Installation of Plant

A borewell was installed at the Tepul site on 15th July 2017 for the second aquifer. An existing civil structure was determined to be used to house the machinery and the tanks. The pipes, pump, starter box, valves, wires, and tanks were procured by the end of July 2017. Some of the photos taken during the borewell installation are shown in Fig. 15.9.

Fig. 15.9 : **a** the manual drilling process; **b** soil sample, silty clay to medium sand; **c** installation of the pump

Acknowledgements This work was financially supported by the Newton UK-India Innovate Grant, Project No 102727, 2016 awarded to Scientific Event Limited, Northern Ireland, Heriot-Watt University, Scotland and IIEST, Shibpur, India.

The authors would like to gratefully acknowledge the immense contribution of Late Prof Kalyan Bhar of IIEST, who was the moving spirit behind this project and Dr Arnab Saha of Heriot-Watt University for his help in drawdown calculation.

Appendix

A1: Design Basis for Tubewell Installation at Tepul

Maximum demand = 12 m^3/day, taking 3 cycles @ 4 m^3/cycle; for future provision; consider 3 times higher output, 36 m^3/day (approx).

The thickness of the semiconfined aquifer = 30 m (local information)

Average groundwater depth: 4 m (winter), 7 m (summer)-local data, cross-checked by the author.

Maximum observed drawdown = 1 m (proposed to reduce the drop in the water level in the nearby dug wells.

The radius of the circle of influence is 25 m (local data)

Permeability = 1.2 × 10–5 m/s (1.0368 $m^3/m^2 \cdot$ day) from third-party data. This value matches with the standard value of 1.0 × 10–5 m/s for aquifer sand (https://structx.com/Soil_Properties_007.html)

Using Dupuit-Thiem Equation (USGS Open File pp.135–140; pubs.usgs.gov/of/1990/ofr90-183/pdf/ofr_90-183_e.pdf)

Q = 2.72 P. t. (drawdown)/[log10 (R/r)],

Where, Q = discharge or yield of a well, m^3/day

p = porosity ratio

k = transmission coefficient.

P = permeability per unit area = **p. k**, m^3/m^2 per day

R = radius of the circle of influence, m

r = radius of well, m (to be calculated)

H = height of water between the static water table and bottom of the well, m

h = depth of water in well, m

H–h = drawdown, m

t = thickness of the aquifer, m

For the proposed well,

Q = 2.72 × 1.0368x 30 (1)/ [log10 (25/r)] = 36 m^3/day

or, (2.72 × 1.0368x30) /Q = log10 (25/r)

36 = (2.72 × 1.0368x30)/ log10 (25/r)

(2.72 × 1.0368x30)/36 = log10 (25/r) = 2.35; 25/r = 102.35 = 224

r = 25 /224 = 0.111 m = 11.1 cm

The tube well should be 4–6 inches in dia (two standard sizes)

A 4-inch tube well will do the job since the future demand was considered as 3 times higher.

Check: tubewell dia 4 inch or 10 cm

Q = 2.72 × 1.0368x 30 (1)/ [log10 (25/0.1)] = 2.72 × 1.0368 × 30 (1)/2.398 = 35.28 m^3/day

Check: tubewell dia 6-inch dia (15 cm)

Q = 2.72 × 1.0368x 30 (1)/ [log10 (25/0.15)] = 2.72 × 1.0368 × 30 (1)/2.2218 = 38.07 m^3/day.

The difference is small in this case because of the low permeability of the aquifer and the low drawdown value. Therefore a 6-inch dia (preferred) or a 4-inch dia well is recommended.

A2: Aeration Rate Calculation (Fig. A1)

Aeration remains fundamental to arsenic and iron oxidation for safe water supply from the well. The tank was designed to maintain the DO level at 4 mg/L.

Tank diameter = 1956 mm (Brand—Patton, widely available)

Tank height = 1982 mm.

Tank depth from shower head = 1800 mm (design parameter)

Area of the tank = $\pi r^2 = 3.141*(0.978)^2 = 3\ m^2$

Hourly pumping rate = 2.7 m^3/hr.

Superficial water velocity = $\frac{Q}{A} = \frac{2.7}{3} = 0.9$ m/hr

Average pumping time = 1.5 h/cycle

Total amount pumped = 1.5 h * 2.7 m^3/hr = 4.05 m^3

Height of the water at the end of the cycle = $\frac{4.05}{3} = 1.35$ m = 1350 mm

At t = 0, the water drop descends 1800 mm.

At t = 1.5 h, the water drop descents = (1800–1350) = 450 mm

Keeping a 150 mm bottom clearance for the exit pipe,

At t = 0, (beginning of the cycle) water drop descends = 1650 mm

At t = 1.5 h, it is (450–150) = 300 mm

On average, a water droplet descends by = $\frac{1650+300}{2} = 975$ mm

Velocity of the water in the 1.5-inch PVC pipe before entering the shower head:

Considering 1.5-inch schedule 80 PVC/HPPE pipe,

diameter = 3.81 cm, radius = 1.905 cm, area = $\pi * (1.905)^2 = 11.398\ cm^2$

Flowrate 2700 l/hr = 45 l/min = 0.75 l/s.

= 750 cm^3/s; since the pipe area = 11.398 cm^2 (as shown above)

V = 65 cm/s.

The total aerator pore area is designed to have a 10% area of 11.398 cm^2 to increase the velocity of the drops.

Therefore, the velocity of the spray is approx. 650 cm/s (10 times the water velocity in the 1.5-inch pipe)

This value matches closely with the ASCE aerator discharge velocity of 692 cm/s.

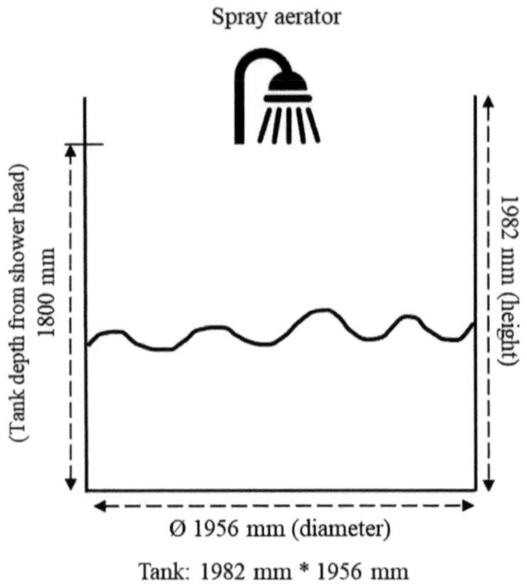

Fig. A1 Spray aerator configuration

The ASCE spray aerator drop trajectory length is taken as 61 cm, in this case, it is 97.5 cm.

Here K value is taken as 6.0 m. d^{-1} (Ref: McWhirter et al., Int. Eng. Chem. Res., 1995, 34, 2644–2654). This is equal to 0.25 m. h^{-1}.

Oxygen balance in the tank (Ref: Banks et al., J. Environ. Eng., 1984, 110(1): 1–14)

$$M = Q * C_r + K * A * (C_s - C)$$

(inflow *DO*) (aeration *DO*)

C_r = Concentration of *DO* in the drop (water entering the tank) = 2 mg/L (design value).

Q = discharge rate into the tank = 2.7 m^3/hr

C_s = saturation value 9.2 mg/L at 20 °C; K = 6.0 m/d or 0.25 m/h.

Amount of *DO* added due to aeration = 0.25 m/hr * 3 m^2 * (9.2 – 2) mg/L × 1.5 h = 8.1 m^3 * mg/L.

DO increase over in the tank = $\frac{8.1}{4.05}$ = 2.0 mg/L (added due to aeration)

Initial *DO* = 2 mg/L at 20 °C

Final *DO* = (2 + 2) mg/L = 4.0 mg/L

Target = 4.0 mg/L at 20 °C

Actual measurement = 4.2 mg/L

Selection of shower head: recommended 2.5 US gallons per minute, shower heads. Such HDPE shower heads made to the US standard are widely available.

2.5 US gal/min = 11.36 L/min; total no of showerheads required = 45.0 / 11.36 = 3.96 or 4 showerhead

Note: It is left to the installation team to adjust the aeration rate to achieve the final DO value of 4 mg/L. Experience shows that for the alluvial aquifers of the North

24 Parganas (West Bengal, India), this DO value is sufficient to achieve the target Eh value of + 200 mV for the pumped water (after 8 weeks) and reach + 300 mV in less than 6 months.

Drawdown estimation by using Theis Equation (https://www.twdb.texas.gov/groundwater/models/analytical/index.asp)

Theis equation (drawdown): It was necessary to check the drawdown for the high withdrawal rate by the proposed installation as well as other installations in the vicinity. A maximum of 16,000.00 L per hour abstraction has been assumed (4 times the max abstraction needed). The maximum drawdown is <1 m (design limit)

Drawdown information (Fig. A2).

Drawdown Calculation using Theis Equation- Distance from the well is 5 meters

Pumping Rate, Q	16000 litres/hour	70.4459 GPM	0.156954 cfs
Aquifer Transmissivity, T	700 m^2/day	7534.74 ft^2/day	0.087208 ft^2/second
Time since pumping began, t	2 hrs	0.0833333 days	
Aquifer storativity, S	0.0003		
Radial distance, r	5 m	16.40 ft	

u = r^2 S$_c$/(4 T t)	0.000032
Well function, W	9.768152404
Drawdown (ho-h) at day 0.0833333: 1.40 ft using series calculation of W(u) to u6	

$$h_o - h = \frac{Q}{4\pi T} W(u) \qquad u = \frac{r^2 S}{4Tt}$$

$$h_o - h = \frac{Q}{4\pi T}\left[-0.5772 - \ln u + u - \frac{u^2}{2\cdot 2!} + \frac{u^3}{3\cdot 3!} - \frac{u^4}{4\cdot 4!} + \ldots\right.$$

Q is the constant pumping rate (L3/T; ft3/day or m3/day)
h is hydraulic head (L; ft or m)
ho is hydraulic head before pumping started (L; ft or m)
ho-h is the drawdown (L; ft or m)
T is aquifer transmissivity (L2/T; ft2/day or m2/day)
t is time since pumping began (T; days)
r is radial distance from the pumping well (L; ft or m)
S is aquifer storavity (dimensionless)
b is aquifer thickness (L; ft or m)

Drawdown over the course of a year

Days	Drawdown (ft)	Drawdown (m)
1	1.75	0.5334
4	1.95	0.5944
7	2.03	0.6187
10	2.08	0.6340
13	2.12	0.6462
16	2.15	0.6553
19	2.18	0.6645
22	2.2	0.6706
25	2.22	0.6767
28	2.23	0.6797
30	2.24	0.6828
60	2.34	0.7132
90	2.4	0.7315
120	2.44	0.7437
150	2.47	0.7529
180	2.5	0.7620
210	2.52	0.7681
240	2.54	0.7742
270	2.56	0.7803
300	2.57	0.7833
330	2.59	0.7894
365	2.6	0.7925

Drawdown over multiple years

Years	Drawdown (ft)	Drawdown (m)
1	2.6	0.7925
2	2.7	0.8230
3	2.76	0.8412
4	2.8	0.8534
5	2.83	0.8626
6	2.86	0.8717
7	2.88	0.8778
8	2.9	0.8839
9	2.91	0.8870
10	2.93	0.8931
11	2.94	0.8961
12	2.96	0.9022
13	2.97	0.9053
14	2.98	0.9083
15	2.99	0.9114
16	3	0.9144
17	3.01	0.9174
18	3.01	0.9174
19	3.02	0.9205
20	3.03	0.9235
21	3.04	0.9266
22	3.04	0.9266

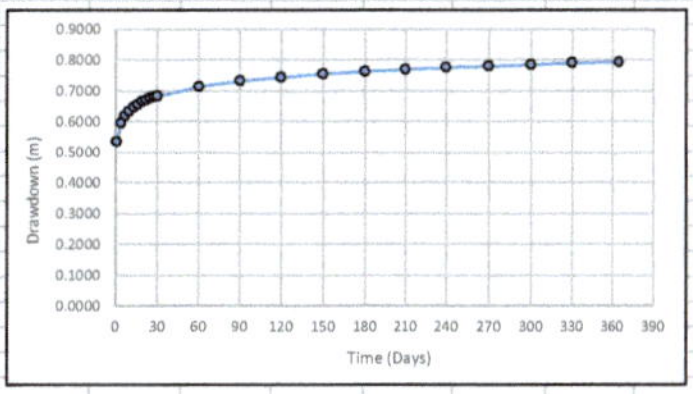

Drawdown Calculation using Theis Equation at 10 m distance

Pumping Rate, Q	16000 litres/hour	70.4459 GPM	0.156954 cfs
Aquifer Transmissivity, T	700 m^2/day	7534.74 ft^2/day	0.087208 ft^2/second
Time since pumping began, t	2 hrs	0.0833333 days	
Aquifer storativity, S	0.0003		
Radial distance, r	10 m	32.81 ft	

u = r^2 S$_c$/(4 T t)	0.000129
Well function, W	8.381954467
Drawdown (ho-h) at day 0.0833333: 1.20 ft using series calculation of W(u) to u6	

Drawdown over the course of a year

Days	Drawdown (ft)	Drawdown (m)
1	1.56	0.4755
4	1.75	0.5334
7	1.84	0.5608
10	1.89	0.5761
13	1.92	0.5852
16	1.95	0.5944
19	1.98	0.6035
22	2	0.6096
25	2.02	0.6157
28	2.03	0.6187
30	2.05	0.6248
60	2.14	0.6523
90	2.2	0.6706
120	2.24	0.6828
150	2.28	0.6949
180	2.3	0.7010
210	2.32	0.7071
240	2.34	0.7132
270	2.36	0.7193
300	2.38	0.7254
330	2.39	0.7285
365	2.4	0.7315

Drawdown over multiple years

Years	Drawdown (ft)	Drawdown (m)
1	2.4	0.7315
2	2.5	0.7620
3	2.56	0.7803
4	2.6	0.7925
5	2.63	0.8016
6	2.66	0.8108
7	2.68	0.8169
8	2.7	0.8230
9	2.72	0.8291
10	2.73	0.8321
11	2.74	0.8352
12	2.76	0.8412
13	2.77	0.8443
14	2.78	0.8473
15	2.79	0.8504
16	2.8	0.8534
17	2.81	0.8565
18	2.82	0.8595
19	2.82	0.8595
20	2.83	0.8626
21	2.84	0.8656
22	2.84	0.8656

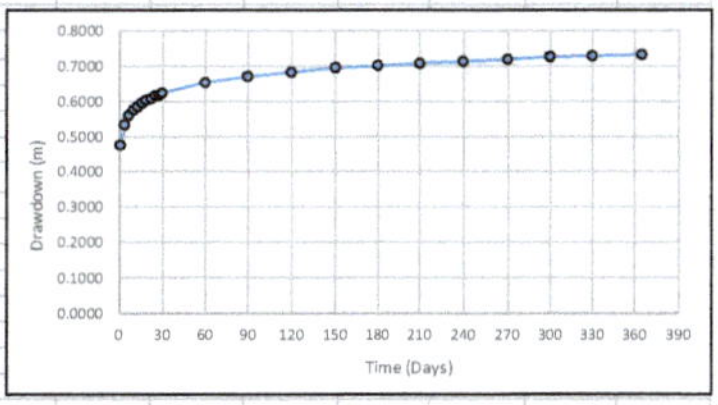

Fig. A2 Drawdown calculations and plots

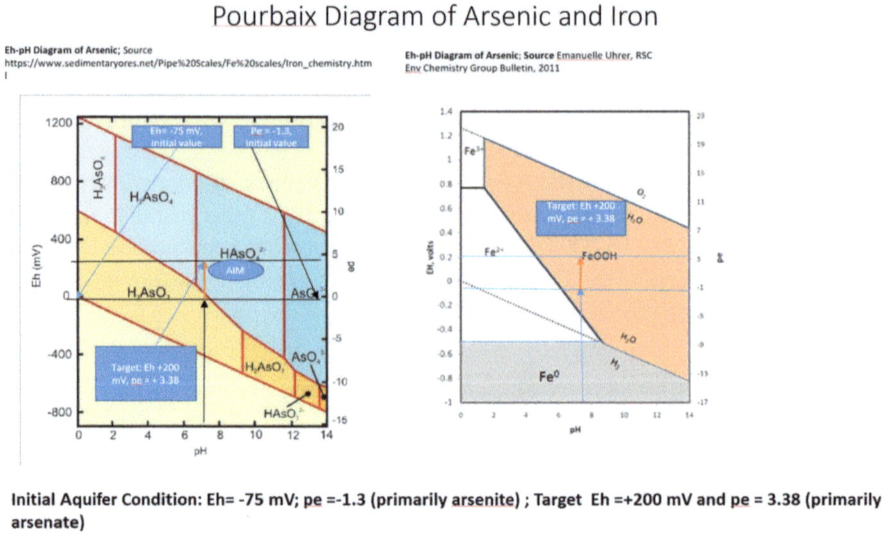

Fig. A3 : Pourbaix diagrams of Fe and As

Operating Conditions

The operating conditions are best described from the Pourbaix diagram of iron and arsenic. ThepH of the water was 6.5 and the initiale H was −75 mV, which changed to + 200 mV upon planned aeration over several weeks. The eH value was kept around + 200 mV through controlled aeration on a daily basis. This operation ensured that arsenic was < 10 μg/L and iron was 0.3 mg/L at all times during operation (Fig. A3).

References

https://www.bbc.co.uk/news/uk-scotland-39647456

Hashim, M. A., et al. (2011). Remediation technologies for heavy metal contaminated groundwater. *Journal of Environmental Management, 92*, 2355–2388.

Sen Gupta, B., et al. (2009). A simple chemical-free arsenic removal method for community water supply—A case study from West Bengal, India. *Environmental Pollution, 157*, 3351–3353.

Sen Gupta, B. (2010). Water treatment goes underground. *The Chemical Engineer (IChemE), 984*, 35–37.

SAR technology: Home (insituarsenic.org)

Chapter 16
Performance Evaluation of a Subterranean Arsenic Removal (SAR) Community Water Treatment Plant: A Sustainable Long-Term Approach to Removing Arsenic from Drinking Water

Bhaskar Sen Gupta, Isita Sen Gupta, Soumyadeep Mukhopadhyay, Sumona Mukherjee, Debra Helen Phillips, Amitava Bandopadhyay, and Arup K. Sen Gupta

Abstract Subterranean arsenic removal (SAR) is a low-cost, zero-waste, and 'easy to operate' process that can remove arsenic and iron from groundwater without using any adsorption bed. The SAR plant creates an aerobic high Eh bed in the aquifer by recharging oxygenated groundwater, thereby supporting the growth of arsenic and iron oxidizing aerobic bacteria. The arsenic and iron are immobilized in the

---

B. S. Gupta (✉)
Institute of Infrastructure and Environment, Heriot-Watt University, Edinburgh Campus, Scotland EH14 4AS, UK
e-mail: b.sengupta@hw.ac.uk

I. S. Gupta · S. Mukhopadhyay
Scientific Event Ltd, 31, Cannon Street, Seton Park, Winchburgh, Broxburn, Scotland EH52 6WN, UK

S. Mukhopadhyay
Commission for Regulation of Utilities, Dublin, Ireland

S. Mukherjee
Dublin City University, Dublin, Ireland

D. H. Phillips
School of Natural and Built Environment, Queen's University Belfast, David Keir Building, BT9 5AG Northern Ireland, UK

A. Bandopadhyay
Centre for the Science and Technology of the Non-Aligned and Other Developing Countries (NAM Centre), Habitat Centre, New Delhi, India

A. K. S. Gupta
Environmental Engineering Program, Department of Civil and Environmental Engineering, Lehigh University, Bethlehem, PA, USA

B. Sen Gupta and N. Martínez-Villegas (eds.), *Arsenic Remediation of Food and Water*,
https://doi.org/10.1007/978-981-97-4764-1_16

aquifer sand in the forms of As (V) and Fe(III), respectively. This work reports the key performance data of a SAR plant installed at Ghetugachi village in West Bengal, India under an increased demand three times above the designed production volume. The naturally occurring water has an As concentration of 154 ppb, which was decreased to 13 ppb for 3000 L per day (LPD) water during the period of 2009 to 2015. However, with an increase in demand of up to 10,000 LPD, due to an increase in local consumer population, the As in the SAR treated water increased to 30 ppb. In order to control the As and Fe levels, a novel HAIX-nano Fe resin bead media filter (Lehigh University) was installed in line with the SAR plant thereby reducing the As and Fe in delivery water to 6 ppb and 240 ppb, respectively. This resulted in low filter cost, no filter clogging over the past 6 months, and a lower maintenance cost of both the SAR plant and HAIX media filter. The combined SAR-HAIX plant has been able to maintain a favourable Eh–pH value of the water in order to immobilize the arsenic and iron consistently over the last 6 months of the study. Locally, ~600 people and two schools depend upon the safe water supplied by this plant and the operating cost comes to about $30.00 US dollars a month to produce 10,000 L of safe water per day.

Keywords Arsenic · Iron · Community water supply · Sustainable in-situ groundwater treatment · Chemical free · Waste free · Subterranean Arsenic Removal (SAR) · Haix resin beads

Introduction

Groundwater arsenic contamination poses a very serious threat to public health globally, especially in the Ganga-Meghna-Brahmaputra basin in Asia, which is home to more than 1 billion people (Adeel, 2002, Mukherjee et al., 2006). Arsenic is a natural constituent of the earth's crust and can be found in water that has flowed through arsenic-rich rocks and minerals. High concentrations (>50 ppb) of arsenic in groundwater are also found in other parts of the world including Argentina, Bangladesh, Chile, China, Hungary, India (West Bengal), Mexico, Nepal, Pakistan, Thailand, USA, and Vietnam (Mukherjee et al., 2006; Ravenscroft, 2007; WHO, 2004). Inorganic arsenic is a documented human carcinogen and 10 ppb was established as a provisional guideline value for arsenic by the USEPA (USEPA, 2001). When arsenic-contaminated water is used for irrigation practices, arsenic enters the food chain through rice and other crops (Biswas et al., 2011; Carbonell-Barrachina et al., 2009; O'Neill et al., 2013). Prolonged intake of arsenic contaminated food and drinking water above 10 ppb may lead to melanosis, keratosis with or without anaemia, conjunctivitis, bronchitis, gastroenteritis, Blackfoot disease and skin cancer (Jomova et al., 2011). The arsenic crisis usually triggers a varied number of societal problems and the poorest fraction of the society are the worst affected (Adeel, 2002; WHO, 2000).

A number of pump and treat technologies have been developed around the world over the past two decades to remove arsenic from contaminated groundwater through domestic and community units using chemical oxidation, chemical co-precipitation, coagulation and filtration, adsorption, and ion exchange processes (Visoottiviseth & Ahmed, 2008). However, all of these technologies encounter several operational problems in the long run, such as safe disposal and stabilization of toxic sludge, back-wash and handling of filters, and high operative costs over the operational lifetime, etc. (Bhattacharyya et al., 2003). This problem which lies within the aquifer cannot be solved by bringing up toxic sludge onto the soil surface poisoning the ecosystem (Leist et al., 2000). A technology that has minimum environmental impact, produces almost zero waste using no chemicals, and can serve the rural population with easy operation and inexpensive maintenance should be the primary treatment option to remove the bulk of the iron and arsenic. Additionally, a 'pump and treat' technology such as an adsorption bed can be used as a secondary treatment option for polishing the treated water before public delivery.

This paper highlights the performance of a Subterranean Arsenic treatment (SAR) plant and a planned retrofit for serving an increasing number of consumers in a village in West Bengal, India. In an earlier publication, the principle and workings of a SAR plant had been described (Sen Gupta et al., 2009). According to UNDP, India's HDI is 0.547, which gives the country a rank of 134 out of 187 countries with comparable data. The HDI of South Asia as a region increased from 0.356 in 1980 to 0.548 currently, placing India below the regional average. The health index (life expectancy) rank is 142 (index value of 0.717) with 1.1% of the total GDP being expended on public health (UNDP, 2011). These statistics validate the urgency of establishing a sustainable and stable system of water treatment for the rural population plagued by the slow poisoning of arsenic, which negatively impacts the already burdened economy with deteriorating public health.

The objective of this work was to fulfil the current demand of 10,000 LPD safe water while maintaining the USEPA maximum permissible limit of 10 ppb from the existing SAR plant. In order to achieve this objective, two separate strategies were adopted: (i) the introduction of three operating cycles of 4 h each—a total of over 12 h of operation a day, and (ii) the addition of an adsorber/polisher media, to compensate for a 10% recharge. If these filtration systems are used as a polisher and fed with water already having lower levels of pollutants, then they will be highly effective over a long duration, without having the necessity to deal with large and frequent build-up of contaminants. With this objective, a novel hybrid anion exchange (Haix-nanoFe) resin media was attached to the SAR in at the site.

Materials and Methods

SAR Plant Retrofitted with a Haix-Column at the Site

Groundwater sources that supplied water to Ghetugachi, a small village in Nadia District of West Bengal, India (Lat 23.07607; Long 88.52548) (Sen Gupta et al., 2009) (Fig. 16.1), contained 160–180 ppb arsenic (varying within this range depending on season and groundwater levels) as compared to 10 ppb which is the USEPA and WHO maximum permissible limit. In 2008, A SAR plant was designed and installed to produce only 3000 L of arsenic-free water a day for 600 people in the village. The schematics of the SAR plant are shown in the Fig. 16.2. The principle and workings of a SAR plant were described by Sen Gupta et al. (2009). Ghetugachi village is located in the Ganga-Meghna-Brahmaputra basin where sediments originating from granitic and metamorphic rock of the Himalayas weathered releasing sulphide minerals, including arsenopyrite. These sediments were deposited in the Ganges delta region (McArthur et al., 2004; Nickson et al., 1998).

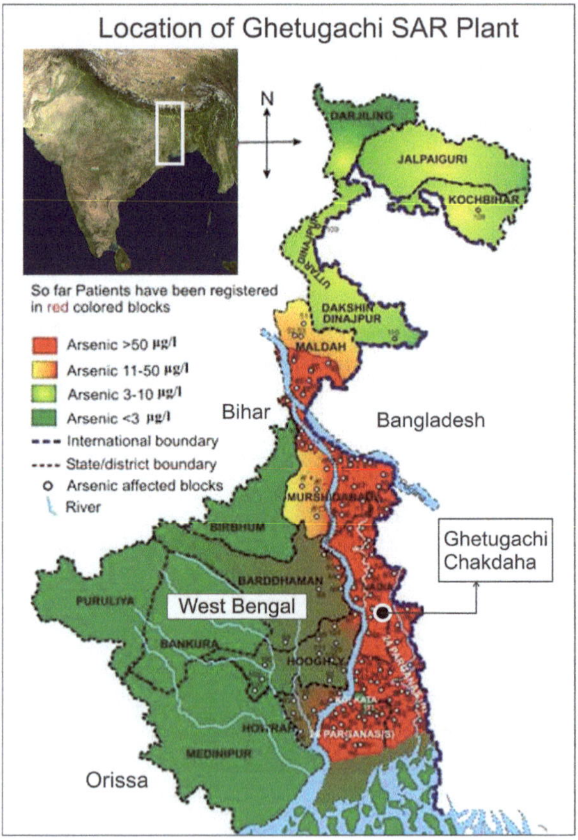

Fig. 16.1 Location of Ghetugachi village in the Nadia District of West Bengal, India

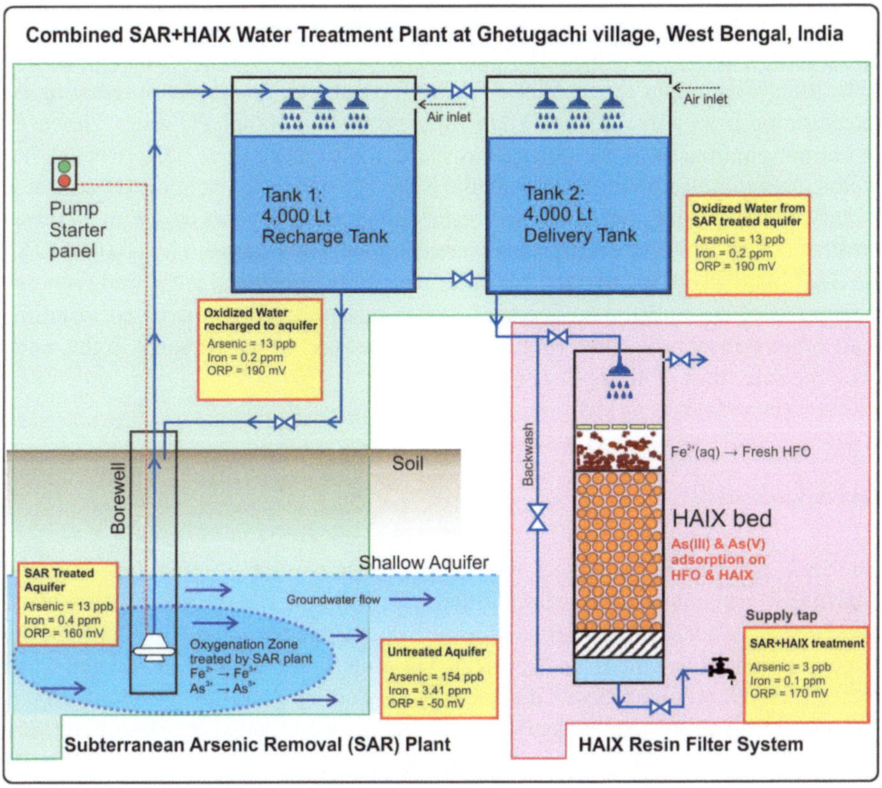

Fig. 16.2 Schematics of the complete SAR-HAIX system

The socio-economic condition of the people did not permit them to buy pure and safe drinking water. Arsenic filters available on the market were too expensive to maintain for regular water supply. Under such circumstances, a small-scale SAR plant was installed at this site, with three conditions in mind (1) construction cost should be low, (2) operation and maintenance should be easy and cheap, and (3) plants should deliver arsenic-free water over long periods (compared to 6 months running time of arsenic filters assisted by backwash). The plant was constructed with locally available materials by local plumbers with financial help from the World Bank for a total cost of $5000.00 US dollars apiece in 2008 (Sen Gupta et al., 2009). Operating costs included an electricity bill which costs < $10.00 US dollars per month and some minor maintenance costs, such as repairing leaking pipes and changing tap washers. Local villagers, trained by the research team, operate the plant. The operators collect a small token fee from the users to cover the cost of electricity. The research team also financially supported the operators.

Arsenic-contaminated water is pumped from the shallow aquifer and sprayed through shower heads into a 4,000L recharge tank (Tank 1) which oxygenates the water. This oxygenated water is then sprayed into a 4,000L delivery tank (Tank 2) and is then pumped back into the aquifer. This comprised one operational cycle and took approximately 4 h to complete. The SAR plant setup had been described in another publication (Sen Gupta et al., 2009). The SAR plant was successful in reducing As concentration from ~160 ppb to 13 ppb while producing 3000 LPD water. However, the current population of the village has increased to more than 1000 over the last 7 years. Additionally, about 20 m from the SAR, plant two new schools (Ghetugachi Primary and Secondary Schools) were established about 5 years ago with a current enrolment of about 400 students, thus increasing the water demand to ~10,000 LPD. The SAR plant struggled to treat the water beyond the designed value and recorded an increase of As to 30 ppb. The trial in this research project involved installing an adsorber/polisher designed by Lehigh University to remove arsenic in the water above the safe limit of 10 mg/L (Fig. 16.2).

HAIX Column

The SAR plant was retrofitted with a community-scale arsenic removal unit consisting of a stainless-steel column filled with 100 L of adsorbent including Haix-nano Fe resin beads. The SAR-treated water was pumped into the HAIX column for further removal of arsenic from the water. The water inlet where SAR-treated water enters the HAIX column is located at the top of the column and designed with a spray head and splash plates so that small water droplets are formed at the entry point. The top part of the column has a large void space with vent connections that allow the air to mix with the water droplets (Fig. 16.2). The dissolved air helps to oxidize traces of ferrous iron (~0.25 ppm) in the SAR-treated water to form fine precipitates of hydrated ferric oxide (HFO) which also helps to remove arsenic and iron from the water (Phillips et al., 2018). Therefore, arsenic (~30 ppb) which was not eliminated by the SAR treatment is removed while the water trickles down to the bottom of the unit from where the HAIX-nano Fe treated water is collected.

Geochemical Sampling and Analysis

Groundwater and treated water samples were collected, stored in refrigeration with addition of a drop of nitric acid, and transported to an accredited laboratory in nearby city of Kolkata for analysis. Total Fe and As was measured by a Perkin-Elmer Atomic Adsorption Spectroscopy (AAS). pH and Eh measurements were carried-out at site using a Fisherbrand™ Handheld pH/ORP meter.

Results and Discussion

SAR-HAIX Plant Performance

Background Borewell

From Dec 2015 to Nov 2016, the background borewell located 75 m from the SAR plant had high total arsenic concentrations ranging from 152–166 ppb and total Fe concentrations ranging from 3500–3750 ppb (Fig. 16.3). This groundwater was noticeably higher in arsenic and iron concentrations compared to water from the SAR borewell, SAR-treated water and Haix-treated water. The groundwater had an ORP that ranged from −55 mV to −60 mV and the pH was slightly acidic, ranging from 6.3 to 6.6 (Table 16.1). Figure 16.4 shows the naturally occurring untreated groundwater is in a reduced state and has an Eh–pH value in the zone of As and Fe mobility which results in the high As and Fe concentrations in the untreated aquifer. There are a number of potential mechanisms for the release of arsenic into groundwater from shallow alluvial sedimentary aquifers such as the one at this site. The oxidation of arsenic-rich pyrite, arsenopyrite, in aquifer sediments and the reductive dissolution of arsenic-rich Fe(III) oxyhydroxides may be the two principal mechanisms (Chowdhury et al., 1999; Harvey et al., 2002). Additional factors such as mobilization of sorbed arsenic by phosphate originating from the intensive use of fertilizers (Acharyya et al., 1999), changes in the sorptive capacity of Fe(III) oxyhydroxides (Smedley & Kinniburgh, 2002), and anaerobic metal-reducing bacteria which reduces As from the + 5 to + 3 state (Islam et al., 2004) may also contribute significantly towards the process.

SAR Borewell and Plant

The SAR plant was able to decrease arsenic concentrations to about to 13 ppb for 3000 LPD water during the period of 2009 to 2015. The SAR plant turned the underground aquifer into a natural biochemical reactor and an oxidized adsorption zone that entraps As(V) along with Fe(III) and Mn(II) at an elevated redox value of groundwater (Eh > 150 mV) in the oxidation zone). Functionally, this technology simply involved pumping out the water from shallow aquifers by a submersible pump, spraying it with shower heads in an overhead tank to oxygenate, and then releasing it back into the aquifer. The bacterial arsenic reduction process is reversed without using any chemicals, by recharging the calculated volume of aerated water (DO $\geq$ 4 mg L^{-1}) in the aquifer to create an oxidized zone. This process boosted the growth of iron oxidizing bacteria and chemoautotrophic As oxidizing bacteria (CAOs) and suppressed the growth of As reducing anaerobic bacteria over a period of six to eight weeks (Sen Gupta et al., 2009). The soil particles developed a surface area of ferric iron hydroxides for adsorption of more ferrous iron and other oxyanions such as arsenate (Appelo & deVet, 2003; Rott et al., 2002). The CAO bacteria oxidizes the

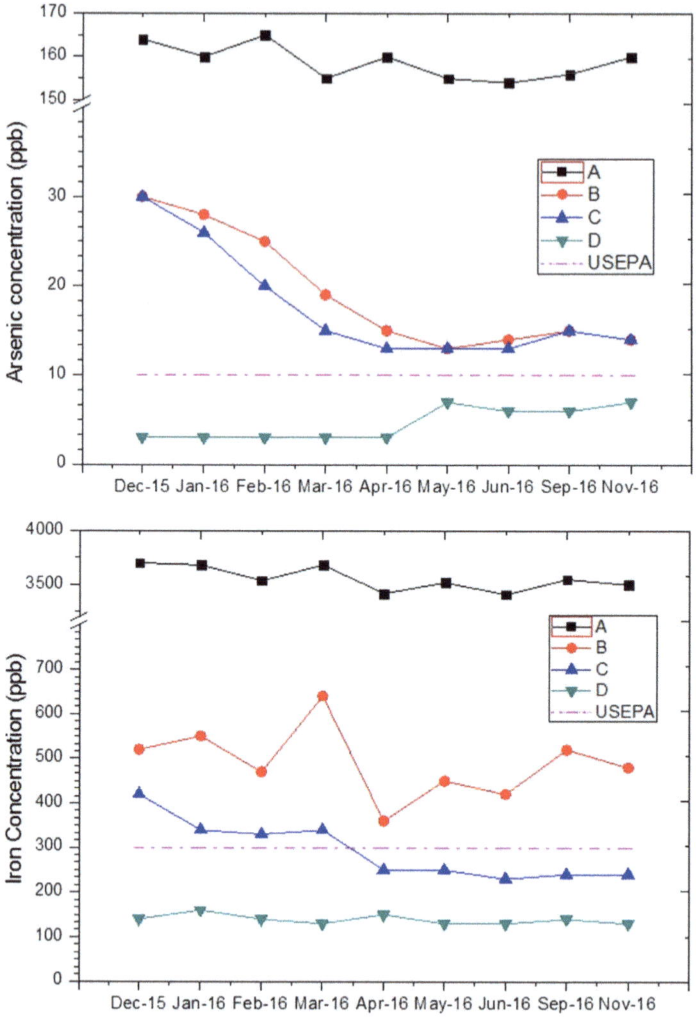

Fig. 16.3 Arsenic and iron levels in water samples during the period December 2015 to Nov 2016. **a** raw water sample from borewell 75 m away from SAR plant; **b** raw water sampled from SAR borewell; **c** water treated by SAR sampled before HAIX filter; **d** water treated by HAIX-nanoFe resin bead adsorber

soluble and more toxic As(III) to less mobile As(V) species by enzymatic oxidation. The success of the process largely depends on controlled precipitation of Fe(III) on the aquifer sand so that the precipitate could acquire a dense goethite or lepidocrocite type structure. Controlled precipitation of Fe(III) also ensured the entrapment of As(V) as it is adsorbed on the aquifer sand and is subsequently oxidized to form a dense and compact structure, without affecting the permeability of the aquifer sand (Sen Gupta et al., 2009).

Table 16.1 ORP (Eh)-pH data during the period December 2015 to Nov 2016 (A) raw water sample from borewell 75 m away from SAR plant; (B) raw water sampled from SAR borewell; (C) water treated by SAR sampled before HAIX filter; (D) water treated by HAIX adsorber

Month	Dec 2015	Jan 2016	Feb 2016	Mar 2016	April 2016	May 2016	June 2016	Sept 2016	Nov 2016
(A) Raw water from Untreated borewell 75 m distance from SAR Plant									
ORP (max) mV	−56	−60	−55	−58	−60	−55	−55	−60	−58
pH	6.4	6.5	6.3	6.6	6.5	6.6	6.4	6.5	6.5
(B) Raw water from Treated borewell of SAR Plant									
ORP (max) mV	145	152	148	155	152	155	160	158	162
pH	6.4	6.7	6.4	6.6	6.4	6.7	6.6	6.7	6.6
(C) Water from SAR plant before HAIX filter									
ORP (max) mV	160	170	170	180	190	195	195	190	190
pH	6.5	6.6	6.3	6.5	6.5	6.6	6.7	6.7	6.6
(D) Water from HAIX adsorber/polisher (Treating SAR produced water)									
ORP (max) mV	170	170	170	170	170	160	160		
pH	6.7	6.8	6.7	6.8	6.7	6.9	6.9	6.9	6.8
USEPA drinking water standards: Arsenic = 10 ppb; Iron: 0.3 ppm, pH = 6.5–8.5									

However, with a local population growth, there was an increase in demand for water production up to 10,000 LPD with a 10% discharge, and as a result As in the SAR treated water increased to the range of 25–32 ppb. From December 2015, the total recharge was increased to 10,000 LPD. A year of water monitoring at the site revealed that the SAR borewell water and the SAR treated water had a similar trend in arsenic concentrations. Arsenic concentrations noticeably decreased from Dec 2015 to May 2016 in water from the SAR borehole and SAR treated water, with arsenic concentrations being lower from Jan 2016 to Apr 2016 after being treated by the SAR plant. From May 2016 to Nov 2016, arsenic concentrations were fairly similar for both the SAR borewell and the SAR treated water ranging between 12–16 ppb. Iron was higher in the SAR borewell ranging from 350 to 650 ppb compared to the iron in the SAR treated water which ranged from 250–400 ppb. From Apr 2016 to Nov 2016 the iron was below the USEPA guideline values.

The ORP was lower in the SAR borehole ranging from 145 to 162 mV compared to the SAR treated water which ranged from 160 to 195 mV. Generally, these ORP values increased over time in the SAR plant. The pH values were similar for the SAR borewell ranging from 6.4 to 6.7 and the SAR treated water ranging from 6.3

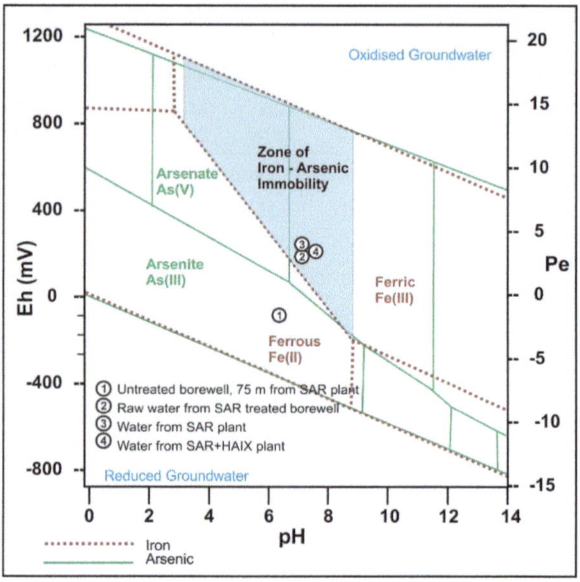

Fig. 16.4 Composite Eh–pH diagram for As and Fe systems showing the position of the water in this study

to 6.7. The treated water from SAR zone and SAR-HAIX plant (discussed below) are oxidized and have Eh–pH values well within the "zone of As-Fe immobility" thereby binding the As and Fe (Fig. 16.4). This Eh–pH is being controlled in the aquifer by SAR process and in the adsorbent bed by HAIX system. Arsenic may occur as a semi metallic element (As°), arsenate (As^{5+}), arsenite (As^{3+}), or arsine (As^{3-}). The double bond oxygen in the arsenate molecule influences its ability to become ionized through the loss of hydrogen ions. A negative charge develops on the arsenate molecule when this dissociation occurs. The double bonded oxygen increases the capacity to delocalize that charge, easing the loss of hydrogen ions (Vance, 1995).

HAIX Column

The HAIX-nano Fe resin bead adsorber bed was installed on one delivery point of the plant in May 2015. Three other delivery points were connected to water treated by subterranean method. Traces of arsenic (~30 ppb) from the SAR treated water was removed while it trickled down to the bottom of the unit from where the HAIX-nano Fe treated water is collected from the water inlet located at the top of the column. The arsenic and iron concentrations of the polished water remained below the USEPA guideline values of <10 ppb and <300 ppb, respectively, for over one year until Nov 2016 for the Haix-nano Fe treated water. The dissolved air helps to oxidize the traces of ferrous iron (~0.25 ppm) in the SAR treated water to form fine precipitates of hydrated ferric oxide (HFO) according to the following reaction. The ORP was steady at 170 mV from Dec 1015 to Apr 2016 decreasing slightly to 160 mV in May to June 2016.

$$\begin{aligned} &4Fe^{2+} = (aq.) + O_2(g) + 10H_2O \Rightarrow 4Fe(OH)_3(s) + 8H^+ \\ &\Delta G_R^0 = -18kJ/mole \end{aligned} \tag{16.1}$$

Oxidation of dissolved Fe(II) to insoluble Fe(III) hydroxide at near-neutral pH (6.7–6.9) similar to the pH values of the treated water in the Haix column is a thermodynamically favorable process due to its relatively high negative free energy of reaction (ΔG^0) at the standard state (Morel & Hering, 1993). Freshly precipitated hydrated Fe(III) oxide (HFO) particle surfaces are considered to be diprotic acid capable of forming the surface functional groups of Fe(OH)2 + and FeOH at circum-neutral pH. HFO particles with these surface groups can selectively bind both arsenites or As(III) and arsenates or As(V) through the formation of bidentate and/or monodentate inner-sphere complexes where Fe(III), a transition metal, serves as electron pair acceptor or Lewis acid (Dzombak & Morel, 1990). A significant portion of influent arsenic is adsorbed by the freshly formed HFO precipitates inside the column. The remaining arsenic in the water is removed by a bed of hybrid anion exchanger (HAIX) impregnated with nano Fe particles of HFO (Phillips et al., 2018; Sarkar et al., 2008) (Fig. 16.5).

The desired treated water flow rate from the arsenic removal unit is 10–12 L/min. The flow rate, however, diminishes over time due to the precipitation of the HFO particles within the bed. Backwashing of the column once a month to drive out the HFO particles is necessary to maintain a good flow rate through the column. The arsenic-laden HFO particles in the waste backwash are trapped on top of a coarse sand filter provided in the same location. Impregnation or dispersion of hydrated Fe(III) oxide or HFO nanoparticles within an anion exchanger posed major challenges because of the associated difficulty in introducing positively charged ferric ions

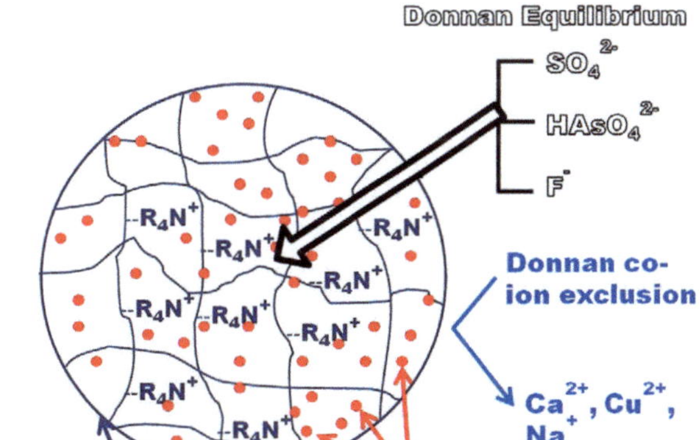

Fig. 16.5 Diagram of an HAIX resin bead with the anion exchange resin support impregnated with nanoparticles of HFOs. Note the transport of counterions (i.e., anions) through the resin per the Donnan membrane equilibrium and rejection of co-ions (i.e., cations) per Donnan co-ion exclusion

(Fe^{3+}) within an anion exchange resin with positively charged quaternary ammonium functional groups. Through several iterations, a reproducible, operationally simple protocol was developed in 2004 and described in detail in United States Patent US 7,291,578 (Sengupta & Cumbal, 2007; Sarkar et al., 2007). Under this patent, HAIX is now commercially available globally as LayneRT™ from Layne Christensen Co. (LayneRT).

Costs

Community participation was ensured at every stage of the installation as well as the operation of the treatment units in this project. A water committee was formed to take care of the management of the treatment unit. The expenditures related to maintenance of the unit, electricity cost, and monthly water testing are covered through the collection of a monthly water tariff from the users in the range of Rs. 15 (equivalent to 30 US cents) per month, a total being ~$30.00 US dollars. The users are regularly updated about the water quality, funds generated, and issues related to the management of the unit. Table 16.2 contains the details of the capital cost of a complete SAR + HAIX plant. The capital cost of the units has so far been provided by international and national funding agencies.

Table 16.2 Capital expenditure for installation of SAR + HAIX plant

	Numbers	Rate (INR)	Total (INR)	Total (US$)
Materials				
Tanks (4000 L)	2	35,000	70,000	1076.92
Submersible pump	1	12,000	12,000	184.62
HAIX unit	1	75,000	75,000	1153.85
Electrical accessories	1	10,000	10,000	153.85
Plumbing accessories	1	20,000	20,000	307.69
Valves	5	1000	5000	76.92
Water meters	3	4000	12,000	184.62
Civil job materials	1	50,000	50,000	769.23
Labor and services (2qww3)				
Transportation			10,000	153.85
Installation and commissioning			10,000	153.85
Plumbing			20,000	307.69
Civil work			20,000	307.69
Electrical work			20,000	307.69
TOTAL			334,000	5138.46

Conclusions and Future Work

In 2008, a SAR plant was installed at Ghetugachi village in West Bengal, India to treat water with an As concentration of 154 ppb. From 2009 to 2015 the arsenic concentrations decreased to 13 ppb for 3000 LPD. However, an increase in consumer population, increased demand to 10,000 LPD resulting in As concentrations increasing to 30 ppb in the SAR-treated water. The SAR plant was retrofitted with a HAIX column containing HAIX-nano Fe resin bead media as a polisher which successfully reduced the concentrations of the residual arsenic and iron remaining in the SAR treated to 6 ppb and 240 ppb, respectively, which are below the USEPA guideline values. The SAR-HAIX plant maintained a favourable Eh–pH value of the water which immobilized the arsenic and iron over the last 6 months of the study. There were low filter costs, no filter clogging over the last 6 months of the study, and a lower maintenance cost for both the SAR plant and HAIX-nano Fe resin bead media. The SAR-HAIX plant is currently supplying 10,000 L of safe water per day water to ~600 people and two schools with a low operating cost of about $30.00 US dollars a month. With regular operation and maintenance, the SAR-HAIX plant is predicted to have a working life of more than 10 years.

Research is underway to develop new HAIX media to better remove both arsenic and fluoride from water. At this site, the fluoride concentration was 3.41 ppm as compared to 1.5 ppm which is the USEPA and WHO maximum permissible limit, therefore, the SAR-HAIX plant would benefit from a HAIX media that could also remove fluoride. This technology has a scope of being replicated for similar resource-poor societies around the world.

Acknowledgements Partial financial support from Water for People (WFP) in Denver, CO, Hilton Foundation, Rotary International, Lehigh University, and several private donors are gratefully acknowledged. The authors also gratefully acknowledge the funding provided by the United States Environmental Protection Agency (USEPA) through the P3 grant award in 2007.

References

Acharyya, S. K., Chakraborty, P., Lahiri, S., Raymahashay, B. C., Guha, S., & Bhowmik, A. (1999). Arsenic poisoning in the Ganges delta. *Nature, 401*(6753), 545–545.

Adeel, Z. (2002). The disaster of arsenic poisoning of groundwater in South Asia—A focus on research needs and UNU's role. *Global Environmental Change, 12*(1), 69–72.

Appelo, C. A. J., & deVet, W. W. J. M. (Eds.). (2003). *Modeling in situ iron removal from groundwater with trace elements such as As.* Arsenic in Groundwater.

Bhattacharyya, R., Chatterjee, D., Nath, B., Jana, J., Jacks, G., & Vahter, M. (2003). High arsenic groundwater: Mobilization, metabolism and mitigation—An overview in the Bengal Delta Plain. *Molecular and Cellular Biochemistry, 253*(1), 347–355.

Biswas, A., Majumdar, J., & Santra, S. C. (2011). Potential arsenic enrichment problems of rice and vegetable crops. *International Journal of Research in Chemistry and Environment, 1*, 29–34.

Carbonell-Barrachina, Á. A., Signes-Pastor, A. J., Vázquez-Araújo, L., Burló, F., & Sengupta, B. (2009). Presence of arsenic in agricultural products from arsenic-endemic areas and strategies to reduce arsenic intake in rural villages. *Molecular Nutrition and Food Research, 53*(5), 531–541.

Chowdhury, T. R., Basu, G. K., Mandal, B. K., Biswas, B. K., Samanta, G., Chowdhury, U. K., Chanda, C. R., Lodh, D., Roy, S. L., Saha, K. C., Roy, S., Kabir, S., Quamruzzaman, Q., & Chakraborti, D. (1999). Arsenic poisoning in the Ganges delta. *Nature, 401*(6753), 545–546.

Dzombak, D. A., & Morel, F. M. M. (1990). *Surface complexation modeling: Hydrous ferric oxide*. Wiley-Interscience.

Harvey, C. F., Swartz, C. H., Badruzzaman, A. B. M., Keon-Blute, N., Yu, W., Ali, M. A., Jay, J., Beckie, R., Niedan, V., Brabander, D., Oates, P. M., Ashfaque, K. N., Islam, S., Hemond, H. F., & Ahmed, M. F. (2002). Arsenic mobility and groundwater extraction in Bangladesh. *Science, 298*(5598), 1602–1606.

Harvey, C. F., Swartz, C. H., Badruzzaman, A. B. M., Keon-Blute, N., Yu, W., Ali, M. A., Jay, J., Beckie, R., Niedan, V., Brabander, D., Oates, P. M., Ashfaque, K. N., Islam, S., Hemond, H. F., & Ahmed, M. F. (2005). Groundwater arsenic contamination on the Ganges Delta: Biogeochemistry, hydrology, human perturbations, and human suffering on a large scale. *Comptes Rendus Geoscience, 337*(1–2), 285–296.

Islam, F. S., Gault, A. G., Boothman, C., Polya, D. A., Chamok, J. M., Chatterjee, D., & Lloyd, J. R. (2004). Role of metal-reducing bacteria in arsenic release from Bengal delta sediments. *Nature, 430*(6995), 68–71.

Jomova, K., Jenisova, Z., Feszterova, M., Baros, S., Liska, J., Hudecova, D., Rhodes, C. J., & Valko, M. (2011). Arsenic: Toxicity, oxidative stress and human disease. *Journal of Applied Toxicology, 31*(2), 95–107.

Katsoyiannis, I. A., & Zouboulis, A. I. (2004). Application of biological processes for the removal of arsenic from groundwaters. *Water Research, 38*(1), 17–26.

Leist, M., Casey, R. J., & Caridi, D. (2000). The management of arsenic wastes: Problems and prospects. *Journal of Hazardous Materials, 76*(1), 125–138.

McArthur, J. M., Banerjee, D. M., Hudson-Edwards, K. A., Mishra, R., Purohit, R., Ravenscroft, P., Cronin, A., Howarth, R. J., Chatterjee, A., Talukder, T., Lowry, D., Houghton, S., & Chadha, D. K. (2004). Natural organic matter in sedimentary basins and its relation to arsenic in anoxic ground water: The example of West Bengal and its worldwide implications. *Applied Geochemistry, 19*(8), 1255–1293.

Morel, F. M. M., & Hering, J. G. (1993). *Principles and applications of aquatic chemistry*. Wiley-Interscience.

Mukherjee, A., Sengupta, M. K., Hossain, M. A., Ahamed, S., Das, B., Nayak, B., Lodh, D., Rahman, M. M., & Chakraborti, D. (2006). Arsenic contamination in groundwater: A global perspective with emphasis on the Asian scenario. *Journal of Health, Population and Nutrition, 24*(2), 142–163.

Nickson, R., McArthur, J., Burgess, W., Matin Ahmed, K., Ravenscroft, P., & Rahman, M. (1998). Arsenic poisoning of Bangladesh groundwater. *Nature, 395*(6700), 338.

O'Neill, A., Phillips, D. H., Kok, S., Chea, E., Seng, B., & Sen Gupta, B. (2013). Arsenic in groundwater and its influence on exposure risks through traditionally cooked rice in Prey Vêng Province, Cambodia. *Journal of Hazardous Materials, 262*, 1072–1079.

Phillips, D. H., Sen Gupta, B., Mukhopadhyay, S., & Sengupta, A. (2018). Arsenic and fluoride removal from contaminated drinking water with Haix-Fe-Zr and Haix-Zr resin beads. *Journal of Environmental Management, 215*, 132–142.

Ravenscroft, P. (2007). *Predicting the global extent of arsenic pollution of groundwater and its potential impact on human health*. UNICEF.

Rott, U., Meyer, C., & Friedle, M. (2002). Residue-free removal of arsenic, iron, mangenese and ammonia from groundwater. *Water Science and Technology: Water Supply, 2*(1), 17–24.

Sarkar, S., Blaney, L. M., Gupta, A., Ghosh, D., & SenGupta, A. K. (2007). Use of ArsenXnp, a hybrid anion exchanger, for arsenic removal in remote villages in the Indian subcontinent. *Reactive and Functional Polymers, 67*(12), 1599–1611.

Sarkar, S., Blaney, L. M., Gupta, A., Ghosh, D., & SenGupta, A. K. (2008). Arsenic removal from groundwater and its safe containment in a rural environment: validation of a sustainable approach. *Environmental Science & Technology, 42*(12), 4268–4273.

Sengupta, A. K., & Cumbal, L. H. (2007). Hybrid anion exchanger for selective removal of contaminating ligands from fluids and method of manufacture thereof. U. S. p. U. 7291578.

Sen Gupta, B., Chatterjee, S., Rott, U., Kauffman, H., Bandopadhyay, A., DeGroot, W., Nag, N. K., Carbonell-Barrachina, A. A., & Mukherjee, S. (2009). A simple chemical free arsenic removal method for community water supply—A case study from West Bengal, India. *Environmental Pollution, 157*(12), 3351–3353.

Smedley, P. L., & Kinniburgh, D. G. (2002). A review of the source, behaviour and distribution of arsenic in natural waters. *Applied Geochemistry, 17*(5), 517–568.

UNDP. (2011). International human development indicators. Retrieved August 10, 2012, from http://hdrstats.undp.org/en/indicators/57506.html

USEPA. (2001). *National primary drinking water regulations; arsenic and clarifications to compliance and new source contaminants monitoring* (pp. 6975–7066). USA

Vance, D. (1995). Arsenic—Chemical behavior and treatment. *National Environmental Journal, 5*(3).

Visoottiviseth, P., & Ahmed, F. (2008). Technology for remediation and disposal of arsenic. *Reviews of Environmental Contamination, Springer, New York, 197*, 77–128.

WHO. (2000). Towards an assessment of the socioeconomic impact of arsenic poisoning in Bangladesh, WHO, Geneva. WHO/SDE/WSH/00.4.

WHO. (2004). Arsenic in drinking water. *IARC Monographs on the Evaluation of Carcinogenic Risks to Humans, 84.*

Part IV
Arsenic Contamination in Water and Food Chain

Chapter 17
Remedial Approaches to Arrest Arsenic in Soil–Plant System to Prevent Its Entry in Rice Grain—a Review

Urvashi Lama, Sharmistha Majumder, Deepanjan Mridha, and Tarit Roychowdhury

Abstract Arsenic (As) contamination has been a world-wide issue since decades which has become a serious problem to the population living both in As-endemic and non-endemic regions. The metalloid, As, has made its way to humans through the inescapable food chain via consumption of As-contaminated food (plants and animals), food products and water. Rice being a staple diet for a large population in the South Asian sub-continent constitutes a significant source of As to the human body due to its efficient As accumulation property. The health risks posed by As in rice are of critical concern, necessitating remedial measures to prevent the entry of As from contaminated soil to plants. This chapter encompasses the different strategies used to reduce or prevent As uptake, translocation and bioavailability by rice crops, highlighting existing research gaps and future research prospects in this field.

Keywords Arsenic · Remediation · Soil–plant system · Organic and inorganic amendments · Biochar · Biofertilizer

Introduction

One of the major environmental issues which people in different parts of the world encounter is arsenic (As) contamination. When exposed to As which is a naturally occurring hazardous metalloid, people may experience both carcinogenic and non-carcinogenic effects (Joardar et al., 2021). The entry of As to human body system is through food chain via consumption of As-contaminated food. Arsenic is a non-essential element that can build up in terrestrial plants through their root system since

U. Lama · S. Majumder · D. Mridha · T. Roychowdhury (✉)
School of Environmental Studies, Jadavpur University, Kolkata 700032, India
e-mail: tarit.roychowdhury@jadavpuruniversity.in; rctarit@yahoo.com

U. Lama
Sovarani Memorial College, Jagatballavpur, Howrah 711408, India

B. Sen Gupta and N. Martínez-Villegas (eds.), *Arsenic Remediation of Food and Water*,
https://doi.org/10.1007/978-981-97-4764-1_17

it is comparable to numerous other elements (Garg & Singla, 2011). The leaves of plants may accumulate atmospheric As, which can then be ingested. These when enter the body through food and water, can cause poisoning in both humans and animals (Shrivastava et al., 2015).

In the case of a developing country such as India, the Ganga-Meghna-Brahmaputra (GMB) river plain with a 569,749 km^2 size and a population of over 500 million, is one of the most severely affected areas in South Asia impacted by As contamination in groundwater (Chakraborti et al., 2018). The contamination of irrigation water sources, which is often unavoidable, consistently results in crop exposure to As. Paddy crops quickly absorb arsenite, the most common type of As present in flooded rice field conditions. Therefore, dietary intake of As-enriched grains can significantly contribute to chronic As poisoning. Half of the world's population eat rice as a staple diet and the build-up of As in rice compromises food security and safety. Therefore, it is necessary to lessen As stress and lower As levels in rice crops. This chapter—(i) discusses the As uptake by rice and its further effects in plants, (ii) reviews the remediation strategies to ameliorate As toxicity in soil–plant system, (iii) emphasises/focuses on environment friendlier approaches, (iii) evaluates the existing gaps in research and points out future prospects in the field of arresting As uptake from soil to plant system which could ultimately reduce health risks and problems associated to As poisoning.

Arsenic in Soil-Paddy System

Arsenic in Soil and Its Uptake by Rice Plants

Arsenic is considered as biologically nonessential. It exists mainly in two forms—inorganic and organic (Zhao et al., 2010). Inorganic forms of As (iAs) are more toxic when compared to the organic (oAs) ones (Jomova et al., 2011) as they get readily absorbed and cause adverse physiological effects. Both soil and water are found to have two major inorganic forms of As species, arsenate (AsV) and arsenite (AsIII). Several forms of oAs such as monomethylarsonous acid (MMAIII), monomethylarsonic acid (MMAV), dimethylarsinous acid (DMAIII), dimethylarsinic acid (DMAV), arsenobetaine (AsB), arsenocholine (AsC), demethylatedarsenosugar, trimethylarsine (TMA), trimethylarsine oxide (TMAO) and tetramethylarsonium ion (TETRA) are also present in the environment but in a lesser degree as compared to iAs (Farooq et al., 2016; Zhao et al., 2013). Figure 17.1 shows the chemical structures of different iAs and oAs.

Uptake of As by roots occurs via phosphate transporters (PHT) which take up As(V) through the phosphate (Pi) transport pathway, whereas As(III) and methylated As species (MMA and DMA) are taken up and deposited in plant tissues by multiple members of the nodulin 26-like intrinsic protein (NIP) subfamily of aquaporins (Li et al., 2009a, b; Ma et al., 2008). As(V) is absorbed as a phosphate analogue

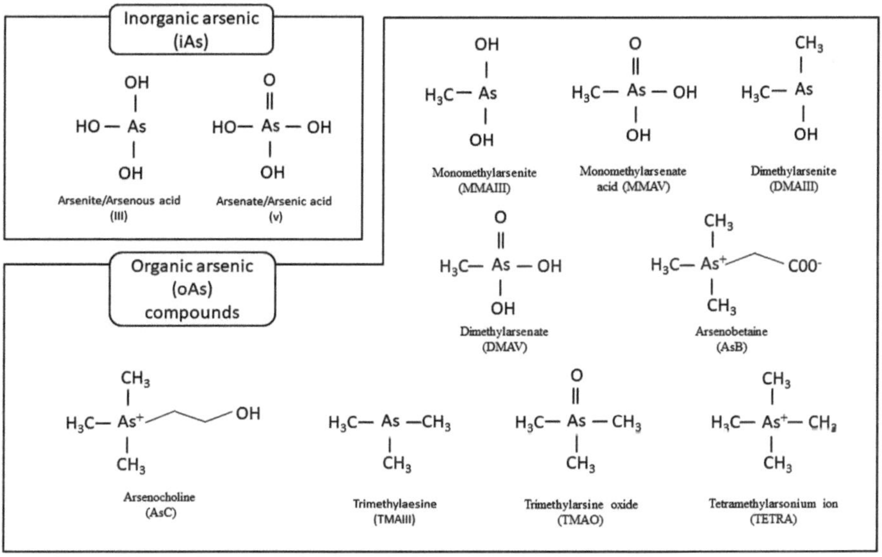

Fig. 17.1 Chemical structures of major arsenic-containing inorganic and organic compounds

by the phosphate absorption system, including *OsPT1*, *OsPT4*, and *OsPT8* in rice (Cao et al., 2017; Kamiya et al., 2013; Wang et al., 2016). On the other hand, As(III) being similar to silicic acid in physiochemical characteristics, rice roots can absorb As(III) via silicon (Si) transporters, such as *OsLsi1* (a member of the aquaporin family) and *OsLsi2* (a putative anion transporter) (Ma et al., 2008). These transporters, *OsLsi1* and *OsLsi2*, are localized in the distal side and the proximal side respectively of the epidermis and endodermis of rice roots (Ma et al., 2006; 2007). Both Si and As(III) have similar sizes and are primarily present as undissociated neutral molecules within the normal physiological and environmental pH range (i.e. pKa = 9.3 & 9.2for Si and As(III) respectively; undissociated at pH 7.0), which facilitates As(III) to use these Si transporters in rice root cells in the shared uptake pathway (Li et al., 2009a). Consecutively, xylem and phloem then mediate the uptake of As in the grain. Figure 17.2 represents the entry pathway of As in plants via roots involving aquaporins and transporters.

Manifestation of Toxicity in Arsenic-Stressed Rice Plants

Exposure to As impacts the physiological and metabolic processes in plants adversely impacting the growth of plants (Chakrabarty et al., 2009; Tripathi et al., 2013). It has been reported that As(III) produces more inhibitory effect on plant growth than As(V) (Mridha et al., 2021). As(III) can interact with the sulfhydryl groups (–SH) of proteins and enzymes in the cytosol changing the working of biochemical

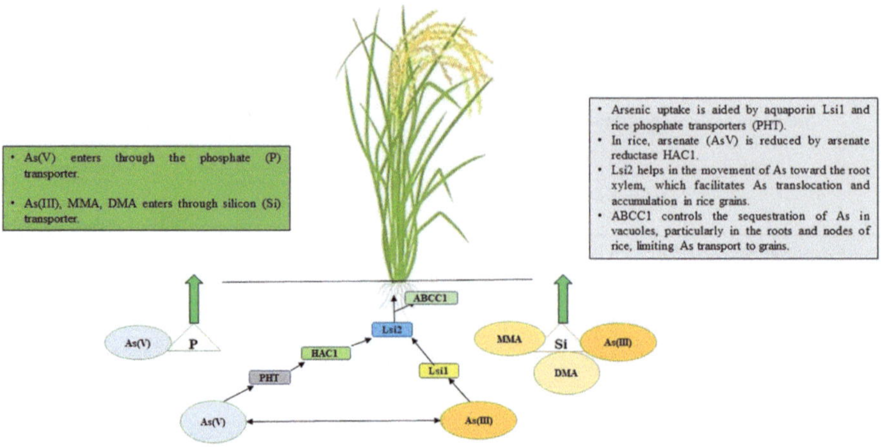

Fig. 17.2 Schematic diagram illustrating the arsenic uptake route

processes involved (Meharg & Hartley-Whitaker, 2002). Arsenic stress produces reactive oxygen species (ROS) which leads to oxygen stress in plants (Requejo & Tena, 2005). Similarly, it was reported that the excessive ROS production brought on by the activation of the ROS-producing enzyme NADPH oxidase and the deterioration of plant antioxidant systems caused by As exposure may have a significant effect on plant metabolic processes (Mishra et al., 2019). Plant cyclins control cell cycle progression in conjunction with highly conserved cyclin-dependent kinases (CDKs). Huang et al. (2012) discovered that a variety of cell cycle-related genes, including five cyclin genes (Orysa;CycD1;1, –D3;2, –D5;2, –D6;1, and –P2;1), were down regulated by As(V).

Plants' Mechanisms to Reduce Arsenic Stress

To combat As toxicity, plants have developed a number of intrinsic defensive systems. Plant cell walls thicken when exposed to stress from heavy metals to shield the cells from damage. Primarily, detoxification of As in rice plants may begin at the roots with the reduction of As(V) by genes which are mostly expressed in the root region. After As(III) is absorbed by the root cells, some of it is immediately released into the rhizosphere, a process that is in part mediated by *Lsi1* acting as a bidirectional channel (Zhao et al., 2010); the remaining As is then either sequestered in the root vacuoles or transported to the shoots, where it is then distributed to various organs. As(V) is converted to As(III) by arsenate reductases such as OsHAC1;1, OsHAC1;2, and OsHAC4 following uptake by rice roots (Xu et al., 2017) which is then transported by OsABCC1 into the vacuoles where it can bind phytochelatin (PC) synthase and be sequestered (Song et al., 2014). Over expression of these arsenate reductases considerably reduced As build-up in the plant and boosted As(III) outflow into the external medium (Shi et al., 2016; Xu et al., 2017). They catalyse a process in the outer cell layer, enabling the outflow of As from the root into the soil. ACR3, an As antiporter, in particular, can extrude some As(III) into the external medium, lowering the As content in plants (Chen et al., 2017).

The production of ROS due to As stress affects the growth and development of plants. To combat the negative impacts induced by ROS, mechanisms are required in plants to scavenge the ROS and balance the enzymatic process and non-enzymatic antioxidants (Upadhyay et al., 2014). Figure 17.3 highlights the production of oxidative stress due to the uptake of As and the simultaneous enhancement of antioxidant system in plants. Stress in plants from As(III) leads to increase in thiol metabolism while that from As(V) results in strengthening of the antioxidant system (Dave et al., 2013; Mishra et al., 2008; Srivastava et al., 2007). On As exposure, rice plants exhibit oxidative stress response. When exposed to As(V), rice seedlings produced hydrogen peroxide (H_2O_2) (Choudhury et al., 2011) while As(III) produced superoxide anion (O^{2-}) and H_2O_2 which caused lipid peroxidation in rice seedlings (Mishra et al., 2011). In rice, JA signalling and lipid metabolism was discovered as responses to As(III) stress by Yu et al. (2012) using Illumina sequencing. Hartley-Whitaker et al. (2001) reported that plant cells manufacture antioxidants including ascorbate peroxidase (APX), catalase (CAT), glutathione (GSH), glutathione S-transferase (GST) and superoxide dismutase (SOD) which can be used to detoxify ROS produced due to As stress. A mechanism to counteract the damage due to the production of ROS is the increased APX activity which reduces H_2O_2 via the ascorbate–glutathione cycle (Dubey et al., 2014). Raab et al. (2005) reported complexation with thiol compounds such as GSH and PC which reduces As mobility sequestering it in vacuole, a method by which plant cells detoxify As on biochemical level. Mishra et al. (2011) found that in seedlings that had been exposed to As, production of ascorbate and glutathione had increased along with the production of PCs and total acid-soluble thiols. Similarly, Tripathi et al. (2012) reported greater levels of PCs due to increased PC-synthesizing activity and coordinated thiol metabolism on exposure to As.

A natural plant growth regulator, 5-aminolevulinic acid (ALA), a crucial biosynthetic precursor of heterocyclic tetrapyrrole molecules, can encourage plant growth and tolerance in a variety of harsh conditions, including those caused by toxic metal stressors (Ali et al., 2013). As defence mechanisms, some studies reported upregulation of several heat shock proteins and enhanced production of proline which could aid in the protection of enzymes in the presence of As(III) (Chakrabarty et al., 2009; Mishra & Dubey, 2006). Di et al. (2021) reported that As(III) stress activates production and signalling of jasmonic acid (JA), considered as a stress hormone, which signifies JA pathway in the plant's response to As(III). Di et al. (2021) suggested that the As(III) stress response of rice may involve not only the JA pathway but also additional phytohormones as ethylene (ET), abscisic acid (ABA), and cytokinin (CK). All of these pathways (JA, ET, ABA, and CK) may work together to restrict plant growth and development in response to As(III) stress, increasing plants' tolerance to As(III). In an earlier study, phytohormones ABA and JA contents were shown to be significantly higher under As stress, according to Srivastava et al. (2013), suggesting

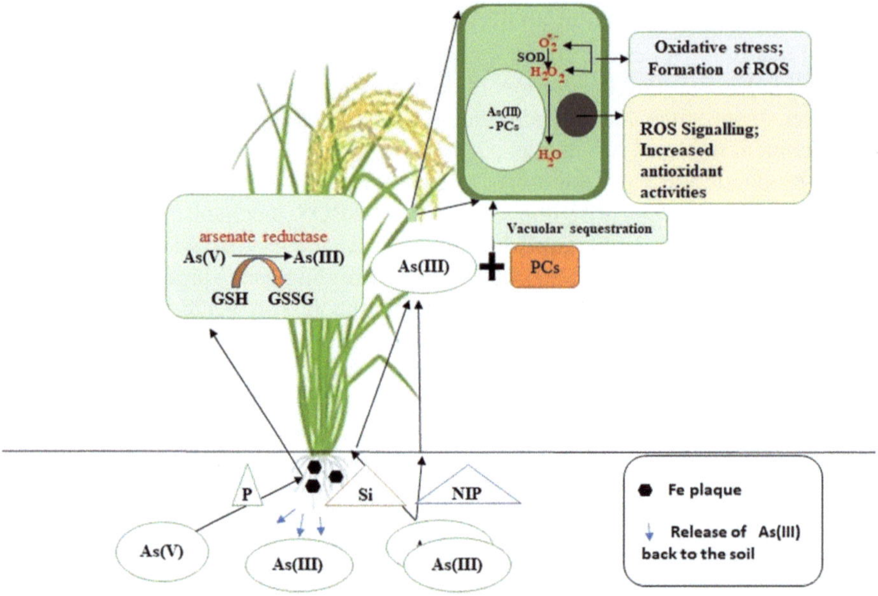

Fig. 17.3 Schematic representation of arsenic uptake, ROS formation and detoxification in rice plants

that ABA and JA may be implicated in the development of protective mechanisms against As toxicity (Bücker-Neto et al., 2017). According to Srivastava et al. (2009), jasmonates were crucial in the perception and reaction processes to As stress.

Rice plants specifically form a coat of reddish iron plaques around the root surface, caused by a reoxidation of Fe(II) to Fe(III) (Becker & Asch, 2005). Studies have shown that iron plaques scavenge As in the soil hindering its uptake by plants resulting to decreased As concentration in the plant (Huang et al., 2012; Wu et al., 2012).

Remedial Approaches Using Different Amendments to Arrest Arsenic in Soil into Plant System

An overall approaches and mechanisms of plants discussed in this chapter to arrest As uptake, reduce its toxicity and accumulation in rice plants is represented in Fig. 17.4.

Use of Inorganic Amendments

Several earlier researchers have reported the application of inorganic amendments to ameliorate As contamination in water, soil and soil–plant system (Miretzky &

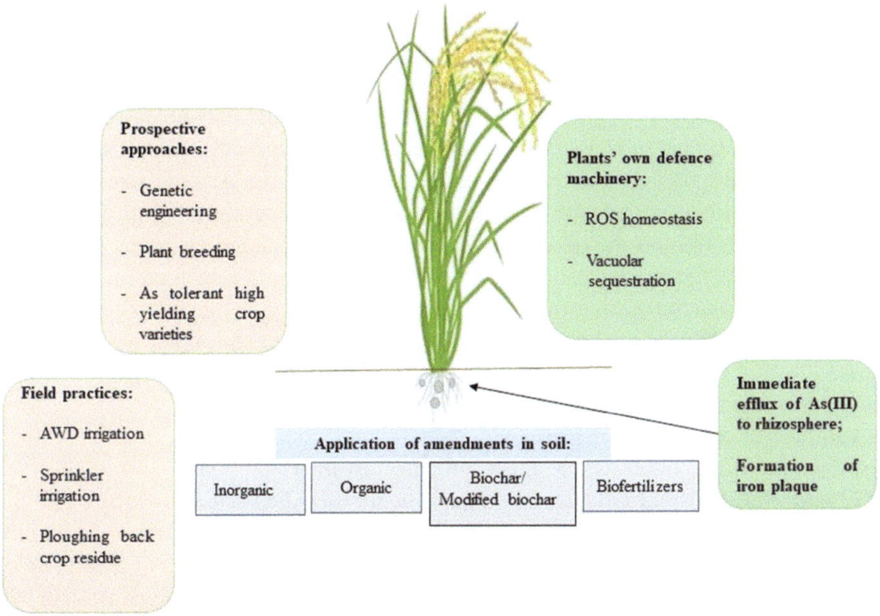

Fig. 17.4 Schematic representation of approaches for reduced uptake and load of arsenic in rice

Cirelli, 2010; Sarkar et al., 2007; Sun et al., 2015; Yu et al., 2017). Studies of iron, phosphorus, silicon, selenium, sulphur and zinc as amendments are discussed below.

It is believed that iron (Fe) in soils immobilises trace elements (Farrow et al., 2015; Lombi et al., 2002). Iron plaque around the roots contributes significantly to the sequestration of As from the soil in anaerobic paddy fields. Weber et al. (2010) found that there is a "coupling of Fe and As," or a significant positive correlation between Fe and As, in the context of the availability of As in soil pore water. Iron compounds have a strong sorption capacity for potentially hazardous elements (Komárek et al., 2013; Qiao et al., 2019). Iron oxide application decreased As bioavailability in soil, reducing its development in crops and subsequent transfer to the food chain, according to Madeira et al. (2012). The study by Yang et al. (2019) provides more evidence for the association between Fe-oxide and As bioaccumulation. Studies by Zhang et al. (2020) using ammonium oxalate revealed a significant amount of As, demonstrating the importance of amorphous Fe-oxide and Al-oxide as a key As sink. According to Yamaguchi et al. (2014), iron plaque, or Fe-oxyhydroxide, has a remarkable capacity to scavenge As. Mlangeni et al. (2020) in a combined investigation of the effectiveness of Fe and irrigation techniques, discovered that the simultaneous use of zero valent iron (Fe°) and alternating wetting and drying (AWD) greatly accelerated methylation of As in rice grains, which would be less hazardous to people. The combined approach also boosted the proportion of productive tillers and grain production compared to continuous flooding (CF).

It was reported that As (V) competes with inorganic phosphorus (Pi) in the cytoplasm, creating unstable ADP-As (V) and disrupting energy fluxes in cells (Hartley-Whitaker et al., 2001). It has been discovered that As stress causes down regulation of phosphate transporters. Begum et al. (2016) discovered that phosphate transporters (*OsPT8*, *OsPT4*, *OsPHO1;2*) were down regulated in rice cultivar BRRI 33 (As-tolerant rice variety) under both high and low P availability. Arsenic stress induces a continuous reduction in tissue P content, increases the expression of phytochelatin synthase (*OsPCS1*) and the activity of phytochelatin-mediated enzymes leading to sequestration of As in root cell vacuoles and restricting As translocation to shoot. Their findings concluded that the signal driving As tolerance mechanisms is generated in the roots. It was reported by Gunes et al. (2009) that adding Pi to As-stressed plants lowered the enzymatic activities of CAT, APX, and lipid peroxidation which resulted in less membrane damage.

Limmer et al. (2018) investigated the effects of silica-rich soil amendments (ground rice husk, charred rice husk, silicate fertilizer) on paddy soil at the plot level and observed that silicate-rich amendments raised Si while lowering As. Si-rich additions increased ferrihydrite in root plaque and organic As in grain. Selenium and As share an antagonistic relationship. It affects the uptake of As in plants. Selenium in the form of selenite or selenate, as demonstrated by numerous research, can aid in lowering the concentration and toxicity of As in plants (Kumar et al., 2014, 2016; Moulick et al., 2016). Depending on the plant type, the amounts of As and Se in the growth medium, the presence of other components, and the duration of As and Se exposure, Se amendment has varying effects on plants growing under As stress (Chauhan et al., 2020; Moulick et al., 2016). Zhou et al. (2017) reported that Se amendment not only reduces As accumulation in grains but also aids in progressing the Se nutrition in plants. It can be attributed to the increase in the concentration of selenomethionine (SeMet) in grains and the decrease in H_2O_2 content and ROS which leads to a reduced oxidative stress and cell injury in crop plants (Pandey & Gupta, 2018). Antagonistically, it promotes the activity of several enzymatic antioxidants like peroxidase (POD), CAT, APX and SOD in crops under abiotic stress (Alam et al., 2020). It causes an increase in the levels of thiols and GSH thereby preventing lipid peroxidation. It enhances the formation of iron plaques on the root surface which further reduces the uptake of As in the plant system. Kaur et al. (2017) reported a reduction of As uptake in crops and an increase in biomass production due to the antagonistic relationship between As and Se. Chauhan et al. (2017) observed that an exogenous application of sodium selenite to the hydroponic solution where rice was grown resulted in a decrease of accumulation of As in the roots and shoots of rice. Moulick et al. (2018a; b) found a similar reduction in As translocation from root to shoot in rice seeds primed with selenite. However, in a study conducted by Moulick et al. (2016), selenite concentrations in growth medium greater than 1.0 mg L^{-1} were found to be toxic to rice. Selenium has a beneficial effect on plant biomass. It increases biomass, which lowers the per unit concentration of As in plants, a process known as the dilution effect. Under As stress, Se application increased shoots and roots length in rice seedlings (Pandey & Gupta, 2015). They found an increase in chlorophyll content in rice seedlings as well. Kumar et al. (2013) found that co-applying P (5.0 mg L^{-1}) and selenite (0.75 mg L^{-1}) significantly reduced As(III) uptake and translocation to shoots while increasing plant biomass in rice.

Sulphur(S) is essential for the production of amino acids, enzymes, and plants' defence mechanisms against biotic and abiotic stress (Shi et al., 2020). It has been reported that under As stress, the sulphate transporter (*OsSultr1;1*) keeps the S pool within the cell stable, allowing the plant to resist the stress (Kumar et al., 2019). Zou et al. (2018) observed that sulphate treatment significantly improved the number, height, and biomass of rice plant tillers in As and Pb co-contaminated soil. Tang et al. (2020) corroborated similar findings with the application of sulphate to rice plants leading to the increased shoot development, biomass and grain weight as compared to the control treatment (without sulphate) when the plant was stressed. Wisawapipat et al. (2021) suggested that the application of S-containing soil amendments can effectively decrease inorganic As accumulation in rice grains grown under the flooded conditions. They highlighted the effect of S amendments on the decrease of soluble and exchangeable As and on enhancing residual As fractions under flooded conditions. Wisawapipat et al. (2021) studied the influence of two S-containing soil amendments viz. chicken manure and elemental S powder in the mitigation of As accumulation in rice grains under flooded conditions, which is the traditional method of rice production, and intermittent flooding conditions. They found that although the intermittent flooding condition was effective in reducing As accumulation (total As, inorganic As, and DMA) in grains, it lowered rice productivity. On the other hand, the traditional method enhanced the productivity of rice while mitigating As in rice grains. Wisawapipat et al. (2021) concluded that in rice crops grown under flooded conditions, the application of chicken manure showed promising results in the mitigation of As and obtaining high yields of rice. They discovered that the key regulators of As bioavailability in soil were the weakly crystalline Fe oxides, which might be the principal dissolvable reactive phases. The decrease in inorganic As concentration in rice grain was relatively due to the decreased time-average concentrations of dissolved As in the pore water as well as the soluble and exchangeable As fraction in the solid phase as a result of S additions. Additionally, a speciation study observed that inorganic As and DMA in rice grain was reduced by 63–71% and 56–71% respectively upon suphate application. The results were consistent with the findings of Dixit et al. (2015) and Srivastava et al. (2016) of reduced translocation factor in rice following S application. As per the study of Mridha et al. (2022), application of sulphate as an amendment to the soil could be advised to produce As safer rice grains.

Zinc is an essential trace element for plant growth and development (Li et al., 2015). Zinc plays a role in reducing As release in soil solution by electrostatically interacting with As and producing an insoluble 'metal-like' complex (zinc-arsenate) in the soil matrix, thereby reducing As toxicity in the soil plant system (Das et al., 2016; Kader et al., 2017). Similar antagonistic relationship of As and Zn was demonstrated when application of Zn to soil (at a rate of 3 mg/L) lowered As availability and uptake in plants (Sanchary & Huq, 2017). Ma et al. (2020)reported reduction of total

As in rice roots (39.5 and 83.3%) and shoots (60.2 and 80.0%) by zinc oxide nanoparticles (ZnONPs) and zinc ions (Zn^{2+}), with the reduction mostly attributable to the decreased inorganic As(III) and organic As species. A recent study by Wang et al. (2022) highlighted the differential activity of various Zn additions on the composition and development of iron plaques, rice production, and As accumulation and speciation in rice grains. Intriguingly, it was found that the ordinary Zn salt showed better benefits in reducing As build-up and the most hazardous As(III) in rice grains than ZnONPs or their bulk analogue.

Use of Organic Amendments

Prospects of using organic amendments to arrest the As uptake and accumulation in plants is growing owing to the sustainability benefits that these amendments provide. Irrespective of the type, both conventional and improved rice types, natural soil additives such as compost, biochar, and biofertilizers lower the concentrations of heavy metals in the soil and rice grains (Yapaa et al., 2022).

Humic substance is a significant part of soil organic matter which contributes in maintaining the fertility of the soil. Organic amendments such as compost, vermicompost, K-humate utilize the humic acids' properties to enrich the soil's fertility while absorbing As through the different functional groups (Mridha et al., 2021). The various functional groups of humic compounds are responsible for both direct and indirect effects on plant nutrition and growth (Zanin et al., 2019). Organo-complexation of As may occur as a result of its binding to the phenolic, carboxylic, amino, hydroxyl, phenolic hydroxyl, aldehyde, and sulfhydryl functional groups of humic chemicals in organic matter (Kumar et al., 2021; Mandal et al., 2019). As a result of adsorption by the various functional groups and the complexation of As with the humic and fulvic substances present in the organic matter, the movement of As in soil is thus constrained (Mandal et al., 2019; Mukhopadhyay & Sanyal, 2004). The humic acids and fulvic acids of vermicompost form very strong complexes with soil As (Sinha & Bhattacharyya, 2012).

Vermicompost is an excellent and environment-friendly soil amendment to increase the nutrients in the soil and reduce As mobility in soil. The humic acids and fulvic acids boost water holding capacity of the soil, improving nutritional availability to the roots entrapping the soil particles (Taha & Osman, 2018). It is reported that they effectively scavenge As(V) from the soil (Ghosh et al., 2012). These substances are present in different molecular sizes. The morphological characteristics of it are relative to factors such as pH, ionic strength and mineral composition of the solution (Durce et al., 2016). A recent study by (Sengupta et al., 2021) evaluated organic amendments (vermicompost, farmyard manure, and mustard cake) and explored irrigation systems (saturation and alternate wetting and drying) and found that the vermicompost amendment along with AWD system of irrigation resulted in the least As accumulation in the edible grains (0.318 mg kg^{-1}) and 25% increased grain yield. Since the effect of organic fertilizer varies on different soil types, it is necessary to choose the amendment wisely according to the soil type and the level of As-contamination.

A list of amendments (both inorganic and organic) for amelioration of As toxicity has been placed in Table 17.1.

Use of Biochars

Biochar is a carbon-rich material produced by slow pyrolysis of organic materials in limited supply or absence of air (Lehmann & Joseph, 2015). The final product, biochar, is being used as a soil amendment to adsorb As (Beesley et al., 2011; Khan et al., 2013). It can remove trace metals via ion exchange and surface complexation (Piscitelli et al., 2018). Earlier research has shown that incorporating biochar to the soil impacts the physiochemical qualities of the soil, microbial activity, soil fertility and immobilizes heavy metals such as As (Kamali et al., 2022; Vithanage et al., 2017). Biochar exhibits a high capacity for adsorption and is known to reduce the uptake of As in plants to some extent by keeping As on its surfaces, but it has little impact on the speciation of As in the soil (Beesley et al., 2011; Khan et al., 2013).

The addition of biochar can have an impact on nutrient conditions, such as the concentration of P, which may engage in the formation of cell wall polysaccharides or thiol-rich peptides and retain heavy metals (Xu et al., 2022; Zhu et al., 2012). In a study by Zama et al. (2022), they sound that the biochar efficiently increased the Ca–As(III) precipitation in the aqueous environment. In another study, Liu et al. (2022) used Ca-rich biochar from food waste digestate and demonstrated that it was affordable as well as the pristine Ca-rich biochar showed to be a viable and sustainable adsorbent for As removal from contaminated water.

However, earlier researchers have shown that biochars have a limited As adsorption capability owing to the negative charge on their surface (Agrafioti et al., 2014; Niazi et al., 2018; Verma & Singh, 2019). Therefore, modifications such as activation by alkaline/acidic solutions, impregnation with zero-valent metals, and surface fictionalization with amides and thiols, have been made to increase the adsorption capabilities of biochar (Cruz et al., 2020; Hussain et al., 2020; Kim et al., 2019; Liu et al., 2019; Wongrod et al., 2018).

Use of Modified Biochar

Additions to biochar have received a lot of attention in recent years as effective measures to control As contamination. The effectiveness of biochar may be influenced by the use of additional organic components (Beesley et al., 2011). Application of biochar and biofertilizers for amelioration of As toxicity has been shown in Table 17.2. Iron-based biochar has grown in popularity as it increases the surface area of the material, the quantity of particular adsorption sites and alter an adsorbent's

Table 17.1 Studies of inorganic and organic amendments for amelioration of arsenic toxicity

Amendments		Substrate	Observation	Reference
Inorganic	P	Soil (pot)	Prevented As uptake in wheat plants.	Pigna et al. (2010)
	Si	Hydroponic system	Si application reduced As uptake and increased photosynthetic characteristics in rice plants.	Sanglard et al. (2016)
	Si	Field soil	Decreased As in rice plants; increased organic As in rice grain and ferrihydrite in root plaque.	Limmer et al. (2018)
	Se	(Seed priming) pot soil	Se primed plants showed enhanced growth and yield; decreased As translocation to shoots; increased translocation of minerals like Fe, Zn and Mn.	Moulick et al. (2018a, b)
	Se	Moist cotton bed	Reduced As stress by reducing H_2O_2, NO, ROS, O^{2-} and cell death; increased the activity of thiol metabolism-related enzymes.	Pandey and Gupta (2018)
	Zn (Zinc oxide nanoparticles, ZnONPs and zinc ions, Zn^{2+})	Soil	Significant reduction of the total As in rice roots and shoots.	Ma et al. (2020)
	Zero valent iron (Fe°)	Soil	Simultaneous use of zero valent iron (Fe°) and alternating wetting and drying (AWD) greatly accelerated methylation of As in rice grains; boosted the proportion of productive tillers and grain production.	Mlangeni et al. (2020)
	S (Na_2SO_4)	Soil (seed bed)	Reduced As in rice grains; encouraged plant growth and yield; reduced uptake and accumulation of As from the soil; decreased oxidative stress in rice plants.	Mridha et al. (2022)
	Zn (ZnONPs, Zn^{2+} or ZnOBPs)	Soil	Only Zn^{2+} effectively reduced total rice grain As.	Wang et al. (2022)

(continued)

Table 17.1 (continued)

Amendments		Substrate	Observation	Reference
Organic	Decomposed paddy husk (PH), vermicompost (VC)	Soil (Pot experiment)	Reduced uptake of As in the order: PH > VC> control.	Kumar et al. (2021)
	Potassium humate	MS medium	Germination and growth of As-stressed rice seedlings enhanced; As buildup and translocation noticeably decreased in the primed seeds.	Mridha et al. (2021)
	Vermicompost, farmyard manure, and mustard cake	Field	Least As accumulation in the edible grains and increased grain yield by co-application of vermicompost with AWD irrigation.	Sengupta et al. (2021)

chemical activity by adjusting the pH which ultimately facilitates electrostatic interactions between As and the adsorbent surface (Asmel et al., 2017; Kong et al., 2018; Vieira et al., 2017; Zhu et al., 2018). Wu et al. (2018) went on to determine how pH, $NaHCO_3$-extractable As concentrations, and As fractions in soils were affected by biochar treated with Fe-oxyhydroxy sulphate, $FeCl_3$, and zero-valent iron. Biochar with Fe-oxyhydroxy sulphate was the most effective among these to remediate As-contaminated soils by considerably reducing the As mobility and bioavailability. Prasad et al. (2022) suggested that biochar's ability to remove As might be increased by modification with calcium salts.

Magnetic biochars from wastes have been utilised extensively for soil–plant system As mitigation (Li et al., 2021; Pan et al., 2021; Wan et al., 2020; Wu et al., 2020). Utilizing magnetised amendments makes the dry magnetic separation easier, allowing for the permanent removal of As from the soil (Wan et al., 2020). Li et al. (2021) used zero-valent iron and Fe_3O_4-loaded magnetic biochar and found that the inclusion of magnetic biochar improved As leachability. The amendments reduced As availability as As became more tightly linked to amorphous Fe oxides. To reduce soil As, Fe-engineered biochars were made from *Platanus orientalis* Linn. Branches and pig carcasses which showed promising outcomes (Pan et al., 2021). Additionally, by increasing bacterial diversity, the Fe-modified biochar was able to reduce the ecotoxic stress of As and Pb while also enhancing soil quality. Earlier, Wan et al. (2020) conducted a novel study in which they followed three magnetised biochar production methodologies from red cedar sawdust (pyrolysis before magnetization, magnetization before pyrolysis, and magnetization in between two pyrolysis process), with the double pyrolysis technique proving to be the most effective in removing As.

Table 17.2 Application of biochar and biofertilizers for amelioration of arsenic toxicity

Amendments		Substrate	Observation	References
Biochar	Waste plant litter biomass biochar (Tectona and Lagerstroemia speciose)	Aqueous solution	Removed As(III) and As(V) from aqueous solutions; in comparison to Lagerstroemia, Tectona biochar showed greater As species eradication effectiveness.	Verma & Singh (2019)
	Iron-modified magnetic biochar	Soil	The overall concentrations of As, Cd, and Pb in soil dropped by 28%, 25%, and 32%, respectively, after the application and separation of the magnetic biochar; the solid fractions of metal(loid)s can also be directly adsorbed to biochar in addition to the soluble and exchangeable fractions being removed.	Wan et al.(2020)
	Calcium-based magnetic biochar	Paddy soil	As and Cd stabilised by Ca-based magnetic biochar; unstable As fraction may become stable; bacterial diversity and abundance in soil boosted.	Wu et al. (2020)

(continued)

Table 17.2 (continued)

Amendments		Substrate	Observation	References
	Zero-valent iron (ZVI) and a Fe_3O_4-loaded magnetic biochar	Soil	Arsenic converted to more stable geochemical components; enhancing plant growth; lowered As phytoavailability; ZVI application considerably lowered As leachability.	Li et al. (2021)
Biofertilizer	Purple nonsulphur bacteria: *Rhodopseudomonas palustris* C1 and *Rubrivivax benzoatilyticus* C31	Hydroponic system	Reduced ROS in plants and enhanced growth of rice plants.	Nookongbut et al. (2018)
	Microalgae (algae and cyanobacteria)	Soil (continuous flooded condition)	Significantly improved rice biomass and uptake of N, P, K, and S in rice grains; greatly decreased the DMA concentration and total As accumulation in rice grains.	Wang et al. (2018)
	Consortium of rhizobacteria (Pseudomonas putida) + algae (Chlorella vulgaris)	Hydroponic system; Hewitt nutrient medium (1/3rd strength), pH 5.5	Alleviated As stress; Iimproved growth of rice seedlings subjected to the stress; Rreduced As accumulation and lessower As in rice grain.	Awasthi et al. (2018, 2021)
	Chlorella sp.	Hewitt nutrient medium (Hewitt, 1966)	iAS absorption and toxicity reduced in rice seedlings due to increased GSH dependent enzyme activity and SAMT iAs methylation.	Ranjan et al. (2021)

Given its success in aqueous experiments, it is hypothesised that magnetic biochar produced by Fe modification, which benefits from the special qualities of both biochar (high sorption capacity for cations, such as Cd) and iron oxide (high sorption capacity for As), can increase its sorption capacity of multi-metal(loid)s from soils (Wu et al., 2018). Wu et al. (2020) used magnetic rice straw and calcium modified magnetic rice straw which was produced from rice straw pyrolyzed along with calcium carbonate and iron oxide to stabilise the contamination of several metals. Calcium-based magnetic biochar was found to stabilise As and Cd.

Tan et al. (2020) concluded that the use of magnetic sorbents in co-contaminated environments would be promising while the maximum capacity for As(V) and NO^{3-} sorption in their study was demonstrated by one-step magnetic biochar (synthesis method). Although, magnetic biochar comes with a lot of advantages; however, it can result in the removal of crucial soil constituents (Wan et al., 2020).

Use of Biofertilizers

A sustainable option to the agricultural community to avoid the pollution from heavy metals such as As and also increase crop output is the use of microorganisms as biofertilizers. Certain microorganisms, such as bacteria, fungus, and cyanobacteria help in promoting plant development and alleviating stress in plants caused by As. Plant growth-promoting rhizobacteria, or PGPRs are known to have both direct and indirect impacts in reducing Astoxicity in plants and fostering their growth (Awasthi et al., 2018). A summarised definition of a biofertilizer as given by Mahanty et al. (2017) is a substance that contains living microorganisms which on application encourages plant growth by increasing the supply of nutrients to the host plant; it does so by colonising the rhizosphere or interior of the plants (Bardi & Malusà, 2012; Malusa & Vassilev, 2014; Vessey, 2003). Plant growth-promoting bacteria, or PGPB can promote growth and relieve As stress in rice plants by different mechanisms such as direct adsorption, absorption, and volatilization as well as indirectly through synthetic siderophores which function as a barrier to prevent hazardous metal entry into the plant and also help plants to acquire iron (Armendariz et al., 2015; Nookongbut et al., 2018).

Awasthi et al. (2018) investigated the bacterial and algal consortium in reduction of As toxicity. A consortium of *Chlorella vulgaris* and *Pseudomonas putida* reduced As build-up in plants and lessened the harmful effects of As on rice. During As exposure, the inoculation of a microbial consortia dramatically reduced the toxicity of As, enhanced the formation of root hairs, and preserved the cellular integrity of the epidermis, exodermis, and the stele area. In continuation, Awasthi et al. (2021) found that As toxicity was reduced by the consortium of *Chlorella vulgaris* and *Pseudomonas putida* via altering the protective mechanisms. It effectively altered transcription factors that increase As absorption and increase tolerance in plants. The application of consortium could serve as a potential amendment in field conditions to achieve high yielding rice plants with reduced As. Nookongbut et al. (2018) investigated the effectiveness of mixed purple non-sulphur bacteria (PNSB) comprising of

Rhodopseudomonas palustris C1 and *Rubrivivax benzoatilyticus* C31 in amelioration of As stress in rice plants. Biochemical responses were reported to be stimulated in rice for increased synthesis of chlorophylls and enzymatic (glutathione reductase, GR, SOD, APX, CAT) and non-enzymatic (carotenoids, lipid oxidation, and nitric oxide) processes to eliminate ROS. Purple non-sulphur bacteria (PNSB) demonstrated significant potential in raising resistance in rice plants to As toxicity through a variety of mechanisms, such as production of siderophores and extracellular polymeric substances (EPS) along with bioactive compounds to form biofilm that plays a crucial role as a barrier by protecting the bacterial communities from toxic substances (Armendariz et al., 2015; Nookongbut et al., 2018). Ranjan et al. (2021) observed reduced As uptake and toxicity in rice seedlings when inoculated with *Chlorella* sp. It is known that plants increase the activities of GSH-dependent enzymes as a defensive strategy against oxidative stress (Hernǎndez et al., 2015). The algal application enhanced the activities of GR and glutathione peroxidase (GPX), GSH-dependent antioxidant enzymes, and S-adenosyl methionine (SAM) gene to combat the effects of iAS. The antioxidant enzymes help in preserving the reduced environment within the cells by either scavenging or neutralizing the produced ROS (Ranjan et al., 2021).

Wang et al. (2018) investigated two green algae and five cyanobacteria to check if microalgal inoculation affected rice growth and rice grain grown under continuously flooded conditions. It is known that microalgal addition aids the plant to fix nitrogen (N) efficiently, increase soil phosphorus (P) and promote plant growth (Rossi et al., 2017). Wang et al. (2018) corroborated growth in rice plants owing to enhanced plant biomass after algal inoculation. Other research using a combination of bacteria and algae (Awasthi et al., 2018; Srivastava et al., 2018) and algae alone (Ranjan et al., 2016; Rossi et al., 2017) all produced similar results.

Field Practices

Apart from the application of external amendments into the soil, implementing good agricultural practices could serve beneficial in reducing As load in the grains. For instance, paddy is grown in water-logged conditions. The reducing conditions of the water-logged soil accentuate the prevalence of As(III) for uptake by rice roots which can be altered by changing the water management practices. Aerobic cultivation results in more availability of As(V) than As(III), as As(V) exhibits a noticeably higher adsorption affinity to soil and mineral surfaces, resulting in lower plant availability and toxicity than As(III) (Yamaguchi et al., 2011). It has however been observed that the aerobic conditions results in reduced yield (Carrijo et al., 2018). Intermittent irrigation serves as a reliable method to compensate for the loss of crop yield while restricting As accumulation in the grains (Shehzad et al., 2022). Intermittent irrigation or intermittent-cum-aerobic irrigation technique rather than conventional irrigation or aerobic irrigation should be encouraged as it also reduces the burden on groundwater. Another approach could be the practice of aerobic cultivation with sprinkler irrigation to decrease As accumulation in rice plant (Majumdar et al., 2021; Moreno-Jiménez et al., 2014). The use of efficient water management techniques and raising the concentration of other mineral nutrients in soil solution can significantly alter the solubility and availability of As to plants (Farooq et al., 2016).

A study by Farooq et al. (2019) highlights that As accumulation and distribution in agricultural soils depend on agricultural practices and crop type apart from the As concentration in irrigation water and amount of water supplied. A practice of ploughing back the residual parts of previous paddy crop grown in the field induce increased levels of organic matter in the soil surface which on decomposition alter the sorption characteristics of soil minerals and affect the dispersion and remobilization of As in paddy soils by causing it to drain from the top soil horizon to the lower soil horizon (Farooq et al., 2019; Lawson et al., 2016).

Future Aspects of Arsenic Remediation in Soil–plant System

Several ways to arrest As in soil–plant system is mentioned in this chapter. However in-depth research and knowledge is still required to apply the approaches according to the soil chemistry, kinetics, level of contamination and other relevant information of the As-affected regions.

(i) The first and foremost is to develop sustainable options which are inexpensive. Considering the farmers' aspect to agriculture which serves as the basis of their livelihood, amendments should be cost effective apart from causing no or little secondary pollution to the environment.
(ii) Biochar as an amendment has received enormous popularity among researchers all over the world. Modified and composite biochar has also shown to be tremendously effective in ameliorating As stress in plants. However, energy efficient pyrolysis method needs to be standardized specific to the type of the precursor material of the biochar.
(iii) Utilization of wastes and waste products to form the material for amendments needs to be encouraged. More such researche studies are required to understand the sorption kinetics and application of the derived amendment effectively.
(iv) An in-depth knowledge is required on mechanisms of microorganisms, which can be considered as biofertilizers, and their interaction with rice and other plants for reducing As toxicity and accumulation in plants. This understanding would help researchers use the full ability of biofertilizers in also mitigating As from soil while replenishing soil with nutrients and fertility.
(v) Functions of certain proteins involved in signalling pathways and transportation of As to seeds and grains need to be critically evaluated.
(vi) A big challenge faced in research is the practical application of the research from laboratory to land or field conditions. Effectiveness of amendments in agricultural fields in amelioration of As contamination in soil, toxicity, uptake and accumulation in crops need to be investigated. However, a proper detailed

survey and study of the area need to be done before applying the amendments in the field.

(vii) Owing to the advancements in the fields related to genetic engineering and plant breeding, there's a huge prospect in developing high yielding As-tolerant crop varieties.

The researchers may come up with new sustainable strategies for crops to combat with As pollution while increasing the productivity which is very much required to suffice the increasing demand from the growing human population. Additionally, the inescapable different manifestations of the changing climate and global warming are a threat to the agricultural sector and food security. The development of the remediation and agronomic practices has to be in line to sustain the conditions of the changing climate patterns.

Acknowledgements The authors acknowledge School of Environmental studies, Jadavpur University for all supports to carry forward the research study.

References

Agrafioti, E., Kalderis, D., & Diamadopoulos, E. (2014). Arsenic and chromium removal from water using biochars derived from rice husk, organic solid wastes and sewage sludge. *Journal of Environmental Management, 133*, 309–314.

Alam, M. Z., Carpenter-Boggs, L., Hoque, M. A., & Ahammed, G. J. (2020). Effect of soil amendments on antioxidant activity and photosynthetic pigments in pea crops grown in arsenic contaminated soil. *Heliyon, 6*(11).

Ali, B., Wang, B., Ali, S., Ghani, M. A., Hayat, M. T., Yang, C., Xu, L., & Zhou, W. J. (2013). 5-Aminolevulinic acid ameliorates the growth, photosynthetic gas exchange capacity, and ultrastructural changes under cadmium stress in *Brassica napus* L. *Journal of Plant Growth Regulation, 32*(3), 604–614.

Armendariz, A. L., Talano, M. A., Oller, A. L. W., Medina, M. I., & Agostini, E. (2015). Effect of arsenic on tolerance mechanisms of two plant growth-promoting bacteria used as biological inoculants. *Journal of Environmental Sciences, 33*, 203–210.

Asmel, N. K., Yusoff, A. R. M., Krishna, L. S., Majid, Z. A., & Salmiati, S. (2017). High concentration arsenic removal from aqueous solution using nano-iron ion enrich material (NIIEM) super adsorbent. *Chemical Engineering Journal, 317*, 343–355.

Awasthi, S., Chauhan, R., Dwivedi, S., Srivastava, S., Srivastava, S., & Tripathi, R. D. (2018). A consortium of alga (*Chlorella vulgaris*) and bacterium (*Pseudomonas putida*) for amelioration of arsenic toxicity in rice: A promising and feasible approach. *Environmental and Experimental Botany, 150*, 115–126.

Awasthi, S., Chauhan, R., Indoliya, Y., Chauhan, A. S., Mishra, S., Agrawal, L., Dwivedi, S., Singh, S. N., Srivastava, S., Singh, P. C., & Chauhan, P. S. (2021). Microbial consortium mediated growth promotion and Arsenic reduction in Rice: An integrated transcriptome and proteome profiling. *Ecotoxicology and Environmental Safety, 228*, 113004.

Bardi, L., & Malusà, E. (2012). Drought and nutritional stresses in plant: alleviating role of rhizospheric microorganisms. *Abiotic stress: New research*, 1–57.

Becker, M., & Asch, F. (2005). Iron toxicity in rice - conditions and management concepts. *Journal of Plant Nutrition and Soil Science, 168*(4), 558–573.

Beesley, L., Moreno-Jiménez, E., Gomez-Eyles, J. L., Harris, E., Robinson, B., & Sizmur, T. (2011). A review of biochars' potential role in the remediation, revegetation and restoration of contaminated soils. *Environmental Pollution, 159*(12), 3269–3282.

Begum, M. C., Islam, M. S., Islam, M., Amin, R., Parvez, M. S., & Kabir, A. H. (2016). Biochemical and molecular responses underlying differential arsenic tolerance in rice (*Oryza sativa* L.). *Plant Physiology and Biochemistry, 104*, 266–277.

Lehmann, J., & Joseph, S. (Eds.). (2015). *Biochar for environmental management: science, technology and implementation.* Routledge.

Bücker-Neto, L., Paiva, A. L. S., Machado, R. D., Arenhart, R. A., & Margis-Pinheiro, M. (2017). Interactions between plant hormones and heavy metals responses. *Genetics and Molecular Biology, 40*, 373–386.

Cao, Y., Sun, D., Ai, H., Mei, H., Liu, X., Sun, S., Xu, G., Liu, Y., Chen, Y., & Ma, L. Q. (2017). Knocking out OsPT4 gene decreases arsenate uptake by rice plants and inorganic arsenic accumulation in rice grains. *Environmental Science & Technology, 51*(21), 12131–12138.

Carrijo, D. R., Akbar, N., Reis, A. F., Li, C., Gaudin, A. C., Parikh, S. J., Green, P. G., & Linquist, B. A. (2018). Impacts of variable soil drying in alternate wetting and drying rice systems on yields, grain arsenic concentration and soil moisture dynamics. *Field Crops Research, 222*, 101–110.

Chakrabarty, D., Trivedi, P. K., Misra, P., Tiwari, M., Shri, M., Shukla, D., Kumar, S., Rai, A., Pandey, A., Nigam, D., & Tripathi, R. D. (2009). Comparative transcriptome analysis of arsenate and arsenite stresses in rice seedlings. *Chemosphere, 74*(5), 688–702.

Chakraborti, D., Singh, S. K., Rahman, M. M., Dutta, R. N., Mukherjee, S. C., Pati, S., & Kar, P. B. (2018). Groundwater arsenic contamination in the Ganga River Basin: A future health danger. *International Journal of Environmental Research and Public Health, 15*(2), 180.

Chauhan, R., Awasthi, S., Tripathi, P., Mishra, S., Dwivedi, S., Niranjan, A., Mallick, S., Tripathi, P., Pande, V., & Tripathi, R. D. (2017). Selenite modulates the level of phenolics and nutrient element to alleviate the toxicity of arsenite in rice (*Oryza sativa* L.). *Ecotoxicology and Environmental Safety, 138*, 47–55.

Chauhan, R., Awasthi, S., Indoliya, Y., Chauhan, A.S., Mishra, S., Agrawal, L., Srivastava, S., Dwivedi, S., Singh, P.C., Mallick, S. & Chauhan, P.S. (2020). Transcriptome and proteome analyses reveal selenium mediated amelioration of arsenic toxicity in rice (*Oryza sativa* L.). *Journal of Hazardous Materials, 390*, 122122.

Chen, Y., Hua, C. Y., Jia, M. R., Fu, J. W., Liu, X., Han, Y. H., Liu, Y., Rathinasabapathi, B., Cao, Y., & Ma, L. Q. (2017). Heterologous expression of *Pteris vittata* arsenite antiporter PvACR3; 1 reduces arsenic accumulation in plant shoots. *Environmental Science & Technology, 51*(18), 10387–10395.

Choudhury, B., Chowdhury, S., & Biswas, A. K. (2011). Regulation of growth and metabolism in rice (*Oryza sativa* L.) by arsenic and its possible reversal by phosphate. *Journal of Plant Interactions, 6*(1), 15–24.

Cruz, G. J., Mondal, D., Rimaycuna, J., Soukup, K., Gómez, M. M., Solis, J. L., & Lang, J. (2020). Agrowaste derived biochars impregnated with ZnO for removal of arsenic and lead in water. *Journal of Environmental Chemical Engineering, 8*(3), 103800.

Das, I., Sanyal, S. K., Ghosh, K., & Das, D. K. (2016). Arsenic mitigation in soil-plant system through zinc application in West Bengal soils. *Bioremediation Journal, 20*(1), 24–37.

Dave, R., Tripathi, R. D., Dwivedi, S., Tripathi, P., Dixit, G., Sharma, Y. K., Trivedi, P. K., Corpas, F. J., Barroso, J. B., & Chakrabarty, D. (2013). Arsenate and arsenite exposure modulate antioxidants and amino acids in contrasting arsenic accumulating rice (*Oryza sativa* L.) genotypes. *Journal of Hazardous Materials, 262*, 1123–1131.

Di, X., Zheng, F., Norton, G. J., Beesley, L., Zhang, Z., Lin, H., Zhi, S., Liu, X., & Ding, Y. (2021). Physiological responses and transcriptome analyses of upland rice following exposure to arsenite and arsenate. *Environmental and Experimental Botany, 183*, 104366.

Dixit, G., Singh, A. P., Kumar, A., Dwivedi, S., Deeba, F., Kumar, S., Suman, S., Adhikari, B., Shukla, Y., Trivedi, P. K., & Pandey, V. (2015). Sulphur alleviates arsenic toxicity by reducing its accumulation and modulating proteome, amino acids and thiol metabolism in rice leaves. *Scientific Reports, 5*(1), 16205.

Dubey, S., Shri, M., Misra, P., Lakhwani, D., Bag, S. K., Asif, M. H., Trivedi, P. K., Tripathi, R. D., & Chakrabarty, D. (2014). Heavy metals induce oxidative stress and genome-wide modulation in transcriptome of rice root. *Functional & Integrative Genomics, 14*(2), 401–417.

Durce, D., Maes, N., Bruggeman, C., & Van Ravestyn, L. (2016). Alteration of the molecular-size-distribution of boom clay dissolved organic matter induced by Na^+ and Ca^{2+}. *Journal of Contaminant Hydrology, 185*, 14–27.

Farooq, M. A., Islam, F., Ali, B., Najeeb, U., Mao, B., Gill, R. A., Yan, G., Siddique, K. H., & Zhou, W. (2016). Arsenic toxicity in plants: Cellular and molecular mechanisms of its transport and metabolism. *Environmental and Experimental Botany, 132*, 42–52.

Farooq, S. H., Chandrasekharam, D., Dhanachandra, W., & Ram, K. (2019). Relationship of arsenic accumulation with irrigation practices and crop type in agriculture soils of Bengal Delta, India. *Applied Water Science, 9*, 1–11.

Farrow, E. M., Wang, J., Burken, J. G., Shi, H., Yan, W., Yang, J., Hua, B., & Deng, B. (2015). Reducing arsenic accumulation in rice grain through iron oxide amendment. *Ecotoxicology and Environmental Safety, 118*, 55–61.

Garg, N., & Singla, P. (2011). Arsenic toxicity in crop plants: Physiological effects and tolerance mechanisms. *Environmental Chemistry Letters, 9*(3), 303–321.

Ghosh, K., Das, I., Das, D. K., & Sanyal, S. K. (2012). Evaluation of humic and fulvic acid extracts of compost, oilcake, and soils on complex formation with arsenic. *Soil Research, 50*(3), 239–248.

Gunes, A., Pilbeam, D. J., & Inal, A. (2009). Effect of arsenic–phosphorus interaction on arsenic-induced oxidative stress in chickpea plants. *Plant and Soil, 314*, 211–220.

Hartley-Whitaker, J., Ainsworth, G., & Meharg, A. A. (2001). Copper-and arsenate-induced oxidative stress in *Holcus lanatus* L. clones with differential sensitivity. *Plant, Cell & Environment, 24*(7), 713–722.

Hernåndez, L. E., Sobrino-Plata, J., Montero-Palmero, M. B., Carrasco-Gil, S., Flores-Cåceres, M. L., Ortega-Villasante, C., & Escobar, C. (2015). Contribution of glutathione to the control of cellular redox homeostasis under toxic metal and metalloid stress. *Journal of Experimental Botany, 66*(10), 2901–2911.

Huang, T. L., Nguyen, Q. T. T., Fu, S. F., Lin, C. Y., Chen, Y. C., & Huang, H. J. (2012). Transcriptomic changes and signalling pathways induced by arsenic stress in rice roots. *Plant Molecular Biology, 80*, 587–608.

Huang, Y., Chen, Z., & Liu, W. (2012b). Influence of iron plaque and cultivars on antimony uptake by and translocation in rice (*Oryza sativa* L.) seedlings exposed to Sb (III) or Sb (V). *Plant and Soil, 352*(1), 41–49.

Hussain, M., Imran, M., Abbas, G., Shahid, M., Iqbal, M., Naeem, M. A., Murtaza, B., Amjad, M., Shah, N. S., Khan, U. H., & Z., & Ul Islam, A. (2020). A new biochar from cotton stalks for As (V) removal from aqueous solutions: Its improvement with H3PO4 and KOH. *Environmental Geochemistry and Health, 42*(8), 2519–2534.

Joardar, M., Das, A., Chowdhury, N. R., Mridha, D., De, A., Majumdar, K. K., & Roychowdhury, T. (2021). Health effect and risk assessment of the populations exposed to different arsenic levels in drinking water and foodstuffs from four villages in arsenic endemic Gaighata block, West Bengal, India. *Environmental Geochemistry and Health, 43*(8), 3027–3053.

Jomova, K., Jenisova, Z., Feszterova, M., Baros, S., Liska, J., Hudecova, D., Rhodes, C. J., & Valko, M. (2011). Arsenic: Toxicity, oxidative stress and human disease. *Journal of Applied Toxicology, 31*(2), 95–107.

Kader, M., Lamb, D. T., Wang, L., Megharaj, M., & Naidu, R. (2017). Zinc-arsenic interactions in soil: Solubility, toxicity and uptake. *Chemosphere, 187*, 357–367.

Kamali, M., Sweygers, N., Al-Salem, S., Appels, L., Aminabhavi, T. M., & Dewil, R. (2022). Biochar for soil applications-sustainability aspects, challenges and future prospects. *Chemical Engineering Journal, 428*, 131189.

Kamiya, T., Islam, R., Duan, G., Uraguchi, S., & Fujiwara, T. (2013). Phosphate deficiency signaling pathway is a target of arsenate and phosphate transporter OsPT1 is involved in As accumulation in shoots of rice. *Soil Science and Plant Nutrition, 59*(4), 580–590.

Kaur, S., Singh, D., & Singh, K. (2017). Effect of selenium application on arsenic uptake in rice (*Oryza sativa* L.). *Environmental Monitoring and Assessment, 189*, 1–8.

Khan, S., Chao, C., Waqas, M., Arp, H. P. H., & Zhu, Y. G. (2013). Sewage sludge biochar influence upon rice (*Oryza sativa* L.) yield, metal bioaccumulation and greenhouse gas emissions from acidic paddy soil. *Environmental Science &Technology, 47*(15), 8624–8632.

Kim, J., Song, J., Lee, S. M., & Jung, J. (2019). Application of iron-modified biochar for arsenite removal and toxicity reduction. *Journal of Industrial and Engineering Chemistry, 80*, 17–22.

Komárek, M., Vaněk, A., & Ettler, V. (2013). Chemical stabilization of metals and arsenic in contaminated soils using oxides–a review. *Environmental Pollution, 172*, 9–22.

Kong, X., Tian, T., Xue, S., Hartley, W., Huang, L., Wu, C., & Li, C. (2018). Development of alkaline electrochemical characteristics demonstrates soil formation in bauxite residue undergoing natural rehabilitation. *Land Degradation & Development, 29*(1), 58–67.

Kumar, N., Mallick, S., Yadava, R. N., Singh, A. P., & Sinha, S. (2013). Co-application of selenite and phosphate reduces arsenite uptake in hydroponically grown rice seedlings: Toxicity and defence mechanism. *Ecotoxicology and Environmental Safety, 91*, 171–179.

Kumar, A., Dixit, G., Singh, A. P., Dwivedi, S., Srivastava, S., Mishra, K., & Tripathi, R. D. (2016). Selenate mitigates arsenite toxicity in rice (*Oryza sativa* L.) by reducing arsenic uptake and ameliorates amino acid content and thiol metabolism. *Ecotoxicology and Environmental Safety, 133*, 350–359.

Kumar, S., Khare, R., & Trivedi, P. K. (2019). Arsenic-responsive high-affinity rice sulphate transporter, OsSultr1; 1, provides abiotic stress tolerance under limiting sulphur condition. *Journal of Hazardous Materials, 373*, 753–762.

Kumar, S. N., Mishra, B. B., Kumar, S., & Mandal, J. (2021). Organo-arsenic complexation studies explaining the reduction of uptake of arsenic in wheat grown with contaminated irrigation water and organic amendments. *Water, Air, & Soil Pollution, 232*(3), 1–10.

Kumar, A., Singh, R. P., Singh, P. K., Awasthi, S., Chakrabarty, D., Trivedi, P. K., & Tripathi, R. D. (2014). Selenium ameliorates arsenic induced oxidative stress through modulation of antioxidant enzymes and thiols in rice (*Oryza sativa* L.). *Ecotoxicology, 23*(7), 1153–1163.

Lawson, M., Polya, D. A., Boyce, A. J., Bryant, C., & Ballentine, C. J. (2016). Tracing organic matter composition and distribution and its role on arsenic release in shallow Cambodian groundwaters. *Geochimica Et Cosmochimica Acta, 178*, 160–177.

Li, R. Y., Ago, Y., Liu, W. J., Mitani, N., Feldmann, J., McGrath, S. P., Ma, J. F., & Zhao, F. J. (2009a). The rice aquaporin Lsi1 mediates uptake of methylated arsenic species. *Plant Physiology, 150*(4), 2071–2080.

Li, R. Y., Stroud, J. L., Ma, J. F., McGrath, S. P., & Zhao, F. J. (2009b). Mitigation of arsenic accumulation in rice with water management and silicon fertilization. *Environmental Science & Technology, 43*(10), 3778–3783.

Li, J., Yu, H., & Luan, Y. (2015). Meta-analysis of the copper, zinc, and cadmium absorption capacities of aquatic plants in heavy metal-polluted water. *International Journal of Environmental Research and Public Health, 12*(12), 14958–14973.

Li, J., Zhang, Y., Wang, F., Wang, L., Liu, J., Hashimoto, Y., & Hosomi, M. (2021). Arsenic immobilization and removal in contaminated soil using zero-valent iron or magnetic biochar amendment followed by dry magnetic separation. *Science of the Total Environment, 768*, 144521.

Limmer, M. A., Mann, J., Amaral, D. C., Vargas, R., & Seyfferth, A. L. (2018). Silicon-rich amendments in rice paddies: Effects on arsenic uptake and biogeochemistry. *Science of the Total Environment, 624*, 1360–1368.

Liu, X., Gao, M., Qiu, W., Khan, Z. H., Liu, N., Lin, L., & Song, Z. (2019). Fe–Mn–Ce oxide-modified biochar composites as efficient adsorbents for removing As (III) from water: Adsorption performance and mechanisms. *Environmental Science and Pollution Research, 26*(17), 17373–17382.

Liu, J., Zhang, W., Mei, M., Wang, T., Chen, S., & Li, J. (2022). A Ca-rich biochar derived from food waste digestate with exceptional adsorption capacity for arsenic (III) removal via a cooperative mechanism. *Separation and Purification Technology, 295*, 121359.

Lombi, E., Zhao, F. J., Fuhrmann, M., Ma, L. Q., & McGrath, S. P. (2002). Arsenic distribution and speciation in the fronds of the hyperaccumulator *Pteris vittata*. *New Phytologist, 156*(2), 195–203.

Ma, J. F., Tamai, K., Yamaji, N., Mitani, N., Konishi, S., Katsuhara, M., Ishiguro, M., Murata, Y., & Yano, M. (2006). A silicon transporter in rice. *Nature, 440*(7084), 688–691.

Ma, J. F., Yamaji, N., Mitani, N., Tamai, K., Konishi, S., Fujiwara, T., Katsuhara, M., & Yano, M. (2007). An efflux transporter of silicon in rice. *Nature, 448*(7150), 209–212.

Ma, J. F., Yamaji, N., Mitani, N., Xu, X. Y., Su, Y. H., McGrath, S. P., & Zhao, F. J. (2008). Transporters of arsenite in rice and their role in arsenic accumulation in rice grain. *Proceedings of the National Academy of Sciences, 105*(29), 9931–9935.

Ma, X., Sharifan, H., Dou, F., & Sun, W. (2020). Simultaneous reduction of arsenic (As) and cadmium (Cd) accumulation in rice by zinc oxide nanoparticles. *Chemical Engineering Journal, 384*, 123802.

Madeira, A. C., De Varennes, A., Abreu, M. M., Esteves, C., & Magalhães, M. C. F. (2012). Tomato and parsley growth, arsenic uptake and translocation in a contaminated amended soil. *Journal of Geochemical Exploration, 123*, 114–121.

Mahanty, T., Bhattacharjee, S., Goswami, M., Bhattacharyya, P., Das, B., Ghosh, A., & Tribedi, P. (2017). Biofertilizers: A potential approach for sustainable agriculture development. *Environmental Science and Pollution Research, 24*(4), 3315–3335.

Majumdar, A., Upadhyay, M. K., Giri, B., Srivastava, S., Srivastava, A. K., Jaiswal, M. K., & Bose, S. (2021). Arsenic dynamics and flux assessment under drying-wetting irrigation and enhanced microbial diversity in paddy soils: A four year study in Bengal delta plain. *Journal of Hazardous Materials, 409*, 124443.

Malusa, E., & Vassilev, N. (2014). A contribution to set a legal framework for biofertilisers. *Applied Microbiology and Biotechnology, 98*(15), 6599–6607.

Mandal, J., Golui, D., & Datta, S. P. (2019). Assessing equilibria of organo-arseniccomplexes and predicting uptake of arsenic by wheat grain from organic matter amendedsoils. *Chemosphere, 234*, 419–426.

Meharg, A. A., & Hartley-Whitaker, J. (2002). Arsenic uptake and metabolism in arsenic resistant and nonresistant plant species. *New Phytologist, 154*(1), 29–43.

Miretzky, P., & Cirelli, A. F. (2010). Remediation of arsenic-contaminated soils by iron amendments: A review. *Critical Reviews in Environmental Science and Technology, 40*(2), 93–115.

Mishra, S., & Dubey, R. S. (2006). Inhibition of ribonuclease and protease activities in arsenic exposed rice seedlings: Role of proline as enzyme protectant. *Journal of Plant Physiology, 163*(9), 927–936.

Mishra, S., Srivastava, S., Tripathi, R. D., & Trivedi, P. K. (2008). Thiol metabolism and antioxidant systems complement each other during arsenate detoxification in Ceratophyllum demersum L. *Aquatic Toxicology, 86*(2), 205–215.

Mishra, S., Jha, A. B., & Dubey, R. S. (2011). Arsenite treatment induces oxidative stress, upregulates antioxidant system, and causes phytochelatin synthesis in rice seedlings. *Protoplasma, 248*(3), 565–577.

Mishra, S., Dwivedi, S., Mallick, S., & Tripathi, R. D. (2019). Redox homeostasis in plants under arsenic stress. *Redox Homeostasis in Plants: From Signalling to Stress Tolerance*, 179–198.

Mlangeni, A. T., Perez, M., Raab, A., Krupp, E. M., Norton, G. J., & Feldmann, J. (2020). Simultaneous stimulation of arsenic methylation and inhibition of cadmium bioaccumulation in rice grain using zero valent iron and alternate wetting and drying water management. *Science of the Total Environment, 711*, 134696.

Moreno-Jiménez, E., Meharg, A. A., Smolders, E., Manzano, R., Becerra, D., Sánchez-Llerena, J., Albarrán, Á., & López-Piñero, A. (2014). Sprinkler irrigation of rice fields reduces grain arsenic but enhances cadmium. *Science of the Total Environment, 485*, 468–473.

Moulick, D., Ghosh, D., & Santra, S. C. (2016). Evaluation of effectiveness of seed priming with selenium in rice during germination under arsenic stress. *Plant Physiology and Biochemistry, 109*, 571–578.

Moulick, D., Santra, S. C., & Ghosh, D. (2018a). Effect of selenium induced seed priming on arsenic accumulation in rice plant and subsequent transmission in human food chain. *Ecotoxicology and Environmental Safety, 152*, 67–77.

Moulick, D., Santra, S. C., & Ghosh, D. (2018b). Seed priming with Se mitigates As-induced phytotoxicity in rice seedlings by enhancing essential micronutrient uptake and translocation and reducing As translocation. *Environmental Science and Pollution Research, 25*(27), 26978–26991.

Mridha, D., Paul, I., De, A., Ray, I., Das, A., Joardar, M., Chowdhury, N. R., Bhadoria, P. B. S., & Roychowdhury, T. (2021). Rice seed (IR64) priming with potassium humate for improvement of seed germination, seedling growth and antioxidant defense system under arsenic stress. *Ecotoxicology and Environmental Safety, 219*, 112313.

Mridha, D., Ray, I., Sarkar, J., De, A., Joardar, M., Das, A., Chowdhury, N.R., Acharya, K. & Roychowdhury, T. (2022). Effect of suphate application on inhibition of arsenic bioaccumulation in rice (*Oryza sativa* L.) with consequent health risk assessment of cooked rice arsenic on human: A pot to plate study. *Environmental Pollution, 293*, 118561.

Mukhopadhyay, D., & Sanyal, S. K. (2004). Complexation and release isotherm of arsenic in arsenic-humic/fulvic equilibrium study. *Soil Research, 42*(7), 815–824.

Niazi, N. K., Bibi, I., Shahid, M., Ok, Y. S., Burton, E. D., Wang, H., Shaheen, S. M., Rinklebe, J., & Lüttge, A. (2018). Arsenic removal by perilla leaf biochar in aqueous solutions and groundwater: An integrated spectroscopic and microscopic examination. *Environmental Pollution, 232*, 31–41.

Nookongbut, P., Kantachote, D., Megharaj, M., & Naidu, R. (2018). Reduction in arsenic toxicity and uptake in rice (*Oryza sativa* L.) by As-resistant purple nonsulphur bacteria. *Environmental Science and Pollution Research, 25*(36), 36530–36544.

Pan, H., Yang, X., Chen, H., Sarkar, B., Bolan, N., Shaheen, S. M., Wu, F., Che, L., Ma, Y., Rinklebe, J., & Wang, H. (2021). Pristine and iron-engineered animal-and plant-derived biochars enhanced bacterial abundance and immobilized arsenic and lead in a contaminated soil. *Science of the Total Environment, 763*, 144218.

Pandey, C., & Gupta, M. (2015). Selenium and auxin mitigates arsenic stress in rice (*Oryza sativa* L.) by combining the role of stress indicators, modulators and genotoxicity assay. *Journal of Hazardous Materials, 287*, 384–391.

Pandey, C., & Gupta, M. (2018). Selenium amelioration of arsenic toxicity in rice shows genotypic variation: A transcriptomic and biochemical analysis. *Journal of Plant Physiology, 231*, 168–181.

Pigna, M., Cozzolino, V., Giandonato Caporale, A., Mora, M. L., Di Meo, V., Jara, A. A., & Violante, A. (2010). Effects of phosphorus fertilization on arsenic uptake by wheat grown in polluted soils. *Journal of Soil Science and Plant Nutrition, 10*(4), 428–442.

Piscitelli, L., Rivier, P. A., Mondelli, D., Miano, T., & Joner, E. J. (2018). Assessment of addition of biochar to filtering mixtures for potential water pollutant removal. *Environmental Science and Pollution Research, 25*(3), 2167–2174.

Prasad, B., Dave, H., Maurya, D. M., Singh, D., Kumari, M., & Prasad, K. S. (2022). Sorptive removal of aqueous arsenite and arsenate ions onto a low cost, calcium modified *Moringa oleifera* wood biochar (CaMBC). *Environmental Quality Management, 31*(4), 461–468.

Qiao, J., Yu, H., Wang, X., Li, F., Wang, Q., Yuan, Y., & Liu, C. (2019). The applicability of biochar and zero-valent iron for the mitigation of arsenic and cadmium contamination in an alkaline paddy soil. *Biochar, 1*(2), 203–212.

Raab, A., Schat, H., Meharg, A. A., & Feldmann, J. (2005). Uptake, translocation and transformation of arsenate and arsenite in sunflower (*Helianthus annuus*): Formation of arsenic–phytochelatin complexes during exposure to high arsenic concentrations. *New Phytologist, 168*(3), 551–558.

Ranjan, K., Priya, H., Ramakrishnan, B., Prasanna, R., Venkatachalam, S., Thapa, S., Tiwari, R., Nain, L., Singh, R., & Shivay, Y. S. (2016). Cyanobacterial inoculation modifies the rhizosphere microbiome of rice planted to a tropical alluvial soil. *Applied Soil Ecology, 108*, 195–203.

Ranjan, R., Kumar, N., Gautam, A., Dubey, A. K., Pandey, S. N., & Mallick, S. (2021). Chlorella sp. modulates the glutathione mediated detoxification and S-adenosylmethionine dependent methyltransferase to counter arsenic toxicity in *Oryza sativa* L. *Ecotoxicology and Environmental Safety, 208*, 111418.

Requejo, R., & Tena, M. (2005). Proteome analysis of maize roots reveals that oxidative stress is a main contributing factor to plant arsenic toxicity. *Phytochemistry, 66*(13), 1519–1528.

Rossi, F., Li, H., Liu, Y., & De Philippis, R. (2017). Cyanobacterial inoculation (cyanobacterisation): Perspectives for the development of a standardized multifunctional technology for soil fertilization and desertification reversal. *Earth-Science Reviews, 171*, 28–43.

Sanchary, I. J., & Huq, S. M. I. (2017). Remediation of arsenic toxicity in the soil-plant system by using zinc fertilizers. *Journal of Agricultural Chemistry and Environment, 6*(01), 30.

Sanglard, L. M., Detmann, K. C., Martins, S. C., Teixeira, R. A., Pereira, L. F., Sanglard, M. L., Fernie, A. R., Araújo, W. L., & DaMatta, F. M. (2016). The role of silicon in metabolic acclimation of rice plants challenged with arsenic. *Environmental and Experimental Botany, 123*, 22–36.

Sarkar, D., Makris, K. C., Vandanapu, V., & Datta, R. (2007). Arsenic immobilization in soils amended with drinking-water treatment residuals. *Environmental Pollution, 146*(2), 414–419.

Sengupta, S., Bhattacharyya, K., Mandal, J., Bhattacharya, P., Halder, S., & Pari, A. (2021). Deficit irrigation and organic amendments can reduce dietary arsenic risk from rice: Introducing machine learning-based prediction models from field data. *Agriculture, Ecosystems & Environment, 319*, 107516.

Shehzad, M. T., Sabir, M., Siddique, A. B., Rahman, M. M., & Naidu, R. (2022). Impact of Water Regimes on Minimizing the Accumulation of Arsenic in Rice (Oryza sativa L.). *Water, Air, & Soil Pollution, 233*(9), 383.

Shi, S., Wang, T., Chen, Z., Tang, Z., Wu, Z., Salt, D. E., Chao, D. Y., & Zhao, F. J. (2016). OsHAC1; 1 and OsHAC1; 2 function as arsenate reductases and regulate arsenic accumulation. *Plant Physiology, 172*(3), 1708–1719.

Shi, G., Lu, H., Liu, H., Lou, L., Zhang, P., Song, G., Zhou, H., & Ma, H. (2020). Suphate application decreases translocation of arsenic and cadmium within wheat (*Triticum aestivum* L.) plant. *Science of the Total Environment, 713*, 136665.

Shrivastava, A., Ghosh, D., Dash, A., & Bose, S. (2015). Arsenic contamination in soil and sediment in India: Sources, effects, and remediation. *Current Pollution Reports, 1*(1), 35–46.

Sinha, B., & Bhattacharyya, K. (2012). Role of organics in reducing arsenic loading in soil-plant system under rice in the lower Gangetic plain of West Bengal. *Journal of Crop and Weed, 8*(1), 79–83.

Song, W. Y., Yamaki, T., Yamaji, N., Ko, D., Jung, K. H., Fujii-Kashino, M., An, G., Martinoia, E., Lee, Y., & Ma, J. F. (2014). A rice ABC transporter, OsABCC1, reduces arsenic accumulation in the grain. *Proceedings of the National Academy of Sciences, 111*(44), 15699–15704.

Srivastava, S., Mishra, S., Tripathi, R. D., Dwivedi, S., Trivedi, P. K., & Tandon, P. K. (2007). Phytochelatins and antioxidant systems respond differentially during arsenite and arsenate stress in Hydrilla verticillata (Lf) Royle. *Environmental Science & Technology, 41*(8), 2930–2936.

Srivastava, S., Srivastava, A. K., Suprasanna, P., & D'souza, S. F. (2013). Identification and profiling of arsenic stress-induced microRNAs in Brassica juncea. *Journal of Experimental Botany, 64*(1), 303–315.

Srivastava, S., Akkarakaran, J. J., Sounderajan, S., Shrivastava, M., & Suprasanna, P. (2016). Arsenic toxicity in rice (*Oryza sativa* L.) is influenced by sulphur supply: Impact on the expression of transporters and thiol metabolism. *Geoderma, 270*, 33–42.

Srivastava, S., Srivastava, S., Bist, V., Awasthi, S., Chauhan, R., Chaudhry, V., Singh, P. C., Dwivedi, S., Niranjan, A., Agrawal, L., & Chauhan, P. S. (2018). *Chlorella vulgaris* and *Pseudomonas putida* interaction modulates phosphate trafficking for reduced arsenic uptake in rice (*Oryza sativa* L.). *Journal of Hazardous Materials, 351*, 177–187.

Srivastava, S., Srivastava, A. K., Suprasanna, P., & D'souza, S. F. (2009). Comparative biochemical and transcriptional profiling of two contrasting varieties of *Brassica juncea* L. in response to arsenic exposure reveals mechanisms of stress perception and tolerance. *Journal of Experimental Botany, 60*(12), 3419–3431.

Suganami, M., Suzuki, Y., Kondo, E., Nishida, S., Konno, S., & Makino, A. (2020). Effects of overproduction of Rubisco activase on Rubisco content in transgenic rice grown at different N levels. *International Journal of Molecular Sciences, 21*(5), 1626.

Sun, Y. Y., Liu, R. L., Zeng, X. B., Lin, Q. M., Bai, L. Y., Li, L. F., Su, S. M., & Wang, Y. N. (2015). Reduction of arsenic bioavailability by amending seven inorganic materials in arsenic contaminated soil. *Journal of Integrative Agriculture, 14*(7), 1414–1422.

Taha, S. S., & Osman, A. S. (2018). Influence of potassium humate on biochemical and agronomic attributes of bean plants grown on saline soil. *The Journal of Horticultural Science and Biotechnology, 93*(5), 545–554.

Talukder, A. S. M. H. M., Meisner, C. A., Sarkar, M. A. R., Islam, M. S., Sayre, K. D., Duxbury, J. M., & Lauren, J. G. (2012). Effect of water management, arsenic and phosphorus levels on rice in a high-arsenic soil–water system: II. *Arsenic Uptake. Ecotoxicology and Environmental Safety, 80*, 145–151.

Tan, G., Mao, Y., Wang, H., & Xu, N. (2020). A comparative study of arsenic (V), tetracycline and nitrate ions adsorption onto magnetic biochars and activated carbon. *Chemical Engineering Research and Design, 159*, 582–591.

Tan, X., Cao, J., Liu, J., Wang, J., Duan, G., Zhang, Y., & Lin, A. (2022). *Characteristics of three organic fertilizers and their influence on the mobility of cadmium and arsenic in a soil-rice (Oryza sativa L.) system.* Pre-Print Version.

Tang, X., Li, L., Wu, C., Khan, M. I., Manzoor, M., Zou, L., & Shi, J. (2020). The response of arsenic bioavailability and microbial community in paddy soil with the application of sulphur fertilizers. *Environmental Pollution, 264*, 114679.

Tripathi, P., Mishra, A., Dwivedi, S., Chakrabarty, D., Trivedi, P. K., Singh, R. P., & Tripathi, R. D. (2012). Differential response of oxidative stress and thiol metabolism in contrasting rice genotypes for arsenic tolerance. *Ecotoxicology and Environmental Safety, 79*, 189–198.

Tripathi, P., Tripathi, R. D., Singh, R. P., Dwivedi, S., Goutam, D., Shri, M., Trivedi, P. K., & Chakrabarty, D. (2013). Silicon mediates arsenic tolerance in rice (*Oryza sativa* L.) through lowering of arsenic uptake and improved antioxidant defence system. *Ecological Engineering, 52*, 96–103.

Upadhyay, A. K., Singh, N. K., & Rai, U. N. (2014). Comparative metal accumulation potential of Potamogetonpectinatus L. and Potamogetoncrispus L.: Role of enzymatic and non-enzymatic antioxidants in tolerance and detoxification of metals. *Aquatic Botany, 117*, 27–32.

Verma, L., & Singh, J. (2019). Synthesis of novel biochar from waste plant litter biomass for the removal of Arsenic (III and V) from aqueous solution: A mechanism characterization, kinetics and thermodynamics. *Journal of Environmental Management, 248*, 109235.

Vessey, J. K. (2003). Plant growth promoting rhizobacteria as biofertilizers. *Plant and Soil, 255*(2), 571–586.

Vieira, B. R., Pintor, A. M., Boaventura, R. A., Botelho, C. M., & Santos, S. C. (2017). Arsenic removal from water using iron-coated seaweeds. *Journal of Environmental Management, 192*, 224–233.

Vithanage, M., Herath, I., Joseph, S., Bundschuh, J., Bolan, N., Ok, Y. S., Kirkham, M. B., & Rinklebe, J. (2017). Interaction of arsenic with biochar in soil and water: A critical review. *Carbon, 113*, 219–230.

Wan, X., Li, C., & Parikh, S. J. (2020). Simultaneous removal of arsenic, cadmium, and lead from soil by iron-modified magnetic biochar. *Environmental Pollution, 261*, 114157.

Wang, P., Zhang, W., Mao, C., Xu, G., & Zhao, F. J. (2016). The role of OsPT8 in arsenate uptake and varietal difference in arsenate tolerance in rice. *Journal of Experimental Botany, 67*(21), 6051–6059.

Wang, Y., Li, Y. Q., Lv, K., Cheng, J. J., Chen, X. L., Ge, Y., & Yu, X. Y. (2018). Soil microalgae modulate grain arsenic accumulation by reducing dimethylarsinic acid and enhancing nutrient uptake in rice (*Oryza sativa* L.). *Plant and Soil, 430*(1), 99–111.

Wang, X., Jiang, J., Dou, F., Li, X., Sun, W., & Ma, X. (2022). Zinc fertilizers modified the formation and properties of iron plaque and arsenic accumulation in rice (*Oryza sativa* L.) in a life cycle study. *Environmental Science & Technology, 56*(12), 8209–8220.

Weber, F. A., Hofacker, A. F., Voegelin, A., & Kretzschmar, R. (2010). Temperature dependence and coupling of iron and arsenic reduction and release during flooding of a contaminated soil. *Environmental Science & Technology, 44*(1), 116–122.

Wisawapipat, W., Chooaiem, N., Aramrak, S., Chittamart, N., Nookabkaew, S., Rangkadilok, N., Satayavivad, J., & Christl, I. (2021). Sulphur amendments to soil decrease inorganic arsenic accumulation in rice grain under flooded and nonflooded conditions: Insights from temporal dynamics of porewater chemistry and solid-phase arsenic solubility. *Science of the Total Environment, 779*, 146352.

Wongrod, S., Simon, S., van Hullebusch, E. D., Lens, P. N., & Guibaud, G. (2018). Changes of sewage sludge digestate-derived biochar properties after chemical treatments and influence on As (III and V) and Cd (II) sorption. *International Biodeterioration & Biodegradation, 135*, 96–102.

Wu, C., Ye, Z., Li, H., Wu, S., Deng, D., Zhu, Y., & Wong, M. (2012). Do radial oxygen loss and external aeration affect iron plaque formation and arsenic accumulation and speciation in rice? *Journal of Experimental Botany, 63*(8), 2961–2970.

Wu, J., Huang, D., Liu, X., Meng, J., Tang, C., & Xu, J. (2018). Remediation of As (III) and Cd (II) co-contamination and its mechanism in aqueous systems by a novel calcium-based magnetic biochar. *Journal of Hazardous Materials, 348*, 10–19.

Wu, J., Li, Z., Huang, D., Liu, X., Tang, C., Parikh, S. J., & Xu, J. (2020). A novel calcium-based magnetic biochar is effective in stabilization of arsenic and cadmium co-contamination in aerobic soils. *Journal of Hazardous Materials, 387*, 122010.

Xu, J., Shi, S., Wang, L., Tang, Z., Lv, T., Zhu, X., Ding, X., Wang, Y., Zhao, F. J., & Wu, Z. (2017). OsHAC4 is critical for arsenate tolerance and regulates arsenic accumulation in rice. *New Phytologist, 215*(3), 1090–1101.

Xu, M., Gao, P., Wu, J., Ma, J., Zhang, X., Yang, G., Long, L., Chen, C., Song, C., & Xiao, Y. (2022). Biochar promotes arsenic sequestration on iron plaques and cell walls in rice roots. *Chemosphere, 288*, 132422.

Yamaguchi, N., Nakamura, T., Dong, D., Takahashi, Y., Amachi, S., & Makino, T. (2011). Arsenic release from flooded paddy soils is influenced by speciation, Eh, pH, and iron dissolution. *Chemosphere, 83*(7), 925–932.

Yamaguchi, N., Ohkura, T., Takahashi, Y., Maejima, Y., & Arao, T. (2014). Arsenic distribution and speciation near rice roots influenced by iron plaques and redox conditions of the soil matrix. *Environmental Science & Technology, 48*(3), 1549–1556.

Yang, Y., Hu, H., Xie, K., Huang, Y., Liu, H., Li, X., Yao, H., & Naruse, I. (2019). Insight of arsenic transformation behavior during high-arsenic coal combustion. *Proceedings of the Combustion Institute, 37*(4), 4443–4450.

Yapaa, N., Dub, W., Madhushanc, A., Yanb, K., Asadd, S., Karunarathnae, S. C., & Bamunuarachchigef, C. (2022). Potential of biofertilizers and natural soil amendments to mitigate heavy metal contents of soil in lowland rice (*Oryza sativa* L.) farming. *Science Asia, 48*(3), 326–334.

Yu, L. J., Luo, Y. F., Liao, B., Xie, L. J., Chen, L., Xiao, S., Li, J. T., Hu, S. N., & Shu, W. S. (2012). Comparative transcriptome analysis of transporters, phytohormone and lipid metabolism pathways in response to arsenic stress in rice (*Oryza sativa*). *New Phytologist, 195*(1), 97–112.

Yu, H. Y., Wang, X., Li, F., Li, B., Liu, C., Wang, Q., & Lei, J. (2017). Arsenic mobility and bioavailability in paddy soil under iron compound amendments at different growth stages of rice. *Environmental Pollution, 224*, 136–147.

Zama, E. F., Li, G., Tang, Y. T., Reid, B. J., Ngwabie, N. M., & Sun, G. X. (2022). The removal of arsenic from solution through biochar-enhanced precipitation of calcium-arsenic derivatives. *Environmental Pollution, 292*, 118241.

Zanin, L., Tomasi, N., Cesco, S., Varanini, Z., & Pinton, R. (2019). Humic substances contribute to plant iron nutrition acting as chelators and biostimulants. *Frontiers in Plant Science, 10*, 675.

Zhang, X., Dayton, E. A., & Basta, N. T. (2020). Predicting the modifying effect of soils on arsenic phytotoxicity and phytoaccumulation using soil properties or soil extraction methods. *Environmental Pollution, 263*, 114501.

Zhao, F. J., McGrath, S. P., & Meharg, A. A. (2010). Arsenic as a food chain contaminant: Mechanisms of plant uptake and metabolism and mitigation strategies. *Annual Review of Plant Biology, 61*, 535–559.

Zhao, F. J., Zhu, Y. G., & Meharg, A. A. (2013). Methylated arsenic species in rice: Geographical variation, origin, and uptake mechanisms. *Environmental Science & Technology, 47*(9), 3957–3966.

Zhou, X., Gao, A., Lai, F., Zhang, C., & Xu, W. (2017). The role of selenium in soil: Effect on the uptake and translocation of arsenic in rice (*Oryza sativa* L). *International Journal of Agriculture and Biology, 19*, 1227–1234.

Zhu, X. F., Lei, G. J., Jiang, T., Liu, Y., Li, G. X., & Zheng, S. J. (2012). Cell wall polysaccharides are involved in P-deficiency-induced Cd exclusion in *Arabidopsis thaliana. Planta, 236*(4), 989–997.

Zhu, F., Cheng, Q., Xue, S., Li, C., Hartley, W., Wu, C., & Tian, T. (2018). Influence of natural regeneration on fractal features of residue microaggregates in bauxite residue disposal areas. *Land Degradation & Development, 29*(1), 138–149.

Zou, L., Zhang, S., Duan, D., Liang, X., Shi, J., Xu, J., & Tang, X. (2018). Effects of ferrous suphate amendment and water management on rice growth and metal (loid) accumulation in arsenic and lead co-contaminated soil. *Environmental Science and Pollution Research, 25*(9), 8888–8902.

Chapter 18
Arsenic in Groundwater: A Threat to Agriculture and Its Mitigation Measures to Protect the Food Chain

Pankaj Kumar Srivastava, Mariya Naseem, and Richa Raghuwanshi

Abstract Arsenic—a non-threshold class-I carcinogen, is mostly of geogenic origin contaminating groundwater worldwide. The use of this arsenic contaminated groundwater for potable and irrigation purposes poses a serious threat to human health. Irrigation of agricultural fields with arsenic-tainted groundwater is adding considerable amounts of arsenic to soils. This build-up of arsenic further contaminates crops and vegetables being cultivated in these fields. Arsenic in agriculture produces puts human health at risk of several diseases when consuming these arsenic-laden eatables. The temporal and spatial distribution of groundwater arsenic in irrigated cultivable fields is a serious concern worldwide, especially in South-East Asian countries. Data such as origin, fate and transfer, distribution of arsenic from aquifers to the soil, bioaccumulation at different levels of the food chain, and its related health impacts on human beings and livestock need to be generated. R&D on alternative sources of irrigation groundwater, mitigation measures in relation to arsenic contamination and toxicity and integration of groundwater arsenic problems with socio-economic concerns are also required. Strategies for stakeholders' involvement, improving legal, policy, institutional and human capacity, and awareness programs for the community at risk are highly warranted. Developing countries should address some specific challenges in the context of improving mitigation approaches for arsenic contamination in groundwater and irrigated ecosystems. The long term solutions and strategies would require joint interdisciplinary R&D, capacity building, and effective awareness through a diverse and integrated regional network and partnership about tackling the groundwater arsenic threat in agriculture.

Keywords Arsenic · Groundwater · Agriculture · Soil · Mitigation

---

P. K. Srivastava (✉) · M. Naseem
Environmental Technologies Division, CSIR-National Botanical Research Institute, Lucknow, India
e-mail: pankajk@nbri.res.in

R. Raghuwanshi
Department of Botany, Banaras Hindu University, Varanasi, India

B. Sen Gupta and N. Martínez-Villegas (eds.), *Arsenic Remediation of Food and Water*,
https://doi.org/10.1007/978-981-97-4764-1_18

Introduction

Arsenic is known to be one of the most important deadly substances present in the environment. According to WHO, Arsenic is a non-threshold class I carcinogen for human health, which is ubiquitous in the environment. Arsenic is a natural geogenicpoisonous metalloid found in the earth's crust with a mean concentration of 3–5 mg kg^{-1}. Arsenic is associated with minerals viz., arsenopyrites (FeAsS), realgar (As_4S_4), and orpiment (As_2S_3 and is released into the environment by natural processes, predominantly the reductive dissolution of soil mineral hydroxides by microbes. Arsenic exists in inorganic forms (arsenate AsV and arsenite AsIII), which are more toxic than the organic forms. Microbial activities are responsible for releasing arsenic into the aquifers. Potential risks of arsenic contamination to ecosystems and human health, particularly in the South Asian region (India and Bangladesh, etc.), have been thoroughly investigated.

Arsenic is a known carcinogen for humans as well as other living organisms and a potent human poison (Hughes et al., 2011). It can cause alteration at the nucleic acid level causing amplification of genes and transforming eukaryotic cells (Bobrowicz et al., 1997). The studies done by Patra et al. (2004) have shown that arsenic is involved in inactivating enzymes, encompassing crucial pathways such as the energy production pathways, replication and repair system of DNA. Arsenic inactivates proteins by interacting with the sulfhydryl group of the protein (Patra et al., 2004); it can substitute P in DNA. Arsenic can also uncouple mitochondrial oxidative phosphorylation. It also produces ROS, which triggers lipid peroxidation and DNA damage. The thiol or sulfhydryl groups present in the human organ tissues having high keratin content are known to bind arsenic (Valko et al., 2005). Consumption of high doses of arsenic either deliberately or unknowingly through contaminated food and water is delineated by the skin condition of the person. Regions where people are exposed to high arsenic concentrations, show various symptoms on their skin such as changes in its colour, spots on hands and feet, and even skin cancer. The other harmful effects of arsenic exposure is hypertension, diabetes, disorders of the reproductive system, and harm to the lungs and kidneys. Chronic exposure to arsenic causes arsenicosis in humans (Mandal & Suzuki, 2002).

Arsenic is known to cause phytotoxicity with arsenite reported to be more fatal than other forms of arsenic (Kumar et al., 2004). The high amount of arsenic is toxic to plant cell membranes, interrupts enzyme specificity, and interferes with cellular functions (Ordonez et al., 2005). It has been reported that arsenic exposure to vegetables can cause a decrease in leaf weight and fruiting which ultimately leads to economic losses; it also causes damage to the root membrane and disturbs upward translocation (Barrachina et al., 1995). The exposure of rice plants (*Oryza sativa*) to Dimethylarsinic acid (DMA) results in a net decrease in photosynthesis and at higher concentrations may lead to a decrease in the leaf areaand dry matter production (Marin et al., 1993). The high arsenic accumulation in bean plants was found to cause a decrease in the leaf gaseous exchange, overall growth, protein content, water potential, and biomass gain of the plant. On contrary, an increase in lipid peroxidation and peroxidase activity as a result of oxidative stress was observed (Stoeva et al., 2005). In rice plants, excess arsenic may cause straight-head disease which is characterized by unfilled panicles. This causes economic losses due to a decrease in the yield. A committee of FAO/WHO on Food Additives allows for aprovisional tolerable weekly intake (PTWI) of inorganic arsenic of 15 $\mu g\ kg^{-1}$ body weight, equivalent to130 $\mu g\ day^{-1}$ for a 60 kg person. Due to the unavailability of toxicological data, such a value has not been set for organic arsenic. Inorganic arsenic is dominant in terrestrialfood plants.

Groundwater Arsenic in Agriculture: Specific Challenges at the National and Global Scale

Arsenic contamination in groundwater through geogenic origin has been reported in the USA, Argentina, Mexico, Japan, Chile, etc. (Casentini et al., 2011; Bundschuh et al., 2012). More than 150 million people worldwide are getting affected by the consumption of contaminated potable and groundwater. Impacts on human health resulting from the consumption of arsenic contaminated groundwater for drinking and irrigation is a global concern, particularly in Asia and Southeast Asia. Groundwater from tube wells used to irrigate agricultural fields is adding large amounts of arsenic to the soil every year, resulting in increased levels of arsenic in crops grown on arsenic contaminated arable soils. For example, the use of 1200–1600 mm of water level containing 100 μg/L of arsenic for paddy cultivation may increase to 40 kg of arsenic per hectare of soil per year. Arsenic accumulates in the edible parts of plants, and subsequently contaminates the food chain, posing a serious health risk to humans. The transfer of arsenic into soil crop systems represents one of the major factors of human exposure to arsenic. The recommendations and observations of the Public-Accounts Committee (2014–15) of the Sixteenth Lok Sabha of Govt. of India on Agriculture have laid the foundation for arsenic mapping in the agriculture domain based on the first published report, which mentioned the arsenic problem in 96 districts of 12 states of India viz., Uttar Pradesh, West Bengal, Assam, Punjab, Bihar, etc. The magnitude of the problem is humongous and unparalleled, with risks of cancer and other diseases involved, affecting 36 million people in the Ganga-Meghna-Brahmaputra basin alone.

The problem of arsenic pollution is widespread in different parts of India with the state of West Bengal being severely affected. The Bengal Delta Plain which encompasses West Bengal in India and Bangladesh has high levels of groundwater as, more than the national drinking water threshold of 50 μg/L (Bhattacharya et al., 2003). The reason for such high arsenic levels in Bengal Delta Plain is the presence of ferric ions on the surface sediments which adsorb ferrous ions during iron oxide reduction. In turn, the adsorbed ferrous ion reduces arsenate to more toxic arsenite. In this way, a more toxic and mobile form of arsenic i.e. arsenite gets introduced into the groundwater of the Bengal Delta Plain (Bhattacharya et al., 2003). In India, many parts of states like Uttar Pradesh, Bihar, Chhattisgarh, Assam, and West Bengal were reported to have high arsenic contamination of soil and sediments (Somani, 2001; Srivastava & Sharma, 2013). The groundwater arsenic contents were found to be in the range of 0.41 to 1.01 mg/L in the West Bengal region. The soil arsenic level was found to be13.12 mg/kg which was beyond the globally acceptable limit of 10.0 mg/kg. The total arsenic content in food crops was found up to 1.464 mg/kg dw (Rahaman et al., 2013). The studies conducted in the different districts of Uttar Pradesh; India showed a high arsenic groundwater contamination (Singh et al., 2015). Out of all the groundwater samples drawn 46.5% had arsenic levels that exceeded the WHO threshold of 10 μg/L for drinking water, 26.7% exceeded 50 μg/L (Indian standard) and 10% had arsenic levels of 300 μg/L which is far beyond acceptable (Ahamed et al., 2006). The arsenic level in groundwater was studied in the Vaishali and Bhagalpur districts of Bihar, India. In both districts, the arsenic level was reported to exceed the WHO standard of 10 μg/L for drinking water. The contents of arsenicin groundwater were found to range from 20 μg/L to143 μg/L, even higher than the FAO standard of 100 μg/L for irrigation water (Singh et al., 2014). The arsenic contaminated groundwater when used for irrigation increases the arsenic level in soil which ultimately increases the chances to enter the food chain and cause health hazards (Singh et al. 2015).

There has been global concern about the toxicity originating from the consumption of arsenic contaminated groundwater for drinking and irrigation purposes, especially in Asia and Southeast Asia (Brammer & Ravenscroft, 2009). The groundwater from tubewells used for irrigating agricultural fields is adding large amounts of arsenic annually, which has led to elevated levels of arsenic in the agricultural soils as well as in the crops grown on these soils (Norton et al., 2012). Compared to other cereals, arsenic accumulation in the paddy crop, a major staple food, is more prevalent owing to its flooded agronomic practice (Hua et al., 2011). Arsenic accumulation in paddy crops can cause many human health risks (Hite, 2013). Rice is the staple food in many regions, and its contamination with arsenic has a profound negative impact on the health and well-being of people residing in the contaminated areas (Bundschuh et al., 2012). In addition to human health risks, high arsenic levels in soil may reduce root growth, plant height, and grain yields in paddy crops along with the occurrence of other physiological disorders like straight head disease in rice (Hua et al., 2011). Arsenic toxicity in paddy leads to a reduction in crop yield as well as toxicity to the food chain (Panaullah et al., 2009). The temporal and spatial distribution of arsenic in paddy soils and paddy crops is subject to the level of arsenic in irrigation groundwater, soil properties, and the rice varieties/genotypes being cultivated (Zheng et al., 2011). Rice is a major staple food of the world, especially the South and South East Asia. India is a major rice-producing nation, where paddy cultivation occupies 23.3% of the gross cropped area. The average intake of cooked rice by an adult human is about 300–600 g per day, especially in Asian countries (Banerjee et al., 2013). Rice cultivation occupies a million hectares in the Indo-Gangetic Plains (IGP) of the South Asian region.

The application of arsenic-tainted groundwater for irrigating cultivable fields is a serious concern worldwide, especially in South-East Asian countries. The most severely affected regions are Bangladesh and India (Hossain, 2006). Arsenic contamination in crops, especially rice owes its presence to the irrigation of paddy agricultural fields with As-tainted groundwater. Arsenic gets accumulated in edible parts of the plant and becomes available to living organisms through the food chain (Meharg & Jardine, 2003) Rice tends to accumulate more arsenic as compared to other crops, because it is cultivated under flooded conditions (Talukder et al., 2012). This result is a subsequent build-up of arsenic in soils and eventually in rice grains (Singh et al., 2014). Generally, millions of people in the world especially in Asia consume rice as a staple food, exposing them to arsenic toxicity (Meharg & Zhao, 2012). Huhmann et al. (2019) reported that the yield of rice crops is reduced due to soil arsenic build-up from irrigation with high-As groundwater. There have been several reports on arsenic uptake and transfer to rice plants being influenced by different mineral nutrients and amendments suggesting nutrient management as an efficient solution to mitigate arsenic uptake in rice (Singh & Srivastava, 2020; Hussain et al., 2021). Therefore, an appropriate arsenic remediating strategy to reduce the arsenic uptake and its translocation from root to grains must be devised for minimizing human health risks. High contents of arsenic ranging from 19 to 489 μg/kg have been observed in vegetables growing in the arsenic contaminated regions of Bangladesh and West Bengal. Das et al. (2004) have reported the range of arsenic (μg/kg) in rice and vegetables as 40–270, and 70–3,990, respectively which became a major issue in their import to European countries. Arsenic contamination in different matrices in developing countries is given in Tables 18.1, 18.2, and 18.3.

Groundwater Arsenic in Agriculture: Goal and Strategies for Mitigation

Soaring incidences of arsenic groundwater contamination worldwide especially in developing countries have been documented as a worldwide issue. Demanding corrective action on a global scale. There is currently a lack of data that challenges the sternness of incorporating arsenic as a mandatory parameter to be analyzed for drinking or irrigation water. This insufficiency of data can be ascribed to the lack of research and monitoring. Data on origin, fate and transfer, distribution of arsenic from aquifers to the soil, bioaccumulation in different levels of the food chain, and its related health impacts on human beings and cattle need to be generated. R&D on alternative sources of arsenic free potable and irrigational groundwater and ways to mitigate arsenic toxicity to humans need to be taken-up on an urgent basis. There is also a need to integrate groundwater arsenic contamination problems with the socio-economic behaviours of the local inhabitants to come up with well-equipped mitigation strategies. Strategies for stakeholders' involvement, improving legal, policy, institutional and human capacity, and awareness programs for the community at risk are highly warranted.

Table 18.1 Arsenic contamination in groundwater

Country	Arsenic concentration	References
Bangladesh	<0.5–> 4600 μg/L	Sandhi, A., Yu, C., Rahman, M. M., & Amin, M. (2022). Arsenic in the water and agricultural crop production system: Bangladesh perspectives. *Environmental Science and Pollution Research*, 1–13
West Bengal, India	1.5–4.0 mg/L (Mean inorganic as)	Koley, S. (2021). Contemporary practices in groundwater arsenic remediation and wastewater management in West Bengal, India: a systematic review. *International Journal of Advanced Technology and Engineering Exploration*, 8(80), 797
Punjab, India	0.42–20.97 μg/L	Sidhu, S. S., Brar, J. S., Biswas, A., Banger, K., & Saroa, G. S. (2012). Arsenic contamination in soil–water–plant (Rice, Oryza sativa L.) continuum in central and sub-mountainous Punjab, India. *Bulletin of environmental contamination and toxicology*, *89*(5), 1046–1050
Afghanistan	10–500 μg/L	Mukherjee, A., Sengupta, M. K., Hossain, M. A., Ahamed, S., Das, B., Nayak, B., … & Chakraborti, D. (2006). Arsenic contamination in groundwater: a global perspective with emphasis on the Asian scenario. *Journal of Health, Population and Nutrition*, 142–163
Ghopuz, Iran	1031 μg/L	Dastgiri, S., Mosaferi, M., Fizi, M. A., Olfati, N., Zolali, S., Pouladi, N., & Azarfam, P. (2010). Arsenic exposure, dermatological lesions, hypertension, and chromosomal abnormalities among people in a rural community of northwest Iran. *Journal of health, population, and nutrition*, *28*(1), 14
Ogun state, Nigeria	40–160 μg/L	Gbadebo, A. M. (2005). Arsenic pollution in aquifers located within limestone areas of Ogun State, Nigeria. In *Natural Arsenic in Groundwater* (pp. 101–108). CRC Press
Namaqualand, South Africa	< 20–253 μg/L	Abiye, T. A., & Bhattacharya, P. (2019). Arsenic concentration in groundwater: Archetypal study from South Africa. *Groundwater for Sustainable Development*, *9*, 100,246

(continued)

Table 18.1 (continued)

Country	Arsenic concentration	References
Greater Giyani, South Africa	0.1–172.52 µg/L	Mudzielwana, R., Gitari, M. W., Akinyemi, S. A., Talabi, A. O., & Ndungu, P. (2020). Hydrogeochemical characteristics of arsenic rich groundwater in Greater Giyani Municipality, Limpopo Province, South Africa. *Groundwater for sustainable development, 10*, 100,336
Northern Chile	400–600 µg/L	Sancha, A. M., & O'Ryan, R. (2008). Managing hazardous pollutants in Chile: arsenic. *Reviews of Environmental Contamination and Toxicology Vol 196*, 123–146
Peru	0.1–193.1 µg/L	George, C. M., Sima, L., Arias, M., Mihalic, J., Cabrera, L. Z., Danz, D., … & Gilman, R. H. (2014). Arsenic exposure in drinking water: an unrecognized health threat in Peru. *Bulletin of the World Health Organization, 92*, 565–572
Cajicá, Zipaquirá, Mosquera, Funza and Soacha(Colombia)	14–255 µg/L	Alonso, D. L., Latorre, S., Castillo, E., & Brandão, P. F. (2014). Environmental occurrence of arsenic in Colombia: A review. *Environmental pollution, 186*, 272–281
San José, Costa Rica	40 µg/L	Kunz, S., Romero-Esquivel, L. G., Otter, P., Feistel, U., Grischek, T., Valverde-Cerdas, J., & Feller, J. (2018). Treatment of arsenic-contaminated water using in-line electrolysis, co-precipitation and filtration in Costa Rica. *Water Science and Technology: Water Supply, 18*(1), 40–48
Brazil: The Iron Quadrangle	< 5–2980 µg/L	Gonçalves, J. A. C., de Lena, J. C., Paiva, J. F., Nalini, H. A., & Pereira, J. C. (2007). Arsenic in the groundwater of Ouro Preto (Brazil): its temporal behavior as influenced by the hydric regime and hydrogeology. *Environmental geology, 53*(4), 785–793
South western Uruguay	0.1–58.0 µg/L	Manganelli, A., Goso, C., Guerequiz, R., Fernández Turiel, J. L., García Vallès, M., Gimeno, D., & Pérez, C. (2007). Groundwater arsenic distribution in South-western Uruguay. *Environmental geology, 53*(4), 827–834
Mixco, Guatemala	15 µg/L	Lotter, J. T., Lacey, S. E., Lopez, R., Socoy Set, G., Khodadoust, A. P., & Erdal, S. (2014). Groundwater arsenic in Chimaltenango, Guatemala. *Journal of Water and Health, 12*(3), 533–542
Cerro Alto, Guatemala	47.5 µg/L	

Table 18.2 Arsenic contamination in soil

Country	Arsenic concentration	References
Bangladesh	20–30 mg/kg	Meharg, A. A., & Rahman, M. M. (2003). Arsenic contamination of Bangladesh paddy field soils: implications for rice contribution to arsenic consumption. *Environmental science & technology*, 37(2), 229–234
West Bengal, India	8.91–13.12 mg/kg	Rahaman, S., Sinha, A. C., Pati, R., & Mukhopadhyay, D. (2013). Arsenic contamination: a potential hazard to the affected areas of West Bengal, India. *Environmental geochemistry and health*, *35*(1), 119–132
Punjab, India	0.26–2.33 mg/kg	Sidhu, S. S., Brar, J. S., Biswas, A., Banger, K., &Saroa, G. S. (2012). Arsenic contamination in soil–water–plant (Rice, Oryza sativa L.) continuum in central and sub-mountainous Punjab, India. *Bulletin of environmental contamination and toxicology*, *89*(5), 1046–1050
Egypt	25–67 mg/kg	Shaheen, S. M., Kwon, E. E., Biswas, J. K., Tack, F. M., Ok, Y. S., &Rinklebe, J. (2017). Arsenic, chromium, molybdenum, and selenium: Geochemical fractions and potential mobilization in riverine soil profiles originating from Germany and Egypt. *Chemosphere*, *180*, 553–563
Río Caritaya, Northern Chile	489 ± 15 mg/kg	Bundschuh, J., Nicolli, H. B., Blanco, M. D. C., Blarasin, M., Farías, S. S., Cumbal, L., … & Fernández Cirelli, A. (2008). Distribución de arsénicoen la regiónsudamericana. In *Distribución del arsénicoen las regionesIbérica e Iberoamericana* (pp. 137–185)
Río Camarones (Illapata) Northern Chile	300 ± 7 mg/kg	
Concepción province, Peru	9.16–112.66 mg/kg	Custodio, M., Peñaloza, R., Orellana, E., Aguilar-Cáceres, M. A., & Maldonado-Oré, E. M. (2021). Heavy metals and arsenic in soil and cereal grains and potential human risk in the central region of Peru. *Journal of Ecological Engineering*, *22*(1)
Ataco Zone, Cerro Copete Colombia	1.0–45.2 mg/kg	Alonso, D. L., Latorre, S., Castillo, E., & Brandão, P. F. (2014). Environmental occurrence of arsenic in Colombia: A review. *Environmental pollution*, *186*, 272–281

(continued)

Table 18.2 (continued)

Country	Arsenic concentration	References
Guayas and Los Ríos provinces, Ecuador	4.48 ± 3 mg/kg	Ochoa, M., Tierra, W., Tupuna-Yerovi, D. S., Guanoluisa, D., Otero, X. L., & Ruales, J. (2020). Assessment of cadmium and lead contamination in rice farming soils and rice (Oryza sativa L.) from Guayas province in Ecuador. *Environmental Pollution, 260*, 114,050
Costa Rica	2.0–21.0 mg/kg	Molinedo Garcia, N. J., & Vargas Azofeifa, I. (2020). Arsenic in soil and rocks of volcanic origin in an AguasZarcas area and La Palmera, northern zone of Costa Rica. *Ciencia y Tecnologia (Online), 36*(1), 23–36
Nova Lima, Brazil	960 mg/kg	Teixeira, M. C., Santos, A. C., Fernandes, C. S., & Ng, J. C. (2020). Arsenic contamination assessment in Brazil–Past, present and future concerns: A historical and critical review. *Science of the Total Environment, 730*, 138,217
Paso de la Laguna, Uruguay	3.62 mg /kg	Carracelas, G., Hornbuckle, J., Verger, M. E. L. I. S. S. A., Huertas, R. A. Q. U. E. L., Riccetto, S., Campos, F., & Roel, A. L. V. A. R. O. (2019). Irrigation management and variety effects on rice grain arsenic levels in Uruguay. *Journal of Agriculture and Food Research, 1*, 100,008
Venezuela	5.75–12.13 mg/kg	Bundschuh, J., Armienta, M. A., Morales-Simfors, N., Alam, M. A., López, D. L., Delgado Quezada, V., … & Ahmad, A. (2021). Arsenic in Latin America: New findings on source, mobilization and mobility in human environments in 20 countries based on decadal research 2010–2020. *Critical reviews in environmental science and technology, 51*(16), 1727–1865

In unexplored geographical areas, arsenic contamination level in the irrigation groundwater sources and in agricultural fields needs to be monitored and assessed (Sidhu et al., 2012). This may help not only in developing new remediation strategies to reduce arsenic uptake by crops such as rice but also in evaluating the potential health risks involved. In India, the Ganga-Meghna-Brahmaputra basin has reported high groundwater arsenic levels (up to 500 $\mu g\ l^{-1}$) in potable water sources (Bhattacharya et al., 2007). The groundwater, which was found unsafe for community

Table 18.3 Arsenic contamination in different food samples

Country	Arsenic concentration	References
Bangladesh	Rice (*Oryza sativa*) 0.058–1.835 mg/kg	Meharg, A. A., & Rahman, M. M. (2003). Arsenic contamination of Bangladesh paddy field soils: implications for rice contribution to arsenic consumption. *Environmental science & technology*, 37(2), 229–234
Bangladesh	Arum leaves (*Colocasia antiquorum*) 0.09–3.99 mg/kg	Das, H. K., Mitra, A. K., Sengupta, P. K., Hossain, A., Islam, F., & Rabbani, G. H. (2004). Arsenic concentrations in rice, vegetables, and fish in Bangladesh: a preliminary study. *Environment International*, 30(3), 383–387
Bangladesh	Potato (*Solanum tuberosum*) 0.07–1.39 mg/kg	Das, H. K., Mitra, A. K., Sengupta, P. K., Hossain, A., Islam, F., & Rabbani, G. H. (2004). Arsenic concentrations in rice, vegetables, and fish in Bangladesh: a preliminary study. *Environment International*, 30(3), 383–387
Bangladesh	Water Spinach (*Ipomoea reptans*) 0.1–1.53 mg/kg	Das, H. K., Mitra, A. K., Sengupta, P. K., Hossain, A., Islam, F., & Rabbani, G. H. (2004). Arsenic concentrations in rice, vegetables, and fish in Bangladesh: a preliminary study. *Environment International*, 30(3), 383–387
Bangladesh	Bean (*Phaseolus vulgaris*) 0.52 mg/kg fresh weight	Islam, M., Ahmed, M., & Habibullah-Al-Mamun, M. (2016). Apportionment of heavy metals in soil and vegetables and associated health risks assessment. *Stochastic environmental research and risk assessment*, 30(1), 365–377
Bangladesh	Lata fish (*Ophicephalus punctatus)* 0.02–0.04 mg/kg	Das, H. K., Mitra, A. K., Sengupta, P. K., Hossain, A., Islam, F., & Rabbani, G. H. (2004). Arsenic concentrations in rice, vegetables, and fish in Bangladesh: a preliminary study. *Environment International,* 30(3), 383–387
West Bengal, India	Potato (*Solanum tuberosum*) 0.456 mg/kg	Rahaman, S., Sinha, A. C., Pati, R., & Mukhopadhyay, D. (2013). Arsenic contamination: a potential hazard to the affected areas of West Bengal, India. *Environmental geochemistry and health*, *35*(1), 119–132

(continued)

Table 18.3 (continued)

Country	Arsenic concentration	References
West Bengal, India	Rice (*Oryza sativa*) 0.429 mg/kg	Rahaman, S., Sinha, A. C., Pati, R., & Mukhopadhyay, D. (2013). Arsenic contamination: a potential hazard to the affected areas of West Bengal, India. *Environmental geochemistry and health, 35*(1), 119–132
Punjab, India	Rice (*Oryza sativa*) 1.01–13.95 µg/kg	Sidhu, S. S., Brar, J. S., Biswas, A., Banger, K., &Saroa, G. S. (2012). Arsenic contamination in soil–water–plant (Rice, Oryza sativa L.) continuum in central and sub-mountainous Punjab, India. *Bulletin of environmental contamination and toxicology, 89*(5), 1046–1050
Taiwan	Rice (*Oryza sativa*) 0.063–0.2 mg/kg	Sandhi, A., Yu, C., Rahman, M. M., & Amin, M. (2022). Arsenic in the water and agricultural crop production system: Bangladesh perspectives. *Environmental Science and Pollution Research*, 1–13
China	Rice (*Oryza sativa*) 0.283–0.725 mg/kg	Xie, Z. M., & Huang, C. Y. (1998). Control of arsenic toxicity in rice plants grown on an arsenic-polluted paddy soil. *Communications in Soil Science and Plant Analysis, 29*(15–16), 2471–2477
Vietnam	Rice (*Oryza sativa*) 0.032–0.465 mg/kg	Phuong, T., Chuong, P., & áTongKhiem, D. (1999). Elemental content of Vietnamese ricePart 1. Sampling, analysis and comparison with previous studies. *Analyst, 124*(4), 553–560
Egypt	Rice (*Oryza sativa*) 0.01–0.58 mg/kg	Meharg, A. A., Williams, P. N., Adomako, E., Lawgali, Y. Y., Deacon, C., Villada, A., … &Yanai, J. (2009). Geographical variation in total and inorganic arsenic content of polished (white) rice. *Environmental science & technology, 43*(5), 1612–1617
Thailand	Rice (*Oryza sativa*) 0.01 0.39 mg/kg	Meharg, A. A., Williams, P. N., Adomako, E., Lawgali, Y. Y., Deacon, C., Villada, A., … &Yanai, J. (2009). Geographical variation in total and inorganic arsenic content of polished (white) rice. *Environmental science & technology, 43*(5), 1612–1617

(continued)

Table 18.3 (continued)

Country	Arsenic concentration	References
La Villa river basin, Panama	Sugar cane (*Saccharum officinarum*) 162.5 mg/kg	Núñez, J. V., Santo Pineda, A., Pérez, J. V., & Zachrisson, I. R. (2021). Assessment of Heavy Metal Concentration in Water, Soils and Sediments of La Villa River Basin, Panama. *Recent Progress in Plant and Soil Research Vol. 4*, 56–
Uruguay	Rice (*Oryza sativa*) 0.015–0.629 mg/kg	Roel, A., Campos, F., Verger, M., Huertas, R., & Carracelas, G. (2022). Regional variability of arsenic content in Uruguayan polished rice. *Chemosphere*, *288*, 132,426
Uruguay	Rice (*Oryza sativa*) 0.07 mg/kg	Carracelas, G., Hornbuckle, J., Verger, M. E. L. I. S. S. A., Huertas, R. A. Q. U. E. L., Riccetto, S., Campos, F., & Roel, A. L. V. A. R. O. (2019). Irrigation management and variety effects on rice grain arsenic levels in Uruguay. *Journal of Agriculture and Food Research*, *1*, 100,008
Venezuela	Rice (*Oryza sativa*) 0.062–0.089 mg/kg	Bundschuh, J., Armienta, M. A., Morales-Simfors, N., Alam, M. A., López, D. L., Delgado Quezada, V., … & Ahmad, A. (2021). Arsenic in Latin America: New findings on source, mobilization and mobility in human environments in 20 countries based on decadal research 2010–2020. *Critical reviews in environmental science and technology*, *51*(16), 1727–1865
San Pedro de Atacama, Chile	Beet (*Beta vulgaris*) 0.520 mg/kg Radish (*Raphanus sativus*) 0.938 mg/kg Cabbage (*Brassica oleracea*) 0.715 mg/kg	Sancha, A. M., & O'Ryan, R. (2008). Managing hazardous pollutants in Chile: arsenic. *Reviews of Environmental Contamination and Toxicology Vol 196*, 123–146

potable purposes in a UNICEF survey in 2005, is still being used incessantly for irrigating cultivable fields in the area. For evaluating arsenic accumulation in paddy crops, several studies have been carried out which include, but are not limited to hydroponics, greenhouse pot experiments using arsenic spiked soils and/or irrigating rice plants with arsenic contaminated waters (Srivastava et al., 2013). Norton et al. (2009) have suggested developing knowledge of paddy grain arsenic content across multiple environments under field conditions. The relationship between soil properties and soil arsenic content is also necessary to be understood for identifying key soil properties influencing soil arsenic content under field conditions. However, data on the arsenic accumulation in the soil and paddy crop with special reference to actual field conditions is not sufficient and is urgently required to reveal the spread of the arsenic problem in the region (Srivastava et al., 2013). None of the studies have included dedicated field studies on the accumulation of arsenic in agricultural soil and its accumulation by different crop varieties in arsenic-tainted regions. As mentioned earlier, the regular use of arsenic-laden groundwater for irrigation was contributing to a significant build-up of arsenic in the paddy fields and contaminating different rice varieties grown in the region. Some key soil properties have a profound effect on the bioavailability of soil arsenic and may regulate soil arsenic content under field conditions. Taking all the considerations in mind the goal should be to find the total and bioavailable arsenic contents of the agricultural soils and the factors governing its bioavailability to crops in order to select appropriate crops that can be grown in the arsenic-contaminated soil. The monitoring of arsenic content in different crops and vegetables being cultivated in the arsenic affected region is also necessary along with the regular monitoring of the arsenic contents in irrigation groundwater sources, assessment of the role the site specific key soil properties influencing arsenic build-up in soils (Singh & Srivastava, 2020).

There is a requirement to identify hot spots in arsenic contaminated regions based on published reports of potable water, piezometric heads of groundwater, and India Mark-II handpumps. Furthermore, extensive grid-based sampling along with geotagging of field sampling sites should be done, which may lead to the development of a web portal of a geo-referenced database of arsenic mapping in the contaminated region. The arsenic reconnaissance in different crop seasons can be done by sampling food crops, vegetables, spices, fruits, and legumes, and any other standing crops in arsenic contaminated agricultural fields and then analyzing them for the total arsenic concentration. The sampling and analysis of the groundwater used for irrigation and the agricultural soil being irrigated should be done in arsenic prone areas. The samples of feedwater, feedstock and bovine produce should also be drawn wherever the groundwater is also being served for animal husbandry. This is equally important as arsenic may enter the food chain through contaminated bovine products also. Detailed soil characterization with respect to physicochemical, microbiological, and enzymatic parameters may give an insight into the factors affecting arsenic loading in soil and its bioavailability to crop plants.

Accumulation of arsenic by rice plants has been associated with several factors viz., soil properties, biogeochemistry, rice cultivars, etc. Estimating the total soil arsenic content is not sufficient and inappropriate to indicate the actual bioavailable fraction of arsenic in the soil and its corresponding toxicity (Huang et al., 2012). Therefore, it is important to understand arsenic fractionation in the soil to better estimate arsenic bioavailable fraction, which is the one responsible for the toxicity in the rice plants (Singh & Srivastava, 2020). The bioavailability of arsenic in soil is influenced by different physico-chemical characteristics of soil viz., pH, clay content, organic matter (OM), redox potential (Eh) (Zou et al., 2018), and presence of hydro(oxide)s of metals, organic carbon content, and phosphate in close association with rice cultivars (Meharg & Zhao, 2012).

In arsenic-contaminated soils, murate of potash, triple superphosphate, zinc sulphate, ferrous sulphate, and ferric chloride can be applied to reduce soil arsenic pollution (Hwang & Neculita 2013). Other management practices including irrigation water management, application of fertilizers, and advocacy to grow low arsenic accumulating (arsenic-safe) rice cultivars are also suggested to mitigate arsenic pollution in soil and crops (Mukherjee et al., 2017). Effects of individual applications of iron, zinc, phosphorus, and silicon on arsenic uptake in rice have also been studied, but mostly at the lab experiment level (Arco-Lázaro et al., 2018).

Properties of soil, such as pH, clay, organic matter, redox potential, phosphorus, and sulphur, are reported as the most important factors influencing arsenic uptake in crops and vegetables. Specifically, the arsenic accumulation in vegetables was observed to be in the subsequent order: leafy > root > tubers. The soil property thresholds for safer cultivation of vegetables in arsenic-contaminated regions may be assessed by using nonlinear models between arsenic bioconcentration factor (BCF) and the specific soil property. Further, variance partitioning analysis (VPA) and partial least squares (PLS) path modeling can be applied to study the correlated effect of the conforming key soil characteristics that influenced most of the BCF of arsenic. While, studying arsenic contamination in crops/vegetables, evaluation of the estimated daily intake and the target hazard quotient as μg/kg/day can be done for children and adults in the affected regions. It is required to find out unsafe crops/vegetables in terms of subsistence maximum daily tolerable dietary intake (MTDI) by human beings according to the regulatory standards of WHO/FAO. Excessive accumulation of arsenic in rice may lead to a loss in crop production and disease like 'straight-head disease' with empty panicle in rice, this may eventually lead to forfeiture of yield and economic losses.

Processes such as bioremediation depend on the natural capabilities of microbes and plants to mitigate pollutants and are accepted as the standard, safe, environmentally compatible practice to restore heavy metal-contaminated soils. Bioremediating agents, on one hand, can carry out their usual life functions in respective stress conditions and may metabolize chemical pollutants as their energy sources by biotransforming these contaminants to harmless end products on other hand. These agents can be used as a substitute to mitigate harmful contaminants through biological activities. Bioremediation is effective compared to conventional methods, and is also eco-friendly and cost-effective (Adams et al., 2015; Singh & Prasad, 2015; Singh et al. 2015). Bioremediation involves (i) biotransformation to lesser harmfulentities; (ii) complete mineralization; and (iii) immobilization. Soil microbes (both bacteria and fungi) are the main agents in several detoxification mechanisms for remediating arsenic contamination e.g., oxidation–reduction, efflux, biosorption, bioaccumulation, and biomethylation (Srivastava et al., 2011). For instance, for *in-situ* bioremediation of heavy metal(loid)s tainted soils, two strategies maybe proposed namely, biostimulation and bioaugmentation (Lebeau et al., 2011). These authors have suggested to conduct field trials using laboratory-tested novel microbial strains. Appropriate technology for *in-situ* arsenic remediation of contaminated soil is urgently needed, especially in the Indian context. Many strategies are being used to mitigate arsenic contamination, which include developing arsenic-tolerant rice varieties through breeding, the development of transgenics, and phytoextraction. However, these methods have their limitations. Therefore, lacunae exist in achieving the desired targets. Different strains of bacteria viz. *Pseudomonas, Xanthomonas, and Flavobacterium* are known forarsenic methylation due tothe presence of a gene encoding arsenite S-adenosyl-L-methionine (SAM) methyltransferase, which catalyzes the transfer of a methyl group from SAM to arsenite, thus converting the more toxic inorganic form to a relatively less toxic organic form (Qin et al., 2006). Microbes such as *Actinobacteria, Microbacterium, Pseudomonas,* and *Rhizobium*with arsenite oxidizing activity as a mode of arsenite detoxification have also been reported (Paul et al., 2014). Fungal species such as *Aspergillus, Aureobasidium, Cladosporium, Funalia, Pleurotus, Ganoderma,* and *Penicillium* were reported for their heavy metal remediation potential (Naseem et al., 2021). Verma et al. (2016) used *S. cerevisiae* as a model organism for validating the methyltransferase gene function isolated from the soil fungi *Westerdykella aurantiaca.* Bakers' yeast, *S. cerevisiae* has been reported for bioremediation of arsenic contaminated water (Roy et al., 2013). The role of plant growth-promoting microbes in enhancing the productivity of crops under contaminated soil conditions has been studied for several crops (Tripathi et al., 2013; Armendariz et al., 2015). The bioremediation project for the biological removal of soil arsenic generally comprises a microbial formulation consisting of rigorously tested bacterial, fungal, and yeast strains with arsenic biotransformation ability for reducing arsenic uptake in rice by reducing its bioavailability in plant-soil interface. Certain nutrient amendments may also limit arsenic uptake in plants by enhancing the bioavailability and accumulation efficiency of some important mineral elements (viz., phosphate, silicon, sulphate, etc.) in turn reducing the uptake/accumulation of arsenic in plants. These can be used to reduce the arsenic uptake in the susceptible crop (Mitra et al., 2017). In general, the bioavailability of arsenic in soil is influenced by different physicochemical characteristics of soil viz., pH, clay content, organic matter (OM), redox potential (Eh) (Zou et al., 2018) and presence of hydro(oxide)s of metals, organic carbon content and phosphate in close association with particular crop cultivars (Meharg & Zhao, 2012).

Developing countries should address some specific challenges in the context such as improving mitigation approaches with reference to arsenic contamination in groundwater and irrigated ecosystems. This will intend state-of-the-art focused research on sustainable integrated solutions to As-contaminated water and food derived from its use for irrigation. The most important challenge is to create awareness regarding the existing and new knowledge/technologies to counter the arsenic problem in groundwater through education, training, and regional networking.

Groundwater Arsenic in Agriculture: Perspectives of Regional Efforts and Roadmap for Solution

Arsenic has emerged as a food chain contaminant. Accumulation of arsenic in soils and crops due to the application of arsenic-contaminated groundwater for irrigation has twisted instant apprehensions worldwide. The building-up of arsenic in agricultural soils may ultimately lead to arsenic accumulation in the edible parts of crops and vegetables, facilitating increased doses of arsenic being ingested by the local inhabitants of the affected region. It is highly warranted to initiate monitoring and assessment programmes for arsenic contamination of irrigation groundwater sources, agricultural fields being irrigated using those sources and the crops grown in those fields. These programmes might be fruitful in exploring the threats of arsenic contamination in the newer areas still unexplored for the purpose. The data from spiked soil and hydroponics experiments cannot be extrapolated as the true representative of the field condition of arsenic contamination of soil and crops, therefore extensive field trials are warranted. An understanding of soil arsenic content and other soil properties (physicochemical, biochemical and microbiological) in relation to arsenic accumulation and toxicity to crops and vegetables is of utmost importance therefore, such studies are needed to be conducted urgently under the field conditions. A good correlation between arsenic uptake in plants/crops and total arsenic content in soils is not always found. It is revealed that soil total arsenic content is not likely a good predictor of arsenic uptake and toxicity under different soil types and micro climatic conditions. It is the bioavailable fraction of total arsenic in specific soil conditions, which is potentially causing a threat to the crops/plants. As per FAO, the arsenic induced phytotoxicity is likely to be observed at 20 mg arsenic kg^{-1} soil. In the case of paddy cultivation with reference to soil arsenic pollution, representative data of arsenic toxicity are required under non-flooded, intermittently flooding and flooded soil conditions to ascertain the role of arsenic in limiting paddy growth and yield under different flooding conditions. It is necessary to sufficiently understand and quantify the factors determining soil arsenic build-up, regulating its bioavailability and toxicity to crops and vegetables in the field. Identification of geographic areas either presently contaminated with arsenic or susceptible to its contamination in near future, is an important step for risk assessment and developing remediation strategies. The mapping of arsenic content which is being transported from soil or irrigation water to edible plant parts is important for protecting the human population against arsenic exposure through contaminated food and water. Assessment of soil arsenic bioavailability may greatly support the scope of remediation options required at the contaminated sites. It is concluded that future R&D should be focused on speciation, biotransformation and bioavailability of arsenic in soil and associated soil properties to reveal tangible latent hazards of soil arsenic contamination to agriculture and related food chain links.

For instance, the arsenic content in rice grains depends on arsenic levels in irrigation groundwater, soils, and rice varieties. Soil arsenic content is influenced by different soil properties, especially soil carbon, phosphorus, sulfur, texture, macronutrients, oxidoreductases, and hydrolases (Hossain et al., 2008; Singh et al., 2022). The arsenic accumulating rice varieties that are currently and popularly grown in the arsenic affected regions might not be a suitable option for human consumption owing to the high arsenic contamination of the grains and switching to low arsenic accumulating rice varieties or arsenic bioremediation of the contaminated soil might be the only option. The magnitude of the arsenic contamination in the water-soil-paddy continuum in the arsenic contaminated region needs to be studied. The arsenic contamination of crops and vegetables in the arsenic contaminated region is affected not only by genotypes of crops and vegetables, but also by the geology, geography, and the site-specific environmental factors. It is suggested that regular arsenic monitoring studies should be conducted by the regulatory agencies to determine the increasing magnitude of arsenic accumulation in the agricultural soils and the crop produces.

However, more studies are needed to evaluate the use of soil amendments in different combinations for minimizing the uptake of arsenic in crop plants (De La Fuente et al., 2010). Given the chemical interaction and affinity of different nutrients with arsenic, it can be postulated that they may synergistically reduce bioavailable fractions of soil arsenic. It can be studied whether various commercially available nutrient amendments could reduce the bioavailability of arsenic in soil individually and in combination to prove the hypothesis. The corresponding decrease in arsenic accumulation in different rice plant parts linked to the reduction in different bioavailable fractions of soil was also studied. The study also aimed to examine the changes in different soil physicochemical properties upon the application of different nutrient amendments, and discuss their effect on the contents of different soil arsenic fractions and plant arsenic contents.

Crops like rice growing on arsenic contaminated soil eventually accumulate arsenic in their edible parts. The amount of arsenic accumulating in the plants is governed by the contents of the bioavailable fraction of soil as. Its bioavailability is governed by various soil properties viz., pH, nutrient availability, and their interactions. Soil nutrients viz., zinc, sulphur, iron, calcium, silicon, potassium, nitrogen, and phosphorus do interact with arsenic chemically and affect its bioavailability to the plant. Arsenic levels in roots of plants might be strongly dependent on the bioavailable content of soil arsenic. The addition of amendments containing these nutrients may reduce the arsenic content in edible as well as in other plant parts. Application of these selected amendments in the appropriate doses may efficiently reduce arsenic contents in plant parts, and subsequently improve plant growth. This will, in turn, alleviate human health risks via the consumption of arsenic contaminated crop/vegetable produces/eatables.

Rhizospheric manipulation of plant roots using microbes with arsenic detoxification potentials can be one of the effective solutions to mitigate contamination of arsenic in the soil–plant interface. For example, fungi may serve as a sink to bioaccumulate and biosorbarsenic along with biomethylation and bioconversion of arsenic. A consortium of novel soil fungal strains having capabilities to detoxify arsenic can be exploited to lower the bioavailable soil arsenic fraction in the plant root zone. This reduction of bioavailable soil arsenic may facilitate a reduction in the contents of arsenic in edible plant parts and may lead to safe crop cultivation in arsenic contaminated agricultural fields under in-situ conditions.

The perspectives of regional collaboration will aim at addressing the arsenic menace in groundwater by linkages among the institutions and stakeholders capable of providing sustainable long term solutions with substantial social impact and relevance. The long term solutions and strategies would require joint interdisciplinary R&D, capacity building, and effective awareness through a diverse and integrated regional network and partnership about tackling groundwater arsenic threats in agriculture.

Conclusion

Worldwide, arsenic is considered the most perilous natural geogenic toxicant due to its highly toxic effects on living organisms including the human population. Markedly, new regions and nations from where the arsenic contamination problem was not been reported previously are now being reported to have arsenic contamination. Accordingly, the number of the affected human population keeps growing. Exploration of the potential sources of highly arsenic contaminated groundwater and tainting agricultural soil and crop production with arsenic above the thresholds is a massive task but highly required as well. There is a need to develop a regional network of stakeholders to deal with scientific, social, economic, and organizational challenges. Not only the spatial and temporal variations of arsenic contents in the groundwater and its correlation to the age of the source but also the intensity of the use of the source needs to be monitored. This may yield a database of probable loadings of arsenic from groundwater to agricultural soils, and further in crops and vegetables. The comprehensive risk assessment and subsequent remediation solutions will keep the human population residing in areas prone to groundwater arsenic contamination and its loading onto the agricultural soil safe. The challenge lies in developing the best solutions which could be suitable for mitigating site-specific problems of groundwater arsenic to agriculture and safeguarding the food chain contamination. The complexity of arsenic fate and transfer in the irrigation groundwater-agricultural soil-crop continuum is high which poses difficulty to offer a common solution to solve the problem. The administration of a combination of solutions would be more feasible, applicable, and acceptable to site-specific social and economic conditions rather than any single approach. The selection of appropriate crop varieties, nutrient amendments, increase in dependence on surface water for irrigation, rainwater management, application of feasible and effective bioremediation measures, etc. are quite a few sustainable approaches to this problem. This highlights the requirement of a regional network for collaboration between academia of the developed, developing, and underdeveloped countries to resolve this mammoth task.

Acknowledgements The author is thankful to the Director, CSIR-National Botanical Research Institute, Lucknow—INDIA and his peer group for the support of the study. Author, Ms. Mariya Naseem is thankful to the CSIR HRDG for the award of senior research fellowship (31/008/(0391)/2019-EMR-I).

References

Adams, G. O., Fufeyin, P. T., Okoro, S. E., & Ehinomen, I. (2015). Bioremediation, biostimulation and bioaugmention: A review. *Int J Environ Bioremediation Biodegradation, 3*(1), 28–39.

Ahamed, S., Sengupta, M. K., Mukherjee, A., Hossain, M. A., Das, B., Nayak, B., Pal, A., Chandra, S. M., Pati, S., Nath, D. R., Chatterjee, G., Mukherjee, A., Srivastava, R., & Chakraborti, D.

(2006). Arsenic groundwater contamination and its health effects in the state of Uttar Pradesh (UP)in upper and middle Ganga plain, India: A severe danger. *Science of the Total Environment, 370*(2–3), 310–322.
Arco-Lázaro, E., Pardo, T., Clemente, R., & Bernal, M. P. (2018). Arsenic adsorption and plant availability in an agricultural soil irrigated with As-rich water: Effects of Fe-rich amendments and organic and inorganic fertilisers. *Journal of Environmental Management, 209*, 262–272.
Armendariz, A. L., Talano, M. A., Oller, A. L., Medina, M. I., & Agostini, E. (2015). Effect of arsenic on tolerance mechanisms of two plant growth-promoting bacteria used as biological inoculants. *Journal of Environmental Sciences, 33*, 203–210.
Banerjee, M., Banerjee, N., Bhattacharjee, P., Mondal, D., Lythgoe, P. R., Martínez, M., Pan, J., Polya, D. A., & Giri, A. K. (2013). High arsenic in rice is associated with elevated genotoxic effects in humans. *Science and Reports, 3*, 2195.
Barrachina, A. C., Carbonell, F. B., & Beneyto, J. M. (1995). Arsenic uptake, distribution, and accumulation in tomato plants: Effect of arsenite on plant growth and yield. *Journal of Plant Nutrition, 18*(6), 1237–1250.
Bhattacharya, P., Welch, A. H., Stollenwerk, K. G., McLaughlin, M. J., Bundschuh, J., & Panaullah, G. (2007). Arsenic in the environment: Biology and Chemistry. *Science of the Total Environment, 379*, 109–120.
Bhattacharyya, R., Chatterjee, D., Nath, B., Jana, J., Jacks, G., & Vahter, M. (2003). High arsenic ground water: Mobilization, metabolism and mitigation_an overview in the Bengal Delta Plain. *Molecular and Cellular Biochemistry, 253*(1–2), 347–355.
Bobrowicz, P., Wysocki, R., Owsianik, G., Goffeau, A., & Ułaszewski, S. (1997). Isolation of three contiguous genes, ACR1, ACR2 and ACR3, involved in resistance to arsenic compounds in the yeast *Saccharomyces cerevisiae. Yeast, 13*(9), 819–828.
Brammer, H., & Ravenscroft, P. (2009). Arsenic in groundwater: A threat to sustainable agriculture in South and South-east Asia. *Environment International, 35*, 647–654.
Bundschuh, J., Nath, B., Bhattacharya, P., Liu, C. W., Armienta, M. A., Moreno López, M. V., Lopez, D. L., Jean, J. S., Cornejo, L., Lauer Macedo, L. F., & Filho, A. T. (2012). Arsenic in the human food chain: The Latin American perspective. *Science of the Total Environment, 429*, 92–106.
Casentini, B., Hug, S. J., & Nikolaidis, N. P. (2011). Arsenic accumulation in irrigated agricultural soils in Northern Greece. *Science of the Total Environment, 409*, 4802–4810.
Das, H. K., Mitra, A. K., Sengupta, P. K., Hossain, A., Islam, F., & Rabbani, G. H. (2004). Arsenic concentrations in rice, vegetables, and fish in Bangladesh: A preliminary study. *Environment International, 30*, 383–387.
De La Fuente, C., Clemente, R., Alburquerque, J. A., Vélez, D., & Bernal, M. P. (2010). Implications of the use of as-rich groundwater for agricultural purposes and the effects of soil amendments on as solubility. *Environmental Science and Technology, 44*, 9463–9469.
Hite, A. H. (2013). Arsenic and rice: A call for regulation. *Nutrition, 29*, 353–354.
Hossain, M. F. (2006). Arsenic contamination in Bangladesh—an overview. *Agriculture, Ecosystems & Environment, 113*, 1–16.
Hossain, M. B., Jahiruddin, M., Panaullah, G. M., Loeppert, R. H., Islam, M. R., & Duxbury, J. M. (2008). Spatial variability of arsenic concentration in soils and plants, and its relationship with iron, manganese and phosphorus. *Environmental Pollution, 156*, 739–744.
Hua, B., Yan, W., Wang, J., & Deng, B. J. Y. (2011). Arsenic accumulation in rice grains: Effects of cultivars and water management practices. *Environmental Engineering Science, 28*, 591–596.
Huang, H., Jia, Y., Sun, G. X., & Zhu, Y. G. (2012). Arsenic speciation and volatilization from flooded paddy soils amended with different organic matters. *Environmental Science and Technology, 46*, 2163–2168.
Hughes, M. F., Beck, B. D., Chen, Y., Lewis, A. S., & Thomas, D. J. (2011). Arsenic exposure and toxicology: A historical perspective. *Toxicological Sciences, 123*(2), 305–332.

Huhmann, B., Harvey, C. F., Uddin, A., Choudhury, I., Ahmed, K. M., Duxbury, J. M., Ellis, T., & van Geen, A. (2019). Inversion of high-arsenic soil for improved rice yield in Bangladesh. *Environmental Science and Technology, 53*, 3410–3418.

Hussain, M. M., Bibi, I., Niazi, N. K., Shahid, M., Iqbal, J., Shakoor, M. B., Ahmad, A., Shah, N. S., Bhattacharya, P., Mao, K., Bundschuh, J., Ok, Y. S., & Zhang, H. (2021). Arsenic biogeochemical cycling in paddy soil-rice system: Interaction with various factors, amendments and mineral nutrients. *Science of the Total Environment, 773*, 145040.

Hwang, T., & Neculita, C. M. (2013). In situ immobilization of heavy metals in severely weathered tailings amended with food waste-based compost and zeolite. *Water, Air, and Soil Pollution, 224*, 1–9.

Kumar, P. R., Chaudhari, S., Khilar, K. C., & Mahajan, S. P. (2004). Removal of arsenic from water by electrocoagulation. *Chemosphere, 55*(9), 1245–1252.

Lebeau, T. (2011). Bioaugmentation for in situ soil remediation: How to ensure the success of such a process. In A. Singh, N. Parmer, & R. C. Kuhad (Eds.), *Bioaugmentation, biostimulation and biocontrol* (pp. 129–186). Springer.

Mandal, B. K., & Suzuki, K. T. (2002). Arsenic round the world: A review. *Talanta, 58*(1), 201–235.

Marin, A. R., Pezeshki, S. R., Masschelen, P. H., & Choi, H. S. (1993). Effect of dimethylarsenic acid (DMAA) on growth, tissue arsenic, and photosynthesis of rice plants. *Journal of Plant Nutrition, 16*(5), 865–880.

Meharg, A. A., & Zhao, F. J. (2012). Arsenic & rice. Springer Science & Business Media, Springer Netherlands.

Meharg, A. A., & Jardine, L. (2003). Arsenite transport into paddy rice (*Oryza sativa*) roots. *New Phytologist, 157*, 39–44.

Mitra, A., Chatterjee, S., Moogouei, R., & Gupta, D. K. (2017). Arsenic accumulation in rice and probable mitigation approaches: a review. *Agronomy, 7*, 67.

Mukherjee, A., Kundu, M., Basu, B., Sinha, B., Chatterjee, M., Bairagya, M. D., ··, U. K., & Sarkar, S. (2017). Arsenic load in rice ecosystem and its mitigation through deficit irrigation. *Journal of Environmental Management, 197*, 89–95.

Naseem, M., Raghuwanshi, R., Verma, P. C., & Srivastava, P. K. (2021). Mycoremediation—Effective strategy to ameliorate arsenic toxicity. In V. K. Sharma, M. P. Shah, S. Parmar, & A. Kumar (Eds.), *Fungi bio-prospects in sustainable agriculture* (pp. 433–458). Academic Press.

Norton, G. J., Duan, G., Dasgupta, T., Islam, M. R., Lei, M., Zhu, Y., Deacon, C. M., Moran, A. C., Islam, S., Zhao, F. J., Stroud, J. L., McGrath, S. P., Feldmann, J., Price, A. H., & Meharg, A. A. (2009). Environmental and genetic control of arsenic accumulation and speciation in rice grain: Comparing a range of common cultivars grown in contaminated sites across Bangladesh, China, and India. *Environmental Science and Technology, 43*, 8381–8386.

Norton, G. J., Pinson, S. R. M., Alexander, J., Mckay, S., Hansen, H., Duan, G. L., Rafiqul Islam, M., Islam, S., Stroud, J. L., Zhao, F. J., McGrath, S. P., Zhu, Y. G., Lahner, B., Yakubova, E., Guerinot, M. L., Tarpley, L., Eizenga, G. C., Salt, D. E., Meharg, A. A., & Price, A. H. (2012). Variation in grain arsenic assessed in a diverse panel of rice (*Oryza sativa*) grown in multiple sites. *New Phytologist, 193*, 650–664.

Ordonez, E., Letek, M., Valbuena, N., Gil, J. A., & Mateos, L. M. (2005). Analysis of genes involved in arsenic resistance in Corynebacterium glutamicum ATCC 13032. *Applied and Environment Microbiology, 71*(10), 6206–6215.

Panaullah, G. M., Alam, T., Hossain, M. B., Richard, H., Loeppert, J., Lauren, G., Meisner, C. A., Ahmed, Z. U., & Duxbury, J. M. (2009). Arsenic toxicity to rice (*Oryza sativa* L.) in Bangladesh. *Plant and Soil, 317*, 31–39.

Patra, M., Bhowmik, N., Bandopadhyay, B., & Sharma, A. (2004). Comparison of mercury, lead and arsenic with respect to genotoxic effects on plant systems and the development of genetic tolerance. *Environmental and Experimental Botany, 52*(3), 199–223.

Paul, D., Poddar, S., & Sar, P. (2014). Characterization of arsenite-oxidizing bacteria isolated from arsenic-contaminated groundwater of West Bengal. *Journal of Environ Sci Health A Tox Hazard Subst Environ Eng, 49*(13), 1481–1492.

Qin, J., Rosen, B. P., Zhang, Y., Wang, G., Franke, S., & Rensing, C. (2006). Arsenic detoxification and evolution of trimethylarsine gas by a microbial arsenite S-adenosylmethionine methyltransferase. *Proceedings of the National Academy of Sciences of the United States of America, 103*, 2075–2080.

Rahaman, S., Sinha, A. C., Pati, R., & Mukhopadhyay, D. (2013). Arsenic contamination: a potential hazard to the affected areas of West Bengal, India. *Environ Geochem Health, 35*(1), 119–132.

Roy, D., Gaur, P., Verma, N., Pathireddy, M., & Singh, K. P. (2013). Bioremediation of arsenic (III) from water using baker yeast *Saccharomyces cerevisiae. Int J Environ Bioremediat Biodegrad, 1*(1), 14–19.

Sidhu, S. S., Brar, J. S., Biswas, A., Banger, K., & Saroa, G. S. (2012). Arsenic contamination in soil-water-plant (rice, *Oryza sativa* L.) continuum in central and sub-mountainous Punjab, India. *Bull Environ Contam Toxicol, 89*, 1046–1050.

Singh, A., & Prasad, S. M. (2015). Remediation of heavy metal contaminated ecosystem: An overview on technology advancement. *International Journal of Environmental Science and Technology, 12*(1), 353–366.

Singh, S. B., & Srivastava, P. K. (2020). Bioavailability of arsenic in agricultural soils under the influence of different soil properties. *SN Applied Sciences, 2*, 153.

Singh, M., Srivastava, P. K., Verma, P. C., Kharwar, R. N., Singh, N., & Tripathi, R. D. (2015). Soil fungi for mycoremediation of arsenic pollution in agriculture soils. *J. App. Microbiol., 119*, 1278–1290.

Singh, S. B., Naseem, M., Raghuvanshi, R., & Srivastava, P. K. (2022). Application of selected nutrient amendments to regulate soil properties for reducing arsenic accumulation in rice. *Soil Sediment Contam, 41*, 1–17.

Singh, S., Ghosh, A., Kumar, A., Kislay, K., Kumar, C., Tiwari, R., Parwez, R., Kumar, N., & Imam, M. (2014). Groundwater arsenic contamination and associated health risks in Bihar, India. *International Journal of Environmental Research, 8*(1), 49–60

Somani, S. M. (2001). *Chemical warfare agents: toxicity at low levels* (pp. 409–435). CRC Press.

Srivastava, P. K., Vaish, A., Dwivedi, S., Chakrabarty, D., Singh, N., & Tripathi, R. D. (2011). Biological removal of arsenic pollution by soil fungi. *Science of the Total Environment, 409*(12), 2430–2442.

Srivastava, P. K., Singh, M., Singh, N., & Tripathi, R. D. (2013). Soil arsenic pollution: a threat to crops. *J. Bioremed. Biodeg., 4*, 7.

Srivastava, P. K., Singh, M., Gupta, M., Singh, N., Kharwar, R. N., Tripathi, R. D., & Nautiyal, C. S. (2015). Mapping of arsenic pollution with reference to paddy cultivation in the middle Indo-Gangetic Plains. *Environmental Monitoring and Assessment, 187*(4), 198.

Srivastava, S., & Sharma, Y. K. (2013). Arsenic occurrence and accumulation in soil and water of Eastern districts of Uttar Pradesh, India. *Environ Monit Assess, 185*(6), 4995-5002.

Stoeva, N., Berova, M., & Zlatev, Z. (2005). Effect of arsenic on some physiological parameters inbean plants. *Biologia Plantarum, 49*(2), 293–296.

Talukder, A. S., Meisner, C. A., Sarkar, M. A., Islam, M. S., Sayre, K. D., Duxbury, J. M., & Lauren, J. G. (2012). Effect of water management, arsenic and phosphorus levels on rice in a high-arsenic soil-water system: II. *Arsenic Uptake. Ecotoxicol. Environ. Saf., 80*, 145–151.

Tripathi, P., Singh, P. C., Mishra, A., Chauhan, P. S., Dwivedi, S., Bais, R. T., & Tripathi, R. D. (2013). Trichoderma: A potential bioremediator for environmental cleanup. *Clean Technologies and Environmental Policy, 15*, 541–550.

Valko, M. M. H. C. M., Morris, H., & Cronin, M. T. D. (2005). Metals, toxicity and oxidative stress. *Current Medicinal Chemistry, 12*(10), 1161–1208.

Verma, S., Verma, P. K., Meher, A. K., Dwivedi, S., Bansiwal, A. K., Pande, V., Srivastava, P. K., Verma, P. C., Tripathi, R. D., & Chakraborthy, D. (2016). A novel arsenic methyltransferase gene of *Westerdykellaaurantiaca* isolated from arsenic-contaminated soil: Phylogenetic, physiological, biochemical studies and its role in arsenic bioremediation. *Metallomics, 8*, 344–353.

Zheng, M. Z., Cai, C., Hu, Y., Sun, G. X., Williams, P. N., Cui, H. J., Li, G., Zhao, F. J., & Zhu, Y. G. (2011). Spatial distribution of arsenic and temporal variation of its concentration in rice. *New Phytologist, 189*(1), 200–209.

Zou, Q., An, W., Wu, C., Li, W., Fu, A., Xiao, R., Chen, H., & Xue, S. (2018). Red mud-modified biochar reduces soil arsenic availability and changes bacterial composition. *Environmental Chemistry Letters, 16*, 615–622.

Chapter 19
Arsenic Contamination in Groundwater and Food and Remediation Measures in Myanmar

Mya Thandar Khin and Phyo Wai

Abstract Few studies have looked into how agricultural or mining activities affect rural groundwater and how industrial/urban activities affect urban groundwater. The Irrawaddy delta area has the highest likelihood of exceeding the WHO Drinking Water Quality Standard (2017) 10 μg/L, and National Drinking Water Quality Standard of Myanmar 50 μg/L for arsenic. In the Bago Region, they also reported 41 and 8% exceeding of 10 and 50 μg/L, respectively. The Gravel Sand Filter system (GSF) is the most common and a possible way to remove arsenic from groundwater which is economic and technologically feasible. Myanmar needs capacity building training of advance arsenic mitigation technologies and equipments for short and long-term training.

Keywords Arsenic · Ayeyarwaddy · Gravel sand filter · Groundwater · Tubewells

Background

The Republic of the Union of Myanmar is bordered on the west by India, Bangladesh, and the Bay of Bengal, on the north by China, on the east by Thailand and Laos, and on the south by the Andaman Sea. Myanmar has a total land area of 676,550sq.km and is divided into one union territory, seven regions and seven states. In 2006, Nay Pyi Taw (union territory) replaced Yangon as the country's capital. The Irrawaddy River, also known as the Burmese Ayeyarwaddy, is Myanmar's main river, flowing through the country's heartland. It travels from the northern highlands down the middle lowlands to the delta, where it splits. The Chindwin River, which borders the

M. T. Khin (✉)
Environmental Conservation Department (ECD), Ministry of Natural Resource and Environmental Conservation (MONREC), Naypyidaw, Myanmar
e-mail: myathanderkhin@gmail.com

P. Wai
Irrigation and Water Utilization Department (IWUMD), Ministry of Agriculture, Livestock and Irrigation (MOALI), Naypyidaw, Myanmar

B. Sen Gupta and N. Martínez-Villegas (eds.), *Arsenic Remediation of Food and Water*,
https://doi.org/10.1007/978-981-97-4764-1_19

Patkai and Naga mountains, is the greatest of its tributaries. The Irrawaddy River is 2,170 km length and 412,500 km^2 total basin area that 372,905 km^2 basin areas are located in Myanmar. The Irrawaddy River has 55.5% of its land area in Myanmar (Fig. 19.1).

Water pollution from urban sources and mining-related operations, as well as deforestation and accompanying increased runoff and soil erosion, area part of Myanmar's major environmental problems. The mineral and other natural resources are plentiful in Myanmar. For example, silver is associated with lead, zinc, and antimony in sulphide minerals in the Namtu-Bawdwin mine of the Shan Plateau. Copper is obtained from the Monywa porphyry copper deposit along the Chindwin River. Nickel deposits are associated with ophiolites that are obtained from Chin State and Katha District in Sagaing Region.

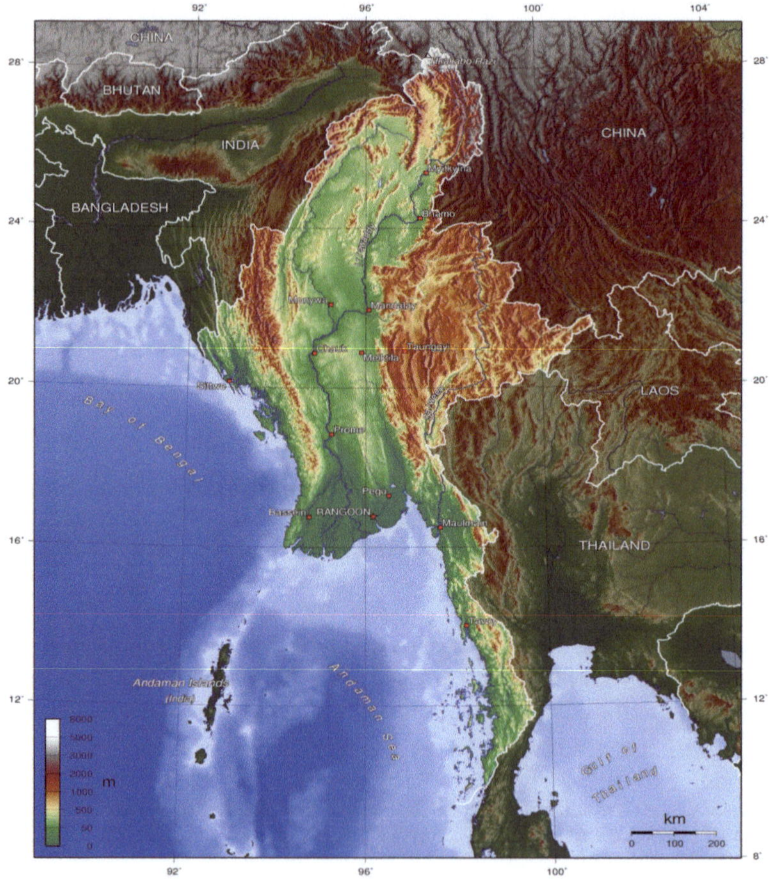

Fig. 19.1: Topographic map of Myanmar (Burma) (*Source* https://commons.wikimedia.org/wiki/ under licensee: CC BY-SS 3.0)

Introduction

The biggest issue with groundwater quality in the lower and middle Irrawaddy alluvial aquifer was contaminated by arsenic. Arsenic has no odour and taste. Therefore, arsenic is difficult to detect if it is present in food, water, or in the air without carrying out specific tests. Chronic health related issues are the most common effects of long-term exposure to high levels of Arsenic in drinking water and food. Arsenic is an extremely hazardous metalloid that raises neonatal mortality, lowers cognitive function in children, and raises adult mortality from cardiovascular disease and lung, liver, and bladder malignancies.

Over the last 50 years, people in South and Southeast Asia have made a cultural shift to drinking groundwater rather than surface water, which has resulted in a drop in the rate of infection caused by water-borne pathogens. However, due to chronic exposure to naturally occurring inorganic hazardous chemicals like arsenic, this change has resulted in an undesirable rise in diseases. To achieve Sustainable Development Goals (Goal 6 in particular) related to "Clean water and sanitation for all", it is critical to reduce exposure to harmful substances.

Testing and Results

According to the study, the Irrawaddy delta area has the highest likelihood of exceeding the WHO Drinking Water Quality Standard (2017) guideline value of 10 μg/L for arsenic. Over the last two decades, a number of groundwater arsenic assessments have been conducted in Myanmar, mainly in the Irrawaddy alluvial and deltaic plain. Water Resources Utilization Department (WRUD) and Unicef (2002) had analyzed the arsenic in groundwater data from the Sagaing, Mandalay, Bago, Ayeyarwaddy, and Yangon regions, as well as the Chin, Kayah, Shan, and Rakhine states (4969 samples), and discovered high concentrations in the Ayeyarwady, Bago, Rakhine, and Shan states. According to Occupational and Environmental Health Division of Myanmar (OEHD, 2017–2018), arsenic values more than 50 μg/L (National Drinking Water Quality Standard of Myanmar) were found in 10% of 2085 tubewell and 0.9% of dug well samples in the Ayeyarwaddy Region. In Bago Region, 41% samples breached 10 μg/L guideline value and 8% of the samples exceeded 50 μg/L (Table 19.1 and Fig. 19.2).

Tun (2003) also analyzed 1912 groundwater samples from shallow tubewells in the Irrawaddy delta, out of which 45% exceeded the WHO guideline limit of 10 μg/L and 21% exceeded 50 μg/L, while 55% had no detectable arsenic. In groundwater from tubewells <60 m deep in the southern Irrawaddy delta, Van Geen et al. (2014) also discovered arsenic in the range <0.05–630 μg/L (median 22.9 μg/L, 55 samples) from <60 m deep tube wells in the southern Irrawaddy delta. They detected minimal amounts of arsenic in less than 10 m deep tube well. WRUD (2005) reported high arsenic contents in groundwater samples (>10 μg/L) from Myaung, Amarapura,

Table 19.1 Arsenic in groundwater from tube wells tested in Ayeyarwaddy alluvial aquifer

Region/state	Township	No. of tube well	>10 μg/L (WHO drinking water quality standard 2017)		>50 μg/L (National drinking water quality standard in Myanmar)	
			No. of tube well	Percentage %	No. of tube well	Percentage %
Sagaing	Chaung-U	500	34	6.8	–	–
	Monywa	221	–	–	–	–
	Shwebo	5,556	30	0.6	1	0.02
	Wetlet	563	91	16.2	–	–
	Sagaing	1,809	264	14.6	9	0.5
	Myinmn	1,781	323	18.1	41	2.3
	Myaung	3,181	877	27.6	145	4.6
Mandalay	Kyauske	2,826	362	12.8	54	1.9
	Myingyan	614	60	9.8	17	2.8
	Mahlaing	500	102	20.4	6	1.2
	Amarapura	500	144	28.8	12	2.4
	Madayar	500	200	40	20	4
	Leway	2,782	809	29.1	63	2.3
	Sintgaing	4,650	765	16.5	82	1.8
	Myittha	6,061	933	15.4	37	0.6
	Tada-U	2,852	452	15.9	12	0.4
Magway	Yesagyo	522	96	18.4	7	1.3

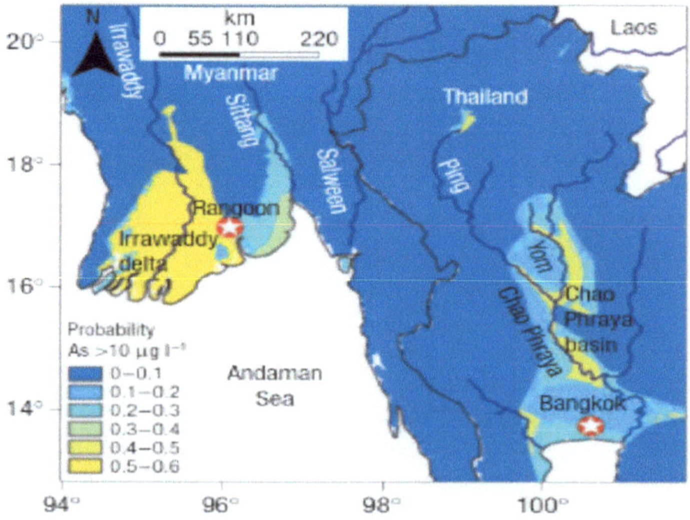

Fig. 19.2 Arsenic in groundwater with concentrations >10 μg/L from the Southern Ayeyarwaddy basin

Madayar, and Leway townships, all of which are located in the middle Irrawaddy alluvial plain.

In 2013, Kyaunggon township was found to be the most severely affected among the 17 townships in Ayeyawaddy region, with 21.56% of water sources having arsenic levels above the 50 μg/L (National Drinking Water Quality Standard of Myanmar). The GSF method, which reduces arsenic levels at a reasonable cost, was developed in this township (Table 19.2 and Fig. 19.3).

Table 19.2 Arsenic concentration data in each village

	As level (μg/L)		Number of tube well				Ratio
	min	Max	total	As≦10	10<As≦50	50<As	0% 25% 50% 75% 100%
Byike	3	21	3	2	1	0	67 / 33 / 0
Kyaunggon	13	272	7	0	5	2	0 / 71 / 29
Wetchaung	0	93	8	2	5	1	25 / 63 / 13
HleSeik	0	188	17	7	7	3	41 / 41 / 18
Chaunggyi	0	13	4	1	3	0	25 / 75 / 0
WetChanGayat	0	87	12	6	4	2	50 / 33 / 17
DoeTan	0	384	9	5	2	2	56 / 22 / 22
LayEain	0	92	7	4	2	1	57 / 29 / 14
DaungGyi	0	268	20	4	12	4	20 / 60 / 20
PaukSeinPe	0	63	9	4	4	1	44 / 44 / 11
total			96	35	45	16	36 / 47 / 17

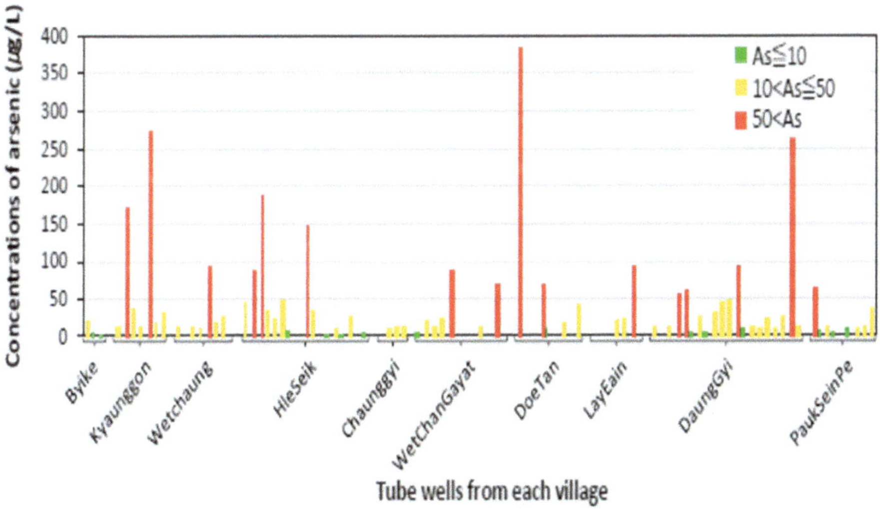

Fig. 19.3 Concentration of arsenic in water from each tube well in Kyaunggon township

In 2015, researchers tested as concentrations in 23 tube wells from Kyonpyaw Township, which is in the lower Irrawaddy. The average arsenic level of 64.7 μg/L was above the 50 μg/L (National Drinking Water Quality Standard of Myanmar).

Challenges

Myanmar needs capacity building training for advanced arsenic remediation technologies and necessary equipments to mitigate both short and long-term exposures. Financial resources are also barriers for arsenic mitigation process. Moreover, Myanmar population, in particular those living in villages require more awareness for arsenic contamination in food and water as well as its mitigation processes. More cooperation with international agencies is required to implement affordable arsenic mitigation technologies.

Conclusion

The study demonstrated the state of arsenic contamination in Myanmar and the importance of 'Gravel Sand Filter' system (GSF) by evaluating the effectiveness of GSF filtration system. It can be concluded that, in Myanmar, advanced arsenic removal technologies for application in drinking water and food production system should be adopted and implemented. More research should be done in finding the sources of arsenic contamination of groundwater and its consequences on human health and the environment.

References

An analysis of arsenic contamination in drinking water source of Ayeyarwaddy Division (UNICEF)

Kumar, A., Chen, Y., Kirby, J., Williams, M., Kookana, R., & O'Neill, C. (2017). *Ayeyarwaddy State of the Basin Assessment (SOBA)*

Bacquart, T., Frisbie, S., Mitchell, E., Grigg, L., Cole, C., Small, C., & Sarkar, B. (2015). *Multiple inorganic toxic substances contaminating the groundwater of Myingyan Township, Myanmar: Arsenic, manganese, fluoride, iron, and uranium*

van Geen, A., Win, K. H., Zaw, T., Naing, W., Mey, J. L., & Mailloux, B. (2014). *Confirmation of elevated arsenic levels in groundwater of Myanmar*

Environmental Health in Myanmar. (2018). *Occupational and environmental health division, department of public health*

Groundwater Quality: Myanmar (British Geological Survey)

Phyu, K. P., Zin, P. W., & Win, K. K. (2020). *Assessment and treatment of arsenic contaminated tube well water from selected areas in kyaunggon township, Ayeyarwady Region, Myanmar*

Water Resources Management for arsenic contamination in Ayeyarwaddy Region, Min KyawHtin, IWUMD, Ayeyarwaddy Region

Mochizuki, H., Phyu, K. P., Aung, M. N., Zin, P. W., Yano, Y., Myint, M. Z., Yamamoto, Y., Hishikawa, Y., Thant, K. Z., Maruyama, M., & Kuroda, Y. (2019). *Peripheral neuropathy induced by drinking water contaminated with low-dose arsenic in Myanmar*
Mukherjee, A., Sengupta, M. K., Hossain, M. A., Ahamed, S., Das, B. (2006). *Arsenic contamination in groundwater: a global perspective with emphasis on the Asian scenario*
Pincetti-Zúniga, G. P., Richards, L. A., Tun, Y. M., Aung, H. P. (2020). *Major and trace (including arsenic) groundwater chemistry in Central and Southern Myanmar*
Republic of the Union of Myanmar: Irrigated Agriculture Inclusive Development Project. (2016).

Chapter 20
Arsenic Toxicity: Contamination Through Groundwater and Food Chain and Role of Genetic Factors

Payel Singh, Subhamoy Bhowmick, Debashis Chatterjee, and Sreemanta Pramanik

Abstract Groundwater is one of our most precious natural resources. Many aquifers around the world have been found to be naturally contaminated with arsenic (As), a toxic metalloid and a potent human carcinogen. Chronic exposure to arsenic is harmful to human health causing skin lesions, hepatotoxicity, vascular disease, diabetes as well as cancers of various organs. High concentrations of arsenic in groundwater, more than the permissible limit (10 μg/L as per World Health Organization, WHO) have been reported from many countries of Asia, Africa, Europe, North and South America and Australia. Around 200 million people, majority of them living in South and South-East Asia, are at risk of arsenic toxicity mainly from contaminated groundwater. Minerals, such as pyrites and arsenopyrites, have a high amount of arsenic content and release it in groundwater upon breakdown under oxic environment. Several hypotheses, such as, pyrite oxidation, iron reduction, competitive exchange and geomicrobial process, have been proposed to elucidate the release mechanism and mobilization of arsenic from sediment to groundwater. Use of groundwater laden with high concentrations of arsenic for irrigation of agricultural crops and entry of arsenic in food chain including crops has been the subject of intense research in the last decade. Some recent reports on the assessment of population's dietary intake suggest that rice and water are significant sources of arsenic. Induction of an oxidative stress through the generation of reactive oxygen species is one of the principal mechanisms of arsenic toxicity. These reactive oxygen species results in DNA damage, lipid peroxidation and loss of antioxidant defense and induction of stress response genes. Polymorphisms in the arsenic metabolism and detoxification, antioxidant defense and DNA repair genes could also contribute

---

P. Singh · S. Bhowmick · S. Pramanik (✉)
Kolkata Zonal Centre, CSIR-National Environmental Engineering Research Institute, i-8 Sector-C, East Kolkata Township, Kolkata 700107, India
e-mail: sr_pramanik@neeri.res.in; sreemanta@gmail.com

P. Singh · S. Bhowmick
Academy of Scientific and Innovative Research (AcSIR), Ghaziabad 201002, India

D. Chatterjee
Department of Chemistry, University of Kalyani, Kalyani, West Bengal 741245, India

B. Sen Gupta and N. Martínez-Villegas (eds.), *Arsenic Remediation of Food and Water*,
https://doi.org/10.1007/978-981-97-4764-1_20

to the inter-individual differences in arsenic biotransformation and consequently, influence inter-individual susceptibility to the health effects of arsenic toxicity.

Keywords Arsenic · Groundwater · Exposure · Toxicity · Food chain · Biotransformation · Genes · DNA damage · Polymorphisms

Introduction

Groundwater is one of the most important natural resources for mankind. A growing global population coupled with increased demands in agricultural and industrial sectors, have led to an ever-increasing demand for groundwater. However, a large number of aquifers which act as natural reservoirs of groundwater, have been found to be contaminated with arsenic in several parts of the world. Arsenic (As) exists as a metalloid in the environment and is often referred to as "king of poisons" because of its extremely toxic nature. It is also known as a potent human carcinogen. Chronic exposure to arsenic, more than the permissible limit 10 μg/L in drinking water (as per WHO), is extremely detrimental to human health causing skin lesions, hepatotoxicity, vascular disease, diabetes as well as cancers of the skin, bladder, lung and liver. Elevated concentrations of arsenic in groundwater have been reported from many countries of Asia, Africa, Europe, North and South America and Australia. More than 200 million people, most of them living in South and South-East Asia, are at risk of arsenic toxicity from groundwater contamination. Minerals, such as pyrites and arsenopyrites, have a high amount of arsenic content and release it in groundwater upon breakdown under an oxic environment. Several hypotheses have been proposed to elucidate the release mechanism and mobilization of arsenic from sediment to groundwater, notable of them being pyrite oxidation, iron reduction, competitive exchange and geomicrobial processes. Many researches around the world have found specific links between consuming water with elevated concentration of arsenic to enhanced risk to human health. Two parameters, namely, daily water consumption and arsenic concentration in water, generally affect how much arsenic is consumed from water, and there is a lot of variation in both these parameters. Use of groundwater laden with high concentrations of arsenic for irrigation of agricultural crops and entry of arsenic in food-chain including crops has been the subject of intense research in the last decade. Some studies have shown that leafy vegetables had a higher concentration of arsenic than non-leafy vegetables and root crops. Rice is a staple diet for the populations of South and South-East Asia and some recent reports on the assessment of population's dietary intake suggest that rice and water are significant sources of arsenic. Studies also have found the link between size and color of rice grains with arsenic content in rice. Now, it has been established that arsenic in cooked rice could be influenced by the cooking method employed. Many studies have documented the genotoxic potential of inorganic arsenic (iAs) in animals and humans. Induction of an oxidative stress through the generation of reactive oxygen species is one of the principal mechanisms of arsenic toxicity. These reactive oxygen species result in DNA damage, lipid peroxidation and loss of antioxidant defense and induction of stress response genes. Polymorphisms in the arsenic metabolism and detoxification, antioxidant defense and DNA repair genes could also contribute to the inter-individual differences in arsenic biotransformation and consequently, influence inter-individual susceptibility to the health effects of arsenic toxicity.

Distribution of Arsenic in Groundwater

Groundwater (hidden resource) is essential for human survival. Globally, per capita fresh water availability, notably drinking water supply in rural areas, is rapidly declining. United Nations has emphasized on ensuring availability and supply of clean water (SDG 6). The safe and sustainable water resource and ecosystem are essential need for the survival of mankind.

Unfortunately, a large number of aquifers in several parts of the globe have now been identified with problems of arsenic contamination, with concentrations usually greater than WHO drinking water guideline values (>10 μg/L) (Chatterjee et al., 2018; Shaji et al., 2021). The occurrence of arsenic in groundwater largely varies from continent to continent and even within countries (Herath et al., 2016). The source of arsenic in groundwater has been reported to be both of geogenic and anthropogenic origin (Bhattacharya et al., 2007). It is important to note that several environments such as reducing, oxidizing, and mixed, are also linked with global distribution of groundwater arsenic occurrences. Recent studies have reported elevated levels of arsenic in groundwater in many countries of Asia, Africa, Europe, North and South America, and Australia (Shaji et al., 2021). Among these countries, the natural contamination of arsenic in groundwater has been significantly high, notably in Asia, Africa, even some parts of Europe and US (Herath et al., 2016). Geothermal, mining and volcanic related arsenic occurrence in groundwater, has also been identified in several parts of the globe.

The majority of the geogenic groundwater arsenic occurrence has been reported from South and South-East Asia (Herath et al., 2016). The most affected countries are Bangladesh, Cambodia, China, India, Pakistan, Taiwan, Thailand, Laos and Vietnam (Herath et al., 2016). Nearly 200 million people in South and South-East Asia are at risk of elevated levels of arsenic exposure (Podgorski & Berg, 2020). Generally, there are several delta plains of South and South-East Asia which have been identified to be severely affected with geogenic arsenic contamination in groundwater. Among these delta plains, the worst affected is the Bengal Delta Plain (BDP), both in terms of geographical area coverage and human exposure (Chatterjee et al., 2018; Mukherjee et al., 2006). Apart from BDP, Mekong Delta Plain (Laos, Cambodia and Vietnam) has also been identified as severely contaminated with arsenic from natural sources (Berg et al., 2007). Shallow aquifers of these plains are mostly affected with arsenic from natural sources (Olson & Chau, 2022; Richards et al., 2018; Shaji et al., 2021). In contrast, China is facing groundwater arsenic problem from both industrial effluent and natural resources (Rodríguez-Lado et al., 2013). Similarly, in Taiwan, both the

South-West and North-East coastal zone have largely been associated with arsenic problem in groundwater (Liu & Wu, 2019). In Taiwan, the "Black Foot" disease is a serious health issue associated with other typical health problems including cancer (Cheng et al., 2016). The occurrences of arsenic in groundwater have been reported from several sources such as natural grey sediments or Holocene aquifers, mining activities and geothermal water (Chatterjee et al., 2018; Shaji et al., 2021). However, there is a critical role of natural environments in releasing arsenic in groundwater. In BDP, the Holocene sediments are associated with a mixed composition of sand-silt, clay including micaceous sands, which are found to be contaminated with arsenic (Bhattacharya et al., 2011; Herath et al., 2016). On the contrary, black shale, fine sand and muds are identified from arsenic source in Taiwan (Tseng et al., 1968). In Vietnam, Holocene sediments, partly Pleistocene sediments unlike BDP are reported with arsenic contamination (Berg et al., 2008).

Several minerals, notably pyrites and arsenopyrites have been identified with high arsenic content; these minerals break down under an oxic environment, and release significant amount of arsenic in groundwater (Bhattacharya et al., 2011; Chatterjee et al., 2018). On the contrary, several metal oxides and hydroxides notably iron oxides and hydroxides, goethite, hydrous ferric oxide, and several Al-containing oxides, especially gibbsite and Mn-oxides are also acting as host for arsenic in sediments and upon favorable condition, release arsenic from sediment and groundwater (Chatterjee et al., 2018).

The arsenic concentration in groundwater varies globally, regionally and even in local village scale. There are several interplaying geochemical processes that are often contributing simultaneously for the release of arsenic in groundwater (Chatterjee et al., 2018; Shaji et al., 2021). Still, there are several co-factors that regulate such processes to large extent (Chatterjee et al., 2018). Few examples are—type of clay and its characteristics (as well as vertical and horizontal extensions); nature, type, and availability of organic matter, anthropogenic activities notably land use pattern, local recharge, sediment color etc. (Bhattacharya et al., 2011; Chatterjee et al., 2018).

Many hypotheses have been put forward to explain the release mechanism and mobilization of arsenic from sediment to groundwater (Fig. 20.1). These mechanisms are principally represented by various models such as pyrite oxidation, iron reduction, competitive exchange and geomicrobial process.

The pyrite oxidation hypothesis was preliminarily proposed to explain rich concentration of arsenic in groundwater in the shallow aquifers, notably in South and South-East Asia (Bhattacharya et al., 2011; Chatterjee et al., 2018; Herath et al., 2016). Oxidation of arsenic bearing sulfidic minerals (notably arsenopyrites, contains nearly 30% arsenic) was thought to be the primary reason for release of arsenic in the aquifers (Herath et al., 2016). It has been further hypothesized that the naturally occurring sulfidic minerals are rapidly oxidizing when they are oxidized due to the influx of oxygen in the aquifer under intensive large-scale groundwater development (Chatterjee et al., 2018; Herath et al., 2016). During the development and movement of oxygenated front downwards which makes the entry of air (atmospheric oxygen) into the aquifer and, ultimately results in the breakdown of pyritic mineral

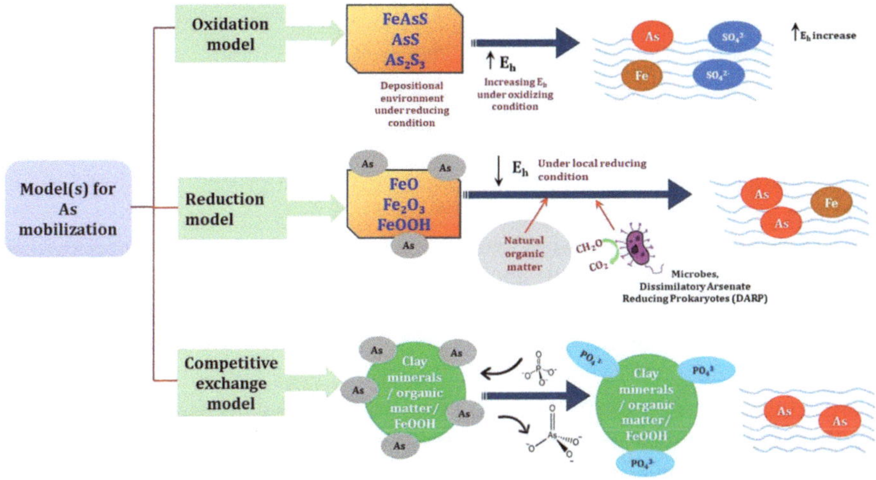

Fig. 20.1 Mechanism of arsenic release and mobilization into groundwater

and thereby, releasing arsenic into the aquifer (Fig. 20.1). According to the proposed hypothesis, the change in groundwater chemistry will occur with high concentration of arsenic [As (III)/As (V)], Fe [Fe (II)/Fe (III)], SO_4^{2-} with decreasing pH towards acidic region. Such phenomenon commonly occurs in oxic condition (positive DO and Eh) (Chatterjee et al., 2018). However, several researchers, particularly in BDP regions, have demonstrated that the nature of arsenic contaminated shallow aquifers is reducing (deficit in dissolved oxygen) with presence of both As (III) and As (V) and low sulphate. Moreover, arsenic total (As_T) and sulphate concentration do not demonstrate any association in the entire BDP, irrespective of high, medium and low occurrences of arsenic in groundwater (Ahmed et al., 2004).

Iron reduction hypothesis has been widely accepted by the researchers to explain arsenic release in groundwater, notably in entire South and South-East Asia and is the reversal hypothesis of pyrite oxidation model (Fig. 20.1). Researcher of KTH (Sweden) and University of Kalyani (India) were one of the first groups to explain arsenic release in groundwater in BDP using this hypothesis (Bhattacharya et al., 1997). In this model, the dissolved arsenic are usually adsorbed/co-precipitated on several metal oxides (Fe, Mn, Al—oxides) as arsenate and are buried in the sediment under oxidizing condition (Herath et al., 2016). These buried oxidized metal oxides along with arsenate later on release arsenic into the groundwater, when there is a local drop in the redox environment of the aquifer (drop in aquifer redox potential). It has been found that the Fe-oxides, hydroxides, oxyhydroxides have a very strong well documented affinity for arsenic, particularly for As (V) (Bhattacharya et al., 2011). In this context, existence of natural organic matter is crucial and plays a key role in the release of arsenic in groundwater. In fact, the Fe-oxides and oxyhydroxides are most common constituents in BDP and are often found as yellow coating on sedimentary sand grains (Bhattacharya et al., 2001; Bhattacharyya et al., 2003a).

Recently, the arsenic mobilization in groundwater and shallow aquifers in BDP, have been widely reported by several workers where reductive dissolution of different Fe-oxides in presence of labile organic matter has been proposed to be the most accepted mechanism to explain release of arsenic in groundwater (Bhattacharya et al., 2011). Moreover, the process of reductive dissolution warrants strong participation of fresh organic matter which has to be supplied to enhance the process to a large extent (Herath et al., 2016).

Another important model is ion-exchange model where basically competitive ion exchange may take place in between PO_4^{3-} and SO_4^{2-}, because they have comparable adsorptive capability (Acharyya et al., 2000). It has been reported that the dissolved PO_4^{3-} is usually coming from agriculture field by surface run-off and ultimately enters the aquifer via local recharge. The large-scale use of fertilizers in paddy fields for maintaining sustainable growth of agricultural production is the most important issue to consider for the ion-exchange model. During competitive exchange, arsenic is released from different adsorbed surfaces (clay, organic matter, weathered mica) and PO_4^{3-} is chemically exchanged to display arsenic during adsorption–desorption consequences (Fig. 20.1) (Appelo et al., 2002; Bauer & Blodau, 2006, 2009). However, researchers have proposed several other competitive exchangers (such as HCO_3^-, SiO_4^{2-}) as they behave equally like PO_4^{3-} in arsenic exchange process (Appelo et al., 2002; Bauer & Blodau, 2006, 2009). Researchers have found several drawbacks to establish the model under field condition. The main drawback is the competitive exchange in ion concentration (As vs. PO_4^{3-}) in groundwater should be inversely proportional to depth whereas the reverse has been reported (Bauer & Blodau, 2006, 2009). Additionally, an elevated concentration of arsenic is supposed to correlate with low HCO_3^- concentration. However, this has been reported reverse in field condition (Bauer & Blodau, 2006, 2009).

The geomicrobial arsenic release process has also been proposed by several workers (Lovely, 1995; Oremland & Stolz, 2003). In this process, the microbes uptake energy from organic matter (proxy substrate) (Islam et al., 2004). Later on, several DIRB (dissimilarity iron reducing bacteria) are reported to strengthen micro-geochemical arsenic mobilization process. During the process, bacteria are assimilating energy by breaking down organic carbon. Both the iron and arsenic will be decoupled by increasing As (III) concentration in groundwater (Oremland & Stolz, 2005). This could be the possible reason to explain the varying ratio of As (III)/As (V) in BDP groundwater (Bhattacharyya et al., 2003b). The geomicrobial process of arsenic release is a complex issue and the control mechanism of arsenic cycling depends on several factors notably the availability, amount, type, and, nature of organic matter, which could be easily degradable by the natural bacteria.

The most notable feature of groundwater arsenic is the spatial heterogeneity, with patchy distribution of groundwater arsenic concentrations; occurrence of low arsenic zones in close vicinity to high arsenic zones (Chatterjee et al., 2018). The spatial heterogeneity is influenced by several points, e.g. land use pattern, availability of organic matter, sediment color, geomorphological features, pumping and local recharge etc. (Chatterjee et al., 2018). Among these factors, local land use pattern along with large-scale ground water improvement for crop production, degraded ponds, water bodies and sewage leakage from sanitation installation are the major contributors in relation to the release of arsenic (Biswas et al., 2014; Nath et al., 2008). At least in local scale, the supply of labile organic matter is the most important as this will trigger the geochemical and geomicrobiological processes [Fe (II)/Fe (III), As (III)/As (V), Mn (II)/Mn (IV)] (Polizzotto et al., 2008). This can also help to regulate TEAPs (terminal electron accepting process) and thereby, control (upscaling/ downscaling) arsenic release in groundwater (Nickson et al., 2000). Researchers have also found a link between sediment color and arsenic concentration in groundwater, with BSA (brown sand aquifer) having low arsenic concentration compared to the GSA (grey sad aquifer) (Biswas et al., 2014). On the other hand, to explain the heterogeneous distribution of arsenic, studies have also proposed multi-release mechanism, where it is viewed that a sole process cannot be responsible for the elevation of groundwater arsenic, but it could be due to a complex combination of two or more processes (Bhowmick et al., 2013).

Sources of Arsenic Exposure

Humans can be exposed to arsenic through a number of different routes, including inhaling dust and other airborne particles, consuming contaminated food and water, and dermal contact with contaminated environment media (Chung et al., 2014). The affected population is mostly exposed to arsenic through water and daily foods, with minor and underestimated supply of arsenic coming from other different sources (Bhowmick et al., 2018).

People all around the world have exploited groundwater as a reliable source of drinking water. Unfortunately, high quantum of arsenic in the water sources has caused a severe arsenic calamity to spread over to these areas. Numerous researchers have found specific links of consuming water with high concentrations of arsenic to enhanced risk to human health (Bhowmick et al., 2018; Chatterjee et al., 2018; Shaji et al., 2021). Daily water consumption rate and arsenic concentration in water are the two parameters that affect how much arsenic is consumed from water and there could be a lot of variations in both these parameters.

There is considerable uncertainty regarding the definite risk from drinking water less than the current WHO drinking water guideline for arsenic (10 μg/L) (Bhowmick et al., 2018). The excess skin cancer risk from drinking water as per WHO drinking water guideline is much higher (approximately 60 times) than that applied to safeguard public health (Smith et al., 2002). Actually, a 0.17 μg/L guideline value for arsenic would have been more appropriate; owing to analytical challenges in measuring arsenic in routine laboratories, a less restrictive guideline value was proposed (Ahmad & Bhattacharya, 2019). The other variable factor is the amount of water consumed daily. When establishing the drinking water quality standard for pollutants, WHO and other regulatory authorities apply value of 2 L per day for persons weighing 60 kg (WHO, 2011). However, these water intake values (2 L/day for a 60 kg adult) could exceed on several occasions, under hot and humid climatic condition with typical manually labour-intensive livelihood (e.g. agriculture practices) (Hossain et al., 2013). In such cases, the water intake may go up to 4 L/day and significant seasonal variations (27%) have also been observed among the rural recruit (Hossain et al., 2013). Special attention is required from arsenic-exposure (indirect) via beverages prepared using arsenic contaminated water (Joseph et al., 2015). Therefore, the combined water intake from both direct and indirect sources will eventually enhance the risk of arsenic exposure.

Arsenic levels in locally irrigated agricultural crops have recently raised concerns. In order to irrigate crops three to four times a year, groundwater has been exploited to meet the irrigation water demand. But because of the heavy arsenic contamination of the groundwater that was extracted for agricultural reasons, there is considerable arsenic loading in the top soil in these regions (Chatterjee et al., 2018). This has resulted in an imminent threat to sustainable crop production and has compromised food security (Table 20.1).

The effects of arsenic-rich groundwater and the increase in arsenic in local irrigated crops have been the subject of numerous studies (Bhowmick et al., 2018). For instance, Roychowdhury et al. (2002) in their study showed that leafy vegetables had a higher concentration of arsenic than non-leafy vegetables and root crops from their survey of total arsenic in food samples which were collected from people in Murshidabad region of West Bengal. Halder et al. (2013) analyzed the distribution of arsenic species in contaminated vegetables and reported that inorganic arsenic species accounts for the majority of the contamination—91.7 and 100% for rice and other vegetables, respectively. This has salient toxicological implications since inorganic arsenic is more toxic than the organic form of arsenic (such as DMA, MMA and arsenobetaine) (Halder et al., 2013). Additionally, the presence of high percentage of arsenic in the leafy vegetables as inorganic arsenic is of serious concern because of the presence of high concentration of folate in leafy vegetables (water soluble vitamin B) and has been recommended as a supplementary diet to increase biomethylation capacity and thereby, decrease blood arsenic concentration level (Gamble et al., 2007). Therefore, the benefit of consuming locally cultivated leafy green vegetables in places with high levels of inorganic arsenic, such as Bangladesh and West Bengal, India, is still a matter of debate. Future research is required to critically investigate and compare the synchronic benefits of addition of folate and the adverse effects caused by inorganic arsenic in these types of foods.

Recent reports on the assessment of population's dietary intake of arsenic suggest that rice and water are significant sources of arsenic (Table 20.2) (Bhowmick et al., 2018; Moulick et al., 2021).

Rice is the staple diet of one-third of the World population (Bhowmick et al., 2018; Biswas et al., 2021; Halder et al., 2013). In fact, the population of entire South and South-East Asia are heavily dependent upon rice intake (Halder et al., 2013). Flooded irrigation, which is the most popular method of rice cultivation makes arsenic highly bioavailable to rice (Majumdar et al., 2021). In contrast to other cereal crops like wheat and maize, rice intrinsically has the effective efficiency by transported signal mechanism to take-up and translocate significant amount of arsenic with high transfer factor (Mishra et al., 2017). On the other hand, the global production of

Table 20.1 Arsenic concentration in various foodstuffs reported from different geographical locations around the world

Category of food stuff	Food Stuff	Area where the study conducted	Reported As concentration (mg/kg)	References
Cereals	Wheat	Bihar, India	0.0077–0.108	Kumar et al. (2016)
	Wheat	Bihar, India	0.00096–0.234	Mondal et al. (2021)
	Wheat	West Bengal, India	0.097–0.211	Rahaman et al. (2012)
	Wheat	Sub-Saharan Africa	0.004–0.017	Jitaru et al. (2019)
	Wheat	Bangladesh	0.037–0.70	Islam et al. (2017a, 2017b)
	Maize	Bihar, India	0.0048–0.043	Kumar et al. (2016)
	Maize	West Bengal, India	0.123–0.342	Rahaman et al. (2012)
	Maize	Bangladesh	0.21–0.59	Islam et al. (2017a, 2017b)
	Millet	Sub-Saharan Africa	0.001–0.019	Jitaru et al. (2019)
Legumes	Green gram	Samastipur, Bihar, India	0.0079–0.049	Kumar et al. (2016)
	Green gram	Malda, West Bengal, India	BDL-0.128	Rahaman et al. (2012)
	Lentils	Malda, West Bengal, India	0.029–0.121	Rahaman et al. (2012)
	Lentils	Bangladesh	0.17–0.49	Islam et al. (2017a, 2017b)
	Black gram	Malda, West Bengal, India	BDL-0.119	Rahaman et al. (2012)
	Lathyrus	Malda, West Bengal, India	BDL-0.121	Rahaman et al. (2012)
	Gram	Malda, West Bengal, India	0.098–0.213	Rahaman et al. (2012)
	Pea	Malda, West Bengal, India	0.093–0.294	Rahaman et al. (2012)
	Peanuts	Sub-Saharan Africa	0.004–0.10	Jitaru et al. (2019)
Leafy vegetables	Leaves of cassava plants	Tanzania	0.60–1.120	Nyanza et al. (2014)
	Leafy vegetables	Sub-Saharan Africa	0.002–0.013	Jitaru et al. (2019)

(continued)

Table 20.1 (continued)

Category of food stuff	Food Stuff	Area where the study conducted	Reported As concentration (mg/kg)	References
	Spinach	West Bengal, India	0.17–0.79	Bhattacharya et al. (2010)
	Spinach	West Bengal, India	0.641–1.39	Biswas et al. (2012)
	Spinach	Three villages of Nadia, West Bengal, India	0.24–0.35	Halder et al. (2013)
	Indian spinach	Three villages of Nadia, West Bengal, India	0.11–0.18	Halder et al. (2013)
	Spinach	Haringhata Block of Nadia, West Bengal, India	0.265 ± 0.155	Samal et al. (2011)
	Red spinach	Mathpara of Gaighata Block of West Bengal, India	0.055–2.801	Joardar et al. (2021)
	Cabbage	Five different blocks of Nadia District, West Bengal, India	0.11–0.30	Bhattacharya et al. (2010)
	Cabbage	Fourteen As prone areas from Nadia, West Bengal, India	0.463–0.572	Biswas et al. (2012)
	Cabbage	Haringhata Block of Nadia, West Bengal, India	0.315 ± 0.069	Samal et al. (2011)
	Cabbage	Ethiopia	5.73 ± 0.37	Gebeyehu and Bayissa (2020)
	Brassica pekinensis (Lour.) Rupr./ Chinese cabbage	Jiangsu Province, China	0.006–0.054	Jiang et al. (2015)
Leafy vegetables	Brassica chinensis Linn/green cabbage	Jiangsu Province, China	0.010–0.032	Jiang et al. (2015)
	Ladies finger	Five different blocks of Nadia District, West Bengal, India	0–0.49	Bhattacharya et al. (2010)

(continued)

Table 20.1 (continued)

Category of food stuff	Food Stuff	Area where the study conducted	Reported As concentration (mg/kg)	References
	Ladies finger	Fourteen As prone areas from Nadia, West Bengal, India	0.314–0.562	Biswas et al. (2012)
	Ladies finger	Haringhata Block of Nadia, West Bengal, India	0.182 ± 0.035	Samal et al. (2011)
	Peas	Five different blocks of Nadia District, West Bengal, India	0.09–0.20	Bhattacharya et al. (2010)
	Pea	Haringhata Block of Nadia, West Bengal, India	0.069 ± 0.022	Samal et al. (2011)
	Basile	Fourteen As prone areas from Nadia, West Bengal, India	0.182–0.225	Biswas et al. (2012)
	Basile	Haringhata Block of Nadia, West Bengal, India	0.331 ± 0.090	Samal et al. (2011)
	Puin leaf	West Bengal, India	0.020–0.508	Joardar et al. (2021)
	Coriander leaves	Fourteen As prone areas from Nadia, West Bengal, India	0.284–0.411	Biswas et al. (2012)
	Coriander leaves	West Bengal, India	0.0339–0.887	Joardar et al. (2021)
	Arum leaf	West Bengal, India	0.049–0.855	Joardar et al. (2021)
Non-leafy vegetables	Brinjal	West Bengal, India	0.049–0.349	Joardar et al. (2021)
	Brinjal	Bangladesh	0.091–0.27	Islam et al. (2017a, 2017b)
	Cauliflower	West Bengal, India	0.00999–0.292	Joardar et al. (2021)
	Tomato	Ethiopia	24.50 ± 0.50	Gebeyehu and Bayissa (2020)

(continued)

Table 20.1 (continued)

Category of food stuff	Food Stuff	Area where the study conducted	Reported As concentration (mg/kg)	References
	Tomato	Bangladesh	0.14–0.53	Islam et al. (2017a, 2017b)
	Tomato	Jiangsu Province, China	0.002–0.022	Jiang et al. (2015)
	Onion	Bangladesh	0.17–0.49	Islam et al. (2017a, 2017b)
	Chili	Bangladesh	0.15–0.38	Islam et al. (2017a, 2017b)
	Luffa cylindrical (L.) Roem./towel gourd	Jiangsu Province, China	0.001–0.060	Jiang et al. (2015)
	Cucumber	Jiangsu Province, China	0.004–0.060	Jiang et al. (2015)
	Capsicum	Jiangsu Province, China	0.001–0.060	Jiang et al. (2015)
Root vegetables	Radish	West Bengal, India	0.067–0.554	Joardar et al. (2021)
	Chinese radish	Jiangsu Province, China	0.005–0.012	Jiang et al. (2015)
	Arum	West Bengal, India	0.015–0.516	Joardar et al. (2021)
	Elephant foot	Malda, West Bengal, India	0.213–0.399	Rahaman et al. (2012)
	Turnip	West Bengal, India	0.0411–0.267	Joardar et al. (2021)
	Carrot	West Bengal, India	0.0149–0.389	Joardar et al. (2021)
	Carrot	Bangladesh	0.15–0.44	Islam et al. (2017a, 2017b)
Root Vegetables	Potato	West Bengal, India	0.0323–0.162	Joardar et al. (2021)
	Potato	Bihar, India	0.0056–0.176	Mondal et al. (2021)
	Potato	Malda, West Bengal, India	0.312–1.464	Rahaman et al. (2012)
	Potato	Jiangsu Province, China	0.003–0.015	Jiang et al. (2015)
	Sweet Potato	Malda, West Bengal, India	0.245–1.032	Rahaman et al. (2012)

(continued)

Table 20.1 (continued)

Category of food stuff	Food Stuff	Area where the study conducted	Reported As concentration (mg/kg)	References
	Potato	Bangladesh	0.17–0.47	Islam et al. (2017a, 2017b)
Fruits	Mango	Malda, West Bengal, India	BDL-0.194	Rahaman et al. (2012)
	Mango	Bangladesh	0.15–0.29	Islam et al. (2017a, 2017b)
	Litchi	Malda, West Bengal, India	BDL-0.174	Rahaman et al. (2012)
	Papaya	Malda, West Bengal, India	0.123–0.276	Rahaman et al. (2012)
	Guava	Malda, West Bengal, India	0.028–0.056	Rahaman et al. (2012)
	Lime (*Citrus* sp.)	Malda, West Bengal, India	0.046–0.103	Rahaman et al. (2012)
	Banana	Bangladesh	0.086–0.67	Islam et al. (2017a, 2017b)
	Jackfruit	Bangladesh	0.079–0.29	Islam et al. (2017a, 2017b)
Others	Mushroom	Bangladesh	0.18–0.91	Rashid et al. (2018)
	Mushroom caps (*L. amethystina)*	Yuxi, Yunnan province, China	135 ± 14	Zhang et al. (2015)
	Mushroom caps (*L. laccata*)	Eshan, Yunnan Province, China	143 ± 8	Zhang et al. (2015)
	Mushroom caps (*L. laccata)*	Yuxi, Yunnan province, China	14.1 ± 0.4	Zhang et al. (2015)
	Mushroom caps (*L. vinaceoavellanea)*	Zhenyuan, Yunnan Province, China	163 ± 4	Zhang et al. (2015)
	Mushroom caps (*L. vinaceoavellanea)*	Yuxi, Yunnan province, China	130 ± 1	Zhang et al. (2015)

rice is geographically concentrated in Western and Eastern Asia which accounts for more than 90% of world output (FAO, 2011). China and India supply over half of the world's rice, with large quotas also coming from Bangladesh, Thailand, Indonesia, Burma, and Vietnam (FAO, 2011). Most of these countries have extensive areas which are contaminated with high arsenic concentration in groundwater and might therefore present a significant public health hazard for the importing countries due to global food trade (Menon et al., 2020; Shraim, 2017).

Rice field and market-based studies have reported large differences in arsenic concentrations among distinct rice varieties (Bhowmick et al., 2018; Halder et al.,

Table 20.2 Arsenic concentration in rice grain from different countries around the world

Category of food stuff	Type of rice	Source	Area where the study conducted	Concentration of As (mg/kg)	References
Rice (cereals)	Rice	Field	Equador	0.042–0.125	Otero et al. (2016)
	Rice	Field	Peru	0.068–0.345	Mondal et al. (2020)
	Rice	Market-based	Sub-Saharan Africa	0.009–0.045	Jitaru et al. (2019)
	Rice	Market-based	Egypt	0.01–0.58	Meharg et al. (2009)
Rice (cereals)	Rice	Market-based	Mexico	0.06–0.3	García-Rico et al. (2020)
	Rice	Collected from local community where crops were grown	Chile	0.06–1.02	Diaz et al. (2015)
	Rice	Market-based	Brazil	0.01–1.39	Fão et al. (2021)
	Rice	Market-based	Uruguay	0.14–0.28	García-Rico et al. (2020)
	Rice	Field	Argentina	0.04–1.31	Oteiza et al. (2020)
	Rice	Market-based	Italy	0.011–0.0174	Cubadda et al. (2016)
	Rice	Market-based	Belgium	0.017–0.363	Ruttens et al. (2018)
	Rice	Market-based	Sweden	0.03–0.177	Kollander et al. (2019)
	Rice	Market-based	Thailand	0.084–0.49	Hensawang and Chanpiwat (2017)
	Rice	Household	Cambodia	0.088–0.578	Phan et al. (2013)
	Rice	Field	West Bengal, India	Aman rice (0.06–0.60); boro rice (0.19–0.78)	Bhattacharya et al. (2010)
	Rice	Household	West Bengal, India	0.01–0.64	Halder et al. (2013)
	Amon rice	Field	West Bengal, India	0.156 ± 0.009	Samal et al. (2011)
Rice (cereals)	Boro rice	Field	West Bengal, India	0.194 ± 0.019	Samal et al. (2011)

(continued)

Table 20.2 (continued)

Category of food stuff	Type of rice	Source	Area where the study conducted	Concentration of As (mg/kg)	References
	Parboiled rice	Household	West Bengal, India	0.105–2.026	Joardar et al. (2021)
	Rice	Household	West Bengal, India	0.291–1.411	Rahaman et al. (2012)
	Rice	Household	Bihar, India	0.00251–3.947	Kumar et al. (2016)
	Rice	Household	Bihar, India	0.015–0.231	Mondal et al. (2021)
	Rice	Household	Bangladesh	0.014–0.680	Islam et al. (2017a, 2017b)
	Rice	Market-based	Bangladesh	0.028–0.68	Islam et al. (2017a, 2017b)
	Rice	Field and market	Hunan Province, China	0.129 ± 0.0492	Ma et al. (2016)
	Polished rice	Field	Different province of China	0.011–0.235	Mu et al. (2019)
	Brown rice	Field	Different province of China	0.020–0.326	Mu et al. (2019)
	Rice	Local markets	Different province of China	BDL-0.3032	Huang et al. (2015)
	Rice	Field	Jiangsu Province, China	0.046–0.247	Jiang et al. (2015)
	Rice	Field	Mekong river delta	0.08–0.56	Nguyen et al. (2019)
Rice (cereals)	Rice	Market-based	Cambodia	0.047–0.766	Gilbert et al. (2015)
	Rice	Market-based	Pakistan	0.00698–0.178	Sarwar et al. (2020)
	Brown rice	Market-based	Switzerland	0.147–0.238	Guillod-Magnin et al. (2018)
	White rice	Market-based	Switzerland	0.0067–0.281	Guillod-Magnin et al. (2018)
	Rice	Market-based	Brazil	0.059–0.782	Ciminelli et al. (2017)
	Rice	Market-based	Bangladesh	0.03–0.30	Williams et al. (2005)
	Rice	Market-based	US	0.11–0.40	Williams et al. (2005)

(continued)

Table 20.2 (continued)

Category of food stuff	Type of rice	Source	Area where the study conducted	Concentration of As (mg/kg)	References
	Rice	Market-based	US	0.005–0.710	Zavala et al. (2008)
	Rice	Market-based	US	0.127–0.464	Shraim (2017)
	Rice	Market-based	Egypt	0.039–0.154	
	Rice	Market-based	India	0.026–0.226	
	Rice	Market-based	UK	0.01–0.37	Menon et al. (2020)

2013; Moulick et al., 2021). Based on rice growing conditions and geographical origin, the inorganic and organic arsenic species distribution in different rice cultivars are also different (Halder et al., 2013). Studies have shown that inorganic arsenic species is mostly found in the rice grown in Bangladesh, India, and Europe, while the organic arsenic form is mostly dominant in the USA rice (Williams et al., 2005; Zavala et al., 2008). Therefore, based on the variation of inorganic and organic arsenic species, Asian and European rice have been classified as inorganic type and USA rice as DMA types.

Studies have linked size and color of the rice grain with amount of arsenic in rice (Halder et al., 2013). Reports shows that shorter rice grains tend to accumulate more arsenic than the longer rice grain (Halder et al., 2013). In a household survey in rural areas of West Bengal, India, Halder et al. (2012) reported maximum amount of arsenic in short-bold brown rice, followed by medium-slender brown rice, long slender brown rice and the lowest arsenic concentration in extra-long slender basmati rice (Halder et al., 2012). The color of the rice is another important characteristic which has shown correlation with arsenic concentration in rice (Halder et al., 2012). It has been reported that brown rice shows higher arsenic content than the white rice (Halder et al., 2013). This is because of the outer bran coating in the brown rice where arsenic tends to accumulate preferentially (Bhowmick et al., 2018). For white rice, this coating is removed by milling and therefore, arsenic content in rice grain decreases (Bhowmick et al., 2018). Such milling not only lowers the arsenic content in rice grain but also reduces the inorganic arsenic content (Naito et al., 2015). It is crucial to note that removing the outer bran layer results in further depletion of various nutrients (Biswas et al., 2021; Šlejkovec et al., 2021; Yim et al., 2017). As rice is a major source of energy and nutrition for many people, it is essential to improve the polishing process, reduce arsenic, and maintain the nutritional quality of rice. Arsenic content in rice grain is also influenced by the rice growing season (Halder et al., 2013). It has been shown that rice grown during hot and dry season (boro rice), tends to have higher arsenic concentration in their rice grain compared to the rice cultivated in the monsoon season (aman rice) (Halder et al., 2013). This is because groundwater that is frequently contaminated with arsenic is heavily relied upon for rice irrigation during the dry season, which increases the arsenic availability for rice grain uptake (Halder et al., 2013; Mandal et al., 2019). The key, however, is still the soil environment, and some edaphic elements like pH, organic matter content, redox state, and presence of phosphate, calcium carbonate, iron, manganese, and aluminum, as well as microbial activity, which dictates the amount of arsenic that is available for plant uptake. Additionally, studies have also shown genetic influence in translocating arsenic to rice grain (Kumarathilaka et al., 2018). In an important study, Ahmed et al. (2011) reported that environmental factors regulated for 69–80% variability of arsenic content in rice grain, while genotype and genotype × environment interaction accounted for 9–10% and 10–21%, respectively (Ahmed et al., 2011).

Finally, rice is consumed in cooked form and therefore, concentration of arsenic in cooked rice dictates the dietary arsenic intake (Bhowmick et al., 2018). Depending upon rice washing method and amount of water used for cooking, rice cooking can be classified as traditional, intermediate and contemporary (Sengupta et al., 2006). The traditional method involves washing rice 5–6 times with water until clear and cooking it in excess water (normally 5–6 times more than the weight of raw rice), and then discarding the gruel. The intermediate method involves washing rice in a manner akin to the traditional method and cooking it in a small amount of water until there is no more water to be discarded. The contemporary method uses similar volume of cooking water as the intermediate method to cook the rice but without washing it first (Sengupta et al., 2006).

Findings regarding washing rice and the decrease in rice grain arsenic have been contradictory. There have been reports which have suggested substantial reduction of arsenic after washing rice (Chowdhury et al., 2020), while field-based studies have documented no such reduction in arsenic content (Halder et al., 2014). Such variations in the degree of arsenic reduction may be influenced by the washing method being used and the amount of arsenic in the washing water (Bhowmick et al., 2018). However, washing rice grains in arsenic-safe water has been shown to lower the amount of arsenic in the washed rice grains and may be adopted as an intervention strategy at kitchen-levels for reduction of cooked rice arsenic levels (Bhowmick et al., 2018; Sengupta et al., 2006).

It has now been established that arsenic in cooked rice is influenced by the cooking method employed (Bhowmick et al., 2018; Sengupta et al., 2006). In one such study, Signes et al. (2008) used same concentration of arsenic in cooking water (40 μg/L) to investigate the impact of cooking process on the final arsenic concentration in cooked rice. Using atab (non-parboiled) and parboiled as the two different rice cultivars with an initial raw rice arsenic concentration of 185 and 315 mg/Kg, the authors reported an increase of 27.6, 37.8 and 42.2% in the cooked rice arsenic content for atab rice following traditional, intermediate and contemporary cooking method, respectively. In case of parboiled rice, there was a reduction of 12.7% in arsenic content in cooked rice using traditional method, but in case of intermediate and contemporary cooking methods, there was an increase of cooked rice arsenic concentration by 15.9% and 23.5%, respectively. Mandal et al. (2019) reported several fold increase in cooked rice arsenic content compared to raw rice from analysis of raw rice and cooked rice samples that were collected from households in rural areas of

West Bengal, India, where cooking water had high arsenic concentration and rice was cooked in traditional method (Mandal et al., 2019). The authors concluded that arsenic concentration in raw rice and cooking water, and rice variety are all potential contributors to the final arsenic content in cooked rice. Additionally, the study also stated that the arsenic content in cooked rice decreased significantly when cooked with arsenic-free water (<10 μg/L). Similar finding was made by Chowdhury et al. (2020), which reported that cooked rice arsenic content significantly decreased when it was cooked with water that contained low (<3 μg/L) to moderate (36–58 μg/L) levels of arsenic contamination, but significantly increased when it was cooked with water that was highly contaminated with arsenic (84–105 μg/L) (Chowdhury et al., 2020). Therefore, researchers have suggested different threshold levels of cooking water arsenic concentration, which would eventually determine whether there would be an increase or decrease in arsenic concentration in cooked rice compared to raw rice. In this context, Halder et al. (2014) proposed that the final concentration of arsenic in cooked rice is actually determined by the transfer co-efficient (i.e., the ratio of arsenic concentration in cooked rice to that of starch water).

In comparison to raw rice, cooking rice generally does not significantly alter the relative distribution of arsenic species (Moulick et al., 2021). Research has shown that when rice is cooked in arsenic-tainted water, the inorganic arsenic in the water tends to be adsorbed into the rice, increasing both the total and inorganic arsenic content (Mandal et al., 2019). However, when surplus arsenic-safe water is used for cooking rice, the inorganic arsenic from rice is leached out at such high temperature and thereby effectively lowers the total as well as the inorganic arsenic content in cooked rice (Carey et al., 2015; Mandal et al., 2019; Mridha et al., 2022). In this respect, an alternative method of cooking has been experimented where rice is cooked in a continuous stream of percolating low-arsenic water near boiling (Carey et al., 2015; Signes-Pastor et al., 2017). Such methods of cooking have shown effective results in decreasing the substantial amount of inorganic arsenic in cooked rice compared to raw rice (Biswas et al., 2021; Chowdhury et al., 2020). Therefore, it can be logically hypothesized that boiling rice in excess water with low arsenic content can be a temporary intervention measure, so as to reduce the dietary exposure to arsenic.

Role of Genetic Factors

Arsenic (As) is a metalloid and exists in nature in both organic and inorganic forms. Seafoods including shellfish and seaweed contain organic forms of arsenic (e.g., arsenobetaine, arsenosugars, arsenolipids) and are considered to be non-toxic. Inorganic arsenic (iAs) falls among class I human carcinogens, however, it is not considered a mutagenic agent (IARC, 2004; Basu et al., 2001). Around 200 million people worldwide are exposed to the toxic effects of arsenic mainly through contaminated groundwater. People can also be exposed to arsenic through foodstuff such as vegetables, rice and other crops irrigated with contaminated groundwater. Chronic exposure to arsenic could result in non-cancerous health outcomes like skin lesions, cardiovascular, respiratory, hepatic and neurological diseases and diabetes as well as cancerous outcomes like skin, lung, kidney and bladder cancers (Brown et al., 1989). Short period of arsenic exposure could result in many kinds of skin lesions like melanosis, pigmentation and keratosis. These kinds of skin lesions are considered as early manifestations of arsenic toxicity and could be a precursor for arsenic-induced skin cancers (Yu et al., 2006). However, many studies have found that only a small percentage of people who were exposed to arsenic could exhibit these kinds of skin lesions (Argos et al., 2011; Chakraborti et al., 2002).

The genotoxic potential of iAs in animals and humans have been demonstrated through many studies. The exact mechanisms by which the developments of these pathological outcomes happen are not very clear and it could be complex in nature. However, induction of a strong oxidative stress through the generation of these reactive oxygen species (ROS) [e.g., superoxide anion, hydroxyl radical, nitrogen species, hydrogen peroxide] is one of the principal mechanisms of arsenic toxicity as has been found by several studies (Kitchin & Ahmad, 2003; Kitchin & Conolly, 2010; Lantz & Hays, 2006). Generation of these reactive oxygen species results in DNA damage, lipid peroxidation and loss of anti-oxidant defense and induction of stress response genes (Fig. 20.2) (Del Razo et al., 2001; Liu et al., 2001; Nesnow et al., 2002). Also, this arsenic-induced ROS generation has been found to alter signaling pathways in the cells and regulate transcription factors that play a crucial role in carcinogenesis (Fig. 20.2) (Tapio & Grosche, 2006). The genotoxic effects of arsenic include a number of modifications in DNA such as chromosomal aberrations, micronuclei formation, sister-chromatid exchange, aneuploidy, DNA strand breaks, inhibition of DNA repair, deletion mutations and DNA–protein cross-linking (Basu et al., 2001; Rossman, 2003).

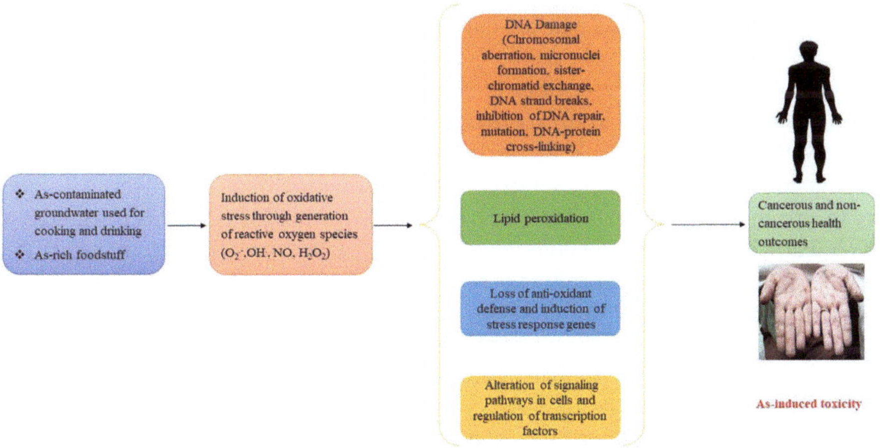

Fig. 20.2 Oxidative stress and its effects on genotoxicity

Inorganic arsenic (iAs) exists in the environment in trivalent (iAs^{III} or arsenite) and pentavalent (iAs^{V} or arsenate) states. These two forms have different toxic potentials and are metabolized differently in animals including humans. Toxicological studies have shown that the trivalent species of arsenic is more toxic than the pentavalent species (Styblo et al., 2000). Arsenic undergoes a biotransformation process on entering the human body. In the classical pathway, pentavalent arsenic (As^{V}) is reduced to trivalent arsenic (As^{III}). Thereafter, oxidative methylation of As^{III} occurs with the help of enzyme, S-Adenosyl methionine (SAM) and mono-, di- and tri-methylated derivatives [e.g., monomethylarsonous acid (MMA), dimethylarsinic acid (DMA)] are generated as end products (Drobna et al., 2009). The end-products of the metabolism are very toxic and their accumulation in the human body disrupts the normal biological processes occurring in the cells and results in genetic and epigenetic changes (Bustaffa et al., 2014). Arsenite methyltransferase (*AS3MT*), glutathione S-transferase omega-1 (*GSTO1*), glutathione S-transferase omega-2 (*GSTO2*), Methylenetetrahydrofolate reductase (*MTHFR*) and purine nucleoside phosphorylase (*PNP*) are the main enzymes of the arsenic biotransformation pathway (De Chaudhuri et al., 2008; Gamble et al., 2005).

Several epidemiological studies have reported inter-individual and ethnic variation in the susceptibility to the toxic effects of arsenic. This variation could be influenced by a number of factors like age, gender, nutritional status, arsenic metabolism and genetic factors (Antonelli et al., 2014; Hernández & Marcos, 2008; Lindberg et al., 2007; Vahter, 2002). All arsenic species like As^{V}, As^{III}, MMA and DMA have been profiled in human urine among several populations and on average, the urinary arsenic levels constitute 60–70% of DMA, 10–20% of MMA and 10–30% of iAs (Vahter, 1999). The distribution of urinary arsenic metabolites varies widely among individuals and populations (Vahter, 2002) and this could lead to the differences in tissue distribution and retention of these toxic metabolites in the body. This variation could explain the differences in the toxic effects of arsenic on human health among individuals and populations who are chronically exposed to arsenic poisoning (Chen et al., 2003; Steinmaus et al., 2006; Tseng et al., 2005). A number of studies have observed that increased percentage of MMA and increased ratio of MMA to DMA is associated with increased toxic effects of arsenic (Chen et al., 2003; Tseng et al., 2005; Vahter, 2002).

A genome is the complete set of genetic information in an organism. All living organisms including humans have their own set of genomes. A gene is a segment of DNA that when transcribed and translated gives rise to a specific protein that carries out a particular function in the human body. After the discovery of non-coding RNAs (ncRNAs) in the early 1990s, ncRNAs are also regarded as genes that can produce functional RNA molecules to regulate gene expression at the transcriptional and post-transcriptional levels. The human genome contains around 3.2 billion base pairs of DNA made up of four types of chemical building blocks, known as nucleotides and denoted by the letters A (adenine), T (thymine), G (guanine) and C (cytosine). The order of the four letters in the double helical structure allows the DNA to function in different ways.

Single nucleotide polymorphisms (SNPs) are single base changes in the DNA where different sequences (referred to as alleles) could exist among normal individuals in populations. SNPs are most common form of sequence variations and account for more than 90% of all variations in the human genome. It is estimated that more than 10 million SNPs are present in the human genome. Since tri or tetra allelic SNPs have very low frequency in humans, SNPs are mostly referred as bi-allelic polymorphisms. SNPs have been extensively used as genetic markers to identify genes associated with complex diseases.

Many studies have reported that polymorphisms in the arsenic metabolism and detoxification, anti-oxidant defense and DNA repair genes could also contribute to the inter-individual differences in arsenic biotransformation and consequently, influence inter-individual susceptibility to the health effects of arsenic toxicity including arsenic-induced skin lesions (Ahsan et al., 2007; De Chaudhuri et al., 2008; Faita et al., 2013; Hernández & Marcos, 2008). Table 20.3 summarizes polymorphisms affecting genes in the arsenic metabolism and detoxification, anti-oxidant defense, DNA repair and some other pathways (e.g. drug metabolism, lipid homeostasis, vascular homeostasis, immunoregulation and tumor suppressor) that have been found to be associated with arsenic toxicity including arsenic-induced skin lesions and cancer.

Arsenite methyltransferase (*AS3MT*) is an S-adenosyl-methionine-dependent (SAM) enzyme that plays an important role in arsenic metabolism by catalyzing the methylation of arsenite species (As^{III}). Variations in the *AS3MT* gene have been found to be associated with arsenic metabolism efficiency to influence arsenic metabolite levels (Vahter et al., 1995). Individuals carrying the variants of the gene could methylate and excrete arsenic more efficiently than other individuals. Meza et al (2005) found that three intronic SNPs in the *AS3MT* gene were associated with DMA/MMA ratios in children among a Mexican population. The Met287Thr polymorphism in *AS3MT* has been extensively studied and correlated with increased MMA percentage of a Central European population (Lindberg et al., 2007). Polymorphisms in the *AS3MT* gene correlated with altered arsenic metabolite patterns in different populations of Bangladesh, Mexico, Argentina and Taiwan (Agusa et al., 2011; Engström et al., 2007; Hsieh et al., 2017). The glutathione S-transferases (*GST*) are a family of enzymes that are involved in the phase II detoxification of xenobiotics through a conjugation reaction with glutathione. Glutathione S-transferase omega-1 (*GSTO1*) and glutathione S-transferase omega-2 (*GSTO2*) take part as rate-limiting enzymes in reduction of methylated arsenic in the biotransformation pathway (Paiva et al., 2010). Paiva et al (2010) found out that Asn412Asp polymorphism in the *GSTO2* correlated with increase in DMA in urine among copper mine workers in Chile. Other genes in the *GST* family like *GSTO1*, *GSTP1*, *GSTM1* and *GSTT1* have also been found to be associated with arsenic-induced skin lesions and arsenic metabolite levels (Hernández & Marcos, 2008). Methylenetetrahydrofolate reductase (*MTHFR*) and 5-methyltetrahydrofolate-homocysteine methyltransferase (*MTR*) genes are involved in SAM metabolism and polymorphisms in these two genes could also affect arsenic metabolism (Engström et al., 2007; Steinmaus et al., 2007). *PNP*, an important enzyme in the arsenic metabolism pathway, plays a role in reducing As^{V} to As^{III}

Table 20.3 Genes in different biological pathways that are associated with arsenic toxicity

Gene	Biological function	Polymorphisms/SNPs in the genes associated with
AS3MT	As metabolism	Arsenic metabolite levels
MTHFR	As metabolism	iAs and arsenic metabolite levels, As-induced skin lesions
MTR	As metabolism	Arsenic metabolite levels
PNP	As metabolism	Arsenic metabolite levels, As-induced skin lesions
GSTO1	As detoxification	iAs level, As-induced skin lesions
GSTO2	As detoxification	iAs and arsenic metabolite levels, major risk for carotid atherosclerosis and urinary transitional cell carcinoma
GSTP1	As detoxification	As-induced skin lesions, major risk for carotid atherosclerosis and bladder cancer
GSTM1	As detoxification	iAs and arsenic metabolite levels, As-induced skin lesions
GSTT1	As detoxification	Arsenic metabolite levels, As-induced skin lesions
CYP1A1	Drug metabolism	Risk for lung cancer
OGG1	DNA repair	8-Oxoguanine levels
APE1	DNA repair	8-Oxoguanine levels, As-induced skin lesions
XRCC1	DNA repair	As-induced skin lesions, risk for bladder cancer
XRCC3	DNA repair	As-induced skin lesions, peripheral neuropathy and conjunctivitis
ERCC2/ XPD	DNA repair	As-induced skin lesions, non-melanoma skin cancer
HO-1/ HMOX1	Anti-oxidant defense	BP regulation and cardiovascular mortality risk
MnSOD	Anti-oxidant defense	Risk for hypertension
MPO	Anti-oxidant defense	Hyperkeratotic skin lesions
CAT	Anti-oxidant defense	Hyperkeratotic skin lesions
APOE	Lipid homeostasis	Risk for carotid atherosclerosis
eNOS/ NOS3	Vascular homeostasis	Risk for hypertension
MCP-1/ CCL2	Immunoregulation	Risk for carotid atherosclerosis
P53	Tumor suppressor	Risk for As-induced keratosis, renal cell carcinoma, skin cancer

in arsenic biotransformation. Three variants of the *PNP* gene were found to render individuals susceptible to arsenic-induced skin lesions in a West Bengal population (De Chaudhuri et al., 2008).

Inhibition of DNA repair process has been implicated as one of the major mechanisms of iAs induced genotoxicity. Nucleotide excision repair (NER) and base excision repair (BER) are two of the major repair processes that repair base damage in DNA induced by ROS after arsenic exposure. While NER is the major pathway for repair of bulky DNA lesions, BER accounts for repair of single strand breaks in DNA. Studies with cultured human fibroblasts have shown that arsenic exposures could cause a reduction in DNA repair capacity (Curnow et al., 2001; Hartwig et al., 1997). Therefore, variants in the DNA repair genes may influence DNA repair capacity by altering the functional properties of DNA repair enzymes and consequently, might modulate susceptibility to cancer (Lunn et al., 1999). Polymorphisms in several genes in the DNA repair pathways such as *XRCC1*, *XRCC3*, *APE1*, *ERCC2* and *OGG1* have been found to influence 8-oxoguanine levels, arsenic-induced skin lesions and bladder and non-melanoma skin cancer (Andrew et al., 2009; Fujihara et al., 2011; Kundu et al., 2011; Minatel et al., 2018).

Occurrence of oxidative DNA damage through generation of reactive oxygen species (ROS) has been proposed as one of the principal mechanisms of arsenic toxicity and arsenic-induced carcinogenicity (Hubaux et al., 2013). Oxidative DNA damage leading to gene mutations and genomic instability has been observed frequently in cancer initiation and progression (Roos et al., 2016). Several polymorphisms in anti-oxidant defense genes such as *HO-1*, *MnSOD*, *MPO* and *CAT*, have been found to increase the risks for hyperkeratotic skin lesions, hypertension and cardiovascular mortality in arsenic toxicity.

Apart from the genes in the arsenic metabolism and detoxification, DNA repair and anti-oxidant defense, several important genes in other biological pathways, such as *CYP1A1* (drug metabolism), *APOE* (lipid homeostasis), *eNOS* (vascular homeostasis), *MCP-1* (immunoregulation) and *P53* (tumor suppressor) have been reported to be associated with arsenic-induced skin lesions, cancer and other diseases like hypertension and carotid atherosclerosis.

Conclusions

Research over the past decades suggests that arsenic is a natural and silent poison that has endangered the lives of millions of people across the globe. Exposure to arsenic is the root cause of a number of human health problems, including cancers of various organs. The seriousness of the problem can be gauzed from the facts that exposure to very low (ultra-trace) levels of arsenic is enough to invoke serious health problems. Moreover, the sources of arsenic exposure are varied and are complicatedly intertwined with the daily habits and lives of the vulnerable populations. But most importantly, arsenic as a potent genotoxic agent, is also capable of inducing genetic changes like DNA damage, lipid peroxidation and loss of anti-oxidant defense and induction of stress response genes. Moreover, many genes in the arsenic metabolism and detoxification, anti-oxidant defense and DNA repair genes could also contribute to the inter-individual differences in arsenic biotransformation.

Therefore, strict policies should be implemented in a time-bound manner to protect the lives of unsuspecting populations who are being exposed to arsenic. Strict regulations should be enforced on the use of groundwater for drinking and irrigation purpose. A robust water-surveillance system should be developed so that the water safety and security plan is not compromised. All efforts should be made to alleviate the quantum of exposure levels and increase the health-status of the common people; thereby achieving many of the goals of Sustainable Development Goals (SDGs) and global priorities of United Nation (UN) and UNESCO, respectively.

Acknowledgements Financial support from the Department of Biotechnology, Government of India (Project No. BT/PR25850/BCE/8/1423/2017) is sincerely acknowledged. KRC, CSIR-NEERI is duly acknowledged for checking the manuscript through the anti-plagiarism software, iThenticate (CSIR-NEERI/KRC/2023/MARCH/KZC/4).

References

Acharyya, S. K., Lahiri, S., Raymahashay, B. C., et al. (2000). Arsenic toxicity of groundwater in parts of the Bengal basin in India and Bangladesh: The role of quaternary stratigraphy and Holocene sea-level fluctuation. *Environmental Geology, 39*, 1127–1137.

Agusa, T., Fujihara, J., Takeshita, H., et al. (2011). Individual variations in inorganic arsenic metabolism associated with AS3MT genetic polymorphisms. *International Journal of Molecular Sciences, 12*(4), 2351–2382.

Ahmad, A., & Bhattacharya, P. (2019). Arsenic in drinking water: Is 10 μg/L a safe limit?. *Current Pollution Reports, 5*, 1–3.

Ahmed, K. M., Bhattacharya, P., Hasan, M. A., et al. (2004). Arsenic enrichment in groundwater of the alluvial aquifers in Bangladesh: An overview. *Applied Geochemistry, 19*(2), 181–200.

Ahmed, Z. U., Panaullah, G. M., Gauch, H., et al. (2011). Genotype and environment effects on rice (Oryza sativa L.) grain arsenic concentration in Bangladesh. *Plant and Soil, 338*, 367–382.

Ahsan, H., Chen, Y., Kibriya, M. G., et al. (2007). Arsenic metabolism, genetic susceptibility, and risk of premalignant skin lesions in Bangladesh. *Cancer Epidemiology Biomarkers & Prevention, 16*(6), 1270–1278.

Andrew, A. S., Mason, R. A., Kelsey, K. T., et al. (2009). DNA repair genotype interacts with arsenic exposure to increase bladder cancer risk. *Toxicology Letters, 187*(1), 10–14.

Antonelli, R., Shao, K., Thomas, D. J., et al. (2014). AS3MT, GSTO, and PNP polymorphisms: Impact on arsenic methylation and implications for disease susceptibility. *Environmental Research, 132*, 156–167.

Appelo, C. A. J., Van der Weiden, M. J. J., Tournassat, C., et al. (2002). Surface complexation of ferrous iron and carbonate on ferrihydrite and the mobilization of arsenic. *Environmental Science & Technology, 36*(14), 3096–3103.

Argos, M., Kalra, T., Pierce, B. L., et al. (2011). A prospective study of arsenic exposure from drinking water and incidence of skin lesions in Bangladesh. *American Journal of Epidemiology, 174*(2), 185–194.

Basu, A., Mahata, J., Gupta, S., et al. (2001). Genetic toxicology of a paradoxical human carcinogen, arsenic: A review. *Mutation Research/Reviews in Mutation Research, 488*(2), 171–194.

Bauer, M., & Blodau, C. (2006). Mobilization of arsenic by dissolved organic matter from iron oxides, soils and sediments. *Science of the Total Environment, 354*(2–3), 179–190.

Bauer, M., & Blodau, C. (2009). Arsenic distribution in the dissolved, colloidal and particulate size fraction of experimental solutions rich in dissolved organic matter and ferric iron. *Geochimica et Cosmochimica Acta, 73*(3), 529–542.

Berg, M., Stengel, C., Trang, P. T. K., et al. (2007). Magnitude of arsenic pollution in the Mekong and Red River Deltas—Cambodia and Vietnam. *Science of the Total Environment, 372*(2–3), 413–425.

Berg, M., Trang, P. T. K., Stengel, C., et al. (2008). Hydrological and sedimentary controls leading to arsenic contamination of groundwater in the Hanoi area, Vietnam: The impact of iron-arsenic ratios, peat, river bank deposits, and excessive groundwater abstraction. *Chemical Geology, 249*(1–2), 91–112.

Bhattacharya, P., Mukherjee, A., & Mukherjee, A. B. (2011). *Arsenic in groundwater of India.*

Bhattacharya, P., Chatterjee, D., & Jacks, G. (1997). Occurrence of arsenic-contaminated groundwater in alluvial aquifers from Delta plains, eastern India: Options for safe drinking water supply. *International Journal of Water Resources Development, 13*(1), 79–92.

Bhattacharya, P., Jacks, G., Jana, J., et al. (2001). Geochemistry of the Holocene alluvial sediments of Bengal Delta Plain from West Bengal, India: Implications on arsenic contamination in groundwater. *Groundwater Arsenic Contamination in the Bengal Delta Plain of Bangladesh, 3084*, 21–40.

Bhattacharya, P., Samal, A. C., Majumdar, J., et al. (2010). Arsenic contamination in rice, wheat, pulses, and vegetables: A study in an arsenic affected area of West Bengal, India. *Water Air Soil Pollution, 213*, 3–13.

Bhattacharya, P., Welch, A. H., Stollenwerk, K. G., et al. (2007). Arsenic in the environment: Biology and chemistry. *Science of the Total Environment, 379*(2–3), 109–120.

Bhattacharyya, P., Tandukar, N., Nekul, A., Valero, A. A., Mukherjee, A. B., & Jacks, G. (2003a, May). Geogenic arsenic in groundwaters from Terai alluvial plain of Nepal. In Journal de Physique IV (Proceedings) (Vol. 107, pp. 173–176). EDP sciences.

Bhattacharyya, R., Jana, J., Nath, B., et al. (2003b). Groundwater As mobilization in the Bengal Delta Plain, the use of ferralite as a possible remedial measure—A case study. *Applied Geochemistry, 18*(9), 1435–1451.

Bhowmick, S., Nath, B., Halder, D., et al. (2013). Arsenic mobilization in the aquifers of three physiographic settings of West Bengal, India: Understanding geogenic and anthropogenic influences. *Journal of Hazardous Materials, 262*, 915–923.

Bhowmick, S., Pramanik, S., Singh, P., et al. (2018). Arsenic in groundwater of West Bengal, India: A review of human health risks and assessment of possible intervention options. *Science of the Total Environment, 612*, 148–169.

Biswas, A., Biswas, S., & Santra, S. C. (2012). Risk from winter vegetables and pulses produced in arsenic endemic areas of Nadia district: Field study comparison with market basket survey. *Bulletin of Environmental Contamination and Toxicology, 88*, 909–914.

Biswas, A., Gustafsson, J. P., Neidhardt, H., et al. (2014). Role of competing ions in the mobilization of arsenic in groundwater of Bengal Basin: Insight from surface complexation modeling. *Water Research, 55*, 30–39.

Biswas, B., Chakraborty, A., Chatterjee, D., et al. (2021). Arsenic exposure from drinking water and staple food (rice): A field scale study in rural Bengal for assessment of human health risk. *Ecotoxicology and Environmental Safety, 228*, 113012–113012.

Brown, K. G., Boyle, K. E., Chen, C. W., et al. (1989). A dose-response analysis of skin cancer from inorganic arsenic in drinking water 1. *Risk Analysis, 9*(4), 519–528.

Bustaffa, E., Stoccoro, A., Bianchi, F., et al. (2014). Genotoxic and epigenetic mechanisms in arsenic carcinogenicity. *Archives of Toxicology, 88*, 1043–1067.

Carey, M., Jiujin, X., et al. (2015). Rethinking rice preparation for highly efficient removal of inorganic arsenic using percolating cooking water. *PLoS One, 10*, 1–12.

Chakraborti, D., Rahman, M. M., Paul, K., et al. (2002). Arsenic calamity in the Indian subcontinent: What lessons have been learned? *Talanta, 58*(1), 3–22.

Chatterjee, D., Mazumder, M., Barman, S., et al., 2018. Arsenic in groundwater: distribution and geochemistry in Nadia District, West Bengal, India. Groundwater of South Asia (pp. 349–373).

Chen, Y. C., Guo, Y. L. L., Su, H. J. J., et al. (2003). Arsenic methylation and skin cancer risk in southwestern Taiwan. *Journal of Occupational and Environmental Medicine*, 241–248.

Cheng, P. S., Weng, S. F., Chiang, C. H., et al. (2016). Relationship between arsenic-containing drinking water and skin cancers in the arseniasis endemic areas in Taiwan. *The Journal of Dermatology, 43*(2), 181–186.

Chowdhury, N. R., Das, A., Joardar, M., et al. (2020). Flow of arsenic between rice grain and water: Its interaction, accumulation and distribution in different fractions of cooked rice. *Science of the Total Environment, 731*, 138937.

Chung, J. Y., Yu, S. D., & Hong, Y. S. (2014). Environmental source of arsenic exposure. *Journal of Preventive Medicine and Public Health, 47*(5), 253.

Ciminelli, V. S. T., Gasparon, M., Ng, J. C., et al. (2017). Dietary arsenic exposure in Brazil: The contribution of rice and beans. *Chemosphere, 168*, 996–1003.

Cubadda, F., D'Amato, M., Aureli, F., et al. (2016). Dietary exposure of the Italian population to inorganic arsenic: The 2012–2014 total diet study. *Food and Chemical Toxicology: An International Journal Published for the British Industrial Biological Research Association, 98*(Pt B), 148–158.

Curnow, A., Salter, L., Morley, N., et al. (2001). A preliminary investigation of the effects of arsenate on irradiation-induced DNA damage in cultured human lung fibroblasts. *Journal of Toxicology and Environmental Health Part A, 63*(8), 605–616.

De Chaudhuri, S., Ghosh, P., Sarma, N., et al. (2008). Genetic variants associated with arsenic susceptibility: Study of purine nucleoside phosphorylase, arsenic (+3) methyltransferase, and glutathione S-transferase omega genes. *Environmental Health Perspectives, 116*(4), 501–505.

Del Razo, L. M., Quintanilla-Vega, B., Brambila-Colombres, E., et al. (2001). Stress proteins induced by arsenic. *Toxicology and Applied Pharmacology, 177*(2), 132–148.

Diaz, O. P., Arcos, R., Tapia, Y., et al. (2015). Estimation of arsenic intake from drinking water and food (raw and cooked) in a rural village of Northern Chile: Urine as a biomarker of recent exposure. *International Journal of Environmental Research and Public Health, 12*, 5614–5633.

Drobna, Z., Styblo, M., & Thomas, D. J. (2009). An overview of arsenic metabolism and toxicity. *Current Protocols in Toxicology, 42*(1), 4–31.

Engström, K. S., Broberg, K., Concha, G., et al. (2007). Genetic polymorphisms influencing arsenic metabolism: Evidence from Argentina. *Environmental Health Perspectives, 115*(4), 599–605.

Faita, F., Cori, L., Bianchi, F., et al. (2013). Arsenic-induced genotoxicity and genetic susceptibility to arsenic-related pathologies. *International Journal of Environmental Research and Public Health, 10*(4), 1527–1546.

FAO. (2011). *Rice market monitor, Seibersdorf, Austria* (Vol. XIV p. 1).

Fão, N., Nascimento, S., de La Cruz, A. H., et al. (2021). Estimation of total arsenic contamination and exposure in Brazilian rice and infant cereals. *Drug and Chemical Toxicology, 44*, 400–408.

Fujihara, J., Soejima, M., Yasuda, T., et al. (2011). Polymorphic trial in oxidative damage of arsenic exposed Vietnamese. *Toxicology and Applied Pharmacology, 256*(2), 174–178.

Gamble, M. V., Liu, X., Ahsan, H., et al. (2005). Folate, homocysteine, and arsenic metabolism in arsenic-exposed individuals in Bangladesh. *Environmental Health Perspectives, 113*(12), 1683–1688.

Gamble, M. V., Liu, X., Slavkovich, V., et al. (2007). Folic acid supplementation lowers blood arsenic. *The American Journal of Clinical Nutrition, 86*(4), 1202–1209.

García-Rico, L., Valenzuela-Rodríguez, M. P., Meza-Montenegro, M. M., et al. (2020). Arsenic in rice and rice products in Northwestern Mexico and health risk assessment. *Food Additives & Contaminants, Part B Surveill, 13*, 25–33.

Gebeyehu, H. R., & Bayissa, L. D. (2020). Levels of heavy metals in soil and vegetables and associated health risks in Mojo area, Ethiopia. *PLoS ONE, 15*, 1–22.

Gilbert, P. J., Polya, D. A., & Cooke, D. A. (2015). Arsenic hazard in Cambodian rice from a market-based survey with a case study of Preak Russey village, Kandal Province. *Environmental Geochemistry and Health, 37*(4), 757–766.

Guillod-Magnin, R., Brüschweiler, B. J., Aubert, R., et al. (2018). Arsenic species in rice and rice-based products consumed by toddlers in Switzerland. *Food Additives & Contaminants: Part A, 35*(6), 1164–1178.

Halder, D., Bhowmick, S., Biswas, A., et al. (2012). Consumption of brown rice: A potential pathway for arsenic exposure in rural Bengal. *Environmental Science & Technology, 46*(7), 4142–4148.

Halder, D., Bhowmick, S., Biswas, A., et al. (2013). Risk of arsenic exposure from drinking water and dietary components: Implications for risk management in rural Bengal. *Environmental Science & Technology, 47*(2), 1120–1127.

Halder, D., Biswas, A., Šlejkovec, Z., et al. (2014). Arsenic species in raw and cooked rice: Implications for human health in rural Bengal. *Science of the Total Environment, 497*, 200–208.

Hartwig, A., Groblinghoff, U. D., Beyersmann, D., et al. (1997). Interaction of arsenic (III) with nucleotide excision repair in UV-irradiated human fibroblasts. *Carcinogenesis, 18*(2), 399–405.

Hensawang, S., & Chanpiwat, P. (2017). Health impact assessment of arsenic and cadmium intake via rice consumption in Bangkok, Thailand. *Environmental Monitoring and Assessment, 189*, 599.

Herath, I., Vithanage, M., Bundschuh, J., et al. (2016). Natural arsenic in global groundwaters: Distribution and geochemical triggers for mobilization. *Current Pollution Reports, 2*, 68–89.

Hernández, A., & Marcos, R. (2008). Genetic variations associated with inter-individual sensitivity in the response to arsenic exposure. *Pharmacogenomics, 9*(8), 1113–1132.

Hossain, M. A., Rahman, M. M., Murrill, M., et al. (2013). Water consumption patterns and factors contributing to water consumption in arsenic affected population of rural West Bengal, India. *Science of the Total Environment, 463*, 1217–1224.

Hsieh, R. L., Su, C. T., Shiue, H. S., et al. (2017). Relation of polymorphism of arsenic metabolism genes to arsenic methylation capacity and developmental delay in preschool children in Taiwan. *Toxicology and Applied Pharmacology, 321*, 37–47.

Huang, Y., Wang, M., Mao, X., et al. (2015). Concentrations of inorganic arsenic in milled rice from China and associated dietary exposure assessment. *Journal of Agricultural and Food Chemistry, 63*(50), 10838–10845.

Hubaux, R., Becker-Santos, D. D., Enfield, K. S., et al. (2013). Molecular features in arsenic-induced lung tumors. *Molecular Cancer, 12*(1), 1–11.

International Agency for Research on Cancer (I.A.R.C.). (2004). Some drinking-water disinfectants and contaminants, including arsenic. Monographs on the evaluation of carcinogenic risks to humans. Lyon, France. *WHO, 84*, 61–96.

Islam, F. S., Gault, A. G., Boothman, C., et al. (2004). Role of metal-reducing bacteria in arsenic release from Bengal delta sediments. *Nature, 430*(6995), 68–71.

Islam, M. S., Ahmed, M. K., Habibullah-Al-Mamun, M., et al. (2017a). Arsenic in the food chain and assessment of population health risks in Bangladesh. *Environment Systems and Decisions, 37*(3), 344–352.

Islam, S., Rahman, M. M., Islam, M. R., et al. (2017b). Geographical variation and age-related dietary exposure to arsenic in rice from Bangladesh. *Science of the Total Environment, 601–602*, 122–131.

Jiang, Y., Zeng, X., Fan, X., et al. (2015). Levels of arsenic pollution in daily foodstuffs and soils and its associated human health risk in a town in Jiangsu Province, China. *Ecotoxicology and Environmental Safety, 122*, 198–204.

Jitaru, P., Ingenbleek, L., Marchond, N., et al. (2019). Occurrence of 30 trace elements in foods from a multi-centre sub-Saharan Africa total diet study: Focus on Al, As, Cd, Hg, and Pb. *Environment International, 133*(Pt B), 105197.

Joardar, M., Das, A., Chowdhury, N. R., et al. (2021). Health effect and risk assessment of the populations exposed to different arsenic levels in drinking water and foodstuffs from four villages in arsenic endemic Gaighata block, West Bengal, India. *Environmental Geochemistry and Health, 43*, 3027–3053.

Joseph, T., Dubey, B., & McBean, E. A. (2015). A critical review of arsenic exposures for Bangladeshi adults. *Science of the Total Environment, 527*, 540–551.

Kitchin, K. T., & Ahmad, S. (2003). Oxidative stress as a possible mode of action for arsenic carcinogenesis. *Toxicology Letters, 137*(1–2), 3–13.

Kitchin, K. T., & Conolly, R. (2010). Arsenic-induced carcinogenesis: Oxidative stress as a possible mode of action and future research needs for more biologically based risk assessment. *Chemical Research in Toxicology, 23*(2), 327–335.

Kollander, B., Sand, S., Almerud, P., et al. (2019). Inorganic arsenic in food products on the Swedish market and a risk-based intake assessment. *Science of the Total Environment, 672*, 525–535.

Kumar, M., Rahman, M. M., Ramanathan, A., et al. (2016). Arsenic and other elements in drinking water and dietary components from the middle Gangetic plain of Bihar, India: Health risk index. *Science of the Total Environment, 539*, 125–134.

Kumarathilaka, P., Seneweera, S., Meharg, A., et al. (2018). Arsenic accumulation in rice (Oryza sativa L.) is influenced by environment and genetic factors. *Science of the Total Environment, 642*, 485–496.

Kundu, M., Ghosh, P., Mitra, S., et al. (2011). Precancerous and non-cancer disease endpoints of chronic arsenic exposure: The level of chromosomal damage and XRCC3 T241M polymorphism. *Mutation Research/Fundamental and Molecular Mechanisms of Mutagenesis, 706*(1–2), 7–12.

Lantz, R. C., & Hays, A. M. (2006). Role of oxidative stress in arsenic-induced toxicity. *Drug Metabolism Reviews, 38*(4), 791–804.

Lindberg, A. L., Kumar, R., Goessler, W., et al. (2007). Metabolism of low-dose inorganic arsenic in a central European population: Influence of sex and genetic polymorphisms. *Environmental Health Perspectives, 115*(7), 1081–1086.

Liu, C. W., & Wu, M. Z. (2019). Geochemical, mineralogical and statistical characteristics of arsenic in groundwater of the Lanyang Plain, Taiwan. *Journal of Hydrology, 577*, 123975.

Liu, S. X., Athar, M., Lippai, I., et al. (2001). Induction of oxyradicals by arsenic: Implication for mechanism of genotoxicity. *Proceedings of the National Academy of Sciences, 98*(4), 1643–1648.

Lovely, D. R. (1995). Microbial reduction of iron, manganese, and other metals. In *Advances in agronomy*, USA.

Lunn, R. M., Langlois, R. G., Hsieh, L. L., et al. (1999). XRCC1 polymorphisms: Effects on aflatoxin B1-DNA adducts and glycophorin A variant frequency. *Cancer Research, 59*(11), 2557–2561.

Ma, L., Wang, L., Jia, Y., et al. (2016). Arsenic speciation in locally grown rice grains from Hunan Province, China: Spatial distribution and potential health risk. *Science of the Total Environment, 557–558*, 438–444.

Majumdar, A., Upadhyay, M. K., Giri, B., et al. (2021). Arsenic dynamics and flux assessment under drying-wetting irrigation and enhanced microbial diversity in paddy soils: A four year study in Bengal delta plain. *Journal of Hazardous Materials, 409*, 124443.

Mandal, U., Singh, P., Kundu, A. K., et al. (2019). Arsenic retention in cooked rice: Effects of rice type, cooking water, and indigenous cooking methods in West Bengal, India. *Science of the Total Environment, 648*, 720–727.

Meharg, A. A., Williams, P. N., Adomako, E., et al. (2009). Geographical variation in total and inorganic arsenic content of polished (white) rice. *Environmental Science & Technology, 43*(5), 1612–1617.

Menon, M., Sarkar, B., Hufton, J., et al. (2020). Do arsenic levels in rice pose a health risk to the UK population? *Ecotoxicology and Environmental Safety, 197*, 110601.

Meza, M. M., Yu, L., Rodriguez, Y. Y., et al. (2005). Developmentally restricted genetic determinants of human arsenic metabolism: Association between urinary methylated arsenic and CYT19 polymorphisms in children. *Environmental Health Perspectives, 113*(6), 775–781.

Minatel, B. C., Sage, A. P., Anderson, C., et al. (2018). Environmental arsenic exposure: From genetic susceptibility to pathogenesis. *Environment International, 112*, 183–197.

Mishra, S., Mattusch, J., & Wennrich, R. (2017). Accumulation and transformation of inorganic and organic arsenic in rice and role of thiol-complexation to restrict their translocation to shoot. *Scientific Reports, 7*(1), 1–13.

Mondal, D., Periche, R., Tineo, B., et al. (2020). Arsenic in Peruvian rice cultivated in the major rice growing region of Tumbes River basin. *Chemosphere, 241*, 125070.

Mondal, D., Rahman, M. M., Suman, S., Sharma, P., et al. (2021). Arsenic exposure from food exceeds that from drinking water in endemic area of Bihar, India. *Science of the Total Environment, 754*, 142082.

Moulick, D., Samanta, S., Sarkar, S., et al. (2021). Arsenic contamination, impact and mitigation strategies in rice agro-environment: An inclusive insight. *Science of the Total Environment, 800*, 149477.

Mridha, D., Gorain, P. C., Joardar, M., et al. (2022). Rice grain arsenic and nutritional content during post harvesting to cooking: A review on arsenic bioavailability and bioaccessibility in humans. *Food Research International*, 111042.

Mu, T., Wu, T., Zhou, T., et al. (2019). Geographical variation in arsenic, cadmium, and lead of soils and rice in the major rice producing regions of China. *Science of the Total Environment, 677*, 373–381.

Mukherjee, A. B., Bhattacharya, P., Jacks, G., et al. (2006). *Groundwater arsenic contamination in India: extent and severity* (pp. 553–593). CSIRO Publishing.

Naito, S., Matsumoto, E., Shindoh, K., et al. (2015). Effects of polishing, cooking, and storing on total arsenic and arsenic species concentrations in rice cultivated in Japan. *Food Chemistry, 168*, 294–301.

Nath, B., Stüben, D., Mallik, S. B., et al. (2008). Mobility of arsenic in West Bengal aquifers conducting low and high groundwater arsenic. Part I: Comparative hydrochemical and hydrogeological characteristics. *Applied Geochemistry, 23*(5), 977–995.

Nesnow, S., Roop, B. C., Lambert, G., et al. (2002). DNA damage induced by methylated trivalent arsenicals is mediated by reactive oxygen species. *Chemical Research in Toxicology, 15*(12), 1627–1634.

Nguyen, T. P., Ruppert, H., Pasold, T., et al. (2019). Paddy soil geochemistry, uptake of trace elements by rice grains (Oryza sativa) and resulting health risks in the Mekong River Delta, Vietnam. *Environmental Geochemistry and Health.*

Nickson, R. T., McArthur, J. M., Ravenscroft, P., et al. (2000). Mechanism of arsenic release to groundwater, Bangladesh and West Bengal. *Applied Geochemistry, 15*(4), 403–413.

Nyanza, E. C., Dewey, D., Thomas, D. S. K., et al. (2014). Spatial distribution of mercury and arsenic levels in water, soil and cassava plants in a community with long history of gold mining in Tanzania. *Bulletin of Environmental Contamination and Toxicology, 93*, 716–721.

Olson, K. R., & Chau, K. M. (2022). Natural and anthropic sources of arsenic in the groundwater and soils of the Mekong Delta. *Open Journal of Soil Science, 12*(11), 541–570.

Oremland, R. S., & Stolz, J. F. (2003). The ecology of arsenic. *Science, 300*(5621), 939–944.

Oremland, R. S., & Stolz, J. F. (2005). Arsenic, microbes and contaminated aquifers. *Trends in Microbiology, 13*(2), 45–49.

Oteiza, J. M., Barril, P. A., Quintero, C. E., et al. (2020). Arsenic in Argentinean polished rice: Situation overview and regulatory framework. *Food Control, 109*, 106909.

Otero, X. L., Tierra, W., Atiaga, O., et al. (2016). Arsenic in rice agrosystems (water, soil and rice plants) in Guayas and Los Ríos provinces, Ecuador. *Science of the Total Environment, 573*, 778–787.

Paiva, L., Hernández, A., Martínez, V., et al. (2010). Association between GSTO2 polymorphism and the urinary arsenic profile in copper industry workers. *Environmental Research, 110*(5), 463–468.

Phan, K., Sthiannopkao, S., Heng, S., et al. (2013). Arsenic contamination in the food chain and its risk assessment of populations residing in the Mekong River basin of Cambodia. *Journal of Hazardous Materials, 262*, 1064–1071.

Podgorski, J., & Berg, M. (2020). Global threat of arsenic in groundwater. *Science, 368*(6493), 845–850.

Polizzotto, M. L., Kocar, B. D., Benner, S. G., et al. (2008). Near-surface wetland sediments as a source of arsenic release to ground water in Asia. *Nature, 454*(7203), 505–508.

Rahaman, S., Sinha, A. C., Pati, R., et al. (2012). Arsenic contamination: A potential hazard to the affected areas of West Bengal, India. *Environmental Geochemistry and Health, 35*(1), 119–132.

Rashid, M. H., Rahman, M. M., Correll, R., et al. (2018). Arsenic and other elemental concentrations in mushrooms from Bangladesh: Health risks. *International Journal of Environmental Research and Public Health, 15*(5), 919.

Richards, L. A., Magnone, D., Boyce, A. J., et al. (2018). Delineating sources of groundwater recharge in an arsenic-affected Holocene aquifer in Cambodia using stable isotope-based mixing models. *Journal of Hydrology, 557*, 321–334.

Rodríguez-Lado, L., Sun, G., Berg, M., et al. (2013). Groundwater arsenic contamination throughout China. *Science, 341*(6148), 866–868.

Roos, W. P., Thomas, A. D., & Kaina, B. (2016). DNA damage and the balance between survival and death in cancer biology. *Nature Reviews Cancer, 16*(1), 20–33.

Rossman, T. G. (2003). Mechanism of arsenic carcinogenesis: An integrated approach. *Mutation Research/Fundamental and Molecular Mechanisms of Mutagenesis, 533*(1–2), 37–65.

Roychowdhury, T., Uchino, T., Tokunaga, H., et al. (2002). Survey of arsenic in food composites from an arsenic-affected area of West Bengal India. *Food and Chemical Toxicology, 40*(11), 1611–1621.

Ruttens, A., Cheyns, K., Blanpain, A. C., et al. (2018). Arsenic speciation in food in Belgium. Part 2: Cereals and cereal products. *Food and Chemical Toxicology, 118*, 32–41.

Samal, A. C., Kar, S., Bhattacharya, P., et al. (2011). Human exposure to arsenic through foodstuffs cultivated using arsenic contaminated groundwater in areas of West Bengal, India. *Journal of Environmental Science and Health. Part A, Toxic/Hazardous Substances & Environmental Engineering, 46*(11), 1259–1265.

Sarwar, T., Khan, S., Yu, X., et al. (2020). Analysis of arsenic concentration and its speciation in rice of different markets of Pakistan and its associated health risk. *Environmental Technology & Innovation*, 101252.

Sengupta, M. K., Hossain, M. A., Mukherjee, A., et al. (2006). Arsenic burden of cooked rice: Traditional and modern methods. *Food and Chemical Toxicology, 44*(11), 1823–1829.

Shaji, E., Santosh, M., Sarath, K. V., et al. (2021). Arsenic contamination of groundwater: A global synopsis with focus on the Indian Peninsula. *Geoscience Frontiers, 12*(3), 101079.

Shraim, A. M. (2017). Rice is a potential dietary source of not only arsenic but also other toxic elements like lead and chromium. *Arabian Journal of Chemistry, 10*, S3434–S3443.

Signes, A., Mitra, K., Burlo, F., et al. (2008). Effect of cooking method and rice type on arsenic concentration in cooked rice and the estimation of arsenic dietary intake in a rural village in West Bengal, India. *Food Additives and Contaminants, 25*(11), 1345–1352.

Signes-Pastor, A. J., Carey, M., Meharg, A. A. (2017). Inorganic arsenic removal in rice bran by percolating cooking water. *Food Chemistry, 234*, 76–80.

Šlejkovec, Z., Gorše, L., Grobler, A., et al. (2021). Arsenic speciation and elemental composition of rice samples from the Slovenian market. *Food Chemistry, 342*, 128348.

Smith, A. H., Lopipero, P. A., Bates, M. N., et al. (2002). Arsenic epidemiology and drinking water standards. *Science, 296*(5576), 2145–2146.

Steinmaus, C., Bates, M. N., Yuan, Y., et al. (2006). Arsenic methylation and bladder cancer risk in case-control studies in Argentina and the United States. Journal of Occupational and Environmental Medicine, pp.478–488.

Steinmaus, C., Moore, L. E., Shipp, M., et al. (2007). Genetic polymorphisms in MTHFR 677 and 1298, GSTM1 and T1, and metabolism of arsenic. *Journal of Toxicology and Environmental Health, Part A, 70*(2), 159–170.

Styblo, M., Del Razo, L. M., Vega, L., et al. (2000). Comparative toxicity of trivalent and pentavalent inorganic and methylated arsenicals in rat and human cells. *Archives of Toxicology, 74*(6), 289–299.

Tapio, S., & Grosche, B. (2006). Arsenic in the aetiology of cancer. *Mutation Research/Reviews in Mutation Research, 612*(3), 215–246.

Tseng, C. H., Huang, Y. K., Huang, Y. L., et al. (2005). Arsenic exposure, urinary arsenic speciation, and peripheral vascular disease in blackfoot disease-hyperendemic villages in Taiwan. *Toxicology and Applied Pharmacology, 206*(3), 299–308.

Tseng, W. P., Chu, H., How, S. W., et al. (1968). Prevalence of skin cancer in an endemic area of chronic arsenicism in Taiwan. *Journal of the National Cancer Institute, 40*(3), 453–463.

Vahter, M. (1999). Methylation of inorganic arsenic in different mammalian species and population groups. *Science Progress, 82*(1), 69–88.

Vahter, M. (2002). Mechanisms of arsenic biotransformation. *Toxicology, 181*, 211–217.

Vahter, M., Concha, G., Nermell, B., et al. (1995). A unique metabolism of inorganic arsenic in native Andean women. *European Journal of Pharmacology: Environmental Toxicology and Pharmacology, 293*(4), 455–462.

WHO. (2011). *Guidelines for drinking-water quality* (4th ed., pp. 164–168). World Health Organization.

Williams, P. N., Price, A. H., Raab, A., et al. (2005). Variation in arsenic speciation and concentration in paddy rice related to dietary exposure. *Environmental Science & Technology, 39*(15), 5531–5540.

Yim, S. R., Park, G. Y., Lee, K. W., et al. (2017). Determination of total arsenic content and arsenic speciation in different types of rice. *Food Science and Biotechnology, 26*, 293–298.

Yu, H. S., Liao, W. T., & Chai, C. Y. (2006). Arsenic carcinogenesis in the skin. *Journal of Biomedical Science, 13*, 657–666.

Zavala, Y. J., Gerads, R., Gürleyük, H., et al. (2008). Arsenic in rice: II. Arsenic speciation in USA grain and implications for human health. *Environmental Science & Technology, 42*(10), 3861–3866.

Zhang, J., Li, T., Yang, Y. L., et al. (2015). Arsenic concentrations and associated health risks in Laccaria mushrooms from Yunnan (SW China). *Biological Trace Element Research, 164*(2), 261–266.

Chapter 21
Transfer of Arsenic in Food Chain Through Groundwater Irrigation: A Threat to Food Safety and Human Health

Shepherd Manhokwe, Victor Nyanhete, Rudo Natasha Mugadza, Ruth Nyoka, and Patience Marume

Abstract *Background*: This paper aims to provide a comprehensive review of the food toxicity caused by transfer of arsenic through groundwater irrigation as well as the associated health complications. The review focuses on various species of arsenic, food systems and irrigation practices. This article presents potential health risks of exposure to arsenic as well as the occurrence and consumption of arsenic in irrigated crops. Additionally, the potential toxicity of arsenic is discussed. *Design/methodology/approach*: Information and data regarding food toxicity from irrigation practices were collected from the scientific sources. *Findings:* In the present scenario, arsenic is available in the food system in the form of inorganic and organic arsenic species. The inorganic forms of arsenic are more toxic as compared to the organic arsenic and the arsenic is incorporated into various food products through different routes. Arsenic becomes exceptionally toxic even at low exposure levels because of its high water solubility and bioaccumulation tendency in different environmental matrices. Contaminated groundwater used for irrigation of crops such as rice and tomatoes, is a major source of exposure to arsenic. Fish and other seafood contain the most of less toxic organic arsenic. *Originality/value*: Prolonged arsenic toxicity leads to carcinogenic and non-carcinogenic health risks such as arsenicosis, cancers, hepatotoxicity, kidney failure, and skin disease. Numerous studies have demonstrated that the consumption of arsenic or arsenic contaminated products can also cause diabetes, anaemia and certain types of cancer. Creating awareness among consumers about the toxicity profile and carcinogenic properties of arsenic is crucial in order to minimize its detrimental effects.

Keywords Arsenic · Underground water · Irrigation · Food system · Toxicity

S. Manhokwe (✉) · V. Nyanhete · R. N. Mugadza · R. Nyoka · P. Marume
Department of Food Science and Nutrition, Midlands State University, P Bag 9055, Gweru, Zimbabwe
e-mail: manhokwes@staff.msu.ac.zw

B. Sen Gupta and N. Martínez-Villegas (eds.), *Arsenic Remediation of Food and Water*,
https://doi.org/10.1007/978-981-97-4764-1_21

Introduction

Access to sufficient amounts of safe and nutritious food is a basic human right and essential to sustaining life and promoting good health (Desye et al., 2023). Humans are exposed to heavy metals and metalloids in the environment, in water and in the entire food system. Groundwater is the most widely distributed freshwater resource on earth. It is significantly important for human beings, industry, agriculture and social development. Arsenic (As) is widely distributed on earth and its occurrence in natural water is dependent on the local geology, hydrogeology, and geochemical characteristics of the aquifer and climatic factors as well as human activities (Herath et al., 2016; Raju, 2022). The content of arsenic in natural water bodies is relatively low, with a content usually less than 1.0 μgL^{-1} (Guo et al., 2014). The recommended arsenic level in drinking water by the World Health Organization (WHO) the regulation limit (as imposed by the respective legislation) of arsenic concentration in drinking water is10 μgL^{-1} (WHO, 2018). High arsenic levels in water bodies can be attributed to high levels of human activities such as mining, industrial operations and hazardous waste dumping (Ahoulé et al., 2015b). High level of arsenic in water bodies are a concern due to high toxicity and carcinogenicity of arsenic (Jan et al., 2015, Styblo et al., 2002).

Sources of Arsenic

Arsenic can be released into the atmosphere by natural or anthropogenic activities (Khosravi-Darani et al., 2022). High concentrations may naturally occur due to erosion of minerals in several areas, which are therefore contaminated by anthropogenic activities (industrial production and uses). Natural activities such as weathering, volcanic eruption, rock erosion, and forest fires may release arsenic into the environment. Particularly, weathering is the main naturally occurring process through which arsenic sulfides present in ores are converted into a highly mobile form of arsenic, which may enter the arsenic cycle as a dust or via dissolution in rain, water reservoirs, or ground water (Smedley & Kinniburgh, 2002). In many parts of Africa, industrial production has been on the rise over the past 10 years, whilst mining of minerals has become a common practice. These activities have therefore contributed to the presence of arsenic in groundwater.

According to a rough estimation, the global annual release of arsenic in the atmosphere is 7.8×10^7 kg/year (Khosravi-Darani et al., 2022). The natural sources are expected to release 1.2×10^7 kg per year, where in volcanoes and microbial volatilization may supplementarily contribute an additional 8.9×10^6 and 2.1×10^7 kg/year, respectively (Lee et al., 2019). This widespread distribution of Arsenic throughout the earth's crust, soil, sediments, water, air and living organisms leads to serious adverse effects on the environment (Smedley & Kinniburgh, 2002; Herath et al., 2016). Activities like topography and rigidity of soil, thereby facilitate for easy dissolution of soil constituents during natural formation of groundwater. Hence contamination can occur by natural runoff and subsequent seepage of arsenical compounds applied to the farm lands, atmospheric deposition, or improperly disposed chemicals (Brunt et al., 2004; Madhukar et al., 2016; Herath et al., 2016). The arsenic concentrations in soils show a similar range to that found in sediments except where contaminated by industry or agricultural activity. The arsenic release potential seems to be correlated with three soils or hydrogeological situations (this does not mean, however, that all soils of these types necessarily carry measurable levels of arsenic) (Alaerts and Khouri, 2004). Peaty or peaty—peaty—clayey soils with humic organic content and with high water table, which contain arsenopyrite crystals. When these soils, often associated with wetlands and marshes, are drained to bring them into production, oxygen penetrates the peat and oxidises the arsenopyrite crystals, thus releasing a sulphate-rish acid as well as dissolved arsenic into ground and drainage water. Arsenic concentrations are usually below 100 μg^{-1}. Such conditions are found for example in South Africa, Central Africa, Congo Basin and many peaty lowlands. However, the potential risk of surface and groundwater contamination by arsenic is associated mainly with the amounts of arsenic present in soils, because arsenic is more highly concentrated in soils than rock (Bhattacharyya et al., 2003; Meharg & Rahman, 2003a, b; Bundschuh et al., 2013). Arsenic may be found in soil as inorganic species as well as binding to organic compounds; however, it exists predominantly as inorganic species in soil. Inorganic arsenic compounds can undergo methylation by microorganisms, producing, under oxidizing conditions, monomethylarsonic acid (MMA), dimethylarsinic acid (DMA), and trimethylarsine oxide (Bhattacharya et al., 2007; Bissen & Frimmel, 2003; Nicolli et al., 2012).

Principle factors affecting the level of arsenic in soils include the parent rock, climate, organic and inorganic components of the soil, and redox potential status (Mandal & Suzuki, 2003; Bissen & Frimmel, 2003; Klump et al., 2006; Kar et al., 2010). The sorption of arsenic in soils is influenced mainly by the bioavailability of subsequent mobility of arsenic in soils, and the presence of inorganic PO_4^{3-}, Ca^{2+}, and Na^+ in different types of soil (Smith, et al., 2002; Bauer & Blodau, 2006; Wang & Mulligan, 2006.) Arsenic concentration in terrestrial soils typically varies from 1.5 to 3 mg/kg, depending on the origin of the soil (Sandhi et al., 2022). A study found that arsenic concentrations could reach up to 20–30 mg/kg in soils that have been irrigated frequently with groundwater (Meharg & Rahman, 2003a; b).

The presence of carbonate and bicarbonate ions in aquifer sediments directly results in a change in pH (Anawar et al., 2004). Carbonate ions can form complexes on the surface sites of iron hydroxide and substitute arsenic from the surface of minerals and sediments, thereby releasing arsenic into groundwater (Wang & Mulligan, 2006; Biswas et al., 2014). The aquifers associated with arsenic usually are linked to the input of dissolved organic matter, which also can greatly facilitate the release of arsenic from soils and sediments to groundwater (Bauer & Blodau, 2006; Kar et al., 2011). In marine ecosystems, complex organic arsenic compounds, such as tetramethylarsonium salts, arsenocholine, arsenobetaine, dimethyl (ribosyl) arsine oxides, and arsenic-containing lipids, have been identified. It is estimated that seawater typically contains 1 to 8 μg/L of arsenic (Bhattacharya et al., 2007). Only a very limited amount of the total arsenic present in the oceans remains in solution, because a large portion of dissolved arsenic can be adsorbed onto suspended particulate materials (Mandal & Suzuki, 2003; Kohlmeyer et al., 2002).

Arsenic also may exist naturally in geothermal systems and springs as a result of the mobilization of arsenic from rocks and mineral phases, and arsenic contaminated geothermal systems and springs are recognized in many areas around the world (Ormachea Munoz et al., 2015; Bundschuh et al., 2013; Bundschuh & Maity, 2015; Lopez et al., 2012).

Arsenic in Irrigated Plants

There has been a gradual increase in the use of groundwater for irrigation over the past two decades. If arsenic contaminated water is used for irrigation, it may create a hazard both in soil environment and in crop quality (Heikens, 2006). Similarly, drinking arsenic contaminated groundwater exposes humans to arsenic toxicity. Concentration of arsenic in irrigation water is accumulated in soil, showing greater accumulation at top layers of soil and it is found in soluble form (Saldaña-Robles et al., 2018). Irrigation with arsenic contaminated groundwater leads to increased concentration of arsenic in cereals, vegetables, and other agricultural products of the arsenic affected areas (Kabir et al., 2016).

Chemical and physical conditions of the aqueous phase of water affect arsenic chemical behavior in the water-soil–plant system (Saldaña-Robles et al., 2018). The relationship between arsenic concentration of water and arsenic concentration absorbed by plants is influenced by parameters such as pH, temperature, soil type and kind of plant (Saldana-Robles et al., 2018). Moreover, the behavior of arsenic is distinctly different under flooded (anaerobic) and non-flooded (aerobic) soil conditions, with flooded conditions being likely the most hazardous in terms of uptake by plants and toxicity (Heikens, 2006). In continuous flooding conditions, reductive dissolution of arsenic containing iron oxyhydroxides (FeOOH) releases arsenic due to reduction of iron (II) accompanied with reduction of arsenate, As (V) to arsenite (Jahan et al., 2021). In general, arsenate (As^{5+}) is the major species in the aerobic or oxygenated soil, whereas arsenite (As^{3+}) prevails in the anaerobic or submerged soil (Sandhi et al., 2022). In water with dissolved oxygen concentrations below 0.8 mg L^{-1} it is more common to find arsenic (III) than arsenic (V). Anions of arsenic (V) that are strongly adsorbed by minerals under conditions of neutral pH whilst under oxidizing conditions, arsenate is stable and the absorption decreases with increasing pH (Saldana-Robles et al., 2018).

Agricultural practices with chemicals, historically used, contribute to the accumulation of arsenic (Romero-Estevez et al., 2023). Some combinations of pesticides and chemical fertilizers were widely applied over long periods of time in the past, owing to arsenic accumulation (Saldaña-Robles et al., 2018). The arsenic contained in the irrigation water, is bioaccumulated in crops, which form part of the diet, and in this way reaches human beings. As a consequence, this bioaccumulation of arsenic has negatively affected human health (Mayorga et al., 2013). The amount of arsenic accumulated by the plant depends on the transfer factor, which is controlled by the plant species and the soil arsenic concentration.

Some plants can accumulate high amount of arsenic and are called as hyperaccumulators (Sandil et al., 2019). Plants have been found to accumulate the highest arsenic concentration in the roots and the lowest in the edible parts. Arsenic accumulation in plants is usually in the order: roots > stem > leaves > edible parts (Lu et al., 2010; Sandil et al., 2021). Stoltz and Greger (2002) stated that some tests showed a high correlation between the concentration of arsenic among varieties of cotton (*Eriophorum angustifolium* and *E. scheuchzeri* Honckenya Hoppe) and arsenic concentration in agricultural soil, reporting maximum concentrations of 8.4 mg kg^{-1} in soil and 276 mg kg^{-1} in root. Sadee et al. (2016) studied arsenic uptake in broad bean (*Viciafaba*) in loamy soil, with soil and irrigation water arsenic concentration 5.32mgkg^{-1} and 1.06μgL^{-1}, respectively. For mangrove plants such as *Aegiceras corniculatum* L. growing in arsenic contaminated soils, the maximum concentration was found in the roots which was 74.54–89.26% of the total arsenic accumulation in the plant and the roots of *Zea mays were found to contain* 14.5–27.4 mg kg^{-1}. (Neidhardt et al., 2015, Wu et al., 2015). The concentration was in the order of root > leaf > pod > stem > bean.

Bioavailable arsenic in the soil is not all phyto available, its uptake depends on plant type and species (Sandil et al., 2021). Interestingly, Arsenic content of irrigation water accumulates in the topsoil (10–15 cm) and is affected by soil pH, redox potential, organic matter (OM), Fe, Mn, P, $CaCO_3$, and clay content (texture) (Sandil et al., 2019, 2021). As (III) and As (V) are taken-up by different mechanisms. As (V) is taken up via the high affinity phosphate uptake system (Meharg, 2004). Arsenic methylation in the soil is controlled by arsenic methyltransferase (arsM). This is a microbial activity that controls the uptake of methylated arsenic by the plant. The methylation process promotes the detoxification process since some end products such as trimethylarsine (TMAs (III) are volatile (Zhu et al., 2014).

Accumulated arsenic in plants affect the growing mechanisms as well as the yield of crops and may negatively affect the health of living beings (Rosas et al., 2016). Excessive arsenic inhibits germination, reduces root and shoot growth, lowers yield, produces reactive oxygen species (ROS), affects photosynthesis, mineral nutrition, and causes necrosis, thus posing a threat to agricultural productivity (Sandil et al., 2021). Plants cultivated in high arsenic concentration have been reported to have decreased biomass and increased arsenic accumulation in the plant parts (Sandil et al., 2019); Sandil et al., 2021). Numerous greenhouse and field studies have revealed that an increase in arsenic in cultivated soils leads to an increase in the levels of arsenic in edible vegetables with many factors affecting bioavailability, uptake, and phytotoxicity of arsenic (Huq, (n.d.). Bioaccumulation in rice has been extensively studied (Dittmar et al., 2010; Sadee et al., 2016; Upadhyay et al., 2020). Several research studies have already proved that the consumption of rice and vegetables cultivated with arsenic elevated groundwater is a potential human health risk (Sadee et al., 2016; Stazi et al., 2016). Table 21.1 shows the concentration of arsenic in food systems.

Table 21.1 Common cereals and vegetable concentration of arsenic

Vegetable/cereal	Concentration	Reference
Broad bean (*Viciafaba*)	1.06 μg L^{-1}	Sadee et al. (2016)
Tomato	0.08–0.11 mg kg^{-1}	Stazi et al. (2016)
Cabbage	7.8 μg kg^{-1}	Huang et al. (2006)
Lettuce	23.3 μg kg^{-1}	Huang et al. (2006)
Rice	0.29–0.95 mg kg^{-1}	Upadhyay et al. (2020)

Human Health Risk Assessment in Arsenic Toxicity

Risk analysis provides a framework to effectively manage, assess and communicate risks in cooperation with the diverse stakeholders involved (Santhini & Lee, 2015). It consists of three components: risk assessment, risk management and risk communication as shown in Fig. 21.1. Risk analysis provides a framework to collect and analyze the best available scientific information on a hazard that presents a risk to people, animals or plants in a certain country, region or globally.

A hazard is a biological, chemical or physical agent in, or condition of, food with the potential to cause an adverse health effect. Inherent property of an agent or

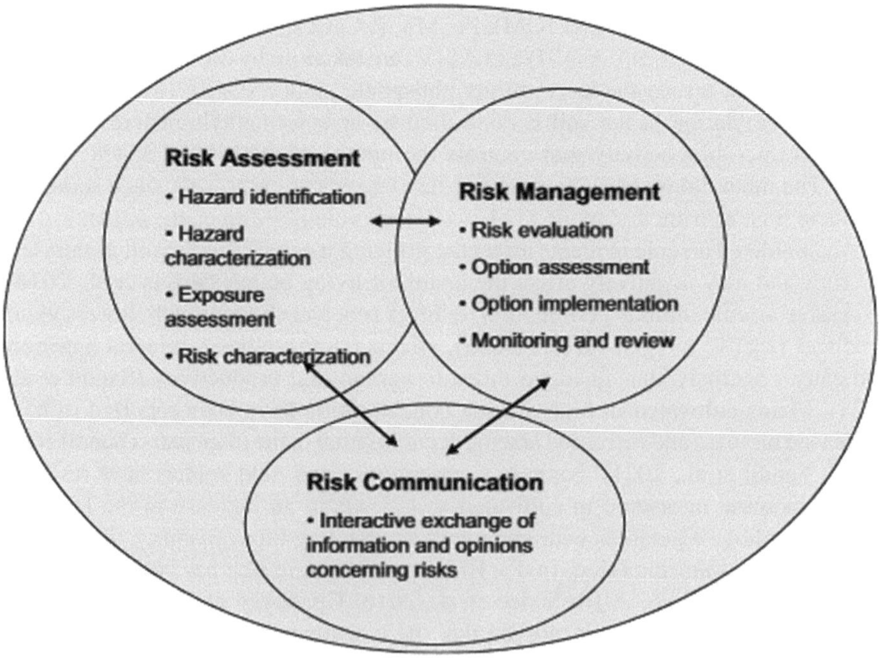

Fig. 21.1 Risk analysis process

situation having the potential to cause adverse effects when an organism, system or (sub) population is exposed to that agent. (IPCS, 2004). Risk, thus, can be defined as function of the probability of an adverse health effect and the severity of that effect, consequential to a hazard(s) in food. The probability of an adverse effect in an organism, system or (sub) population caused under specified circumstances by exposure to an agent (IPCS, 2004).

Risk assessment is the central scientific component of the risk analysis process which characterizes the potential hazards associated with specific risk of life due to exposure for a certain period (Ostrom & Wilhelmsen, 2019). Human health risk assessment is considered as the characterization of the potential adverse health effects of humans as a result of exposures to environmental hazards (USEPA, 2012). It is a scientifically based process consisting of four steps viz hazard identification, hazard characterization, exposure assessment and risk characterization. The risk assessment process uses scientific tools to identify and measure a hazard, to determine possible routes of exposure, and to calculate the potential risk (Lushenko, 2010). The risk to be calculated is usually classified as carcinogenic or non- carcinogenic and this classification determines the procedure to be followed when calculating risk (Koki et al., 2015). Non-carcinogenic chemicals have a threshold value below which no adverse health effects will be observed and an essential part of the dose–response in risk assessment makes use of a reference dose (RfD) (Koki et al., 2015). Exposure to carcinogens has no threshold since someone may get cancer even by being exposed to the lowest doses.

There are two main exposure pathways for arsenic which are intake of the metals through water consumption and by skin absorption (Koki et al., 2015). Approaches for doing exposure assessment include indirect and direct techniques, covering measurement of environmental concentrations and personal exposures, as well as biomarkers (Wambaugh et al., 2019). Assessment of non-carcinogenic risks can be achieved by estimating the hazard quotient (HQ). If $HQ > 1$ there is a high risk associated with the metal. If $HQ < 1$ it means the adverse non cancer effects are unlikely.

Arsenic Metabolism and Toxicity

The intricacy of arsenic metabolism originates from its yield of diverse intermediates/products, each of which has its own distinct behavior and interactions within the human body (El-Ghiaty & El-Kadi, 2023). Ingestion of arsenic activates metabolic pathways for excretion, resulting in a number of metabolites, some of which are more potent and toxic than the originally ingested inorganic form of arsenic (iAs), including mono-and dimethylated arsenicals (Tyler & Allan, 2014). In general, inorganic As (V) is one tenth as toxic as inorganic As (III), while methylarsonic acid (MMA) and dimethylarsinic acid (DMA) are less toxic than inorganic As (V). Once inside a biosystem, iAs undergoes biotransformation, via methylation, into more complex organic arsenic. After oral ingestion, both iAs (III) and iAs (V) are highly absorbed, extensively metabolized, and finally excreted, primarily in urine

(El-Ghiaty & El-Kadi, 2023). This pollutant is associated with increased risk of developing chronic diseases, including metabolic diseases. Metabolic syndrome (MS) is a complex pathology that results from the interaction between environmental and genetic factors. This condition increases the risk of developing type 2 diabetes, cardiovascular diseases, and cancer.

Metabolism of arsenic in humans involves two processes. After entering a cell, arsenate is reduced to arsenite. Arsenite is then methylated to form MMA and DMA. Metabolism of arsenic takes place in the liver where the first step is methylation. Methylated trivalent inorganic arsenic in the liver is released into the bloodstream and excreted in urine via transport through cell membranes (Ro et al., 2022). The presence of monomethylarsenic acid (MMAV) and dimethylarsenic acid (DMAV) indicates the methylation of arsenic in bile and urine (Kulshrestha et al., 2014). Pentavalent methylated species, specifically the dimethylated DMAV, are the dominant excretory forms of arsenic and are more prevalent than trivalent ones in human urine (El-Ghiaty & El-Kadi, 2023).

The toxicity of an arsenical can be generally predicted by its oxidation state and degree of methylation (El-Ghiaty & El-Kadi, 2023). Arsenic cytotoxicity is completely dependent on its oxidation state and chemical structure. During the detoxification of iAs (III), the compound is biomethylated to mono- and dimethylated (in some cases trimethylated) arsenicals of varying toxicity (Khairul et al., 2017). Arsenic is metabolized by a series of reduction and methylation reactions via 1-carbon metabolism, in which methyl groups are transferred from S-adenosylmethionine (SAM) to arsenic in its trivalent state (Vahter, 2007). The severity of arsenic poisoning is closely related to changes in metabolic enzyme activities. Therefore, exploring changes in metabolic enzyme activity during arsenic poisoning may help to identify individuals who are particularly vulnerable to arsenic toxicity (Muhetaer et al., 2022).

Arsenic causes toxicity via oxidative stress by affecting the antioxidant enzymes. Mechanistically, toxic interactions of trivalent, both inorganic and organic, species can be generally ascribed to their affinity to sulfhydryl-containing biomolecules, especially peptides and proteins (El-Ghiaty & El-Kadi, 2023). The properties of arsenic binding to proteins are one of the most important factors that have been found critical in influencing the sensitivity and resistance of a specie towards arsenic exposure (Khairul et al., 2017). Heavy metals react with sulfhydryl group of cysteine residue found frequently in the enzymes. They replace the hydrogen atoms and make covalent bond with sulphur. They induce the alterations in enzymatic confirguration resulting in changed active site subsequently no enzymatic activity (Sabir et al., 2019). Generally speaking, trivalent arsenicals have high binding affinity to sulfhydryl groups of proteins, while pentavalent arsenicals found to be not binding with sulfhydryl groups of proteins (Khairul et al., 2017).

Arsenic alters the activity of a number of enzymes associated with carbohydrate metabolism viz. glucogenesis, glycolysis, gluconeogenesis and pentose phosphate pathway (Sabir et al., 2019). Trivalent arsenicals are moderately effective inhibitors of numerous enzymes involved in glucose metabolism such as succinyl Co-A synthase, α-ketoglutarate dehydrogenase and pyruvate dehydrogenase (PDH).

Inhibition of PDH leads to decreased generation of ATP and energy resulting in cell damage and cell death (Kulshrestha et al., 2014). Arsenic stimulates the production of reactive oxygen species (ROS) which results in the induction of adverse health effects (Kulshrestha et al., 2014). Arsenic exposure generally induces the production of reactive oxygen species that can lead to the production of antioxidant metabolites and numerous enzymes involved in antioxidant defense. Increased ROS causes cellular and metabolic impairment through oxidative damage, which results in physiological abnormalities and deleterious chronic disorders (Kulshrestha et al., 2014). Arsenic interferes with gluconeogenesis in muscle by inhibiting glucose transporters and suppressing regulatory genes in glucose metabolism. Thus, arsenic can substantially interfere with glycolytic and mitochondrial energy generation (Ro et al., 2022).Arsenic can be therefore considered a significant risk factor for developing type 2 diabetes (T2D). Insulin promotes glycolysis and suppresses the glucogenesis, gluconeogenesis, protein catabolism and lipolysis. Impaired insulin secretion, due to pancreatic damage by arsenic and cadmium, results in reduced rate of glycolysis and promotes the glucogenesis, gluconeogenesis, protein catabolism and lipolysis (Sabir et al., 2019). Insulin-independent diabetes is the common form of diabetes mellitus among people chronically exposed to iAs (Kulshrestha et al., 2014). Arsenic binds with disulphide bridge present in insulin and insulin receptor and makes them non-functional. Arsenic impairs the translocation of glucose transporter (GLUT) by decreasing the phosphorylation of protein kinase B (PKB)/Akt. In fact, Akt is necessary for GLUT4 translocation and glucose transport. Therefore, loss of dual phosphorylation of Akt due to the effects of arsenic and its metabolites, causes significant inhibition of insulin-dependent glucose uptake hence results in hyperglycemia (Khairul et al., 2017). In addition, Arsenic competes with Pi and binds with ADP in order to make ADP-arsenate that is subsequently used in the formation of glucose-6-arsenate. Arsenic also causes the elevation of oxidative stress level through which it increases the expression of pro-inflammatory cytokines (TNF–α and IL-6) and decreases the expression of peroxisome-proliferator-activated receptor gamma (PPARγ) and these two factors play chief role in the development of insulin resistance (Sabir et al., 2019).

Oxidative stress has been identified as an important mechanism of arsenic toxicity and carcinogenicity. Accumulated arsenic in the body can generate reactive oxygen species (ROS) and epigenetic dysregulation, which promotes carcinogenesis (Ro et al., 2022). Promotion of oxidative stress by arsenic compounds is one of the main causes for arsenic-induced carcinogenicity (Khairul et al., 2017). During carcinogenesis, arsenic exposure induces a loss of DNA methylation, which increases gene expression (Ro et al., 2022). In particular, arsenic induces oxidative DNA damage and lipid peroxidation (Vahter, 2007). Oxidative carbon metabolism, amino acid and protein relationships, and nitrogen and sulfur assimilation pathways are also impacted by arsenic exposure (Finnegan & Weihua, 2012). ROS can damage proteins, amino acids, purine nucleotides and nucleic acids and cause peroxidation of membrane lipids However, under stronger stresses, such as arsenic exposure, where ROS generation increases, these defense mechanisms may be overwhelmed, leading to cellular damage. This damage can therefore lead to cell death (Finnegan & Weihua, 2012).

Health Problems Associated with Arsenic Poisoning

Humans can be exposed to arsenic in a variety of ways which include water, contactand air. Inorganic arsenic forms are reported to be more toxic than organicarsenic, Pal (2015). The ingestion of arsenic through drinking water, using contaminated water in food preparation, irrigation of crops, food or beverage industrial processes and eating contaminated food are considered to be the primary exposure pathways (Tsuda et al., 1992). Drinking As-contaminated water is the most typical route to take arsenic through ingestion, and can lead to acute and chronic poisoning.

Toxicity of arsenic to humans depends on factors such as age, gender, concentration and the form of arsenic, duration and dose of arsenic exposure and nutritional status of the individual (Jovanovic & Rasic-Milutinovic, 2017). Low selenium intake, smoking and genetic factors, also enhance the development of arsenicosis (Deb et al., 2013; Chen et al., 2001; Gamble et al., 2007; Miyazaki et al., 2005; Lamm et al., 2006). Chronic exposure to inorganic Arsenic increases the risk of diabetes mellitus, adverse pregnancy outcomes, even cancers of the skin, lungs, and urinary bladder (Paul et al., 2007; Vahter, 2009; Wang et al., 2014, Jovanovic & Rasic-Milutinovic, 2017). Arsenic exposure during gestation period has ill-effects on fetus development through irreversible impairment of thioredoxin reductase, methyltransferases and DNA repair enzymes. Arsenic is the first metalloid directly linked with adverse pregnancy after-effects (Jovanovic & Rasi-Milutinovic, 2017; Vahter, 2009). A close relationship of high Arsenic levels with elevated blood sugar level, which in turn affects insulin signal transduction and insulin sensitivity was demonstrated by Paul et al. (2007). Acute arsenic poisoning has no distinct symptoms. Weariness, colitis, loss of reflexes, weight loss, weakness, anorexia, gastritis, hair loss, and anorexia are a few of the non-specific symptoms of chronic poisonin (Fatoki & Badmuys, 2022). Long-term arsenic exposure has been associated with development of skin lesions such as hyperkeratosis and pigmentation changes and black-foot disease, cancer of the bladder, lung, kidney and liver, circulatory disorders, developmental effects, cardiovascular diseases, neurotoxicity and diabetes (Steinmaus et al., 2013; Martinez et al., 2011; Gbaruko et al., 2010 and Smith et al., 1992). Vomiting, abdominal pain, diarrhea, numbness and tingling, muscle pain and cramps, as well as death in extreme cases are reported as the main health symptoms of elevated arsenic intake (Zeng & Zhang, 2020). Villagers with high arsenic ground water in Ghana and Burkina Faso were affected by conditions which include skin lesions which are characterized by melanosis, keratosis, and ulceronecrotic. (COWI, 2005; Smedley et al., 2007).

In 1993, the World Health Organization (WHO) reduced provisionally the guideline value from 50 to 10 $\mu g\ L^{-1}$ which means any water with arsenic level above this value unfit for human consumption. Africa studies do not usually make a direct link between health problems and arsenic in drinking water (Ahoulé et al., 2015a, b). However adverse effects of arsenic on human health have been reported in some countries and are associated to mine smoke and food crops grown in mining-contaminated areas (Ahmad et al., 2018).

Asante et al. (2007) found high concentrations of arsenic in human urine samples from residents of the mining town of Tarkwa in Ghana and from non-mining town of Accra and suggested that human contamination of arsenic has another source, possibly foods. On the other hand, water samples from Tarkwa had very high concentrations of arsenic comparatively to those from Accra. However, although most scholars relate arsenic poisoning from other sources such as food, drinking water remains the significant human exposure way of inorganic arsenic contamination (Irund et al., 2022; Ahoule et al., 2015a; b). Smedley et al. (2007) found a link between the concentrations of inorganic urinary arsenic from sample populations and drinking water in Ghana. Therural stream-water-drinking community was found to have a mean value of 42 $\mu g\,L^{-1}$ of urinary arsenic whilst a suburb of Obuasi community using groundwater for potable supply has 18 $\mu g\,L^{-1}$. Health surveys carried in Ouahigouya area, northern Burkina Faso, found that skin lesions characterized by melanosis, keratosis, and ulceronecrotic tumor were found in patients from three villages with high arsenic in groundwater samples (COWI, 2005; Smedley et al., 2007). Another study by Somé et al. (2012) identified that 29.26 and 46.34% of the population (240 persons) in 20 villages in Yatenga province, Burkina Faso, were affected by melanosis and keratosis, respectively a public health problem which are linked to arsenic is due to drinking water.

Nzihou et al. (2013) focused on the health risk related to medium- and long-term exposure to drinking arsenic contaminated water. They built an exposure factor based on the arsenic concentration in drinking water. The results have shown that for arsenic concentration ranging from 1 up to 87.8 $\mu g\,L^{-1}$, exposes the majority of the population at risk of arsenic-related diseases (cancerogenic or non-cancerogenic). Although there is no enough evidence to link arsenic in water to risk among populations, the occurrence of arsenic in significant concentrations in African water remains a threat to human health (Ahoulé et al., 2015a, b).

Arsenic Toxicity Studies in Africa

The review of studies on arsenic in African waters shows that arsenic can be found in high concentrations in both surface water and groundwater. High concentrations of arsenic water were reported in the groundwater by in Burkina Faso (1.6 mg/l) and in Ghana (1.760 mg/l) with its origin from weathered rocks of the underlying geology (Ahoulé et al., 2015a; b). In Burkina Faso, arsenic in ground water is predominantly present as arsenate and concentrations can exceed 1.0 mg/l (Clair-Caliot et al., 2021). In addition, high arsenic water was reported in some parts of Zimbabwe, especially in the gold mining areas of Mashonaland Central, Mashonaland East, Mashonaland West and Masvingo whereby a range of 0.15—1.19 mg/l of arsenic water was noted from different sources of water which included borehole water, dam as well as municipal water (Mhandire et al., 2017a; b). The contamination of groundwater by naturally occurring arsenic in West Africa has received little attention compared to the widespread As-endemic areas in Asian lowlands (Clair-Caliot et al,. 2021).

According to Sanou et al. (2020) about 560,000 people in rural communities are estimated to rely on water from arsenic-contaminated tube wells and this has been associated with health problems such as skin, liver, lungs, bladder and kidneys cancers as well as dermatological effects such as melanosis and keratosis. In a study by Lyndon et al. (2023) in Kumasi, Ghana, the average concentration of Arsenic (As) in vegetables from all sampling sites exceeded the recommended maximum limit of (RML), 0.10 mg/kg. The highest average concentration of arsenic for lettuce recorded at a farm was 6.58 mg/kg and the highest arsenic concentration in spring onions was 4.84 mg/kg. In another study in Tanzania by Mng'ong'o et al. (2021), determination of arsenic in different irrigation schemes enabled determination of the spatial distribution of arsenic in agroecosystem with highly intensified and commercialized irrigation schemes recording significantly high arsenic values. Irrigation schemes such as *Kapunga* (1279.24 μg/kg), *Ihahi* (1536.31 μg/kg), *(Igalako* 1172.36 μg/kg), *Chimala (*1606.02 μg/kg), (*Ilaji*1790.71 μg/kg) and *Mahongole* (1700.95 μg/kg) had higher arsenic values. In Kenya, levels of arsenic in the raw materials used to make the brews were determined. Maize, millet and sorghum used in the preparation of *busaa* had high levels of arsenic, 2.38 ± 0.440, 1.50 ± 0.190 and 1.33 ± 0.217 μg/kg respectively (Masime et al., 2013). Kapwata et al. (2023) investigated relations between personal exposure to elevated concentrations of arsenic in water and soil and blood arsenic levels amongst people living in rural areas in Limpopo, South Africa. Significant percentages of drinking water, soil and blood samples from the exposed sites were above the internationally recommended guidelines (namely, 10 μg/L, 20 mg/kg and 1 μg/L, respectively). Meck et al. (2019) studied arsenic levels in vegetables consumed in Zimbabwe. Arsenic concentrations were found to be above the FAO/WHO Codex maximum permissible levels (0.3 mg kg^{-1}), in all vegetable samples from the greenstone substrate of Zimbabwe, as well as in the tomatoes (4.03 mg kg^{-1}), pumpkin leaves (5.20 mg kg^{-1}), cabbage (5.80 mg kg^{-1}) and okra (1.00 mg kg^{-1}) in the granitic substrate of Zimbabwe and hence, possible health risks from associated toxicity.

Regulation and Control

The World Health Organization (WHO) developed health-based guidelines to assist nations with a baseline for developing their own enforceable water-quality standards. According to the WHO drinking water quality should have a provisional value of 0.01 mg/L for arsenic content (World Health Organization, 2018). A number of European countries have adopted the WHO provisional guideline of 0.01 mg/L as their standard whilst some developing countries such as Bangladesh, China and India have set their standard for arsenic in drinking water remains at 0.05 mg/L. Particularly in India, drinking water quality standards are specified by the Bureau of Indian Standards (BIS) which has two types of values for general physico-chemical parameters and toxic elements in drinking water namely maximum desirable limit and permissible limit in the case of absence of alternative drinking water sources.

BIS specification for drinking water (IS 10500: 2003) specified both the values as 0.05 mg/L. However, in 2012, BIS reduced the maximum desirable limit of arsenic from 0.05 mg/L to 0.01 mg/L. The permissible limit (in the case of absence of alternative sources) for arsenic remains unchanged as 0.05 mg/L (Shrivastava, 2016). Such regional standards are useful for sharing and transferring technology among countries to meet respective set quality standards. Most importantly these standards are crucial in surveillance and quality assurance as well as ensuring food safety of arsenic water. However, Africa has not developed such regional standards and possess a potential risk of arsenic poisoning to its population.

Conclusion

The inorganic forms of arsenic are more toxic as compared to the organic arsenic and the arsenic is incorporated into various food products through different routes. Arsenic becomes exceptionally toxic even at low exposure levels because of its high water solubility and bioaccumulation tendency in different environmental matrices. Contaminated groundwater used for irrigation of crops such as rice and tomatoes, is a major source of exposure to arsenic. Fish and other seafood contain the most of less toxic organic arsenic. Prolonged arsenic toxicity leads to carcinogenic and non-carcinogenic health risks such as arsenicosis, cancers, hepatotoxicity, kidney failure, and skin disease. Creating awareness among consumers about the toxicity profile and carcinogenic properties of arsenic is crucial in order to minimize its detrimental effects. Regulation of arsenic water is paramount to ensure safety of water and promote public health.

Acknowledgements The authors would like to thank Centre for Science and Technology of the Non-Aligned and Other Developing Countries (NAM S&T Centre) for the financial support to attend the International Training Programme on "Low Cost Technologies for Arsenic Removal from Groundwater" in Cairo, Egypt

Conflict of Interest Authors declare no conflict of interest.

References

Ahmad, S. A., Khan, M. H., & Haque, M. (2018). Arsenic contamination in groundwater in Bangladesh: Implications and challenges for healthcare policy. *Risk Management Health Care Policy, 11*, 251–261.

Ahoulé, D. G., Lalanne, F., Mendret, J., Brosillon, S., & Maïga, A. H. (2015a). Arsenic in African waters: A review. *Water, Air, & Soil Pollution, 226*, 1–13.

Ahoulé, D. L., Mendret, F., Brosillon, S., & Hama, M. A. (2015b). Arsenic in African waters: A review. *Water, Air, and Soil Pollution, 226*. https://doi.org/10.1007/s11270-015-2558-4

Anawar, H. M., Akai, J., & Sakugawa, H. (2004). Mobilization of arsenic from subsurface sediments by effect of bicarbonate ions in groundwater. *Chemosphere, 54*(6), 753–762.
Asante, K. A., Agusa, T., Subramanian, A., Ansa-Asare, O. D., Biney, C. A., & Tanabe, S. (2007). Contamination status of arsenic and other trace elements in drinking water and residents from Tarkwa, a historic mining township in Ghana. *Chemosphere, 66*(8), 1513–1522.
Bauer, M., & Blodau, C. (2006). Mobilization of arsenic by dissolved organic matter from iron oxides, soils and sediments. *Science of the Total Environment, 354*, 179–190.
Bhattacharya, P., Welch, A. H., Stollenwerk K.G., McLaughlin, M. J., Bundschuh, J., & Panaullah, G. (2007). Arsenic in the environment: biology and chemistry. *Sci Total Environ, 379*, 109–120.
Bhattacharyya, R., Jana, J., Nath, B., Sahu, S., Chatterjee, D., & Jacks, G. (2003). Groundwater as mobilization in the Bengal Delta Plain, the use of ferralite as a possible remedial measure—a case study. *Applied Geochemistry, 18*, 1435–1451.
Bissen, M., & Frimmel, F.H. (2003). Arsenic-a review. Part I: occurrence, toxicity, speciation, mobility. *ActaHydrochimHydrobiol, 31*, 9–18.
Biswas, A., Gustafsson, J. P., Neidhardt, H., Halder, D., Kundu, A. K., Chatterjee, D., ... & Bhattacharya, P. (2014). Role of competing ions in the mobilization of arsenic in groundwater of Bengal Basin: Insight from surface complexation modeling. *Water research, 55*, 30–39.
Brunt, R., Vasak, L., & Griffioen, J. (2004). International Groundwater Resources Assessment Centre.
Bundschuh, J., Maity, J. P, Nath, B., Baba, A., Gunduz, O., & Kulp, T. R. (2013). Naturally occurring arsenic in terrestrial geothermal systems of western Anatolia, Turkey: potential role in contamination of freshwater resources. *J Hazard Mater*, *262*, 951–959.
Bundschuh, J., & Maity, J. P. (2015). Geothermal arsenic: Occurrence, mobility and environmental implications. *Renew SustEnerg Rev., 42*, 1214–1222.
Chen, M., Ma, L. Q., Hoogeweg, C. G., & Harris, W. G. (2001). Arsenic background concentrations in Florida, USA surface soils: Determination and interpretation. *Environmental Forensics, 2*(2), 117–126.
Clair-Caliot, G., Marks, S. J., Hug, S. J., Bretzler, A., Tihe, S. F. K., & Lalanne, F. (2021). Uptake of arsenic by irrigated vegetables and cooked food products in Burkina Faso. *Frontiers in Water, 3*, 667308.
Cowi. (2005). *Etude approfondiedans 8 villages de la région Nord ayant des forages à tauxd'arsenicélevés.* COWI Engineering A/S.
Deb, D., Biswas, A., Ghose, A., Das, A., Majumdar, K. K., & Mazumder, D. N. G. (2013). Nutritional deficiency and arsenical manifestations: A perspective study in an arsenic-endemic region of West Bengal, India. *Public Health Nutrition, 16*(9), 1644–1655.
Desye, B., Tesfaye, A. H., Daba, C., & Berihun, G. (2023). Food safety knowledge, attitude, and practice of street food vendors and associated factors in low-and middle-income countries: A systematic review and meta-analysis. *PLoS ONE, 18*(7), e0287996. https://doi.org/10.1371/journal.pone.0287996
Dittmar, J., Voegelin, A., Maurer, F., Roberts, L. C., Hug, S. J., Saha, G. C., & Kretzschmar, R. (2010). Arsenic in soil and irrigation water affects arsenic uptake by rice: complementary insights from field and pot studies. *Environmental Science & Technology, 44*(23), 8842–8848. https://doi.org/10.1021/es101962d
El-Ghiaty, M. A., & El-Kadi, A. O. S. (2023). Annual review of pharmacology and toxicology the duality of arsenic metabolism: impact on human health. *Annu Rev Pharmacol Toxicol, 63*, 341–358.https://doi.org/10.1146/annurev-pharmtox-051921020936
Fatoki, J. O., & Badmus, J. A. (2022). Arsenic as an environmental and human health antagonist: A review of its toxicity and disease initiation. *J Hazard Mater Adv, 5*, 100052.
Finnegan, P. M., & Weihua, C. (2012). Arsenic toxicity: the effects on plant metabolism. *Frontiers in Physiology, 3*. https://doi.org/10.3389/fphys.2012.00182
Gamble, M. V., Liu, X., Slavkovich, V., Pilsner, J. R., Ilievski, V., Factor-Litvak, P., ... & Graziano, J. H. (2007). Folic acid supplementation lowers blood arsenic. *The American Journal of Clinical Nutrition, 86*(4), 1202–1209.

Gbaruko, B. C., Ana, G., & Nwachukwu, J. K. (2010). Ecotoxicology of arsenic in the hydrosphere: Implications for public health. *African Journal of Biotechnology, 7*(25).

Guo, H., Wen, D., Liu, Z., Jia, Y., & Guo, Q. (2014). A review of high arsenic groundwater in Mainland and Taiwan, China: Distribution, characteristics and geochemical processes. *Applied Geochemistry, 41*, 196–217.

Heikens, A. (2006). *Arsenic contamination of irrigation water, soil and crops in Bangladesh: Risk implications for sustainable agriculture and food safety in Asia, Food and Agriculture Organization of the United Nations Regional Office for Asia and the Pacific Bangkok*

Herath, I., Vithanage, M., & Bundschuh, J. (2016). Natural arsenic in global groundwaters: distribution and geochemical triggers for mobilization. *Curr Pollution Rep, 2*, 68–89. https://doi.org/10.1007/s40726-016-0028-2

Huang, R. Q., Gao, S. F., Wang, W. L., Staunton, S., & Wang, G. (2006). Soil arsenic availability and the transfer of soil arsenic to crops in suburban areas in Fujian Province, southeast China. *Science of the Total Environment*. https://doi.org/10.1016/j.scitotenv.2006.03.013

Huq, S. M. I. (n.d.). Fate of arsenic in irrigation water and its potential impact on the food chain. *Arsenic Contamination of Groundwater*, 23–49. https://doi.org/10.1002/9780470371046.ch2

International Programme on Chemical Safety & Organisation for Economic Co-operation and Development. (2004). *IPCS risk assessment terminology*. World Health Organization. https://iris.who.int/handle/10665/42908

Jahan, I., Abedin, M. A., Islam, M. R., Hossain, M., Hoque, T. S., Quadir, Q. F., Hossain, M. I., Gaber, A., Althobaiti, Y. S., & Rahman, M. M. (2021). Translocation of soil arsenic towards accumulation in rice: magnitude of water management to minimize health risk. *Water, 13*, 2816. https://doi.org/10.3390/w13202816

Jan, A. T., Azam, M., Siddiqui, K., Ali, A., Choi, I., & Haq, Q. M. R. (2015). Heavy metals and human health: Mechanistic insight into toxicity and counter defense system of antioxidants. *International Journal of Molecular Sciences, 16*(12), 29592–29630.

Jovanovic, D., & Rasic-Milutinovic, Z. (2017). Public health effects of arsenic exposure. In P. Bhattacharya, D. Polya, & D. Jovanovic (Eds.), *Best practice guide on the control of arsenic in drinking water* (pp. 25–32). IWA Publishing Alliance House.

Kabir, M. S., Salam, M. A., Paul, D. N. R., Hossain, M. I., Rahman, N. M. F., Aziz, Abdullah, & Latif, M. A. (2016). Spatial variation of arsenic in soil, irrigation water, and plant parts: a microlevel study. *The Scientific World Journal*, 1–14. https://doi.org/10.1155/2016/2186069

Kapwata, T., Wright, C. Y., Reddy, T., Street, R., & KuneneA, Z. M. (2023). Relations between personal exposure to elevated concentrations of arsenic in water and soil and blood arsenic levels amongst people living in rural areas in Limpopo. *South Africa, Environmental Science and Pollution Research, 30*, 65204–65216. https://doi.org/10.1007/s11356-023-26813-9

Kar, S., Maity, J. P., Jean, J.-S., Liu, C.-C., Nath, B., & Lee, Y.-C. (2011). Role of organic matter and humic substances in the binding and mobility of arsenic in a Gangetic aquifer. *Journal of Environmental Science and Health, 46*, 1231–1238.

Kar, S, Maity, J. P., Jean, J- S., Liu, C.-C., Nath, B., & Yang, H –J. (2010). Arsenic-enriched aquifers: occurrences and mobilization of arsenic in groundwater of Ganges Delta Plain, Barasat, West Bengal, India. *Applied Geochemistry 25*, 1805–1814.

Khairul, I., Wang, Q. Q., Jiang, Y. H., Wang, C., & Naranmandura, H. (2017). Metabolism, toxicity and anticancer activities of arsenic compounds. *Oncotarget, 8*(14), 23905–23926.

Khosravi-Darani, K., Rehman, Y., Katsoyiannis, I. A., Kokkinos, E., & Zouboulis, A. I. (2022). Arsenic exposure via contaminated water and food sources. *Water, 14*(12), 1884. https://doi.org/10.3390/w14121884

Klump, S., Kipfer, R., Cirpka, O., Harvey, C. F., Brenwald, M., Ashfaque, K., Badruzzaman, A., Hug, S., Imboden, D. M. (2006). Groundwater dynamics and arsenic mobilization in Bangladesh assessed using noble gases and tritium. *Environmental Science & Technology, 40*, 243–250.

Kohlmeyer, U., Kuballa, J., & Jantzen, E. (2002). Simultaneous separation of 17 inorganic and organic arsenic compounds in marine biota by means of high-performance liquid chromatography/inductively coupled plasma mass spectrometry. *Rapid Communications in Mass Spectrometry, 16*, 965–974.

Koki, I. B., Bayero, A. S., Umar, A., & Yusuf, S. (2015). Health risk assessment of heavy metals in water, air, soil and fish. *African Journal of Pure and Applied Chemistry, 9*(11), 204–210.

Kulshrestha, A., Jarouliya, U., Prasad, G. B. K. S., Flora, S. J. S., & Bisen, P. S. (2014). Arsenic-induced abnormalities in glucose metabolism: Biochemical basis and potential therapeutic and nutritional interventions. *World Journal of Translational Medicine 3*(2), 96–111. https://doi.org/10.5528/wjtm.v3.i2.96

Lamm, S. H., Engel, A., Penn, C. A., Chen, R., & Feinleib, M. (2006). Arsenic cancer risk confounder in southwest Taiwan data set. *Environmental Health Perspectives, 114*(7), 1077–1082.

Lee, K., Han, C., Hong, S. B., Jun, S. J., Han, Y., Xiao, C., Du, Z., Hur, S. D., Lee, J. I., & Boutron, C. F. (2019). A 300-year high-resolution Greenland ice record of large-scale atmospheric pollution by arsenic in the Northern Hemisphere. *Environmental Science and Technology, 53*, 12999–13008.

López, D. L., Bundschuh, J., Birkle, P., Armienta, M. A., Cumbal, L., & Sracek, O. (2012). Arsenic in volcanic geothermal fluids of Latin America. *Science of the Total Environment, 429*, 57–75.

Lu, Y., Dong, F., Deacon, C., Chen, H., Raab, A., & Meharg, A. A. (2010). Arsenic accumulation and phosphorus status in two rice (*Oryza sativa* L.) cultivars surveyed from fields in South China. *Environmental Pollution, 158*, 1536–1541.

Lushenko, M. A. (2010). *A risk assessment for ingestion of toxic chemicals in fish from Imperial beach.* San Diego State University.

Madhukar, M., Murthy, S., & Udayashankara, T. H. (2016). Sources of arsenic in groundwater and its health significance-a review. *Nature Environment & Pollution Technology, 15*(3).

Mandal, B. K., & Suzuki, K. T. (2003). Arsenic round the world: A review. *Talanta, 58*, 201–235.

Masime, J. O., Wanjau, R., Murungi, J., & Onindo, C. (2013). Analysis of the levels of arsenic in home-made brews, spirits, in water and raw materials using Hgaas in Nairobi county. *African Journal of Pure and Applied Chemistry, 7*(8), 291–301.

Mayorga, P., Moyano, A., Anawar, H. M., & Garcia, S. A. (2013). Uptake and accumulation of arsenic in different organs of carrot irrigated with as-rich water. *Clean–Soil, Air, Water, 41*(6), 587–592

Meck, M. L., Mudimbu, D., & Davies, T. C. (2019). Accumulation of potentially harmful elements in edible parts of vegetables grown on two different geological substrates in Zimbabwe. *Journal of Geochemical Exploration* (2019). https://doi.org/10.1016/j.gexplo.2019.106392

Meharg, A. A. (2004). Arsenic in rice–understanding a new disaster for South-East Asia. *Trends in Plant Science, 9*(9), 415–417.

Meharg, A. A., & Rahman, M. M. (2003). Arsenic contamination of Bangladesh paddy field soils: implications for rice contribution to arsenic consumption. *Environ Sci Tech., 37*(2), 229–234. https://doi.org/10.1021/es0259842

Meharg, A. A., & Hartley-Whitaker, J. (2002). Arsenic uptake and metabolism in arsenic resistant and nonresistant plant species. *New Phytologist, 154*, 29–43.

Meharg, A. A., & Rahman, M. M. (2003a). Arsenic contamination of Bangladesh paddy field soils: Implications for rice contribution to arsenic consumption. *Environmental Science and Technology, 37*, 229–234.

Mhandire, K., Matarira, H. T., & Chin'ombe, N. (2017b). Environmental arsenic levels in Zimbabwe. *The BioMedical Journal, Zimbabwe, 1*(1), 1–2.

Mhandire, K., Matarira, H. T., & Chin'ombe, N. (2017). Environmental arsenic levels in water sources in Zimbabwe. *TheBioMedical Journal, Zimbabwe, 1*(1), 2

Miyazaki, K., Watanabe, C., Mori, K., Yoshida, K., Ohtsuka, R. (2005). The effects of gestational arsenic exposure and dietary selenium deficiency on selenium and selenoenzymes in maternal and fetal tissues in mice. *Toxicology, 208*(3), 357–365. https://doi.org/10.1016/j.tox.2004.11.030

Mng'ong'o, M., Comber, S., Munishi, L. K., Blake, W., Ndakidemi, P. A., & Hutchinson, T. H. (2021). Assessment of arsenic status and distribution in Usangu agro-ecosystem-Tanzania. *Journal of Environmental Management*. https://doi.org/10.1016/j.jenvman.2021.113012

Muhetaer, M., Yang, M., Xia, R., Lai, Y., & Wu, J. (2022). Gender difference in arsenic biotransformation is an important metabolic basis for arsenic toxicity. *Bmcpharmacology and Toxicology, 23*, 15. https://doi.org/10.1186/s40360-022-00554-w

Neidhardt, H., Kramar, U., Tang, X., Guo, H., & Norra, S. (2015). Arsenic accumulation in the roots of *Helianthus annuus* and *Zea mays* by irrigation with arsenic-rich groundwater: Insights from synchrotron X-ray fluorescence imaging. *Chemie der. Erde—Geochem., 75*, 261–270.

Nicolli, H. B., Bundschuh, J., Blanco, M. C., Tujchneider, O. C., Panarello, H. O., Dapena, C., et al. (2012). Arsenic and associated trace-elements in groundwater from the Chaco-Pampean plain, Argentina: Results from 100 years of research. *Science of the Total Environment, 429*, 36–56.

Nzihou, J. F., Bouda, M., Hamidou, S., & Diarra, J. (2013). Arsenic in drinking water toxicological risk assessment in the north region of Burkina Faso. *Journal of Water Resource and Protection, 5*, 46.

Ormachea Muñoz, M, Bhattacharya, P, Sracek, O, Ramos Ramos, O, Aguirre, J. Q., & Bundschuh, J. (2015). Arsenic and other trace elements in thermal springs and in cold waters from drinking water wells on the Bolivian Altiplano. *Journal of South American Earth Sciences, 60*, 10–20

Ostrom, L. T., & Wilhelmsen, C. A. (2019). *Risk assessment: Tools, techniques, and their applications*. John Wiley & Sons.

Pal, P. (2015). *Groundwater arsenic remediation* (pp. 303–312). ISBN 9780128012819. https://doi.org/10.1016/B978-0-12-801281-9.09999-9

Paul, D. S., Hernández-Zavala, A., Walton, F. S., Adair, B. M., Dedina, J., Matousek, T., et al. (2007). Examination of the effects of arsenic on glucose homeostasis in cell culture and animal studies: Development of a mouse model for arsenic-induced diabetes. *Toxicology and Applied Pharmacology, 222*, 305–314. https://doi.org/10.1016/j.taap.2007.01.010

Raju, J. N. (2022). Arsenic in the geo-environment: A review of sources, geochemical processes, toxicity and removal technologies. *Environmental Research, 203*. ISSN 0013-9351.

Ro, S.-H., Bae, J., Jang, Y., Myers, J. F., Chung, S., Yu, J., Natarajan, S. K., Franco, R., & Song, H.-S. (2022). Arsenic toxicity on metabolism and autophagy in adipose and muscle tissues. *Antioxidants, 11*, 689. https://doi.org/10.3390/antiox11040689

Romero-Estévez, D., Yánez-Jácome, G. S., & Navarrete, H. (2023). Non-essential metal contamination in Ecuadorian agricultural production: A critical review. *Journal of Food Composition and Analysis, 115*, 104932.

Rosas, C. J., Portugal, L., Ferrer, L., Hinojosa, R. L., Guzmán, M. J. L., Hernández, R. A., & Cerd, V. (2016). An evaluation of the bioaccessibility of arsenic in corn and rice samples based on cloud point extraction and hydride generation coupled to atomic fluorescence spectrometry. *Food Chemistry, 204*, 475–482.

Sabir, S., Muhammad Sajid Hamid, A., Fareeha, F., Uzma, S., Hassan, M. M., & Kanwal, R. (2019). Role of cadmium and arsenic as endocrine disruptors in the metabolism of carbohydrates: Inserting the association into perspectives. *Biomedicine & Pharmacotherapy, 114*, 108802. https://doi.org/10.1016/j.biopha.2019.108802

Sackey, L. N., Markin, K., Kwarteng, A., Ayitey, I. M., & Kayoung, P. (2023). Presence and levels of potential trace elements in lettuce and spring onion grown in Kumasi, Ghana. *Kuwait Journal of Science, 2023*. https://doi.org/10.1016/j.kjs.2023.10.010

Sadee, B. A., Foulkes, M. E., & Hill, S. J. (2016a). A study of arsenic speciation in soil, irrigation water and plant tissue: A case study of the broad bean plant, *Viciafaba. Food Chemistry, 210*, 362–370. https://doi.org/10.1016/j.foodchem.2016.04.066

Sadee, B. A., Foulkes, M. E., & Hill, S. J. (2016b). An evaluation of extraction techniques for arsenic in staple diets (fish and rice) utilising both classical and enzymatic extraction methods. *Food AdditContam Part AChem Anal Control Expo Risk Assess., 33*(3), 433–441. https://doi.org/10.1080/19440049.2015.1132479

Saldaña-Robles, A., Abraham-Juárez, M. R., Saldaña-Robles, A. L., Saldaña-Robles, N., Ozuna, C., & Gutiérrez-Chávez, A. J. (2018). Arsenic contamination in agriculture: effects on irrigation water, soils and crops, state of art. *Applied Ecology and Environmental Research, 16*(2), 1533–1551. https://doi.org/10.15666/aeer/1602_15331551

Sandhi, A., Yu, C., Rahman, M., & Amin, N. (2022). Arsenic in the water and agricultural crop production system: Bangladesh perspectives. *Environmental Science and Pollution Research, 29*, 51354–51366.

Sandil, S., Dobosy, P., Kröpfl, K., Füzy, A., Óvári, M., & Záray, G. (2019). Effect of irrigation water containing arsenic on elemental composition of bean and lettuce plants cultivated in calcareous sandy soil. *Food Production, Processing and Nutrition, 1*, 14. https://doi.org/10.1186/s43014-019-0014-3

Sandil, S., Óvári, M., Dobosy, P., Vetési, V., Endrédi, A., Takács, A., & Záray, G. (2021). Effect of arsenic-contaminated irrigation water on growth and elemental composition of tomato and cabbage cultivated in three different soils, and related health risk assessment. *Environmental Research, 197*, 111098. https://doi.org/10.1016/j.envres.2021.111098

Sanou, Y., Kabore, R., & Pare, S. (2020). Adsorption of arsenic and phosphate from groundwater onto a calcined laterite as fixed bed in column experiments. *French-Ukrainian Journal of Chemistry, 8*(2), 227–243.

Santhini, R., & Lee, J. S. (2015). Arsenic risk assessment. In S.J.S. Flora (ed.), *Handbook of arsenic toxicology* (pp. 95–120). ISBN 9780124186880. https://doi.org/10.1016/B978-0-12-418688-0.00004-6

Shrivastava, B. K. (2016). Policy intervention for arsenic mitigation in drinking water in rural habitations in India: Achievements and challenges. *Water Health, 14*(5), 827–838.

Smedley, P. L., & Kinniburgh, D. G. (2002). A review of the source, behavior and distribution of arsenic in natural waters. *Applied Geochemistry, 17*, 517–568. http://dx.doi.org/10.1016/S0883-2927(02)00018-5

Smedley, P. L., Knudsen, J., & Maiga, D. (2007). Arsenic in groundwater from mineralised Proterozoic basement rocks of Burkina Faso. *Applied Geochemistry, 22*(5), 1074–1092.

Smith, A. H., Hopenhayn-Rich, C., Bates, M. N., Goeden, H. M., Hertz-Picciotto, I., Duggan, H. M., Wood, R., Kosnett, M. J., & Smith, M. T. (1992). Cancer risks from arsenic in drinking water. *Environmental Health Perspectives, 97*, 259–267.

Smith, N. M., Lee, R., Heitkemper, D. T., Cafferky, K. D., Haque, A., & Henderson, A. K. (2006). Inorganic As in cooked rice and vegetables from Bangladeshi households. *Science of the Total Environment, 370*, 294–301. https://doi.org/10.1016/j.scitotenv.2006.06.010

Smith, E., Naidu, R., & Alston, A. M. (2002). Chemistry of inorganic arsenic in soils. *Journal of Environmental Quality 31*, 557–563.

Somé, I., Sakira, A., Ouédraogo, M., Ouédraogo, T., Traoré, A., Sondo, B., & Guissou, P. (2012). Arsenic levels in tube-wells water, food, residents' urine and the prevalence of skin lesions in Yatenga province. *Burkina Faso. Interdisciplinary Toxicology, 5*(1), 38–41.

Stazi, S. R., Cassaniti, C., & Marabottini, R. (2016). Arsenic uptake and partitioning in grafted tomato plants. *Horticulture, Environment, and Biotechnology, 57*, 241–247. https://doi.org/10.1007/s13580-016-0036-6

Stoltz, E., & Greger, M. (2002). Accumulation properties of As, Cd, Cu, Pb and Zn by four wetland plant species growing on submerged mine tailings. *Environmental and Experimental Botany, 47*, 271–280.

Styblo, M., Drobna, Z., Jaspers, I., Lin, S., & Thomas, D. J. (2002). The role of biomethylation in toxicity and carcinogenicity of arsenic: A research update. *Environmental Health Perspectives, 110*(suppl 5), 767–771.

Tsuda, T., Ogawa, T., Babazono, A., Hamada, H., Kanazawa, S., Mino, Y., ... & Kurumatani, N. (1992). Historical cohort studies in three arsenic poisoning areas in Japan. *Applied Organometallic Chemistry, 6*(4), 333–341.

Tyler, C. R., & Allan A. M. C. (2014). The effects of arsenic exposure on neurological and cognitive dysfunction in human and rodent studies: a review. *Current Environmental Health Reports, 1*, 132–147. https://doi.org/10.1007/s40572-014-0012-1

Upadhyay, M. K., Shukla, A., Yadav, P., & Srivastava, S. (2019). A review of arsenic in crops, vegetables, animals and food products. *Food Chemistry, 276*, 608–618. https://doi.org/10.1016/j.foodchem.2018.10.069

Upadhyay, M. K., Arnab, M., Jisha, S. K., & Sudhakar, S. (2020). Arsenic in rice agro-ecosystem: solutions for safe and sustainable rice production. *Frontiers in Sustainable Food Systems, 4*, 53. https://doi.org/10.3389/fsufs.2020.00053

USEPA. (2012). *Waste and cleanup risk assessment*. http://www2.epa.gov/risk/waste-and-cleanup-risk-assessment

Vahter, M. E. (2007). Interactions between Arsenic-Induced toxicity and nutrition in early life. *The Journal of Nutrition, 137*(12), 2798–2804. https://doi.org/10.1093/jn/137.12.2798

Vahter, M. (2009). Effects of arsenic on maternal and fetal health. *Annual Reviews of Nutrition, 29*, 381–399. https://doi.org/10.1146/annurev-nutr-080508-141102

Wambaugh, J. F., Bare, J. C., Carignan, C. C., Dionisio, K. L., Dodson, R. E., Jolliet, O., & Isaacs, K. K. (2019). New approach methodologies for exposure science. *Current Opinion in Toxicology, 15*, 76–92.

Wang, X. M., Zhang, J., Wu, L. H., Zhao, Y. L., Li, T., Li, J. Q., et al. (2014). A minireview of chemical composition and nutritional value of edible wild-grown mushroom from China. *Food Chemistry, 151*, 279–285. https://doi.org/10.1016/j.foodchem.2013.11.062

Wang, S., & Mulligan, C. N. (2006). Effect of natural organic matter on arsenic release from soils and sediments into groundwater. *Environmental Geochemistry and Health, 28*, 197–214.

Water Sources in Zimbabwe. *BioMedical Journal, Zimbabwe, 1*(1), 2.

World Health Organization. (2018). *WHO guideline values for drinking-water quality: chemical contaminants II*.

Wu, G. R., Hong, H. L. & Yan, C. L. (2015). Arsenic accumulation and translocation in Mangrove (*Aegiceras corniculatum* L.) Grown in arsenic contaminated soils. *International Journal of Environmental Research and Public Health 12*, 7244–7253

Zeng, Q., & Zhang, A. (2020). Assessing potential mechanisms of arsenic-induced skin lesions and cancers: Human and in vitro evidence. *Environ Pollut*ion, *260*, 113919

Zhu, Y.-G., Yoshinaga, M., Zhao, F. J., & Rosen, B. P. (2014). Earth abides arsenic biotransformations. *Annual Review of Earth and Planetary Sciences, 42*(1), 443–467

Part V
Effect of Arsenic Contamination on Human Health

Chapter 22
Arsenic Contamination of Groundwater and Its Impacts on Health

Kunal Kanti Majumdar

Arsenic is a metalloid which occurs naturally in Earth's crust. It is the twentieth most abundant element and is a component of more than 245 minerals. While groundwater is relatively safe in regard to microbiological contamination and other impurities, however, it is prone to chemical contaminants like arsenic. Arsenic in drinking water has been recognized as a major public health problem in several countries of the world. Arsenic contamination of groundwater may occur in two ways: (i) drawing of water from aquifers that naturally contain arsenic due to geogenic or natural causes and (ii) contamination from anthropogenic activities like mining, agriculture, and certain industrial activities.

Arsenic can combine with both metals and non-metals to form inorganic and organic compounds. The inorganic forms are toxic to human health and consist mostly of arsenite and arsenate compounds. The organic forms are comparatively less toxic and are present in sea foods, but presence of arsenic in sea foods and other food items pose a lesser threat than drinking water. Humans are exposed to arsenic primarily from air, food, and water. Naturally, arsenic is introduced into soil and groundwater during weathering of rocks and minerals followed by subsequent leaching and runoff. Drinking water may be contaminated with arsenic from arsenical pesticides, natural mineral deposits or improperly disposed of arsenical chemicals. Arsenic poisoning is a medical condition that occurs due to elevated levels of arsenic in the body. If arsenic poisoning occurs over a brief time symptoms may include severe vomiting, abdominal pain, encephalopathy, muscle cramps, and watery diarrhea that contains blood. This is known as acute toxicity. Long-term exposure can result in thickening of the skin (keratosis), darker skin (melanosis), abdominal pain, chronic cough,

K. K. Majumdar (✉)
Department of Community Medicine, KPC Medical College & Hospital, Jadavpur, Kolkata, India
e-mail: kunalkm2003@yahoo.com; kunalmajumdar1@gmail.com

UNICEF, Kolkata, India

WHO (SEARO), New Delhi, India

B. Sen Gupta and N. Martínez-Villegas (eds.), *Arsenic Remediation of Food and Water*,
https://doi.org/10.1007/978-981-97-4764-1_22

respiratory distress, diarrhea, heart disease, feeling of numbness, leg pain, and cancer. This is known as a chronic toxicity. The most common reason for long-term exposure is contaminated drinking water.

Arsenic pollution in groundwater used for drinking purposes has been envisaged as a problem of global concern. Population exposed to arsenic from all sources is around 178 million globally, and arsenic cases have been identified in around 66 countries and territories in all regions of the world like Taiwan, China, Argentina, Chile, Mexico, Cambodia, Thailand, Myanmar, Nepal, and the USA (World Health Organization, 2004) but the severity of this contamination in India and Bangladesh is unprecedented. In India, significant arsenic contamination in groundwater was detected in 1983 in West Bengal, when some villagers were diagnosed to be suffering from arsenicosis due to drinking of arsenic contaminated water. Today, in West Bengal, arsenic contamination in groundwater has been detected in 83 blocks in 8 districts. It is suspected that 6 million people are exposed to arsenic contaminated groundwater ($>50\ \mu g/l$). In India, occurrences of arsenic in groundwater have also been reported from Bihar, Jharkhand, Chhattisgarh, Uttar Pradesh Assam, and Manipur (World Health Organization, 2004; Mazumder, 2008; Mazumder et al., 1998a, 1999, 2000, 2003, 2009, 2010; NRC, 1999; Caussy, 2005; Tondel et al., 1999).

Metabolism

After absorption, arsenic is rapidly cleared from the blood and it reaches liver, where it is detoxified by conversion into MMA (monomethyl arsenic acid), DMA (dimethyl arsenic acid), and tri-methylated products. When biomethylation is saturated, a portion of absorbed arsenic is deposited in the skin, hair, and nails firmly bound to keratin. About 45–75% of the administered dose is eliminated through urine within a few days to a week.

Trivalent arsenic reacts readily with vicinal sulphydryl groups of enzyme and protein and accumulates in keratin rich tissues like skin, hairs and nails. Trivalent species are also believed to be carcinogenic i.e., responsible for causing cancer by inducing chromosomal abnormalities, including changes in structure and the number of chromosomes and sister chromatid exchanges. The exact molecular mechanism of arsenic induced carcinogenesis or cancer occurrence is less understood.

Pentavalent arsenic can uncouple mitochondrial oxidative phosphorylation by substituting for inorganic phosphate in the synthesis of ATP, the main energy source of cellular metabolism (Caussy, 2005).

Clinical Features of Chronic Arsenic Toxicity (CAT) (Mazumder, 2008; NRC, 1999; Mazumder et al., 1998a, 1998c, 1999, 2000, 2003, 2009, 2010, 2012; Caussy, 2005; Tondel et al., 1999; Rahman et al., 2019; Deb et al. 2013)

Human health effects of arsenic toxicity depend on the dose, the modality, duration of exposure as well as source and type of arsenic. Chronic arsenic toxicity is designated by the term arsenicosis, to imply a chronic disease caused by prolonged exposure of arsenic to non-lethal doses of 0.005–0.09 mg/kg/day in humans (Caussy, 2005). In the field guide for detection, management and surveillance relating to arsenicosis cases, the condition of chronic arsenic toxicity or arsenicosis has been defined by WHO as "a chronic health condition arising from prolonged ingestion of arsenic above the safe dose for at least six months, usually manifested by characteristic skin lesions of melanosis and keratosis, occurring alone or in combination with or without the involvement of internal organs (Caussy, 2005)". The symptoms of chronic arsenic toxicity (arsenicosis) are insidious in onset. The appearance of lesions usually follows a temporal progression, beginning with hyperpigmentation and then progressing to palmar-plantar keratosis. The condition takes around 6–9 years to develop and is dependent on the magnitude of the dose and the duration of the exposure. Most reports of chronic arsenic exposure in humans focus attention on skin manifestations because of their diagnostic specificity. However, data derived from population-based studies, clinical case series, and reports relating to the intake of inorganic arsenic in drinking water, medications, or occupational and environmental exposure show that chronic arsenic exposure adversely affects multi-organ systems of the human body as mentioned below as non-dermatological manifestations.

The Skin Manifestations

The skin manifestations of chronic arsenic toxicity can be either non-malignant (non-cancerous) or malignant (cancerous). Increased pigmentation (melanosis) and hardening of the skin (keratosis) are common non-malignant manifestations, whereas the most common type of cancer or malignant lesions following chronic exposure is skin cancer. Also, the cancers of the lungs, bladder, kidney, and liver have been reported in studies of arsenic contaminated drinking water. The hyper-pigmentation or changes in the coloration of the skin appears in a finely freckled "rain drop" pattern that is particularly seen in the trunk and extremities and is symmetrically distributed bilaterally (Figs. 22.1, 22.2 and 22.3). Other patterns include diffuse hyper-pigmentation (melanosis), localized and patchy pigmentation, and leukomelanosis (Fig. 22.2) in which hypo pigmented papules take spotty, white appearance. Arsenical hyperkeratosis predominantly appears on palms and planter aspect of feet, although dorsum of extremities and trunk has also been described. It starts with indurated grit like character and later progresses to punctuated warts like keratosis and nodules with cracks and fissures (Fig. 22.4). Keratosis can be mild, moderate, and severe depending on the size of the nodules (Mazumder, 2008; Caussy, 2005; Tondel et al., 1999).

Prevalence of keratosis and hyperpigmentation per 100 for males and females in relation to arsenic exposure in West Bengal. Test for each sex: $P < 0.001$.

Skin abnormalities such as pigmentation (changes in the color of the skin) change and keratosis (thickening and roughening of skin) have been known to be hallmark signs of chronic arsenic exposure and is a serious health problem of public concern, because advanced forms of keratoses are painful and debilitating (Mazumder et al., 1998a; Mazumder, 2008; NRC, 1999). It must be realized that all arsenicosis cases

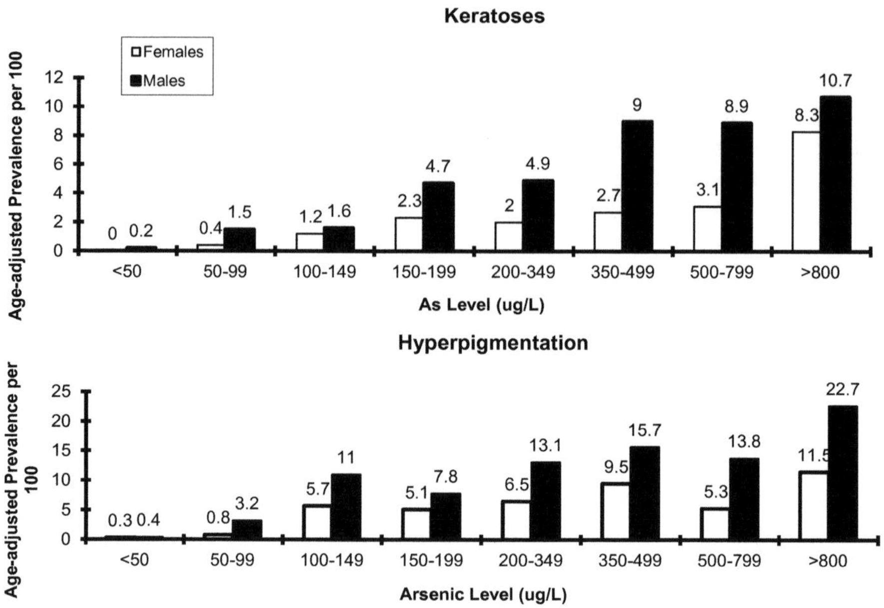

Fig. 22.1 Dose response relationship between prevalence of skin manifestations and arsenic exposure. (*Source* Mazumder et al., 1999)

may not always show dermal manifestations () but dermal manifestation are generally used for diagnosis and identification of patients as these manifestations are somewhat specific and the patients with these manifestations usually seek medical care (Mazumder, 2008; Caussy, 2005) (Fig. 22.5).

Other Systemic Manifestations

The reported clinical manifestations resulting from ingestion of arsenic-contaminated drinking water not only include dermatological manifestations like pigmentation and keratosis of skin but also systemic manifestations due to involvement of other internal organs. The most common systemic manifestations include neurological, hematological, gastrointestinal, and respiratory complications.

The main general manifestations which commonly occur in arsenicosis patients are anemia, weakness, fatigue, and conjunctival congestion (Mazumder, 2008; Caussy, 2005).

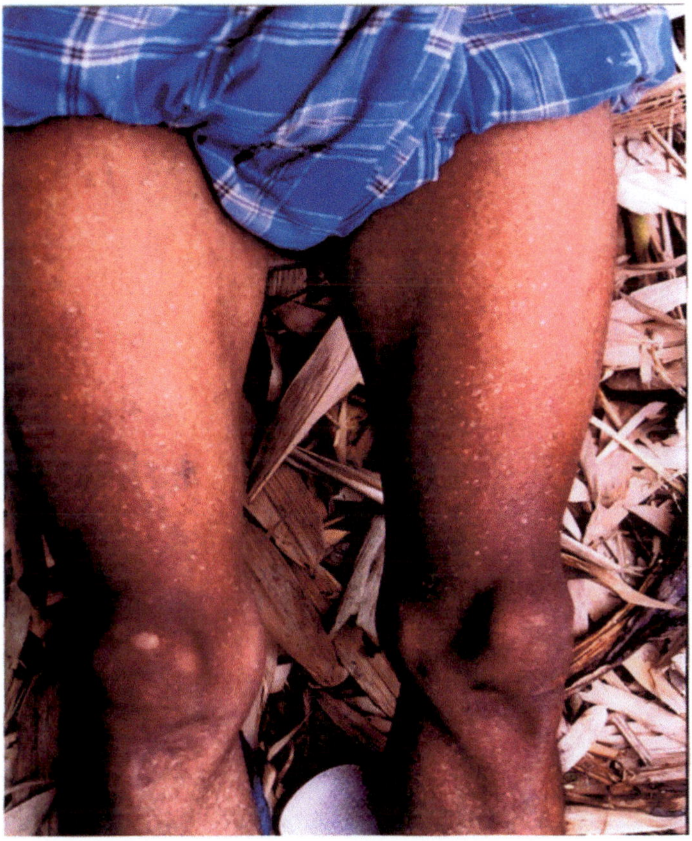

Fig. 22.2 Spotty pigmentation in arsenicosis. (*Courtesy DNGM Research Foundation*)

Respiratory System

Chronic cough and respiratory distress usually occurs more commonly in arsenicosis patients due to chronic lung disease such as chronic bronchitis, chronic obstructive pulmonary disease, and bronchiectasis (Fig. 22.6). Lung cancer can also occur as long term complication (Caussy, 2005; Mazumder et al., 1999, 2000).

Gastrointestinal System and Liver

Liver diseases such as enlargement of liver (Fig. 22.7), portal hypertension, non-cirrhotic portal fibrosis, cirrhosis of liver, ascites, and liver cancer are more commonly seen in arsenicosis patients (Mazumder et al., 1999, 2008; Caussy, 2005).

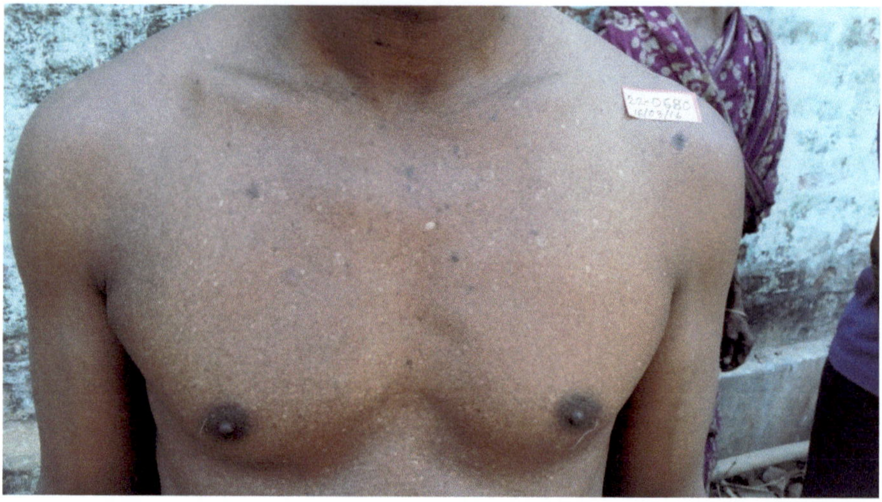

Fig. 22.3 Raindrop pigmentation in arsenicosis. (*Courtesy DNGM Research Foundation*)

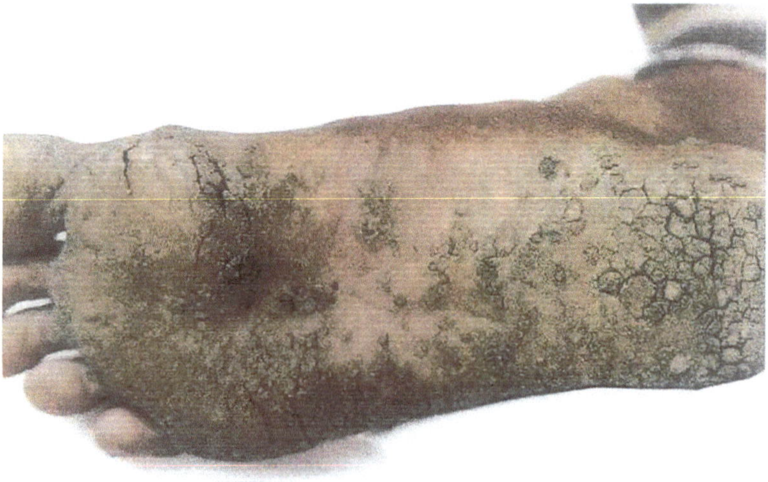

Hyperkeratosis

Fig. 22.4 Hyperkeratosis in arsenicosis. (*Courtesy DNGM Research Foundation*)

Cardio Vascular System

Cardio vascular diseases like peripheral vascular disease, gangrene of toes/limbs, atherosclerosis, hypertension, ischemic heart disease and cerebrovascular diseases like stroke are more commonly seen in arsenicosis patients. Studies from Taiwan have documented the presence of black foot disease, a unique peripheral arterial

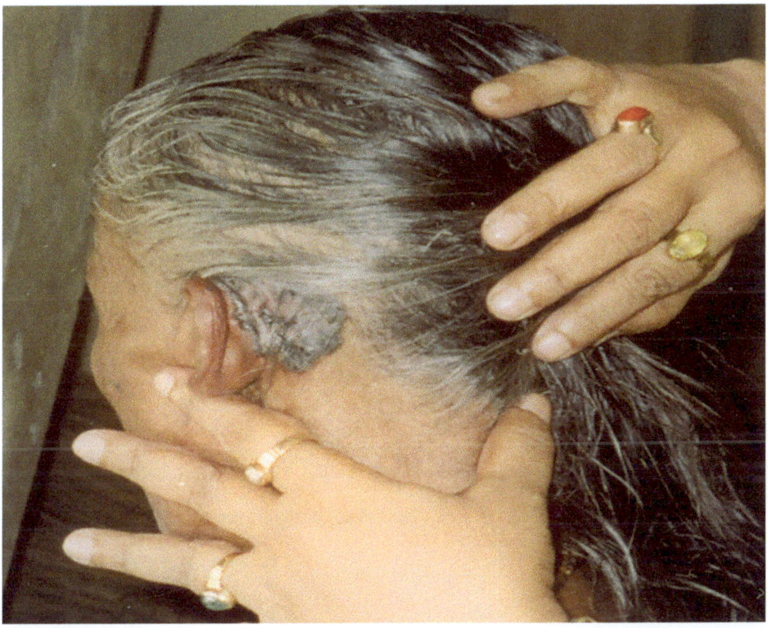

Fig. 22.5 Cancerous lesion due to arsenicosis. (*Courtesy DNGM Research Foundation*)

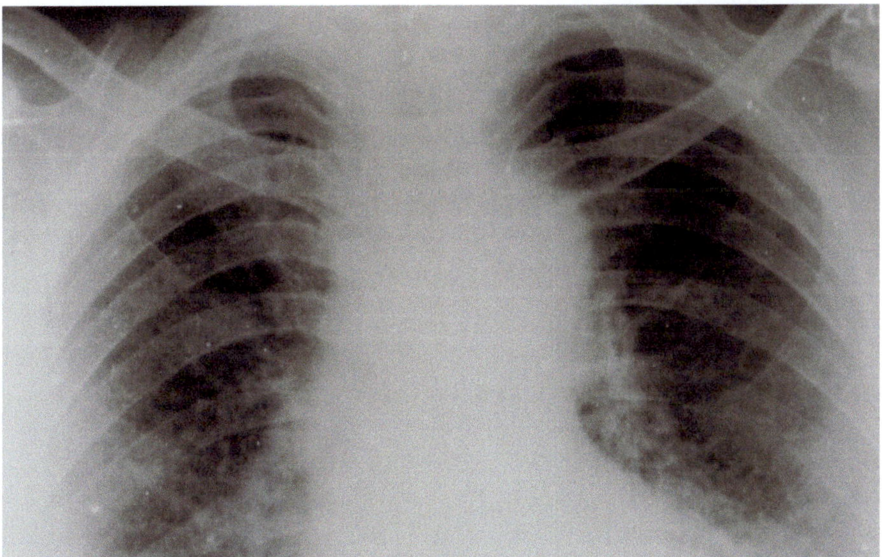

Fig. 22.6 Chest X ray showing lungs involvement in arsenicosis patient. (*Courtesy DNGM Research Foundation*)

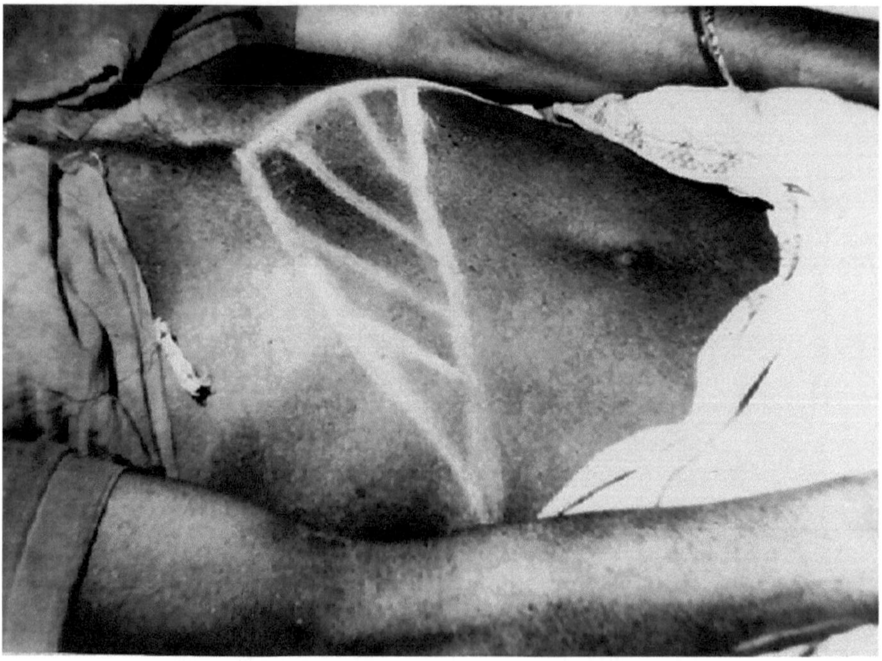

Fig. 22.7 Showing liver enlargement in arsenicosis patient. (*Courtesy DNGM Research Foundation*)

disease with dry gangrene and spontaneous amputations. Leg pain and intermittent cramp in the leg muscles is common, but dry gangrene is less frequently seen in Indo gangetic basin (Mazumder et al., 1999; Mazumder, 2008; Caussy, 2005).

Nervous System

Neurological manifestations like sensor motor, peripheral neuropathy, and polyneuropathy, characterized by the feeling of paresthesia and numbness in the extremities are more commonly seen in arsenicosis patients (Mazumder et al., 1999; Mazumder, 2008; Caussy, 2005).

Urinary System

Passing out urine with blood (haematuria), due to kidney and bladder cancer are more commonly seen in arsenicosis patients (Mazumder et al., 1999; Mazumder, 2008; Caussy, 2005).

Table 22.1 Clinical features of arsenicosis: Experience in West Bengal, India—hospital based study (n = 248)

Symptoms	Percent
Pigmentation spotty	94.35
Keratosis	65.32
Chr. Lung disease	30.24
Hepatomegaly	76.61
Portal hypertension	02.06
Spleenomegaly	29.43
Weakness	65.73
Dyspepsia	66.53
Conjunctivitis	66.53
Non pitting edema	09.27
Periph Vasc dis	01.20
Polyneuropathy	29.83
Diabetes	–
Ischaemic Heart Disease (IHD)	–

Source Mazumder et al. (1999)

Others Manifestations

It has been found in many studies that diabetes and solid oedema (swelling) of limbs especially lower limbs are more commonly seen in arsenicosis patients (Mazumder et al., 1999; Mazumder, 2008; Caussy, 2005).

A hospital based study in West Bengal (Mazumder et al., 1999) showed that among all manifestations skin manifestations, hepatomegaly and splenomegaly (enlargement of liver and spleen), weakness, dyspepsia and chronic lung disease were the most frequent manifestations of arsenicosis (Mazumder et al., 1999) (Table 22.1).

Arsenicosis and Adverse Reproductive Outcomes

There is a dose-dependent relationship between maternal exposure to arsenic during pregnancy and infant (<1yr of age) mortality, meaning that infants born to women exposed to higher concentrations of arsenic or exposed for longer periods, have a higher mortality rate. Studies have shown that mothers exposed to high levels of arsenic during antenatal period had higher incidences of stillbirths and negative reproductive outcomes (Caussy, 2005; Rahman et al., 2019).

Arsenicosis and Cancer

The International Agency for Research on Cancer (IARC) has classified arsenic as a carcinogen (capable of causing cancer) although the exact molecular mechanism of arsenic induced carcinogenesis is still less understood (World Health Organization, 2004). However, there are various theories and hypotheses related to causation for arsenic induced cancer like oxidative stress, alteration of DNA methylation/repair, inhibition of activity, and mutation of p53 human tumor suppressor gene and alteration of structure and a number of chromosomes, etc. (Mazumder, 2008; Caussy, 2005; Mazumder et al., 1999).

Arsenic increases the risk of cancer related to skin, lung, liver, bladder, and kidney (Fig. 22.5) and the increased risk of cancer may result even without the skin lesions being present.

Arsenicosis and Malnutrition

Several human studies have identified associations between malnutrition and arsenic-induced skin lesions, skin cancer, and cardiovascular effects. Literature surveys indicate arsenic resistance and its relation with nutritional status. A cross-sectional study was conducted in Chakdah and Haringhata block of Nadia district of West Bengal to determine the dietary intake of nutrients and its correlation with manifestations of arsenicosis. The study revealed that low socio-economic status along with less dietary intake of calories, protein, and micronutrients may have a definite role in increasing the risk of development of arsenicosis. In the above study, it was evident that calorie consumption and consumption of most of the macro and micronutrients was less than RDA (Recommended Dietary Allowances) in both cases and controls of the study population. But there was a significant difference in intake of protein along with micronutrients like thiamine, riboflavin, niacin, zinc, and choline in the diet of arsenicosis cases compared to exposed controls in both the sexes, indicating the probable role of above nutrients in development of diseases. There is increasing evidence of elevated arsenic levels in rice grains, in the regions of West Bengal and Bangladesh, where paddy-fields are irrigated with arsenic-rich water. On the other hand, arsenic content in rice, grown in regions without groundwater contamination was found to be low. Arsenic contamination of vegetables grown on soils irrigated with arsenic contaminated water has also been reported (Mazumder et al., 2010; Deb et al., 2013; Mazumder et al., 2012).

Some of the skin changes as well as other systemic manifestations associated with arsenic may appear the same or similar to those encountered in the other medical conditions. Clinicians and paramedical personnel practicing in primary care-settings can be trained to screen patients for the possible presence of characteristic arsenic-related skin lesions, but the recommendation from an experienced dermatologist or other physicians with relevant expertise is needed for confirmation of the diagnosis (Mazumder, 2008; Caussy, 2005).

Lesions which are present since birth or with a history less than a month or related to drug intake or occupation, lesions with an elevated margin of hyper-pigmentation or with scaling may not be due to arsenicosis, so the opinion of an experienced dermatologist or a trained physician is often necessary.

Incidence of various clinical manifestations of chronic arsenicosis was found to be significantly higher among patients with higher level of Arsenic exposure (³50 μg/l) in comparison to patients with lower level of arsenic exposure (<50 μg/l) as evident from a epidemiological study conducted in one of the affected districts of West Bengal (24 Parganas South) I995-96 (Mazumder et al., 1998a, 1999) (Table 22.2). A dose response relationship between prevalence of hyperpigmentation and keratosis with arsenic level in water was also found indicating a causal association between level of arsenic in drinking water and occurrence of manifestations of arsenicosis (Mazumder et al., 1998a, 1999) (Fig. 22.1).

Laboratory Diagnosis for Determining the Exposure History of Arsenic (Caussy, 2005; Mazumder et al. 2012)

Testing of water that is currently consumed can detect the recent or past history of exposure to arsenic. This is because consumption of water with an arsenic concentration in excess of the prevailing national standards for at least six months can lead to the development of arsenicosis. The spectrophotometric/colorimetric procedures based on silver diethyl dithiocarbamate complex are used for the estimation of arsenic, but atomic absorption spectrometry (AAS) especially hydride-generation AAS and graphite furnace AAS are used, which are highly sensitive and accurate but costly techniques for the detection of arsenic and are considered as the standard reference method (gold standard).

Table 22.2 Incidence of various clinical manifestations of chronic arsenicosis: Epidemiology study in one of the affected districts of West Bengal (24 parganas-south) 1995–1996

Symptoms	As level < 50 μg/l	As level ≥ 50 μg/l	P value
	(n = 3467)	(n = 4216)	
	(%; n value)	(%; n value)	
Pigmentation	0.34;12	8.82;372	< 0.001
Keratosis	0.11;4	3.64;154	< 0.001
Hepatomegaly	2.99;104	10.21;431	< 0.001
Weakness	1.37;48	4.99;211	< 0.001
Abdominal pain	31.81; 1103	27.84;1174	NS
Nausea	0.31;11	0.74;31	< 0.02
Lung disease	7.74;269	11.68;493	< 0.001
Neuropathy	2.73;95	4.70;198	< 0.001

Source Mazumder et al. (1999)

ICP-MS (inductively coupled plasma-mass-spectrometry), a technique to determine low-concentrations (range: ppb = parts per billion = μg/l) and ultra-low-concentrations of arsenic (range: ptt = parts per trillion = ng/l) is often used. This technique is often used for the determination of arsenic in water and urine with greater speed, precision, and sensitivity.

Reliable test kits are often used for mass screening under field conditions, especially in situations where no laboratory facility exists.

Testing of Biomarkers (Biological Materials) for Arsenic

(a) Testing of hair and nails is usually done by wet digestion or by neutron activation analysis. Presence of arsenic above the guidelines value (mentioned below) in these samples indicates exposure to an unsafe dose of arsenic within the preceding 9–11 months.
(b) Testing of urine and saliva samples is done by AAS or high-performance liquid chromatography-inductively coupled plasma-mass spectrometry (HPLC–ICPMS) for speciation. The presence of arsenic above the guidelines value (mentioned below) in these samples indicates recent exposure to an unsafe dose of arsenic.

The current WHO guideline value of arsenic in laboratory samples (Caussy, 2005).

(a) Water—0.01 mg/L (or 10 ppb or 10 micro g/L)
(b) Urine—0.01- 0.05 mg/lit
(c) Hair—below 1 mg/kg, nail–below 1.3 mg/kg
(d) Blood—0.0015–0.0025 mg/lit. *Blood is of no value in establishing chronic arsenic exposure because of the short half-life of arsenic in blood (WHO).*

The arsenic concentration should always be determined using a validated method performed by trained personnel in a laboratory, meeting national standards and practicing standard operating procedures (Caussy, 2005).

Prevention and Management of Arsenicosis (Mazumder et al., 1998b, 2000, 2003, 2012, 2018; Rahman, et al., 2019; Deb et al., 2013)

Till date, there is no specific treatment for arsenicosis, so the remedy of chronic arsenic toxicity is to consume arsenic-free water. A cohort follow-up study was conducted by Guha Mazumder et al. among 1079 people (arsenic exposed people 623, control population 451) in 2000 in South 24 Parganas District of West Bengal. After five years of the original clinical examination the same population was followed up, and it was found that skin lesions were cleared in 49.7% of people after consuming arsenic-free water (Mazumder et al., 1999, 2003).

So the key approaches for the management of arsenicosis are: (i) cessation of exposure to drinking water or other items with elevated concentration of arsenic; (ii) administration of drugs or nutrients for hastening recovery or preventing disease complications and progression; (iii) provision of non-specific supportive care to improve physical symptoms or treat selected complications; (iv) secondary prevention of hidden effects through medical surveillance, and; (v) counseling and education to address psycho-social effects of the illness and provision of appropriate rehabilitation (Caussy, 2005).

However, arsenic-free safe water sources may not always be available in all the arsenic affected villages. Hence, people need to be aware of the safe water options in their villages and should be counseled about the consequences of consuming arsenic contaminated water. Standard information, education, and communication strategies need to be developed to create awareness. The strategies for safe water supply may include treatment of the surface water, deep tube well, rainwater harvesting, groundwater recharge and the use of arsenic removal filters both at the domestic and community level. But the technologies for the filters should be cost effective, user friendly, and robust, so that villagers can use them. Community participation by forming water committees, promotion of user fees, etc. is essential for creating community ownership, acceptability, and sustainability of such filters, especially community filter.

Despite the magnitude of this potentially fatal toxicity, there is no effective therapy for this disease; patients once affected may not recover, even after remediation of the arsenic-contaminated water. The need for an effective therapy for chronic arsenicosis is obvious.

Chelation therapy for chronic arsenic toxicity is thought to be the specific therapy for the relief of systemic clinical manifestations and reduction of arsenic stores in the body and thereby reducing subsequent risks of developing cancer. Chelation therapy is presumed to be more effective with early features of toxicity, as severe manifestation of polyneuropathy, chronic lung and liver disease, swelling of hands and legs are less likely to respond to this therapy.

Chelating agents like meso-2, 3-dimercaptosuccinic acid (DMSA), sodium 2,3-dimercapto-1-propane sulfonic acid (DMPS) and d-penicillamine have frequently been considered for the treatment of chronic arsenic toxicity (Mazumder et al., 1998b). However, their usefulness as standard method of treatment is yet to be established. Symptomatic treatment for patients with keratosis includes use of keratolytic agents like 5–10% salicylic acid and 10–20% urea cream based ointment (Caussy, 2005).

Antioxidants and vitamins like beta carotene, vitamin E and C are also being tried by some workers for the treatment of arsenicosis (Caussy, 2005; Tondel et al., 1999) for hastening recovery or averting disease progression. But no authentic evidence on the basis of placebo controlled trial has been available substantiating the efficacy.

Poor underweight people with several nutrient deficiencies were found to be more susceptible to arsenical skin lesions in the arsenic-endemic regions. Significantly lower intakes of protein, thiamin, niacin, Mg, Zn, and choline were observed in male and female cases compared with the respective controls (Deb et al., 2013). So dietary advice to increase the consumption of animal protein, roots, tubers, locally available fruits, and other green leafy vegetables along with increasing the consumption of cereals and milk may help reduce the occurrence of arsenical skin lesions.

Besides health, a wide range of psycho-social, behavioral, and economic consequences were observed associated with chronic arsenic toxicity in the society. A study on disease burden, and psychosocial, behavioral, and economic issues related to chronic arsenic exposure were done in Malda, an arsenic affected district of West Bengal. Information on psychosocial, behavioral, and economic issues was assessed amongst the surveyed population. Prevalence of arsenicosis was found to be 8.3% amongst 5355 people surveyed. The majority (62%) of patients who were aware of their illness felt embarrassed, many having anxiety, tension, and depression. Most (93.6%) patients felt that their status had changed within the community. A small number (3%) of patients volunteered for marital problem. Not many (18.9%) patients consulted doctors for their treatment. The majority of patients had disability leading to change of occupation or loss of jobs. A small number (8.7%) of participants considered arsenicosis as a communicable disease. Social exclusion was reported among a small (1.4%) number of participants. Counseling and addressing of these psycho-social issues of the illness are needed, and further rehabilitation needs are to be provided, if necessary (Majumdar et al., 2018).

Secondary prevention of latent effects should be done through medical surveillance, where regular follow-up of patients is needed for early detection of complications like cancer (Caussy, 2005).

Conclusion

There is an urgent need to make arrangements for availability of safe water sources among the arsenic affected community in the endemic areas. Many of the people in the affected villages are not aware of the contamination of their home tube-wells with arsenic. Awareness creation and sensitization among people to test their drinking water sources for arsenic contamination and environmental interventions like rainwater harvesting, groundwater recharge, and restricting excessive use of groundwater for domestic, agricultural and other purposes are also important to prevent further exposure to arsenic.

It can be concluded that an attempt to mitigate chronic arsenic toxicity requires an integrated approach for limiting of the entry of arsenic into the food chain system and in drinking water for consumption.

Further, arsenic affected people with severe skin lesions and systemic manifestations like lung disease, neuropathy, severe skin disease, ulcer, peripheral vascular disease, etc. have unbearable sufferings. These people are very poor and live in distant villages where the hospital facilities are not easily available. The presence of arsenicosis in these patients causes serious socio-economic problems in their families. Arrangements for free treatments of these patients in state referral hospitals and free transport facilities from their villages could help a lot in alleviating the sufferings of these people. Primary prevention by increased levels of awareness about the signs and symptoms of arseniocosis and the available interventions and options of safe water sources among primary care providers of the local region will definitely help to manage and mitigate this important but somewhat neglected public health problem.

References

Deb, D., Majumdar, K. K., & Mazumder, D. N. G. (2013). Arsenicosis and dietary intake among men and women. *Proceedings of the National Academy of Sciences, India; Section B: Biological Sciences, Sci* 83, 405–413.

Mazumder, D. N. G. (2008). Chronic arsenic toxicity and human health. *Indian Journal of Medical Research, 128*, 436–447.

Mazumder, D. N. G., Haque, R., Ghosh, N., et al. (1998a). Arsenic levels in drinking water and the prevalence of skin lesions in West Bengal, India. *Int J Epidemio, 27*, 871–877.

Mazumder, D. N. G., Ghosal, U. C., Saha, J., Santra, A., Dc, B. K., Chatterjee, A., Dutta, S., Angle, C. R., & Centeno, J. A. (1998b). Randomized Placebo-Controlled Trial of 2,3—Dimercaptosuccinic acid in therapy of chronic arsenicosis due to drinking arsenic contaminated subsoil water. *Clinical Toxicology, 36*(7), 683–690.

Mazumder, D. N. G., Ghosh, A., Majumdar, K. K., et al. (2010). Arsenic contamination of ground water and its health impact on population of District of Nadia, West Bengal, India. *Ind J of Community Medicine., 35*, 331–338.

Majumdar, K. K., Ghosh, A. K., Majumder, P. K., Mukherjee, S., & Mazumder, D. N. G. (2018). Arsenical disease burden and psychosocial, behavioral & economic issues associated with chronic arsenic exposure: a study in Malda District, West Bengal. *International Journal of Research & Review 5*(12), 156–165 (2018). (www.ijrrjournal.com)

Majumder, K. K., Guha Mazumder, D. N., Ghose, N., & Lahiri, S. (2009). Systemic manifestations in chronic arsenic toxicity in absence of skin lesions in West Bengal. *Indian Journal of Medical Research, 129*, 75–82.

Mazumder, D. N. G., Das Gupta, J., Santra, A., Pal, A., Ghose, A., Sarkar, S. (1998c). Chronic arsenic toxicity in West Bengal—The worst calamity in the world. *J Indian Med Assoc, 96*, 4–7.

Mazumder, D. N. G., De, B. K., Santra, A., Das Gupta, J., Ghosh, N., Roy, B. K., Chosal, U. C., Saha, J., Chatterjee, A., Dutta, S., Haque, R., Smith, A. H., Chakraborti, D., Angle, C. R, Centreno, J. A., Chappell, W. R., Abernathy, C. O., Calderon, R. L., (eds.) Arsenic exposure and health effects; chronic arsenic toxicity: epidemiology, natural history and treatment (pp. 335–347). Elsevier.

Mazumder, D. N. G., Haque, R., Ghosh, N., De, B. K., Santra, A., Chakraborti, D., & Smith, A. H. (2000). Arsenic in drinking water and the prevalence of respiratory effects in West Bengal, India. *International Journal of Epidemiology 29*, 1047–1052.

Mazumder, D. N. G., Ghose, N., Mazumder, K., Santra, A., Lahiri, S., Das, S., Basu, A., & Smith, A. H. (2003). Natural history following arsenic exposure": a study in an arsenic endemic area of West Bengal, India. In W. R. Chappell, C. O. Abernathy, R. L. Calderon & D. J. Thomas (eds.), *Arsenic Exposure and Health Effects*. Published by Elsevier B. V.

Mazumder, D. N. G., Deb, D., Biswas, A., Saha, C., Nandy, A., Ganguly, B., Ghose, A., Bhattacharya, K., Majumdar, K. K. (2012). Evaluation of dietary arsenic exposure and its biomarkers: A case study of West Bengal, India. *Journal of Environmental Science and Health Part A. Taylor & Francis 48*(8), 896–904.

NRC. (1999). *Arsenic in drinking water*. National Academic Press.

Rahman, A., et al. (2019). Arsenic exposure and risk of spontaneous abortion, stillbirth and infant mortality. *Epidemiology 21*(6), 797–804.

Caussy, D., (ed.) (2005). *A field guide for detection, management and surveillance of arenicosis cases. Clinical aspects and case definition* (Vol. 30, pp. 5–15). WHO Technical Publication; WHO Regional Office for South East Asia.

Tondel, M., Rahman, M., Magnuson, A., Chowdhury, O. A., Faruquee, M. H., & Ahmad, S. A. (1999). The relationship of arsenic levels in drinking water and the prevalence rate of skin lesions in Bangladesh. *Environ; Health Perspect., 107*, 727–729.

WHO. (2004). *Some drinking-water disinfectants and contaminants, including Arsenic; Human Exposure, Arsenic in Drinking Water; IARC Monographs on the Evaluation of Carcinogenic risks to Humans* (Vol. 84, pp. 60–188). World Health Organization.

Correction to: Polymer Nanofilm Composite Membranes for Ionic and Molecular Separation: History, Challenges and Future Perspectives

Pulak Sarkar and Santanu Karan

Correction to:
Chapter 9 in: B. Sen Gupta and N. Martínez-Villegas (eds.), *Arsenic Remediation of Food and Water*, https://doi.org/10.1007/978-981-97-4764-1_9

The original version of the chapter "Polymer Nanofilm Composite Membranes for Ionic and Molecular Separation: History, Challenges and Future Perspectives" was inadvertently published with incomplete PRIS number which has now been updated as PRIS number (CSIR-CSMCRI 117/2024). The book has been updated with the changes.

The updated version of this chapter can be found at
https://doi.org/10.1007/978-981-97-4764-1_9

B. Sen Gupta and N. Martínez-Villegas (eds.), *Arsenic Remediation of Food and Water*,
https://doi.org/10.1007/978-981-97-4764-1_23

MIX
Papier aus verantwortungsvollen Quellen
Paper from responsible sources
FSC® C105338

If you have any concerns about our products,
you can contact us on
ProductSafety@springernature.com

In case Publisher is established outside the EU,
the EU authorized representative is:
Springer Nature Customer Service Center GmbH
Europaplatz 3, 69115 Heidelberg, Germany

Printed by Libri Plureos GmbH
in Hamburg, Germany